BEHAVIOR OF DEEP FOUNDATIONS

A symposium sponsored by ASTM
Committee D18 on Soil and Rock
for Engineering Purposes
AMERICAN SOCIETY FOR
TESTING AND MATERIALS
Boston, Mass., 28 June 1978

ASTM SPECIAL TECHNICAL PUBLICATION 670
Raymond Lundgren, Woodward-Clyde Consultants,
editor

04-670000-38

ASTM
AMERICAN SOCIETY FOR TESTING AND MATERIALS
1916 Race Street, Philadelphia, Pa. 19103

Library of Congress Catalog Card Number: 78-72475
ISBN 0-8031-0291-7

NOTE

The Society is not responsible, as a body,
for the statements and opinions
advanced in this publication.

Printed in Baltimore, Md.
Dec. 1979
Second Printing, Mars, Pa.
July 1985

William S. Housel, 1901–1978

Dedication

ASTM STP 670, Behavior of Deep Foundations, *is dedicated with profound appreciation to William S. Housel as a founding member of ASTM Committee D18 on Soil and Rock for Engineering Purposes and for his continued commitment and professional contribution to that committee. As professor of civil engineering at the University of Michigan for 46 years, as research consultant to the Michigan Department of State Highways for 36 years, and as a consultant to industry over his entire professional career, Bill Housel has made significant contributions to the design and testing of deep foundations. His vision and interest in developing and sharing knowledge in soil and rock for engineering purposes, particularly in deep foundations, are greatly appreciated.*

Foreword

The symposium on Behavior of Deep Foundations was held during the June Committee Week of the American Society for Testing and Materials, 25–30 June 1978, in Boston, Mass. ASTM Committee D18 on Soil and Rock for Engineering Purposes sponsored the symposium. The symposium Committee members were: Raymond Lundgren, Roy Bell, R. D. Darragh, M. I. Esrig, L. C. Reese, M. T. Davisson, and F. M. Fuller. Lundgren also served as editor of this publication.

Related ASTM Publications

Dynamic Geotechnical Testing, STP 654 (1978), 04-654000-38

Dispersive Clays, Related Piping, and Erosion in Geotechnical Projects, STP 623 (1977), 04-623000-38

Soil Specimen Preparation for Laboratory Testing, STP 599 (1976), 04-599000-38

Underwater Soil Sampling, Testing, and Construction Control, STP 501 (1972), 04-501000-38

A Note of Appreciation to Reviewers

This publication is made possible by the authors and, also, the unheralded efforts of the reviewers. This body of technical experts whose dedication, sacrifice of time and effort, and collective wisdom in reviewing the papers must be acknowledged. The quality level of ASTM publications is a direct function of their respected opinions. On behalf of ASTM we acknowledge their contribution with appreciation.

ASTM Committee on Publications

Editorial Staff

Contents

Introduction

With the advent of larger and better construction equipment, including high energy pile driving hammers, and with the competitive desire of industry to design pile members to their full structural capacity, there has been a move toward higher capacity pile foundations. In response to this demand, the material suppliers (concrete, wood, and steel) have been influencing building officials to raise allowable pile stresses in codes and engineers to design for higher stresses than had customarily been used. This has forced the foundation engineer to reassess the true ultimate capacity of piles and the effects of high energy driving on the integrity of the pile member itself.

To address these concerns, a Symposium on the Behavior of Deep Foundations was held on 28 June 1978 under the sponsorship of ASTM Committee D18 on Soil and Rock for Engineering Purposes. The symposium was divided into four sections, each with a state-of-the-art presenter, as follows:

I. "Design and Evaluation of Load Tests on Piles and Caissons," by Professor L. C. Reese, University of Texas;
II. "Soil Capacity for Supporting Deep Foundation Members," by Dr. M. I. Esrig and R. C. Kirby, Woodward-Clyde Consultants, presented by Dr. Esrig;
III. "Stresses in Pile Members-Long Term Performance and Short Term Performance During Driving," by Professor M. T. Davisson, University of Illinois; and
IV. "Design Practice-Present and Proposed-Including Considerations of Standards and Codes," by Frank M. Fuller, Raymond International Builders, Inc.

The symposium chairman introduced the symposium and chaired the morning session, which contained Sections I and II, and R. D. Darragh chaired the afternoon session, which contained Sections III and IV. In addition to the state-of-the-art speakers, there was a panel for each session, consisting of the following panelists:

Morning Session, Sections I and II:
M. Bozozuk, National Research Council of Canada

D. M. Holloway, Woodward-Clyde Consultants
C. P. Wroth, University of Cambridge, England
J. A. Focht, Jr., McClelland Engineers, Inc.

Afternoon Session, Sections III and IV:
William Gamble, University of Illinois
Thomas Dismuke, Bethlehem Steel Corporation
R. M. Armstrong, University of Illinois
W. F. Swiger, Stone & Webster Engineering Company
W. A. Norum, National Forest Products Association

Each panelist made a short presentation, which was followed by a discussion between the panel members and audience.

Although not presented at the symposium, 31 papers were accepted for publication and are included in this volume. The four state-of-the-art papers are based on the personal experiences and knowledge of the state-of-the-art speakers as well as material contained in the papers submitted for each topic. In several cases, discussions on a particular paper were offered for publication and the author of the original paper was given an opportunity to furnish a closure statement. These discussions and closures are also included in this publication. One of the panelists, W. F. Swiger, did not present a paper for publication but his remarks as a panelist are published herein.

The organization of the papers in this volume is as follows: (1) state-of-the-art papers, in the order in which they were presented; followed by (2) papers, in alphabetical order according to first author's surname; and (3) discussions and closures following the paper to which they pertain.

This symposium volume meets its intended purpose. It has considered: new and innovative testing methods; data from full scale load testing and how these data are interpreted; information gained from the latest experience; advancements made in soil mechanics and how the foundation materials perform both during construction and under long-term loading; recent developments in knowledge of material properties; and how an appropriate specification or standard can be formulated.

Discussions and closures concerning the use of higher stresses are included to provide the reader with information on both sides of this issue, on which complete agreement has not been reached.

ASTM Committee D18 is honored to dedicate this volume to the late William S. Housel, Professor Emeritus of Civil Engineering at the University of Michigan and a pioneer in soil mechanics engineering who had a lifetime interest in deep foundations. In 1936, Bill Housel was made Chairman of an ASTM steering committee to organize the newly authorized Committee on Soil for Engineering Purposes. As a founder of Committee D18, he served as First Vice-Chairman of the Committee from 1940 to

1960, as a member of the Executive Committee from 1960 to 1968, and was chairman of the Deep Foundations Committee for a number of years. He received many honors from his colleagues in ASTM, among them the ASTM Award of Merit in 1966, a Special Award of Committee D18 in 1968, and election to Honorary Membership in Committee D18 in 1971. Bill Housel's energy and dedication to ASTM and to the technology of deep foundations have enriched the profession immeasurably, and we are pleased to add this volume to his list of honors.

We hope this publication will be useful to geotechnical engineers, design engineers, building officials, and material suppliers in giving each a better understanding of the views of the others.

It is the hope and belief of the symposium organizers that the volume will provide the latest update on the technology of deep foundations.

Raymond Lundgren

Regional Managing Principal, Executive Vice President, Woodward-Clyde Consultants, San Francisco, Calif. 94111; editor.

L. C. Reese[1]

Design and Evaluation of Load Tests on Deep Foundations

REFERENCE: Reese, L. C., **"Design and Evaluation of Load Tests on Deep Foundations,"** *Behavior of Deep Foundations, ASTM STP 670,* Raymond Lundgren, Ed., American Society for Testing and Materials, 1979, pp. 4–26.

ABSTRACT: The use of methods of prediction of the performance of deep foundations is strongly encouraged. The paper presents models for both axially loaded and laterally loaded deep foundations and shows examples of the use of the models. Internal instrumentation that can be used in future load tests is discussed, along with load test procedures. Information from load tests of instrumented deep foundations will allow improvements in the methods of prediction of the behavior of deep foundation under various kinds of loading.

KEY WORDS: deep foundations, piles, drilled shafts, load tests, instrumentation, prediction models

Load tests on deep foundations are performed for two purposes—to prove a design or to gain information on the interaction between the foundation and the supporting soil. The proof test is frequently inexpensive, and little information is gained beyond the limited purpose. Load tests aimed at developing fundamental information on soil-structure interaction normally require sophisticated instrumentation, are time-consuming in planning and prosecution, and are expensive. However, this latter type of test can be enormously beneficial in allowing the design of foundations at a site to be optimized and in adding to the store of knowledge that is currently meager.

Terzaghi [*1*][2] made a statement that is pertinent to the load test on full-scale, instrumented deep foundations: "Our theories will be superceded by better ones, but the results of conscientious observations in the field will remain as a permanent asset of inestimable value to the profession."

[1] T. U. Taylor Professor of Civil Engineering and Associate Dean of the College of Engineering, The University of Texas at Austin, Austin, Tex.

[2] The italic numbers in brackets refer to the list of references appended to this paper.

While proof tests of deep foundations are frequently inexpensive, they usually do not add information relating foundation behavior to construction procedures, pile geometry, and soil properties. This paper adopts the thesis that when possible, a deep foundation should be subjected to failure loading, that internal instrumentation should be employed in the test pile in a sizable number of cases, and that in nearly all instances, a prediction should be made of the behavior of the deep foundation prior to the application of load.

Single foundations will be discussed that are subjected to two types of loading—axial and lateral. The thrust of the following paragraphs is the development of methods of prediction that can be used to design load tests and to interpret the resulting data.

Deep Foundations Subjected to Axial Loads

Model of Axially Loaded Deep Foundation

Figure 1 shows the model that has been selected to represent the deep foundation under axial load. The stiffness of the spring representing the foundation can easily be obtained from the geometry of the foundation and from the stress-strain properties of the foundation material. The stiffness of the foundation may be nonlinear with load and with depth.

The response of the soil in transmitting load is represented by a series of mechanisms. The details of the mechanisms that are selected, a cantilever spring and a friction block, are unimportant; the mechanisms merely show graphically that the transfer of load from pile to soil can be nonlinear with depth and with pile movement.

The discrete mechanisms shown in the figure are adequate in many cases

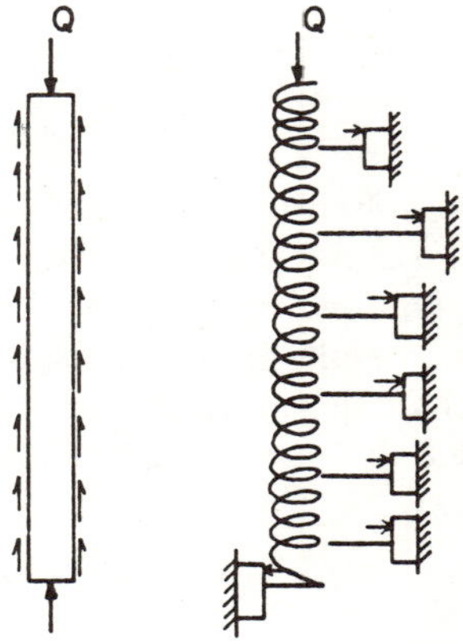

FIG. 1—*Model of axially loaded deep foundation.*

to describe the soil response; however, a modified model can readily be developed if interactions between the various components of a system can be expressed numerically. Thus, the mechanism representing the behavior of the soil at a particular point along the pile could be modified stepwise with time or with pile-load level to reflect the influence on that mechanism of time or of stress level throughout the system. Such a model would be a rigorous representation of the pile–soil system and could be used to deal with locked-in stresses, behavior of a pile in a settling or swelling soil, and group effects. Much research will be necessary to allow the mechanisms to be described numerically; however, the concept is thought to be worthwhile. This paper will deal only with the case of the foundation that is initially unstressed and subjected to a short-term axial load.

The differential equation that must be solved to obtain the behavior of the model is

$$\frac{d^2 z_x}{dx^2} - \eta \beta_x z_x = 0 \tag{1}$$

where

x = axial coordinate measured from top of pile,
z_x = axial movement of pile at point x,
η = C_x/E_xA_x,
C_x = circumference of pile at point x,
E_x = secant to stress–strain curve of the pile material at point x,
A_x = cross-sectional area of pile material at point x,
β_x = s_x/z_x,
s_x = unit load transfer at an axial movement of z_x.

The load Q_x in the pile at any depth x can be obtained by

$$Q_x = E_xA_x \frac{dz_x}{dx} \tag{2}$$

Equation 1 can readily be solved by numerical procedures if the data are at hand as indicated and if a curve is available showing the load at the tip of the pile as a function of axial movement [*2–6*].

Any application of load results in an axial movement of the pile. If the pile material if stiffer than the surrounding soil, one can say intuitively that the bottom of the pile will move when a load is applied at the top.

Analyzing Data from Tests of Instrumented Deep Foundation Under Axial Load

The data that are obtained from a test of an instrumented deep foundation under axial load are shown in Fig. 2. Figure 2*A* shows a load versus settlement curve for the top of the foundation, and Fig. 2*B* shows a family of curves giving the distribution of load as a function of depth.

Figure 2*C* illustrates the steps in the analysis. A particular depth, x_k is selected, along with one of the load-distribution curves. The slope of the curve dQ/dx divided by the circumference of the pile at that point yields the unit load transfer s. The integration of the area under the load distribution curve divided by the appropriate values of axial stiffness yields the compression of the pile over the length selected. That compression subtracted from the movement of the top of the pile for the load being considered yields the vertical movement of the pile z. Thus, a point can be plotted in Fig. 2*D* giving load transfer s as a function of the movement z. Repeating the computations for each of the load-distribution curves produces a curve showing load transfer versus movement for the depth in question. The family of curves shown in Fig. 2*D* can be obtained by performing the analyses for several depths. Using similar procedures, a curve

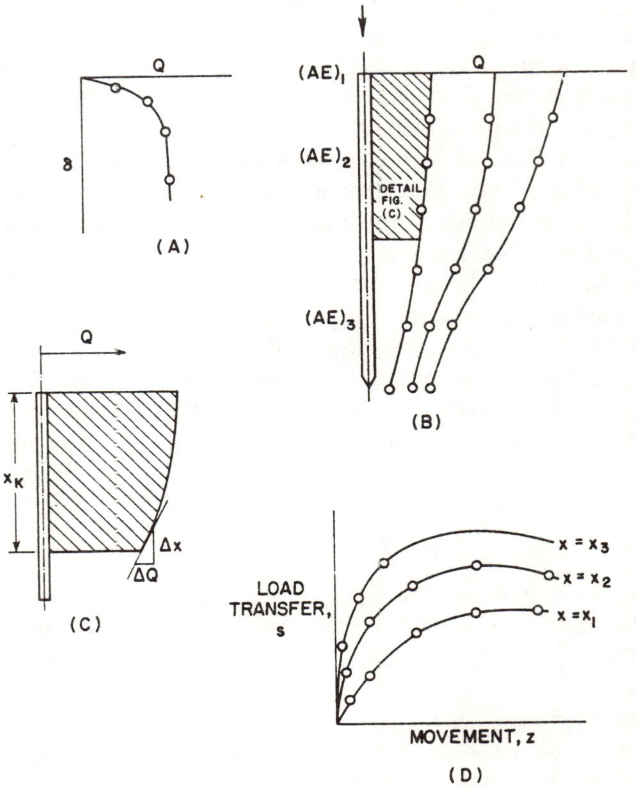

FIG. 2—*Analysis to obtain curves showing load transfer versus pile movement.*

can be developed showing the load at the tip of the pile as a function of the downward movement of the tip.

The data that are obtained on soil responses from a load test of an instrumented deep foundation can be correlated with construction method and with soil properties [*2, 7–12*]. The curves are frequently presented as *T–z* curves, where *T* is the ratio of the load transfer s to the shear strength.

The performance of a number of load tests on deep foundations, instrumented so that axial load can be determined as a function of depth, and the subsequent analyses of data from such tests can provide much useful information. However, there are other kinds of investigations that are needed in order to gain a better understanding of the effects of pile installation on soil properties.

For example, if a pile is driven into a saturated clay, the volume of the clay will change somewhat, but there will be a heave of the ground surface [*13–17*]. The pile will displace the soil and there will be remolding of the clay, with a consequent effect on the shear strength [*10, 13, 17, 18*]. There will be an increase in total stress and pore water stress in the vicinity of the pile and a subsequent decay of these stresses [*17, 19–21*]. The measurement of total stress and pore water stress at the pile wall, along with the sampling and testing of soil at the pile wall, can be helpful in developing prediction methods for properties of the clay. Predicting the properties of clay is essential to the prediction of *T–z* curves for clay.

The effects on the properties of sand of pile installation are no less significant than for clay. Pile driving will usually cause a densification of the sand in the vicinity of the pile unless the relative density is high [*22, 23*]. *In situ* soil testing and tests of instrumented piles will be necessary to gain information to allow the prediction of the properties and the state of stress of sand at the interface of a pile and the sand.

Research has been performed, and some current studies are under way that should be helpful in developing better methods for making rational predictions of the behavior of axially loaded piles. A number of these studies are related to developing prediction methods based on effective stresses [*24–28*].

Instrumenting an Axially Loaded Deep Foundation

To obtain the data shown in Fig. 2*A*, it is necessary, of course, to measure load and deflection at the top of the pile. The procedures set forth by ASTM Testing Piles Under Axial Compressive Load (D 1143) are generally applicable. Load cells should be employed in order to improve the accuracy of load measurement.

To obtain the data shown in Fig. 2*B*, it is necessary to employ instrumentation along the length of the pile. There is no standard approach to this problem. Two possibilities are to employ a number of unstrained rods (telltales) or remote-reading instruments [*29*]. An example of the latter device is shown in Fig. 3 [*30*]. A number of such devices are installed on the rebar cage of a drilled shaft, the rebar cage is placed in the excavation, and concrete is placed with a tremie. Mustran cells have been employed in over 20 load tests of drilled shafts.

There are other approaches to obtaining information on load transfer rather than installing internal instrumentation. Spanovich and Garvin [*31*] describe a procedure for design of load tests of deep foundations that involves no internal instrumentation but yields information on the load transfer in skin friction and in end bearing. Eleven drilled shafts, ranging in diameter from 457.2 mm (18 in.) to 762 mm (30 in.,) were installed at a site where the overburden consists of from 4.6 m (15 ft) to 6.4 m (21 ft) of randomly placed fill underlain by silty shale bedrock. The upper 1 ft of the shale is decomposed. Three of the shafts were terminated on the surface of the shale bedrock, three were socketed 0.9 m (3 ft) into the shale with a void at the base, three were socketed 1.5 m (5 ft) into the shale with

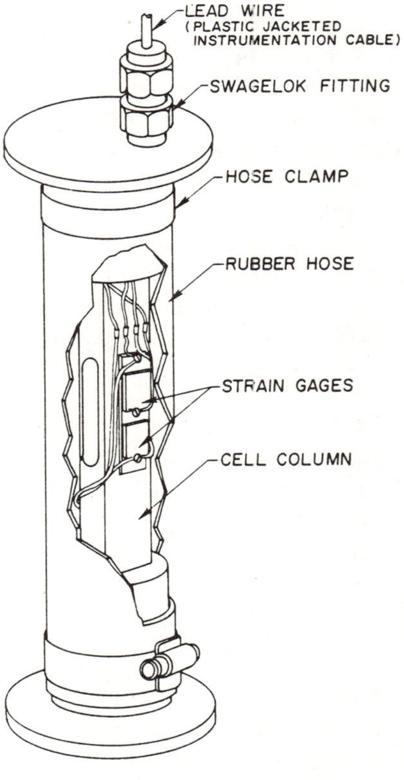

FIG. 3—*Mustran cell.*

a void at the base, one was terminated in fill above the bedrock, and one was socketed 2 m (6.5 ft) into the shale, but lined with a casing to prevent bonding within the shale socket. The results of the load tests were used to make computations for the load supported in skin friction by the overburden, in skin friction in the socket in the shale, and in end bearing in the shale. The settlements necessary to develop the load transfer were also reported.

There are also methods that are based on the use of limited internal instrumentation. Brierley et al [*32*] describe case histories of 14 full-scale test piles and present a method of estimating the behavior of the pile tip. The method is based on the use of a single telltale that extends to the pile tip and on the following assumptions:

1. The amount of skin friction support for the pile is limited by some maximum value, and this value of skin friction is substantially mobilized with pile movement considerably less than that required to cause a bearing capacity failure.
2. The bearing stratum will deflect linearly until the applied load approaches its ultimate bearing capacity.

With data from the telltale, the authors were able to plot two curves, a curve showing elastic compression in the pile as a function of applied load and a curve showing the downward movement of the pile tip as a function of applied load. From the first of the two plots, it was possible to find out whether or not the pile was behaving principally as a friction pile, as an end-bearing pile, or in some other manner. From the second curve, making use of the stated assumptions, the authors were able to construct a curve for the tip behavior. Thus, it was possible to separate any applied load into skin friction and end bearing. The method appeared to yield excellent results for several case studies that were presented.

Test Procedures for Axial Loading

The procedures that should be employed in performing a load test are set forth by ASTM D 1143. The physical arrangement that is frequently employed is shown in Fig. 4. One of the problems that must be solved is the spacing between the anchors. The ASTM procedure states,

> Install a sufficient number of anchor piles or suitable anchoring devices a clear distance from the test pile or pile group, at least five times the maximum diameter of the anchors but not less than 2 m (7 ft).

If a weighted box or platform is to be employed in providing the reaction, the ASTM procedure states that the supports for the dead weight shall be "as far from the test pile as practicable, preferably so that there is a clear distance of at least 2 m (7 ft) between the test pile and the supports to provide adequate working space."

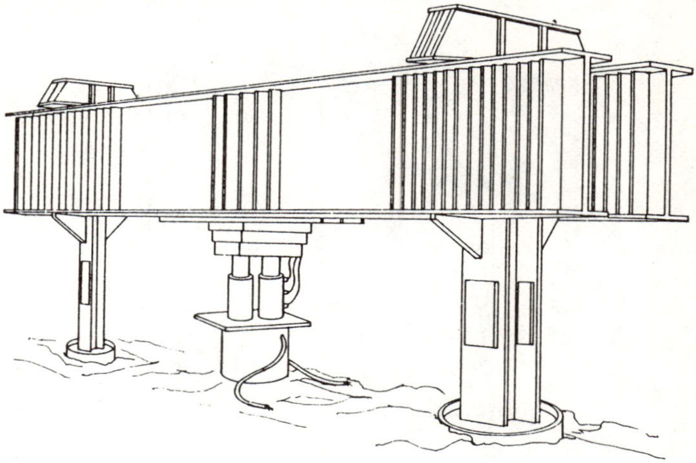

FIG. 4—*Example of arrangement for performing test of deep foundation under axial load.*

The ASTM standard allows several procedures for loading the pile. The writer prefers the constant rate of penetration method or the quick load test method because of the desirability of correlating the load test results with the undrained shear strength of clays [*33*]. No load test can be continued a sufficient length of time to account for creep and consolidation; therefore, such effects must be dealt with by analyses.

Example Results from Testing of Instrumented Deep Foundations Under Axial Load

Figures 5 through 9 are presented to show the kinds of results that can be obtained from the use of internal instrumentation in a series of tests of drilled shafts [*30*]. Figure 5 shows load versus settlement curves for a 760-mm (30-in.) outside diameter drilled shaft that was installed in overconsolidated clay to a penetration of 13.7 m (45 ft). Load transfer versus settlement curves were developed for a number of points along the foundation; however, presentations such as shown in Fig. 5 are instructive to the designer.

The results shown in Figs. 6 through 9 indicate the scatter that can be expected in practice. Much of the scatter is due to inability to assess properly the engineering characteristics of the soil. Other portions of the scatter are due to improper response of the instruments and to variations in the construction procedures that were used. There is probably another portion of the scatter due to the theoretical differences in behavior of piles of varying length [*25,34*]. In spite of the considerable amount of scatter, the results that are shown in the figures, along with other such results that are ob-

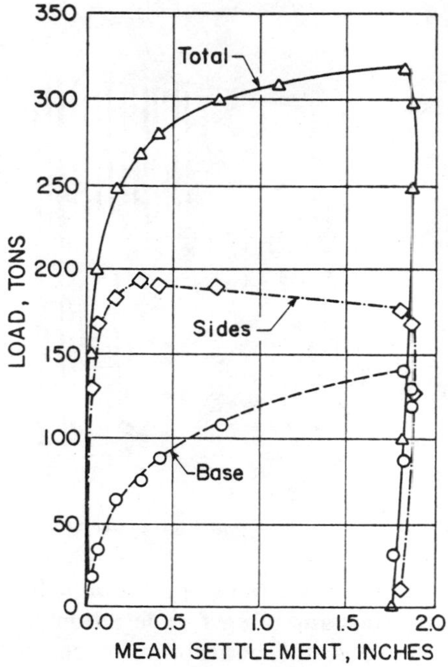

FIG. 5—*Load-settlement curves demonstrating the relative development of side and base resistance (after O'Neill and Reese, 1970).*

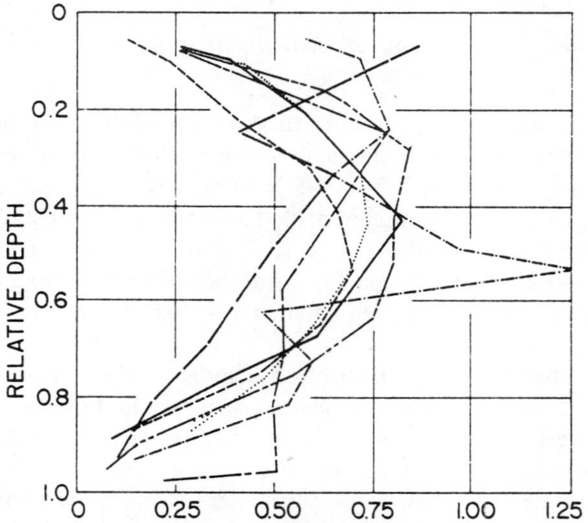

FIG. 6—*Relative maximum load transfer for clay as a function of relative depth.*

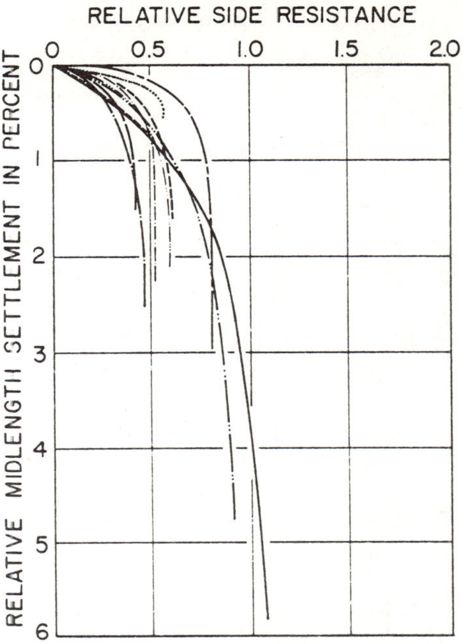

FIG. 7—*Relative side resistance for clay versus relative midlength settlement.*

tained, can lead to a better understanding of the behavior of deep foundations and improved predictive techniques. (The terms used in Figs. 6 through 9 are defined in the Appendix).

Discussion of Models for Axially Loaded Deep Foundations

In the interpretation of results from load tests, there are a number of useful analytical methods that are based on the characteristic responses of axially loaded piles. Leonards and Lovell [*35*] make use of a number of such suggestions. They describe the testing of three high-capacity driven piles, two 355.6-mm (14-in.) pipe piles, and one 355.6-mm (14-in.) H-pile. The piles were driven to a penetration of 33.5 m (110 ft) to 36.6 m (120 ft) into glacial till that overlies hard dolomite bedrock at a depth of more than 36.6 m (120 ft). Telltales were placed at the tip of the three piles to allow both tip and butt settlements to be measured during the load testing. A methodology for interpreting load transfer was presented and was applied to the interpretation of the load tests that were described. The authors stated that the method was not exact but does give considerable insight into the mechanics of load transfer. They concluded that for piles 305 mm (12 in.) to 457 mm (18 in.) in diameter, the ultimate shaft friction is not necessarily mobilized before ultimate load is reached, that loading and

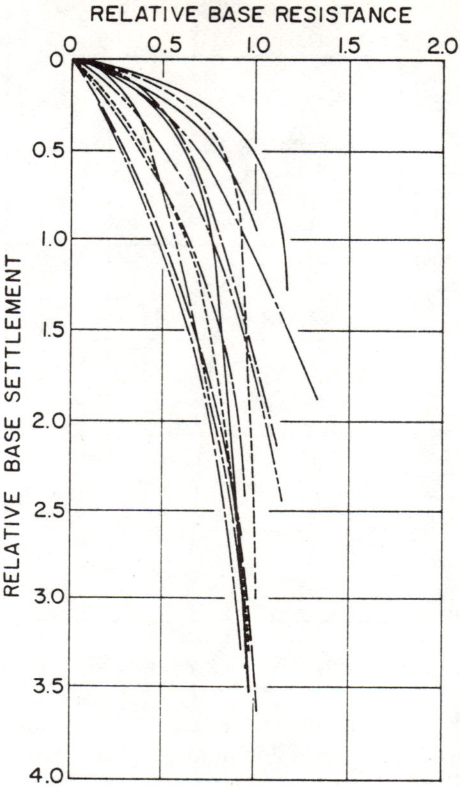

FIG. 8—*Relative base resistance for clay versus relative base settlement.*

unloading a pile can cause irreversible changes in the distribution of effective lateral pressures, that deflection of a pile tip can occur without any load being applied due to strains induced in the soil below the tip by shaft friction forces, and that load transfer in piles is sensitive to small changes in soil strain or pile compression.

An interesting study is described by Randolph and Wroth [*36*] in which the soil is treated as an elastic material with its modulus varying with depth. Design curves are presented by which the settlement of a pile can be obtained as a function of axial load. Excellent agreement between their predictions and experimental results was obtained for the early portion of several load-settlement curves. The authors suggest that the method of solution that is presented can be employed to advantage in the preparation of design charts in cases where single piles must sustain some specified load without excessive deflection. The authors note that their analytical model may be used for the analysis of pile groups.

If possible, predictive methods should be based on a comprehensive

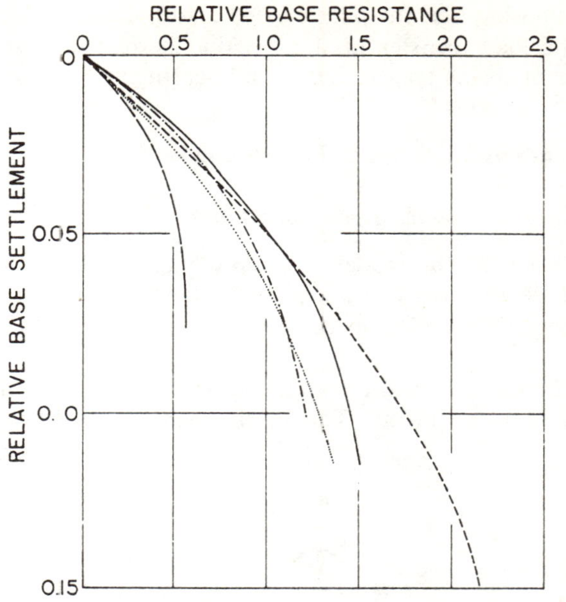

FIG. 9—*Relative base resistance for sand versus relative base settlement.*

description of the soil properties. However, there are a number of methods that are based on the results of *in situ* tests. Cheeks [*37*] describes a test program in which several types of piles were driven at a single site and loaded to failure. Twenty-three piles were driven in the test program, ten were loaded in axial compression, and six in tension. Three static methods and two dynamic methods were used in predicting the pile capacities. The author concludes that a predictive method under development involving the use of the Dutch cone penetration test gave the best results. The model that is presented is believed to be useful in interpreting the results of axial load tests of instrumented deep foundations. Such tests can lead to improved predictive methods.

Montgomery [*38*] used a procedure related to that presented in this paper to evaluate the load-transfer mechanisms for several kinds of piles. Seventeen piles—H-piles, Raymond step taper piles, augercast piles, and steel pipe piles—were tested in compression. Deflection was measured at the butt and by use of telltales at the tip and at one or two intermediate levels. The soils at the site consisted principally of silts and sands. Results from direct shear tests were used to obtain load-transfer relationships in skin friction and results from triaxial tests were used to obtain load versus deflection relationships for the pile tip. Good agreement was obtained between computed and measured movements at the butt, midpoint, and tip for several of the cases. Three families of load-distribution curves were presented, showing that the piles carried load both in skin friction and end bearing. The load transfer in skin friction developed first with the final load increments being transferred in end bearing.

Deep Foundations Subjected to Lateral Loads

Model of Laterally Loaded Deep Foundation

Figure 10 depicts the model that represents the deep foundation under lateral load. With regard to the pile, the conditions are assumed to hold that are normally satisfied in deriving the two-dimensional beam-column equation.

As with the case of axial loading, the response of the soil is represented by a series of mechanisms. The mechanisms shown are not a rigorous

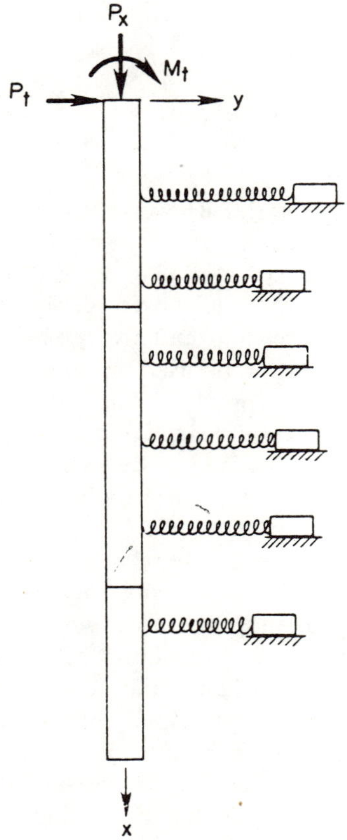

FIG. 10—*Laterally loaded model of deep foundation.*

representation of the soil response but show graphically that the load transfer from pile to soil is nonlinear with depth and with deflection. As with axial behavior, it is assumed that the soil can be replaced by a series of discrete elements; however, the method of solution can be extended to the case where there is interaction between the elements, with the only real limitation being a lack of knowledge concerning the possible interaction. If such interaction can be expressed numerically, the model becomes rigorous; however, as shown later in this paper, the use of the discrete mechanisms to represent the soil yields good results.

The differential equation that must be solved to obtain the behavior of the model is

$$\mathrm{EI}\,\frac{d^4y}{dx^2} + P_x\,\frac{d^2y}{dx^2} + E_s y = 0 \tag{3}$$

where

EI = flexural stiffness of foundation, and
E_s = soil modulus.

Equation 3 can readily be solved by numerical procedures [*39–43*].

Analyzing Data from Test of Instrumented Deep Foundation Under Lateral Load

The data that are obtained from a test of an instrumented deep foundation under lateral load are shown in Fig. 11. Figure 11*A* shows the pile-head deflection and rotation under lateral load, and Fig. 11*B* shows a family of curves giving bending moment in the pile as a function of depth.

Figure 11*C* shows the curves that are desired, a family of curves giving soil resistance p as a function of deflection y. The curves are best obtained by numerical integration and differentiation.

$$y = \iint M_x \tag{4}$$

$$p = \frac{d^2M_x}{dx^2} \tag{5}$$

The integration process can be carried out without difficulty, but the differentiation leads to serious errors unless very accurate bending moment curves are obtained. However, curve fitting and other techniques [*44*] have led to good results. The quality of derived p-y curves can be checked by back computations to see that computed curves are nearly the same as the field data.

The experimental p-y curves can be employed in the development of

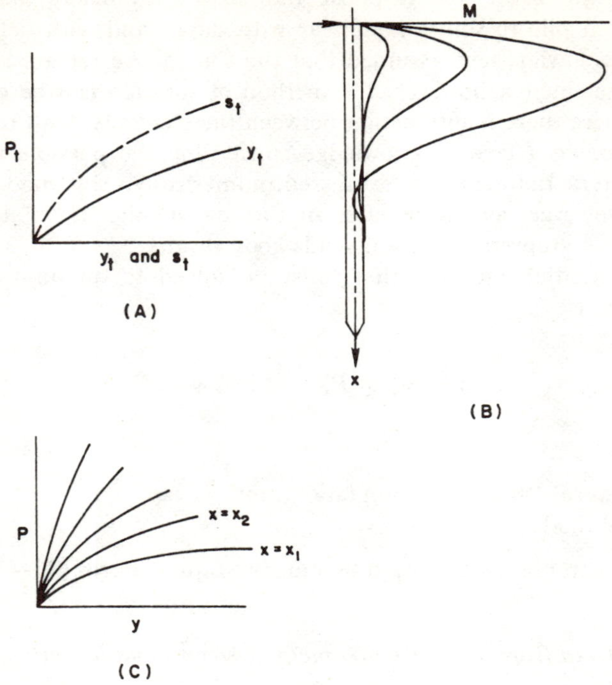

FIG. 11—*Analysis of curves to obtain soil resistance versus lateral deflection.*

correlations with pile geometry; kind of loading that is applied, short-term or cyclic; and with soil properties [*45–48*]. Such correlations can then be employed in making predictions of the behavior of piles under lateral loading.

Instrumenting a Laterally Loaded Deep Foundation

The lateral load, pile-head deflection, and pile-head rotation can be obtained by use of standard procedures. The rotation can easily be obtained by using two or more dial gages at different distances above the groundline. There are a number of other devices, manual or electronic, that can be used to obtain pile-head rotation [*48,49*].

The family of curves giving bending moment in the pile as a function of depth can be obtained by instrumenting a metal pipe with electrical resistance strain gages. The pipe can be the pile itself [*49*] or can be an instrumented core in a drilled shaft [*48*]. In either case, some technique should be employed to calibrate the deep foundation, because the desired accuracy is such that use only of material properties and strain-gage characteristics is inadequate.

A method has been proposed [*44*] such that *p–y* curves can be derived from a knowledge of only the pile-head deflection and rotation, along with the corresponding loading conditions and pile geometry. While the method is useful, there will be a lack of confidence in the curves so derived. The inclinometer can be employed to obtain the pile rotation as a function of depth [*50*]. Such measurements are principally useful in obtaining deflection. The three differentiations of the rotation curves to obtain soil response lead to serious inaccuracies.

Test Procedures for Lateral Loading

ASTM currently has no recommended standard for the performance of tests of piles under lateral loading. A method has been under discussion, however, for some time. One of the principal difficulties is that of establishing pile-head conditions in the field to stimulate those of the foundation that is being designed. A pile with a head that is free to rotate will deflect much more under a given load than a pile with a head that is fixed against rotation. The piles supporting many structures have heads that are neither fixed nor free, but somewhere between.

The above facts argue strongly against the development of a standard for proof-testing a laterally loaded deep foundation, but argue for the use of the prediction method that is indicated. If it is undesirable to use internal instrumentation in the load test, the pile-head deflection (and rotation) can be compared with those predicted by theory and appropriate adjustments can be made in the final design.

Aside from the pile-head conditions, the manner of applying the lateral load presents difficulties. In practice, loads can be dynamic, as from blasts or earthquakes; sustained, as from an earth-retaining structure; cyclic, as from winds, waves or traffic; and possibly short-term. The difficulties of simulating the piles in a particular structure that is subjected to sustained loading or dynamic loading by a field load test are obvious. Thus, the loading procedures have been of two types—short-term static and cyclic. The behavior of a pile under sustained loading or under dynamic loading would start from a prediction of its behavior under a simpler loading and the complexities would be treated analytically as well as the present state-of-the-art allows.

Example Results from Testing of Instrumented Deep Foundations Under Lateral Load

Figure 12 presents a family of *p–y* curves that were obtained from field experiments with an instrumented pipe pile [*47*]. The pile was 610 mm (24 in.) in diameter, had an embedment length of 15.2 m (50 ft), and was installed in overconsolidated clay. Two piles were installed, one was sub-

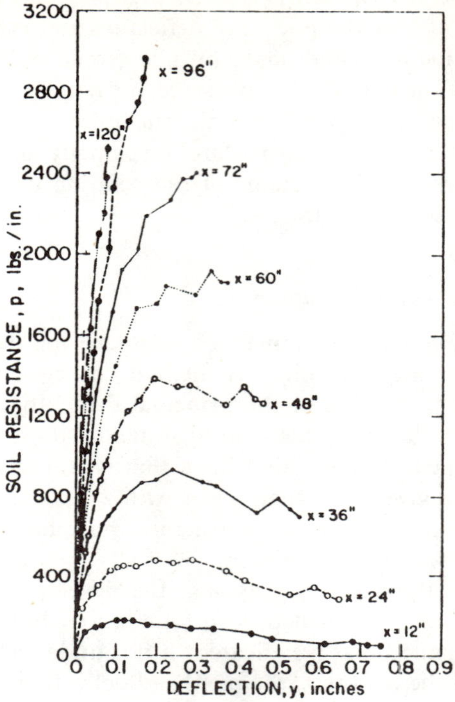

FIG. 12—*Experimental curves giving lateral soil resistance as a function of pile deflection (after Reese, et al, 1975).*

jected to short-term static and the other to cyclic loading. The curves shown are for behavior under static loading.

While there is some scatter in the data, the general quality of the experimental information is excellent. Unfortunately, the amount of such data is very limited.

The important step following the acquisition of data is to develop analytical techniques for predicting *p-y* curves. The predictions must take into account pile geometry; nature of loading, whether short-term static or cyclic; and soil properties. The small number of experiments that have been performed give reasonably good guidance in making most designs; however, a number of other instrumented load tests in a range of soils with a range of pile geometries and stiffnesses is highly desirable.

Discussion of Model for Laterally Loaded Deep Foundation

The model described herein has been used rather extensively for the analysis of piles under lateral load. Agreement between experimental and analytical results has generally been good or the analyses have been somewhat conservative. Two examples are presented in Figs. 13 and 14.

Price [*51*] described a series of field tests on tubular steel piles that were subjected to lateral loading. The piles were 150 mm (6 in.) in diameter and were embedded in stiff clay to a depth of 5.1 m (16.73 ft). Three cases were considered: a single pile, a row of three piles, and a row of three piles with a cap. Static loads, both axial and lateral, were applied in all three cases, and the single pile was subjected to cyclic lateral loads. The tests were designed and conducted with considerable care, and the instrumentation was extensive. The main conclusions were that low levels of cylic horizontal loading cause small adverse changes in the behavior of vertical piles, that the interaction between piles under horizontal load is considerally less than under vertical load, that a cap significantly reduces the horizontal movements but has little influence on vertical movements, and that the experimental techniques that were employed were successful in detecting small changes in pile-soil behavior.

The analytical method described earlier in this paper was employed

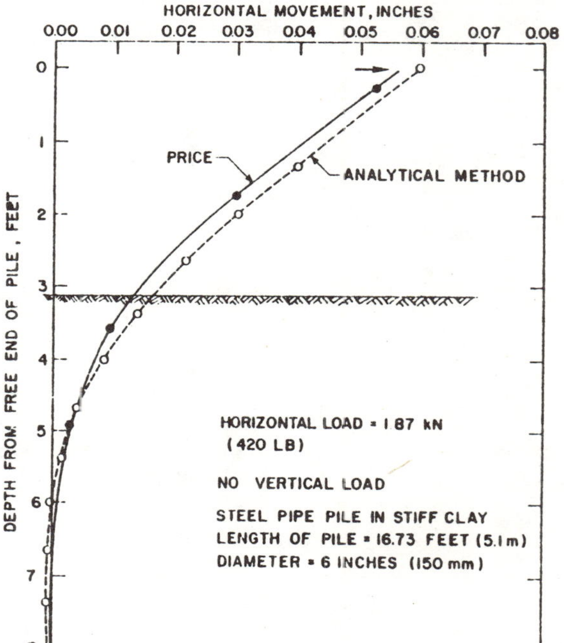

FIG. 13—*Comparison of experimental and analytical results for lateral loading of a pipe pile.*

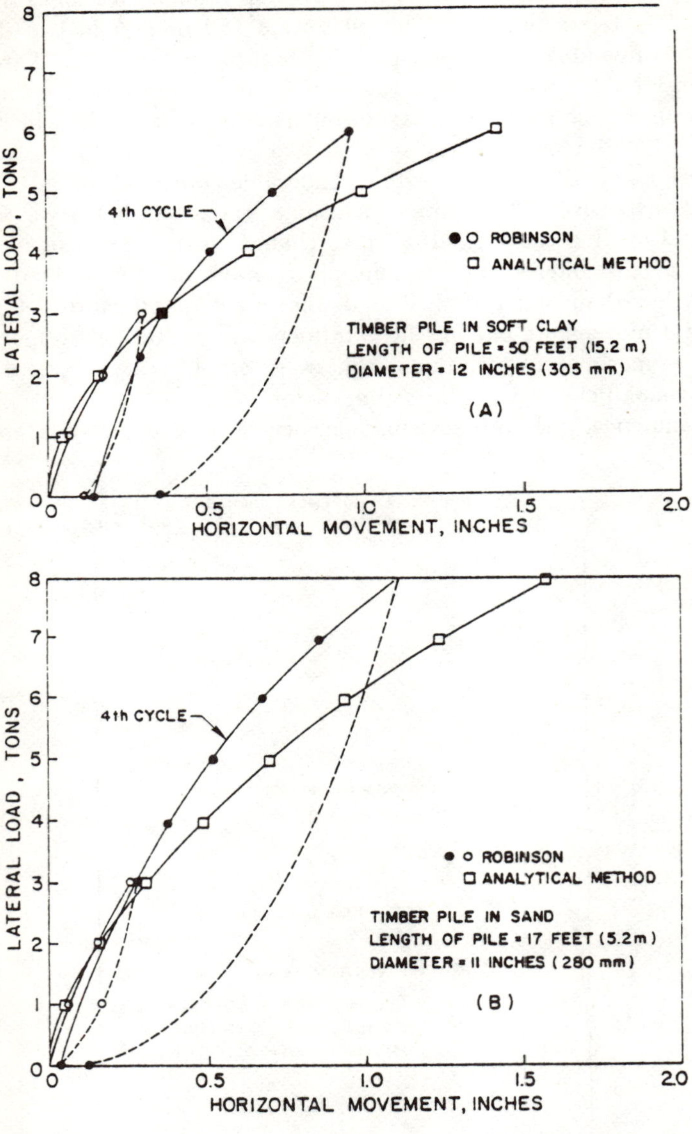

FIG. 14—*Comparison of experimental and analytical results from lateral loading of timber piles; (a—top) timber pile in soft clay, (b—bottom) timber pile in sand.*

to analyze one of the experiments performed by Price. Figure 13 shows a comparison between results of experiment and computations; the agreement is good, with the analytical model being slightly conservative.

Robinson [*52*] presented results from the testing under lateral loading of a number of timber piles. The piles were installed into a variety of soil conditions. Measurements for deflection and slope of the piles were compared with predictions based on elastic theories. The author concluded that existing theoretical techniques were suitable if realistic values for the horizontal subgrade reaction were used. Some values proposed in the literature were found to be conservative.

The analytical method described earlier in this paper was employed to analyze two of the cases reported by Robinson. Figure 14 (*top*) shows the comparison for a timber pile in soft clay, and Fig. 14 (*bottom*) shows the comparison for a timber pile in sand. As may be seen, there is good agreement between experiment and theory for the early portion of the curves; the analytical method is somewhat conservative for the higher loads.

On the basis of the information presented herein, the model for the behavior of laterally loaded piles appears to give reasonably good results. However, the performance of a number of additional tests on instrumented deep foundations is desirable.

Conclusions

1. The use of internal instrumentation in deep foundations during axial load tests is desirable. While the use of instruments that allow the determination of the distribution of load along the length of the deep foundation is desirable, it has been shown that the use of a telltale to the base of the deep foundation yields data that can be interpreted to give considerable insight into the behavior of the pile-soil system.

2. The model presented herein to represent a single pile under axial load, employing discrete but nonlinear mechanisms to simulate the soil, can be employed in the analyses of instrumented load tests. The model is useful in making designs in some cases. However, there is a need for a significant amount of research, both analytical and experimental, in order to represent the soil behavior more faithfully. The research needs to deal with the effects of installation and subsequent loading on soil properties, the interaction between all components of the pile-soil system, and the use of effective stress concepts in developing prediction techniques.

3. It is not practical to develop a general procedure for the performance of proof tests for piles that are to be subjected to lateral load because of the difficulty in simulating the pile-head conditions with accuracy.

4. The model presented herein to represent a pile under lateral load can be employed with considerable success to analyze the results of load tests where no internal instrumentation is employed. Such analyses can

yield information that will allow laterally-loaded piles to be designed with considerable assurance. There is a need, however, to perform a number of lateral-load tests on piles that are instrumented internally in order to gain more information on soil response. Research is needed in order to develop prediction techniques that are based on effective stress parameters. Furthermore, research on laterally-loaded piles is needed in order to develop methods for predicting pile behavior where the loads are sustained or where the loads are related to blasts or earthquakes.

APPENDIX

Definition of Terms

Relative depth is the depth, measured from the ground surface, of a point along the deep foundation divided by the total penetration of the foundation below the ground surface. In the case of a drilled shaft with an underream, the total depth (penetration) was measured to the top of the bell.

Shear strength reduction factor for clay is the unit load transfer divided by the undrained shear strength of clay.

Relative midlength settlement for clay is the downward movement of the midlength of the shaft that is required to develop the maximum skin friction divided by the diameter of the shaft.

Relative side resistance for clay is the maximum load that was carried in skin friction divided by the integral of the undrained shear strength times the differential area over which that shear strength acted.

Relative base settlement for clay is the settlement of the base of the shaft divided by $2B\epsilon_{50}$, where B is the diameter of the base and ϵ_{50} is the strain at one-half of the compressive strength of the soil.

Relative base resistance for clay is the stress developed at the base of a shaft divided by nine times the undrained shear strength of the clay at the base of the shaft. Thus, the relative base resistance would have been unity if the bearing capacity factor had actually been nine, assuming the undrained shear strength was measured correctly.

Relative base settlement for sand is defined as the settlement of the base divided by the diameter of the base.

Relative base resistance for sand is the stress developed at the base of the shaft divided by the ultimate bearing stress at failure. The ultimate bearing stress at failure was assumed to be the stress that existed at a base settlement of 5 percent of the base diameter. This stress, as determined from a small number of tests, was correlated with results from the standard penetration test as follows: $0 \lesseqgtr N_{spt} \lesseqgtr 60$, ultimate base resistance equal to $N_{spt}/1.5$ in tons per square foot; $N_{spt} > 60$, ultimate base resistance equal to 40 tons/ft^2 (3830 kN/m^2).

References

[1] Terzaghi, K. in *From Theory to Practice in Soil Mechanics,* Wiley, New York, 1960, p. 65.

[2] Coyle, H. M. and Reese, L. C. *Journal of the Soil Mechanics and Foundations Division*, American Society of Civil Engineers, Vol. 92, No. SM2, March 1966.
[3] Vijayvergiya, V. N., *Proceedings*, The Ports 1977 Conference, Long Beach, Calif., 10 March 1977.
[4] O'Neill, M. W., Ghazzaly, O. I., and Ha, H. B., "Analysis of Three Dimensional Pile Groups with Nonlinear Soil Response and Pile-Soil-Pile Interaction," Ninth Offshore Technology Conference, Houston, Tex., OTC Paper 2838, May 1977.
[5] O'Neill, M. W. and Peterson, E. H., "Model Studies of Skin Friction in Long Pipe Piles in Clay," Department of Civil Engineering, University of Houston, Oct. 1977.
[6] Matlock, H., Meyer, P. L., and Holmquist, D. V., "AXCOL 3: A Program for Discrete-Element Solution of Axially Loaded Members with Linear or Nonlinear Supports," Report to The American Petroleum Institute, March 1976.
[7] Whitaker, T. and Cooke, R. W. "An Investigation of the Shaft and Base Resistances of Large Bored Piles in London Clay," Symposium on Large Bored Piles, London, 1966.
[8] Poulos, H. G. and Davis, E. H., *Geotechnique*, Vol. 18, No. 3, 1968.
[9] Vesić, A. S., *Journal of the Soil Mechanics and Foundations Division*, American Society of Civil Engineers, Vol. 96, No. SM2, March 1970, pp. 561-584.
[10] Tomlinson, M. J., "*Adhesion of Piles in Stiff Clays*," Construction Industry Research and Information Association, Report 26, Nov. 1970.
[11] Bozozuk, M., Keenan, G. H., and Pheeney, P. E., "Analysis of Load Test on Instrumented Steel Test Piles in Compressible Silty Soil," this volume.
[12] Horvath, R. G., "Field Load Test Data on Concrete-to-Rock 'Bond' Strength for Drilled Pier Foundations," University of Toronto, Publication 78-07, July 1978.
[13] Cummings, A. E., Kerkhoff, G. O., and Peck, R. B., *Transactions*, American Society of Civil Engineers, Vol. 115, 1950, pp. 275-285.
[14] Hagerty, D. J. and Peck, R. B., *Journal of the Soil Mechanics and Foundations Division*, American Society of Civil Engineers, Vol. 97, No. SM1, 1971, pp. 1513-1532.
[15] Massarsch, K. R., "Soil Movements Caused by Pile Driving in Clay," Royal Swedish Academy of Engineering Sciences, Report 51, Stockholm, 1976.
[16] Massarsch, K. R. and Broms, B. B., *Proceedings*, Ninth International Conference on Soil Mechanics and Foundations Engineering, Tokyo, Japan, Vol. 1, 1977, pp. 197-200.
[17] Bozozuk, M., Fellenius, B. H., and Samson, L., *Canadian Geotechnical Journal*, Vol. 15, No. 3, Aug. 1978.
[18] Orrje, O. and Broms, B. B., *Proceedings*, American Society of Civil Engineers, Vol. 93, 1967, No. SM5, Proc. Paper 5415, Sept., 1967, pp. 59-73.
[19] Bjerrum, L. and Hohannessen, I., *Proceedings*, Conference Pore Pressure Suction Soils, London, England, 1960, pp. 108-111.
[20] Hagerty, D. J. and Garlanger, J. E. *Proceedings*, American Society of Civil Engineers, Special Conference on Performance of Earth and Earth-Supported Structures, Purdue University, 1972, Vol. 1, Part 2, pp. 1207-1222.
[21] Holtz, R. D. and Boman, P., *Canadian Geotechnical Journal*, Vol. 11, No. 3, Aug. 1974, pp. 423-430.
[22] Plantema, G. and Nolet, C. A., *Proceedings*, Fourth International Conference on Soil Mechanics, London Vol. 2, 1957, p. 52.
[23] Meyerhof, G. G., *Journal of the Soil Mechanics and Foundations Division*, American Society of Civil Engineers, Vol. 85, No. SM6, 1963, pp. 1-29.
[24] Vesić, A. S., *Journal of the Goetechnical Engineering Division*, American Society of Civil Engineers, March 1972, pp. 265-290.
[25] Esrig, M. I., Kirby, R. C., and Bea, R. G., *Proceedings*, Offshore Technology Conference, May 2-5, Houston, Texas, Vol. 3, 1977, pp. 495-506.
[26] Kirby, R. C. and Wroth, C. P., *Proceedings*, 1977 Offshore Technology Conference, May 2-5, Houston, Tex., Vol. 3, 1977, pp. 483-494.
[27] Randloph, M. F. and Wroth, C. P., "Analytical Solution for the Consolidation Around a Driven Pile," University of Cambridge, Cambridge, England, CUED/C, Soil TR 50, 1978.
[28] Carter, J. P., Randloph, M. F., and Wroth, C. P., "Stress and Pore Pressure Changes in Clay during and after the Expansion of a Cylindrical Cavity," University of Cambridge, CUED/C, Soil TR 51, 1978.

[29] Buttling, S., *Geotechnique*, Vol. 26, No. 1, 1976.
[30] Reese, L. C., *Journal of the Geotechnical Engineering Division*, American Society of Civil Engineers, Vol. 104, No. GT1, Jan. 1978, pp. 91-116.
[31] Spanovich, M. and Garvin, R. G., "Field Evaluation of Caisson-Shale Interaction," this volume.
[32] Brierley, G. S., Thompson, D. E., and Eller, C. W., "Interpreting End-Bearing Pile Load Tests Results," this volume.
[33] Fuller, F. M. and Hoy, H. E., "Pile Load Tests Including Quick-Load Tests Methods, Conventional Methods, and Interpretation," Highway Research Record, Number 333, Pile Foundations, Washington, D.C., 1970, pp. 74-86.
[34] Vijayvergiya, V. N. and Focht, J. A., Jr., "A New Way to Predict the Capacity of Piles in Clay," *Proceedings*, Fourth Annual Offshore Technology Conference, Houston, Tex., May 1972.
[35] Leonards, G. A. and Lovell, D., "Interpretation of Load Tests on High-Capacity Driven Piles," this volume.
[36] Randloph, M. F. and Wroth, C. P., "A Simple Approach to Pile Design and the Evaluation of Pile Tests," this volume.
[37] Cheeks, J. R., "Analytical Methods to Predict Pile Capacities," this volume.
[38] Montgomery, M. W., "Pile Load Tests to Evaluate Load Transfer Mechanisms," this volume.
[39] Gleser, S. M. in *Symposium on Lateral Load Tests on Piles, ASTM STP 154* American Society for Testing and Materials, Philadelphia, 1953, pp. 75-101.
[40] Focht, J. A., Jr. and McClelland, B., *The Texas Engineer*, Texas Section, American Society of Civil Engineers, 1955.
[41] Matlock, H. and Reese, L. C., *Proceedings*, 5th International Conference, International Society of Soil Mechanics and Foundation Engineering, 3B/14, Paris, France, July 1961.
[42] Matlock, H. and Ingram, W. B., *Proceedings*, 2nd Pan-American Conference on Soil Mechanics and Foundation Engineering, Paper No. 32, Brazil, July 1963.
[43] Reese, L. C., *Journal of the Geotechnical Engineering Division*, American Society of Civil Engineers, Vol. 103, No. GT4, Apr. 1977, pp. 287-305.
[44] Reese, L. C. and Cox, W. R., in *Performance of Deep Foundations, ASTM STP 444*, American Society of Testing and Materials, Philadelphia, Vol. 123, 1968, pp. 161-176.
[45] Matlock, H., *Proceedings*, 2nd Annual Offshore Technology Conference, Paper No. OTC 1204, Houston, Tex., 1970, pp. 578-588.
[46] Reese, L. C., Cox, W. R., and Koop, F. D., *Proceedings*, 6th Annual Offshore Technology Conference, Vol. 2, Paper No. OTC 2080, Houston, Tex. 1974, pp. 473-483.
[47] Reese, L. C., Cox, W. R., and Koop, F. D., *Proceedings*, 7th Annual Offshore Technology Conference, Vol. 2, Paper No. OTC 2312, Houston, Tex., 1975, pp. 671-690.
[48] Reese, L. C. and Welch, R. C., *Journal of the Geotechnical Engineering Division*, American Society of Civil Engineers, Vol. 101, No. GT7, February 1975, pp. 633-649.
[49] Matlock, H. and Ripperger, E. A., *Proceedings*, 8th Texas Conference on Soil Mechanics and Foundation Engineering, Special Publication No. 29, Bureau of Engineering Research, The University of Texas, Austin, 1956.
[50] Fellenius, B. H., *Canadian Geotechnical Journal*, Vol. 9, 1972, pp. 25-32.
[51] Price, G., this volume.
[52] Robinson, K. E., this volume.

M. I. Esrig[1] *and R. C. Kirby*[1]

Soil Capacity for Supporting Deep Foundation Members in Clay

REFERENCE: Esrig, M. I. and Kirby, R. C., "**Soil Capacity for Supporting Deep Foundation Members in Clay,**" *Behavior of Deep Foundations, ASTM STP 670,* Raymond Lundgren, Ed., American Society for Testing and Materials, 1979, pp. 27-63.

ABSTRACT: A review and evaluation of the commonly used (state-of-practice) methods for analysis of ultimate capacity, load-deflection behavior, and driveability for driven piles in clay is presented. The evaluation suggests that significant progress has been made in modeling those portions of the problem amenable to conventional methods for structural analysis, but that remarkably little progress has been made toward realistically modeling soil behavior. A general effective stress method (state-of-art) for estimation of ultimate pile capacity is also presented and evaluated. The evaluation suggests that the general effective stress method has promise but is still in the process of development.

KEY WORDS: pile foundations, clay, ultimate capacity, effective stress

The interaction between a deep foundation element subjected to axial loads and the supporting soil is related to at least the following factors:

1. The engineering properties of the soil mass prior to pile installation.
2. The length, stiffness, and displacement characteristics of the pile.
3. The method of pile installation.
4. The number and spacing of piles within a group.
5. The changes in stress and engineering properties produced by pile installation.
6. The changes in stress and engineering properties occurring with time after installation.
7. The loading conditions: long-term loading, transient loading, repeated loading, tension loading, compression loading.

Engineering analysis progresses from very simple idealization (models) of

[1]Vice president and project engineer, respectively, Woodward-Clyde Consultants, Clifton, N.J. 07012.

the real problem to more complete models. For complex situations, such as deep foundations, analytical models are commonly developed for individual elements of performance. The models for each aspect of the problem may be based on different assumptions and are generally at different levels of development.

A discussion of the analytical models presently being used for engineering analysis of pile foundations and an evaluation of the adequacy for these models are presented. The evaluation suggests that significant progress has been made in modeling those portions of the problem amenable to conventional methods of structural analyses but that remarkably little progress has been made toward realistically modeling soil behavior. There is little soil mechanics in the engineering of a pile foundation.

Esrig et al [*1*][2] described the initial development of a general effective stress method for the prediction of axial capacity of driven piles in clay. The very nature of an effective stress methodology necessitates consideration of the mechanics of soils and soil behavior. This work is briefly reviewed, and related recent developments reported in the published literature are discussed. The effective stress methodology is still under development and is not yet ready to replace existing methodologies for the prediction of pile capacity. Nevertheless, the initial work suggests that a general effective stress design method for axial pile capacity is within reach.

State of the Practice—Engineering Analysis for Deep Foundations

Ultimate capacity and load-deflection behavior are the major engineering questions for any foundation element. Added concerns for deep foundations are the feasibility of pile installation, the influence of installation procedures on capacity and load-deflection behavior, group effects, and, for driven piles, the use of performance during pile driving as an indicator of ultimate capacity and load-deflection behavior.

Analytical models for deep foundations have generally been developed to permit some specific aspect of pile performance to be considered. In the following paragraphs, the analytical models for ultimate axial capacity, load-deflection behavior, and pile driveability are briefly described and evaluated. Laterally loaded piles and group effects are excluded from the discussion.

Ultimate Capacity of Axially Loaded Piles

A free body of an axially loaded pile is shown in Fig. 1. The ultimate axial load Q is shared by the soil resistance along the pile shaft Q_s and the soil resistance at the pile tip Q_p.

Point Resistance—The capacity of the tip of a pile bearing in clay is nor-

[2] The italic numbers in brackets refer to the list of references appended to this paper.

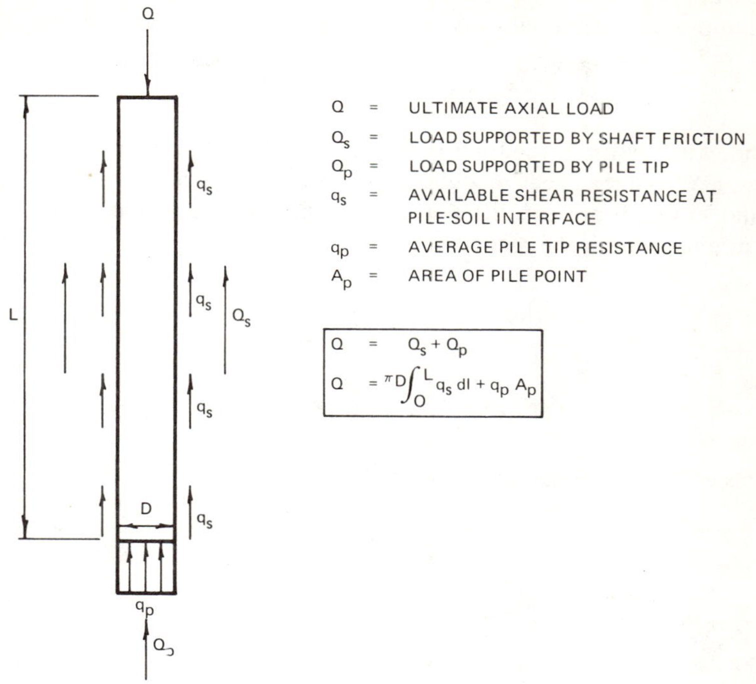

FIG. 1—*Free body of axially loaded pile.*

mally a relatively small portion of the ultimate capacity. As failure under undrained conditions is considered the critical failure mode for design, the unit tip capacity in clay q_p is

$$q_p = S_u N_c \tag{1}$$

in which S_u is the undrained shear strength at the pile tip and N_c is the bearing capacity factor for cohesion. N_c is generally approximated as 9, based on the results of model tests reported by Skempton [*2*] and theoretical analyses and model tests reported by Meyerhof [*3*].

Predictions of point resistance using Eq 1 with N_c equal to 9 and S_u taken as an average value of undrained shear strength are generally sufficiently accurate for engineering purposes because of the small contribution of tip capacity to ultimate pile capacity. However, drilled piers may derive a large portion of their capacity from the pile tip. For these situations, uncertainties in S_u and N_c may be important.

Meyerhof [*3*] and Vesic [*4*] suggest that the value of N_c for a pile can be approximated from the analysis of the limit pressure for expansion of a spherical cavity in an ideal elastic–plastic medium. This problem, for a $\phi = 0$ incompressible material, was solved by Bishop et al [5] who found

$$P_u = {}^4/_3 S_u[\ln (G/S_u) + 1] \tag{2}$$

As indicated by Eq 2, the limit pressure P_u is related to the undrained shear strength S_u and the shear modulus G of the material prior to yield. Meyerhof [3] and Vesic [4] present different approximations for N_c based on the limit pressure for spherical expansion:

$$\text{Meyerhof:} \qquad N_c = \frac{q_p}{S_u} = \frac{P_u}{S_u} + 1 \tag{3}$$

$$\text{Vesic:} \qquad N_c = \frac{q_p}{S_u} = \frac{P_u}{S_u} + 1 + \frac{\pi}{2} \tag{4}$$

Values of N_c for a range of values of G/S_u are

	G/S_u					
	10	50	100	200	400	800
N_c (Eq 3)	5.4	7.5	8.4	9.4	10.3	10.9
N_c (Eq 4)	7.0	9.1	10.0	11.0	11.9	12.8

Values of G/S_u back calculated from measurements[3] of structures founded on normally consolidated and lightly overconsolidated clays with a plasticity index less than 30 are reported by D'Appolonia et al [6] to be about 400. The values were found to be lower (about 200) for clays with a plasticity index greater than 30 and substantially lower (about 40) for a single case history involving a slightly organic plastic clay. Laboratory data were also presented that suggest that G/S_u decreases with increasing overconsolidation ratio (OCR); the value of G/S_u at an OCR of 8 was found to be about one-third the value of G/S_u for normally consolidated specimens.

The application of the analysis for expansion of a spherical cavity in an elastic–plastic medium to the tip capacity of piles is certainly only an approximation to the real problem. Furthermore, estimates of G/S_u from field measurements of initial settlements under surface loadings should be applied with caution to the spherical cavity expansion problem. Nevertheless, these analyses provide the following insights to the range in values for N_c:

1. For a given soil, N_c for overconsolidated clays is expected to be lower than N_c for normally consolidated clays.

[3] The values of G were back calculated at high factors of safety; the values of S_u were obtained from high-quality laboratory tests or field vane tests.

2. For a given OCR, N_c for a clay of low plasticity is expected to be higher than N_c for a clay of high plasticity.

Shaft Resistance: *Total Stress Approach*—The shaft capacity of a pile bearing in clay is normally a very large portion of the ultimate pile capacity. As failure under rapid loading is considered the critical failure mode for design, much of the research on shaft capacity has focused on the undrained shear strength at the pile-soil interface. It is obvious that pile driving displaces and severely distorts the clay located in the path of the penetrating pile. The displacements caused by pile driving in clay produce a large increase in total stress, which is accompanied by an increase in pore pressure and change in effective stress. The pore pressures generated by installing a pile dissipate with time, and the clay reconsolidates. Pile loading shears the soil at the pile-soil interface. This shearing probably occurs at constant volume, but some investigators (for example, Burland [*7*]) believe that drainage can occur in a narrow zone around the pile.

Terzaghi and Peck [*8*] indicate that the soil adjacent to a pile driven into soft clay becomes stronger than the surrounding soil during reconsolidation and suggest that the available shear resistance is approximately equal to the undrained shear strength of the soil as determined by unconfined compression tests on samples recovered prior to pile driving. Tomlinson [*9*] and Woodward et al [*10*] showed that the available shear resistance would be less than the undrained shear strength for piles driven into firm to very stiff clays.

Figures 2 and 3 compare available shear resistance, as back calculated from load tests, to the undrained shear strength of the undisturbed soil. Figure 2 shows results (Flatte and Selnes [*11*]) of such comparisons for 30 pile load tests on timber piles driven into normally consolidated clay. The assumption that the available shear resistance is equal to the undrained shear strength generally overestimates available shear resistance by about 30 percent and the scatter in the correlation is larger. Figure 3 shows the results from load tests evaluated by Vesic [*4*]. As with the data presented by Flatte and Selnes [*11*], the Vesic [*4*] summary shows clear trends but a large amount of scatter about the average values. The amount of scatter in data such as those summarized by Vesic led Tomlinson [*12*] to conclude that the method of installating a pile and the sequence of strata through which a pile penetrates has an important effect on the relationship between available shear resistance and undrained shear strength.

Shaft Resistance: *Simplified Effective Stress Approach*—Starting in about 1960 (Zeevaert [*13*], Eide et al [*14*], Chandler [*15*], Burland [*7*], and others), many researchers began attacking the problem of shaft resistance for piles in clay using an effective stress approach. Few would argue with the principle that shear strength of clay is controlled by effective stress and that the available shear resistance at the pile-soil interface is related to the radial effective stress on the pile *at failure*, the effective stress friction angle, and ef-

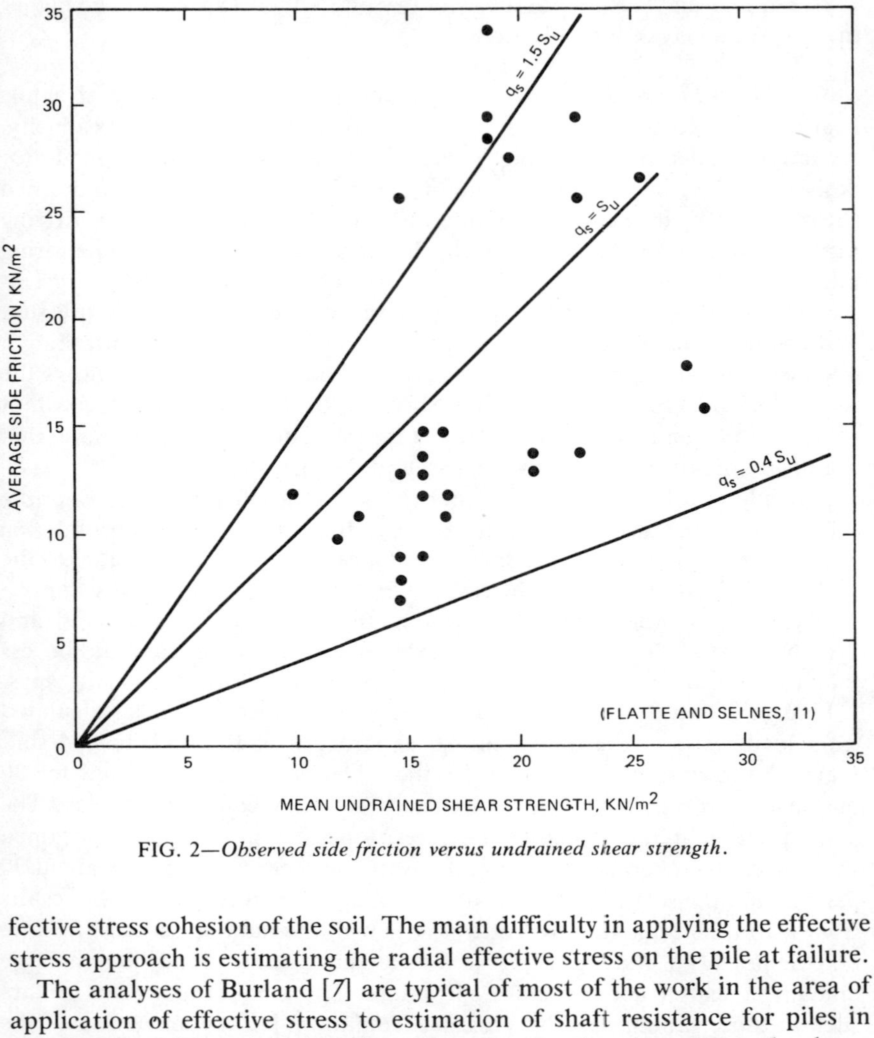

FIG. 2—*Observed side friction versus undrained shear strength.*

fective stress cohesion of the soil. The main difficulty in applying the effective stress approach is estimating the radial effective stress on the pile at failure.

The analyses of Burland [7] are typical of most of the work in the area of application of effective stress to estimation of shaft resistance for piles in clay. Burland [7] makes three important assumptions in order to develop a simple equation for available shear resistance in terms of effective stress:

1. The remolding during pile driving of the soil adjacent to the pile reduces the effective stress cohesion intercept to zero.
2. After dissipation of pore pressures generated by pile driving, the radial effective stress acting on the pile surface is at least equal to the horizontal effective stress ($\bar{\sigma}_{h0}$) prior to pile installation.
3. The major shear distortion during pile loading is confined to a relatively thin zone around the pile shaft and drainage of this thin zone occurs rapidly during loading.

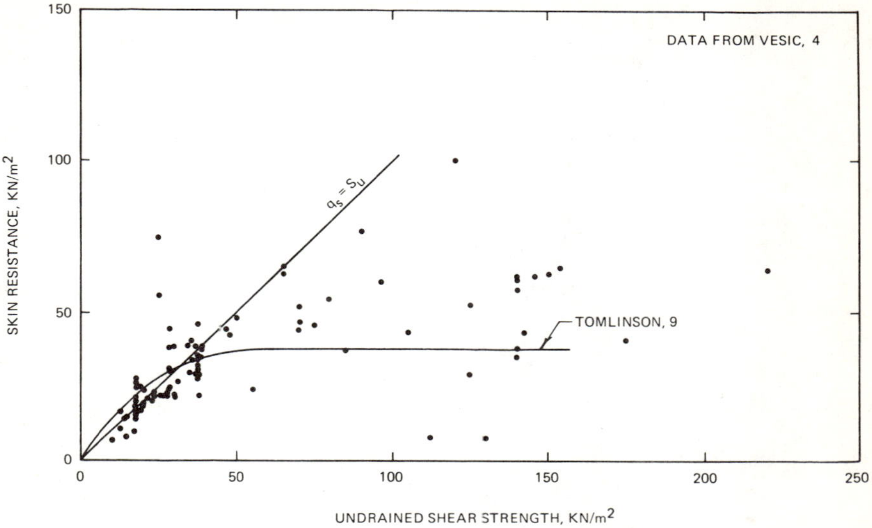

FIG. 3—*Comparison between skin resistance of piles in clay and undrained shear strength.*

The more controversial of these assumptions is that loading of a pile in clay is a drained loading and the implicit assumption that a drained loading would cause no change in radial effective stress. Meyerhof [*16*] generally adopts the assumptions made by Burland [*7*] but defends assumption 3 based on field measurements that indicate that the excess pore pressures generated by pile loading are only about 20 to 50 percent of the undrained shear strength of the clay.

The simple design equation developed by Burland [*7*] can be written as

$$q_s = K_0 \bar{\sigma}_{v0} \tan \bar{\phi} = \beta \bar{\sigma}_{v0} \qquad (5)$$

where q_s is the available unit shear resistance, K_0 is the coefficient of earth pressure at rest ($\bar{\sigma}_{h0}/\bar{\sigma}_{v0}$), $\bar{\sigma}_{v0}$ is the vertical effective stress prior to pile driving, $\bar{\phi}$ is the effective stress friction angle, and β is $K_0 \tan \bar{\phi}$. For normally consolidated clays, K_0 is approximately equal to $1 - \sin \bar{\phi}$ and Burland [*7*] notes that the product of $(1 - \sin \bar{\phi})$ and $\tan \bar{\phi}$ only varies from 0.24 to 0.29 for values of $\bar{\phi}$ between 20 and 30 deg. This surprising result indicates that β should be reasonably constant for normally consolidated clays.

The validity of this conclusion is indicated by the data summarized in Figs. 4 and 5. Figure 4 contains data presented by Burland [*7*] that indicate values of β backcalculated ($\beta = q_s/\bar{\sigma}_{v0}$) from load tests on driven piles in normally consolidated clay. The β values range from 0.25 to 0.4 with an average value of 0.32. Figure 5 contains data presented by Flatte and Selnes [*11*] that indicate values of β back calculated from the same pile load tests that were

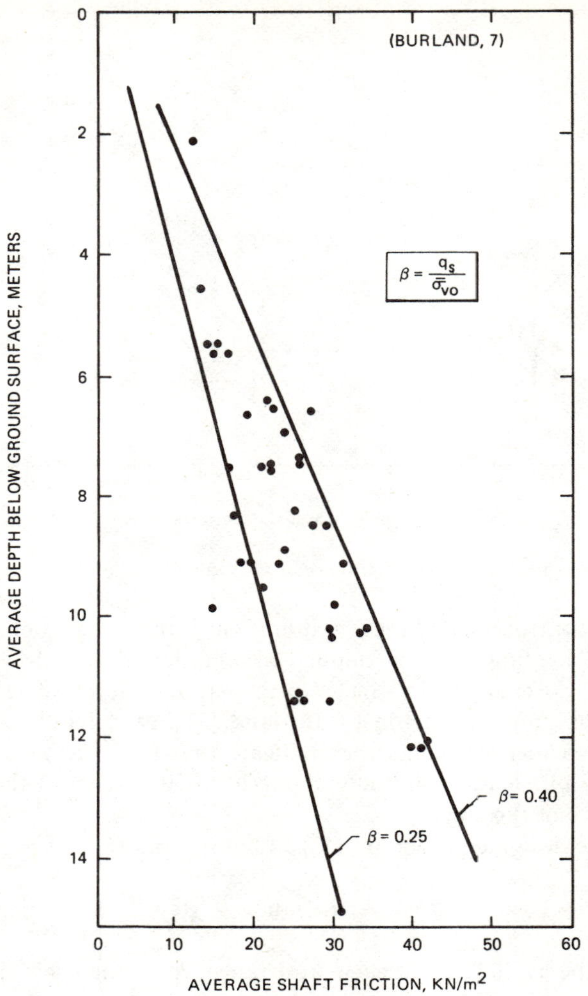

FIG. 4—*Average shaft friction versus average depth for driven piles in soft clay.*

used to prepare Fig. 2. The β values for these tests range from 0.2 to 0.4, with an average value of about 0.32. A comparison of the scatter in data shown in Figs. 2 and 5 indicates that available shear resistance for piles in normally consolidated clay is better correlated with vertical effective stress than with undrained shear strength.

The value of K_0 in Eq 5 is a function of overconsolidation ratio (OCR).

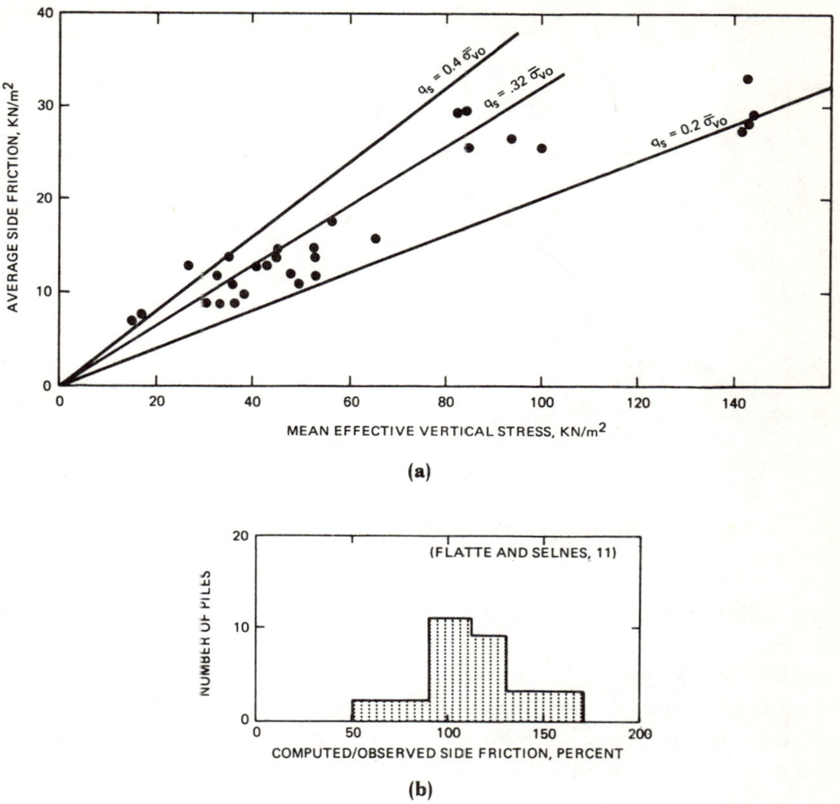

FIG. 5—(a) *Observed side friction versus effective vertical stress.* (b) *Frequency curve for the quotient of calculated to observe side friction* $q_s = 0.32\bar{\sigma}_{v0}$.

Meyerhof [*16*] suggests that the results of laboratory determinations of K_0 can be roughly approximated by the expression

$$K_0 = (1 - \sin \bar{\phi})\,\text{OCR}^{1/2} \tag{6}$$

The accuracy of this expression is discussed later in this report.

Comparisons of available shear resistance predicted using Eq 5 and Eq 6 with available shear resistance back calculated from pile load tests in lightly overconsolidated clays have been made by Flatte and Selnes [*11*]. These comparisons suggest that $\beta = 0.32\,(\text{OCR})^{1/2}$ gives good predictions of available shear resistance as demonstrated by Fig. 6. However, Burland [*7*] reports that the use of a similar equation for driven piles in heavily overconsolidated London clay generally underestimates available shear resistance, often by as much as 50 percent. On the other hand, the available shear resistance for

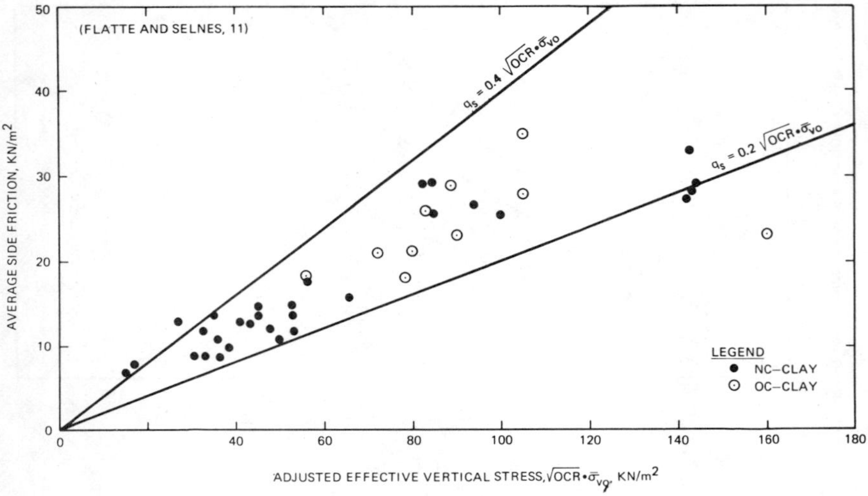

FIG. 6—*Observed side friction versus adjusted effective vertical stress.*

bored piles in London clay is predicted tolerably well by the Burland approach.

Shaft Resistance: *The* λ *Correlation*—Vijayvergiya and Focht [*17*] summarized data from 42 load tests and related available shear resistance to vertical effective stress and undrained shear strength using an empirically determined correlation factor λ,

$$q_s = \lambda(\overline{\sigma}_{v0} + 2S_u) \tag{7}$$

The values of q_s, $\overline{\sigma}_{v0}$, and S_u are average values over the embedded length of the pile. The correlation coefficient λ was plotted against depth of pile penetration and a well-behaved variation of λ with depth developed (see Fig. 7*a*).

A possible danger in the use of such empirically determined correlation coefficients is that they may be surprisingly influenced by the makeup of the data base. The data from short piles included in Fig. 7*a* (penetration less than 50 ft) are generally from piles driven into heavily overconsolidated clay, whereas the data from very long piles are generally piles driven into lightly overconsolidated to underconsolidated clays. A comparison of measured and predicted available shear resistance for the pile load tests reported by Flatte and Selnes [*11*] is shown in Fig. 7*b*. The λ method generally overpredicted the capacity of these piles, and the flatness of the histogram indicates very poor correlation between the measured and predicted values. This is not surprising, since the tests reported in Fig. 7*b* were generally performed on (30-to

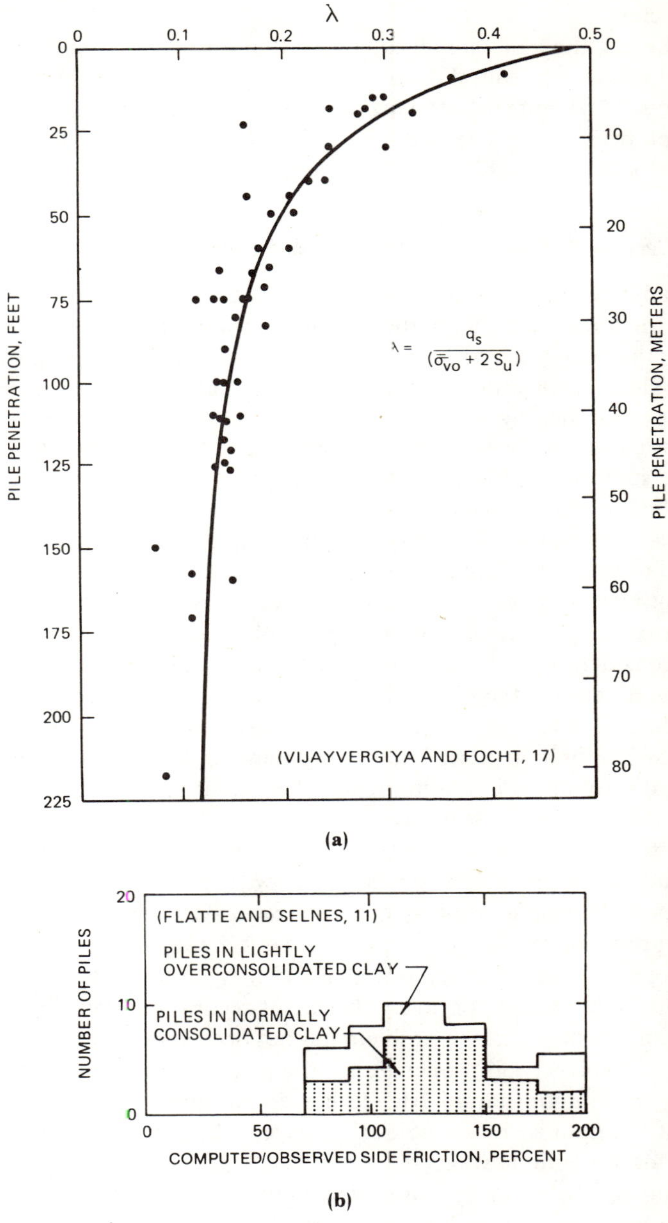

FIG. 7—(a) *Frictional capacity coefficient* λ *versus pile penetration*. (b) *Frequency curve for the quotient of calculated to observed side friction for* $q_s = \lambda(\bar{\sigma}_{v0} + 2S_u)$.

50-ft)-long timber piles driven into normally consolidated or lightly overconsolidated clay.

Nevertheless, the λ method does suggest that the available shear resistance for piles in normally consolidated clays may decrease with increasing pile penetration. Meyerhof [*16*] presents data to suggest that β decreases with pile penetration from about 0.25 to 0.4 for piles up to about (75 ft) long, to about 0.1 to 0.2 for piles from (200 to 300 ft) long.

Load-Deflection Behavior of Axially Loaded Piles

The analytical methods for analyses of load-deflection behavior can be classified in accordance with the method used to idealize soil behavior. Modeling the pile-soil system as an elastic rod supported by an array of nonlinear shear springs along the shaft of the pile and by a point spring at the pile tip was proposed by Seed and Reese [*18*]. Modeling the pile–soil system as an elastic rod supported in an elastic medium has also been investigated and chart solutions published [*19*]. The finite element procedure has also been applied to axially loaded piles (for example, [*20*]).

Analyses of an elastic pile in an elastic medium provide valuable insights to the effects of relative pile length (length/diameter) and relative pile stiffness (equivalent pile modulus/soil modulus) on load transfer and settlement. Recent approximations developed by Randolph and Wroth [*21*] allow economical consideration of piles in layered elastic media. However, relative displacement between the pile and the soil (slip) is not allowed in these solutions, and this is a serious shortcoming, because slip occurs at relatively high factors of safety for long piles. Another problem that cannot be accounted for by the elastic solutions is the influence of residual loads in the pile on the predicted load-deflection behavior.

The finite element method can (in principle) model slip at the pile–soil interface and nonlinear stress–strain behavior and can take into account residual loads in the pile prior to load application. However, the soil properties appropriate for pile loading depend on both the stress conditions and soil properties prior to pile driving, as well as the changes in stress and soil properties due to pile installation and subsequent reconsolidation. Since the installation and reconsolidation problem has not been considered in depth, it is likely that the correct soil properties for the pile loading cannot be selected. Moreover, meaningful parametric studies do not appear to be forthcoming because of the number of variables that have to be considered and the very high cost for the computations.

Modeling soil behavior by equivalent nonlinear springs that support an elastic pile has proven to be a valuable analytical method. The main advantages of this analysis are: (*a*) postslip behavior is modeled, (*b*) the soil springs can be modified to account for residual stresses, and (*c*) the computations are relatively inexpensive and parametric studies can be generated at modest cost. The difficulty with the method is the selection of properties for the soil springs. Guidelines for selection of these properties have been presented based on instrumented pile load tests, but the general validity of these guidelines is uncertain.

Pile Driveability

An overview of the use of wave equation analysis is presented by Peck et al [*22*]. Holloway et al [*23*] describe recent advances in predicting pile driveability and discuss the capabilities of the computer programs that are available. The analytical model idealizes the hammer-pile system as a series of lumped masses connected by springs. The resistance along the shaft of the pile and at the pile point is lumped into springs and dashpots distributed along the pile. The recent refinements to the wave equation analysis have been directed toward (*a*) better simulation of the manner in which diesel hammers supply energy to the pile and (*b*) analyses of the pile response under a "complete" hammer blow, which then permits the simulation of a series of hammer blows and provides a prediction of the residual driving stresses [*24*]. The prediction of residual driving stresses is extremely important to the analysis of load-deflection behavior.

The major shortcoming of the wave equation model is that the simplified rheological model for soil response, especially the dashpot, is not relatable to any measurable engineering soil property. Values of the damping coefficient are experimentally determined by matching the predicted and measured driving resistance and pile capacity. This shortcoming is likely not serious for selection of a suitable hammer-pile system or prediction of peak driving stresses. However, the use of the wave equation analyses to predict the ultimate bearing capacity at the time of pile driving is sensitive to the damping coefficient. The most difficult application of the wave equation is the prediction of pile driveability. This prediction requires estimation of the ultimate bearing capacity at the time of driving and selection of the correct damping coefficient. The damping coefficient is estimated from experience. The ultimate bearing capacity at the time of pile driving must be predicted from soil properties. Unfortunately, little progress has been made toward development of an analytic method that will yield reliable predictions of pile capacity at the time of driving.

Summary and Evaluation of Analytical Methods

Analytical methods for prediction of axial pile capacity, load-deflection behavior, and behavior of piles during driving have been briefly discussed. Summary and evaluation of these analytical methods follow.

Axial Capacity—The analysis of axial capacity for piles in clay involves the development of empirical relationships between the available shear resistance at the pile-soil interface and preconstruction stresses, soil properties, or both. The method for relating available shear resistance to undrained shear strength is well established. However, the scatter in the comparisons between available shear resistance and undrained shear strength is large.

Recently, several investigators have proposed simplified effective stress analyses and correlated available shear resistance with initial vertical effective stress and overconsolidation ratio. These correlations appear to have less scatter than the correlations with undrained shear strength. The simplified effective stress analyses are based on several important and controversial assumptions. It is quite possible that these simplified analyses appear to work because of compensating errors.

A general effective stress analysis for axial capacity of piles in clay is in the process of development. The initial development of this method is described below, and the method appears promising.

In summary, the analysis of axial pile capacity is a very complex soil mechanics problem. The prediction methods that have been developed to date are imperfect, and such predictions should generally be considered as preliminary design estimates. Local experience and load testing to confirm design pile capacity, or a conservative design, are generally required for heavily loaded piles.

Load-Deflection Behavior—Significant progress has been made in the analysis of load-deflection behavior of piles supported by idealized soils. These analytical methods provide important insights to some aspects of pile behavior. However, little progress has been made toward the development of reliable methods to predict the soil characteristics necessary for the analysis. The soil characterization for load-deflection behavior is more demanding than that required for axial capacity, because deformability, as well as strength, is important.

Pile Driveability—The status of development of analytical methods for pile driveability is quite similar to that for load-deflection behavior. Significant progress has been made in analysis of driveability for piles driven into idealized soils. These analytical methods provide important insights to those aspects of pile performance that are insensitive to the soil idealization. Examples of this are selection of a hammer-pile system that will deliver maximum energy and still maintain reasonable levels of driving stresses and thereby minimize pile damage.

However, those aspects of predicted performance that are based on soil properties are frequently unreliable. The correlation between driving resistance and static capacity at the time of driving are sensitive to soil damping. This is not really a soil property, but a correction factor back calculated from driving records and load tests that appears to work. The factors affecting soil damping are not well understood.

A prediction of pile driveability requires an estimate of the static capacity at the time of pile driving and of the soil damping. This is probably one of the most uncertain predictions in foundation engineering.

State of the Art—General Effective Stress Analysis for Axial Capacity

During the past several years, progress has been made in the development of a general effective stresss method for the prediction of the skin frictional resistance of piles driven into clay. The premise of effective stress analysis is that the shearing resistance on the failure surface is a function of the normal effective stress on that surface at failure and the effective stress shear strength parameters $\bar{c}$ and $\bar{\phi}$. It is commonly believed that the critical failure surface is at or close to the pile-soil interface and that the major problem lies in determining the effective stresses at the interface at failure; it is assumed that the appropriate shear strength parameters can be measured.

Determination of the effective stresses at the pile-soil interface can be viewed as a problem of addition; add to (*a*) the initial state of stress in the ground prior to pile driving, (*b*) the change in stress due to pile driving, (*c*) the change in stress due to reconsolidation after pile driving, (*d*) the change in stress due to pile loading to yield the effective stress at failure. The state of our ability to determine each of these stresses and stress changes is described in the next sections.

Initial State of Stress and Stress History

The initial state of effective stress and stress history of an element of soil may be characterized by three parameters—the vertical effective stress $\bar{\sigma}_{v0}$, the earth pressure coefficient K_0, and the overconsolidation ratio OCR.

The vertical effective stress is obtained from measurements of soil unit weight and pore water pressure. Determination of pore water pressures may require the installation of piezometers. Nevertheless, a value of $\bar{\sigma}_{v0}$ for which there is a high degree of confidence can be obtained.

The horizontal stress $\bar{\sigma}_{h0}$ is equal to $K_0\bar{\sigma}_{v0}$. Although K_0 can be measured directly in the laboratory using specialized equipment and techniques, it is most often inferred from correlations with available test data. However, those correlations are functions of OCR, which is defined as

$$\text{OCR} = \bar{\sigma}_{v(\max)}/\bar{\sigma}_{v0} \tag{8}$$

where

$\bar{\sigma}_{v(\max)}$ = maximum past vertical effective stress to which the point in the soil mass has been subjected, and

$\bar{\sigma}_{v0}$ = current vertical effective stress.

Determination of the maximum past vertical effective stress (preconsolidation pressure) to which a soil mass has been subjected has been one of the important bits of information obtained from the results of one-dimensional consolidation tests. Most often, the preconsolidation pressure is determined by the empirical methods suggested by Casagrande [*25*] or Schmertmann [*26*], although other methods (for example, [*27*]) are also in use.

It is recognized that sample disturbance may affect significantly the determination of $\bar{\sigma}_{v(\max)}$ from consolidation tests. Therefore, indirect methods for estimating OCR are often used to verify the results of consolidation tests. The use of liquidity index is one such method.

Computation of $\bar{\sigma}_{v0}$ and liquidity index I_l for a particular soil specimen yields a data point that can be plotted on Fig. 8 as point A. The curve shown in Fig. 8 is a roughly unique relationship between liquidity index and vertical effective stress that pertains to normally consolidated clays. The preconsolidation pressure $\bar{\sigma}_{v(\max)}$ is approximately given by the intersection between the appropriate rebound curve (shown in the figure) that passes through point A and the curve for normally consolidated clays. OCR determined by these means is subject to substantial error as $\bar{\sigma}_{v(\max)}$ increases.

Alternatively, the undrained shear strength S_u may be used as an indicator of OCR through the relationship shown as Fig. 9 [*28*]. In order to use the curve, the undrained shear strength of the soil measured in the laboratory or the field is divided by $\bar{\sigma}_{v0}$ to find $(S_u/\bar{\sigma}_{v0})$. For normally consolidated soils tested in unconfined compression or using the field vane shear apparatus, the ratio is expected to be approximated by

$$(S_u/\bar{\sigma}_{v0})_{nc} = 0.11 + 0.0037 I_p \tag{9}$$

where I_p is the plasticity index of the clay.

As the OCR increases, the normalized ratio $(S_u/\bar{\sigma}_{v0})/(S_u/\bar{\sigma}_{v0})_{nc}$ is believed to increase as shown in Fig. 9. Therefore, determination of this ratio on the

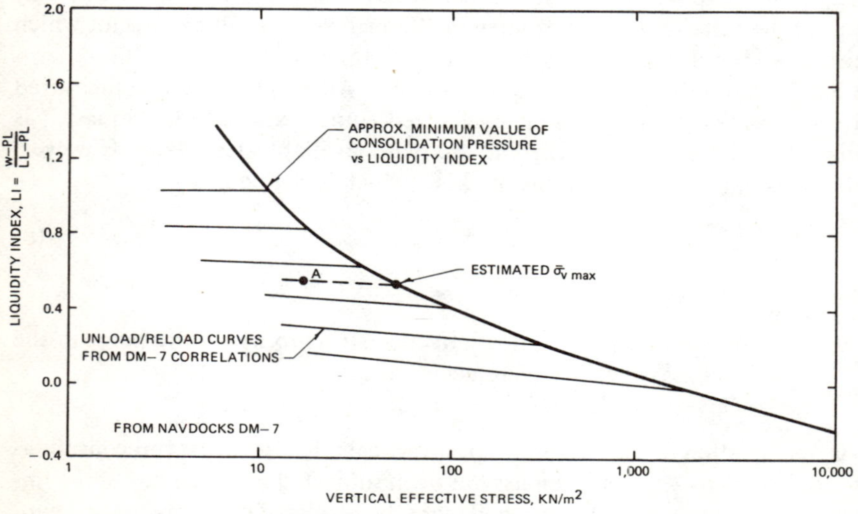

FIG. 8—*Correlation between liquidity index and effective vertical stress showing family of unloading–reloading curves and example calculation.*

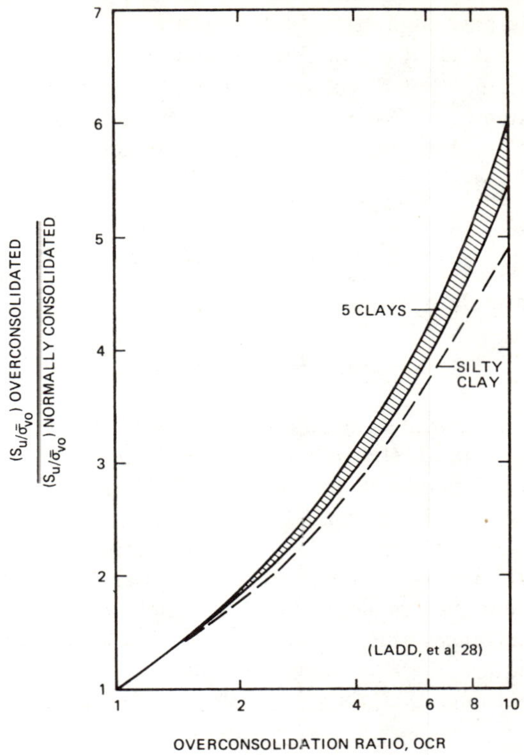

FIG. 9—*Variation in undrained shear strength with overconsolidation ratio.*

basis of laboratory tests permits entry into Fig. 9 to estimate OCR. Alternatively, OCR may be estimated from the relationship

$$\frac{(S_u/\bar{\sigma}_{v0})_{0c}}{(S_u/\bar{\sigma}_{v0})_{nc}} = \text{OCR}^m \tag{10}$$

where

$m = 0.85$ for OCR < 3
$m = 0.8$ for OCR $= 3$ to 6
$m = 0.75$ for OCR > 6

The effects of sample disturbance on the measured value of S_u lead to uncertainties in determination of the normalized ratios. Furthermore, uncertainty about the generality of the relationship in Fig. 9 increases uncertainties associated with its use. Still, the OCR determined by this means, which is believed to be an underestimate of OCR in situ, provides valuable information on which to base engineering judgment.

Determination of OCR permits, within limits, definition of K_0 by indirect methods. Assuming that direct measurement is not possible, correlations from which estimates of K_0 can be obtained are available.

Ladd et al [*28*] have compiled data that indicate that for normally consolidated soils

$$(K_0)_{nc} = (1 - \sin \overline{\phi}) \pm 0.05 \tag{10a}$$

and for overconsolidated soils

$$(K_0)_{oc} = (K_0)_{nc}\,\mathrm{OCR}^n \tag{10b}$$

The power n is recognized as being a function of plasticity index I_p. The relationship between n and I_p shown in Fig. 10 has been developed from relationships presented independently by Ladd et al [*28*] and Gardner [*29*].

It is emphasized by Ladd et al [*28*] and others that Eq 10*b* for $(K_0)_{oc}$ is only valid for soils subjected to a single unloading. Upon reloading, K_0 decreases rapidly and is always smaller than that obtained from Eq 10*b*. Therefore, for complex geologic histories, such as those associated with several glacial advances and retreats, $(K_0)_{oc}$ must be measured by the best available techniques.

With the definition of $\overline{\sigma}_{v0}$, OCR, and K_0, the initial state of stress and the

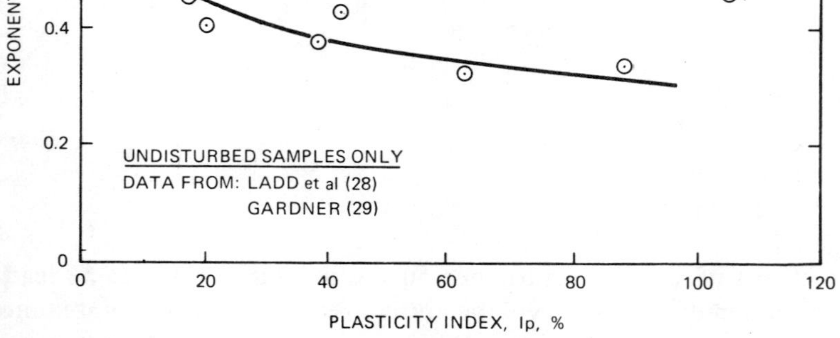

FIG. 10—*Variation* K₀ *exponent* n *with plasticity index.*

stress history prior to pile driving are fully characterized. The stress changes caused by driving the pile must next be investigated.

Stress Changes Due to Pile Driving

The stress changes resulting from pile driving can be separated into several categories—those occurring along the shaft and at the tip of the driven pile due to its installation, and those occurring as a result of driving adjacent piles.

Cooke [*30*] shows by measurements obtained as a pile was jacked into the ground what Tomlinson [*12*] had inferred from observations on piles excavated after installation and others had inferred from model tests (for example, [*31*,*32*])—major downward displacements of soil occur adjacent to the pile shaft and at the pile tip and extend into the soil mass a relatively small distance from the pile. These displacements appear large enough to fully remold the soil at the pile-soil interface as Tomlinson's [*12*] observations suggest and appear to generate large enough shear strains to bring to failure, in the z–r plane, a mass of soil within a distance from the pile shaft of at least one pile diameter. Of particular importance, in the Cooke [*30*] data, is the observation that, upon unloading of the force causing pile penetration, the net vertical displacements near the shaft, for a major portion of the length of the pile are almost zero. That is, as a first approximation, soil displacements at the pile shaft are largely radial over much of the length of the pile.

This observation does not deny the fact that soil heave occurs near a driven pile due to the volume changes required for installation, but rather it indicates that, near the pile shaft, the downward displacement of soil adjacent to the pile observed during driving is essentially counterbalanced by the rebound of the pile on release of the driving force. Pile heave has been described in some detail by Hagerty [*33*] and by Cooke and Price [*34*] as is believed to be a major contributor to the pile problems reported by Clark [*35*].

However, the observation of essentially radial displacement near the surface of the pile for most of its length suggests that it is likely a reasonable first approximation to model the installation of the shaft of a driven pile as the expansion of a cylindrical cavity. Such an analysis yields an estimate of the state of stress and pore water pressures developed at the pile-soil interface as a result of pile driving. Many analyses of this type have been made during the past decade [*1*,*32*,*36*,*37*,*38*]. All have used the assumption of plane strain cavity expansion, none have given consideration to the effects of cyclic vertical movements during pile driving (which may result in large residual stresses on the boundary of the expanding cylinder), and all but Carter et al [*38*] have modeled the soil as an incompressible elastic-plastic material.

The solution by Carter et al [*38*], which utilizes the concept of the critical state and the Cam-clay model of plastic soil behavior, shows that the relatively simple elastic-plastic analysis yields stresses and pore water pressures similar to those obtained from the more complex models. Although it is to be expected that the presence of residual shear stresses at the pile-soil boundary will alter the horizontal stress at the boundary, it is not likely to alter the mean normal effective stress immediately after pile installation nor the distribution of pore water pressure generated by the installation of the pile.

Despite the apparent limitations of the elastic-plastic plane strain model for pile installation, studies performed to date suggest that it leads to reasonable predictions of the magnitude and distribution of pore water pressure generated by pile driving and to reasonable predictions of the change in radial stress at the pile-soil interface. The geometry of the elastic-plastic analysis, the assumed soil behavior, and the governing equations are shown in Fig. 11.

The assumption of elastic behavior precludes the generation of pore water pressures prior to failure, the point at which the change in radial total stress equals the undrained shear strength of the soil. Thereafter, all increases in stress to expand the cylinder are reflected as changes in pore water pressure; that is, the pore water pressure generated by expansion is equal to the change in radial total stress minus the undrained shear strength. The predicted change in radial total stress $\Delta\sigma_{r(\max)}$ and the predicted change in pore water pressure $\Delta u_{\max}$ at the face of the expanding cavity are tabulated below for a variety of assumed ratios of the undrained Young's modulus of the soil E_u to the undrained shear strength S_u.

E_u/S_u	R/ρ	$\Delta\sigma_{r(\max)}/S_u$	$\Delta u_{\max}/S_u$
2000	25.8	7.5	6.5
1000	18.2	6.8	5.8
500	12.9	6.1	5.1
250	9.1	5.4	4.4

The predicted changes in radial total stress and pore water pressure are seen to be relatively insensitive to modulus ratio. As a first approximation, the model predicts

$$\begin{aligned}\Delta\sigma_{r(\max)} &= (7.5 \pm 1)S_u \\ \Delta u_{\max} &= (6.5 \pm 1)S_u\end{aligned} \tag{11}$$

For normally consolidated and lightly overconsolidated soils, where $S_u/\bar{\sigma}_{v0}$ may range between about $^1/_4$ and $^1/_2$, the model predicts $\Delta u_{\max}$ of $1.5\bar{\sigma}_{v0}$ to $3\bar{\sigma}_{v0}$.

The predicted $\Delta\sigma_{r(\max)}$ of $7.5S_u$ may be compared with the direct measurements of Butterfield and Johnston [*39*], who jacked a 100-mm diameter by 4-m long steel pipe into heavily overconsolidated London clay for which E_u/S_u in compression may be 100 to 300 and reported radial stresses of between $6S_u$ and $8S_u$ near the pile tip and between $4S_u$ and $6S_u$ along the pile shaft.

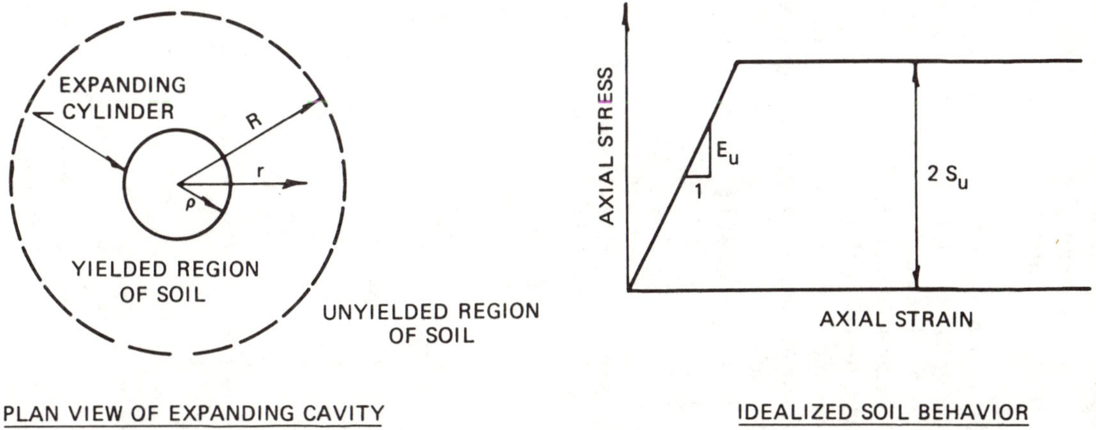

SIZE OF YIELDED REGION : $R/\rho = \sqrt{\dfrac{E_u}{3\,S_u}}$

CHANGE IN RADIAL TOTAL STRESS : $\Delta\sigma_r = S_u\,[1 + 2\ln R/r]$; FOR $\rho \leqslant r \leqslant R$ $\quad \Delta\sigma_r = S_u\,[R^2/r^2]$; FOR $r \geqslant R$

FIG. 11—*Summary of expression for change in radial total stress due to plane strain cylindrical cavity expansion.*

The predicted linear distribution of stress and excess pore pressure with the logarithm of dimensionless radial offset is shown in Fig. 12 for the case of E_u/S_u equal to 1200. The approximate distribution of excess pore pressure obtained from one set of field measurements [*39*] is illustrated in Fig. 13, where it can be seen that the excess pore pressure at the pile-soil interface is approximately $5S_u$.

It is concluded from this discussion that a reasonable first approximation to the distribution of pore water pressure and stress along the shaft of the pile due to pile driving may be obtained from a simple analysis of an expanding cylindrical cavity in elastic-plastic soil that remains at constant volume. Knowledge of this distribution is necessary if the rate of reconsolidation and the stresses at the completion of reconsolidation are to be determined.

The fact that the radial stresses at the pile tip reported by Butterfield and Johnston [*40*] are larger than those along the shaft is not surprising. Tip stresses of $9S_u$ are predicted by conventional bearing capacity theory and may be larger or smaller than that, depending on the modulus ratio E_u/S_u, if the expanding spherical cavity analysis discussed above represents the conditions at the tip. Of special interest, however, in the region of the tip is the stress reduction that appears to develop as the tip stress is applied to the soil. This stress reduction is predicted by the Mindlin [*40*] solution for stresses at a

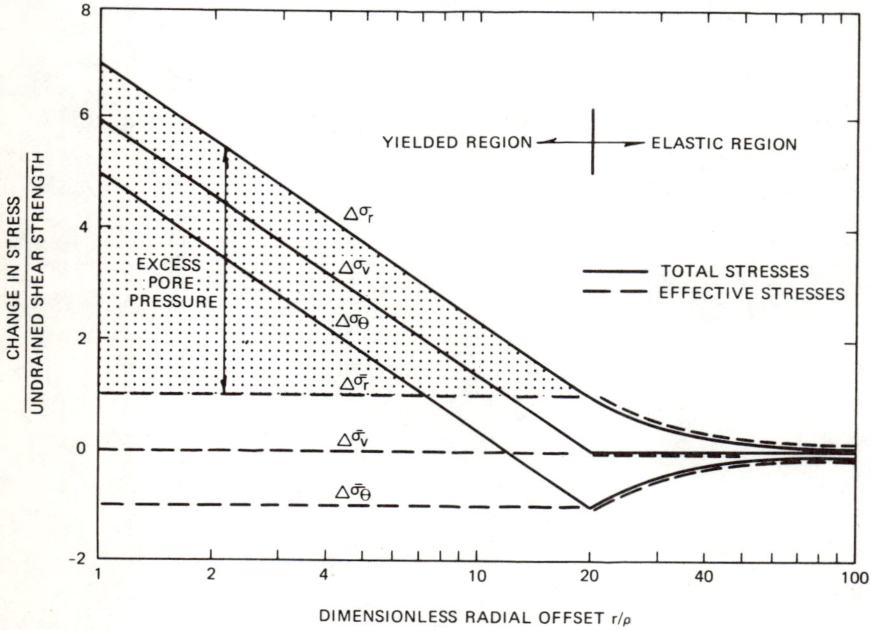

FIG. 12—*Predicted variation of total stress, effective stress, and pore pressure; full displacement pile*, $E_u/S_u = 1200$.

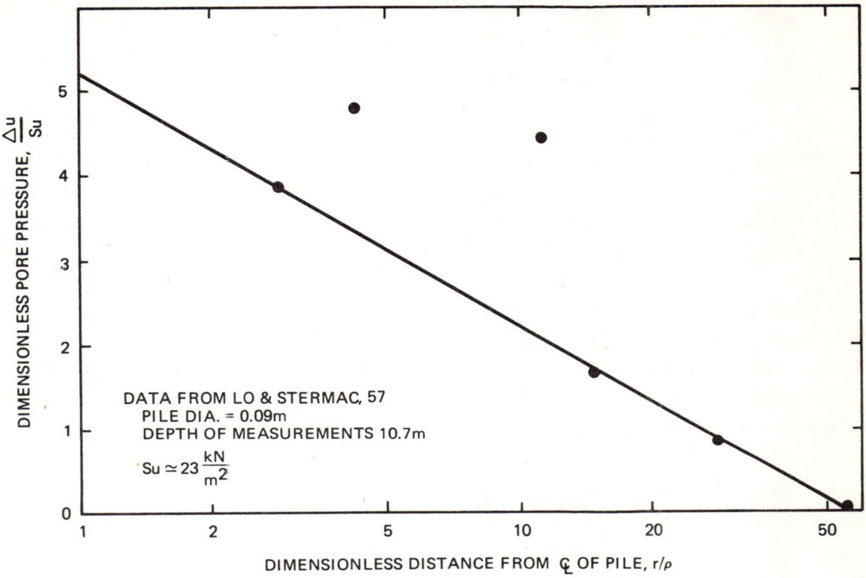

FIG. 13—*Excess pore pressure due to pile driving versus dimensionless radial offset.*

point within an elastic body and may be illustrated by the soil fracturing reported by Clark [*36*]. Butterfield and Ghosh [*42*] report tensile tip stresses on application of working loads to their model piles. Clearly, the state of stress at the pile tip is complex but only affects a relatively small portion of a pile of normal length to diameter ratio.

Before proceeding to the analysis of reconsolidation, it is necessary to further explore the state of effective stress immediately upon pile installation and to introduce the concept of the critical state.

The application of the concept of the critical state to the prediction of axial capacity for driven piles in clay has been described by Kirby and Wroth [*42*]. The critical state concept is considered particularly appropriate for the pile problem, because it is concerned with the behavior of soils after large deformations (soil remolding) have occurred. By definition, when a soil is at the critical state, further soil deformations can occur without further changes in void ratio or mean normal effective stress. Critical state soil mechanics utilizes the familiar Mohr-Coulomb failure criterion but expresses the failure condition in an unfamiliar manner, using terms related to invariants. The terms chosen to describe stress conditions in the triaxial test are $\bar{p}$, the mean normal effective stress, and q, the deviator stress.

$$\bar{p} = {}^1/_3(\bar{\sigma}_1 + \bar{\sigma}_2 + \bar{\sigma}_3) = \bar{\sigma}_{oct} \quad (12a)$$

$$q = \sigma_1 - \sigma_3 = \frac{3}{\sqrt{2}} \tau_{oct} \tag{12b}$$

where

$\bar{\sigma}_1$ = major principal effective stress
$\bar{\sigma}_2$ = intermediate principal effective stress
$\bar{\sigma}_3$ = minor principal effective stress
$\bar{\sigma}_{oct}$ = octahedral normal effective stress, and
τ_{oct} = octahedral shear stress.

The q–$\bar{p}$ stress space is shown as Fig. 14*a*. Also shown is the failure envelope, which is the locus of all points that have achieved failure in accordance with the Mohr–Coulomb failure criterion. This failure envelope rises at an inclination M so that $q = M\bar{p}$. The inclination M can be expressed in terms of the angle of shearing resistance of the soil $\bar{\phi}$ as follows:

$$M_1 = \frac{6 \times \sin \bar{\phi}}{3 - \sin \bar{\phi}} \quad \text{when} \quad \bar{\sigma}_2 = \bar{\sigma}_3 \tag{13a}$$

$$M_2 = \frac{6 \times \sin \bar{\phi}}{3 + \sin \bar{\phi}} \quad \text{when} \quad \bar{\sigma}_1 = \bar{\sigma}_2 \tag{13b}$$

At present, only M_1 is believed to be of concern for the pile problem.

Shown in Fig. 14*b* and designated as the critical state line (CSL)[4] is the locus of all combinations of void ratio e and mean normal effective stress $\bar{p}$ when a soil is being sheared at the critical state. To be at the critical state, a point must be on the CSL in both q–$\bar{p}$ and e–$\bar{p}$ space.

The stress paths shown in Fig. 14 illustrate the stress conditions believed to be associated with pile driving. The initial state of stress at a particular elevation in the ground is given by point 1 in both halves of the figure.[5] It has been assumed that the soil is normally consolidated and that point 1 is on the virgin compression line (VCL). Driving the pile remolds the soil and expands the cylindrical cavity. The effective stress path is from points 1 to 2′ to 3′, following the solid line. The difference between the total stress p and the effective stress $\bar{p}$ is the excess pore water pressure.

Thus, if the distance between point 1 ($\bar{p}_{nc}$) and point 3′ ($\bar{p}_{cs}$) is known, the effective stress after pile driving is also known. As the CSL and the VCL are believed to be parallel in e–log $\bar{p}$ space, the ratio $\bar{p}_{cs}/\bar{p}_{nc}$ is a constant, which when combined with $\bar{p}_{nc}$, defines $\bar{p}_{cs}$.

[4]The CSL is a curve in the arithmetic space shown, but is a straight line in e–log $\bar{p}$ space.

[5]The stresses plotted in Fig. 14 have been reduced by the ambient pore pressure. Consequently, the initial mean normal effective stress and the mean normal total stress can be represented by point 1.

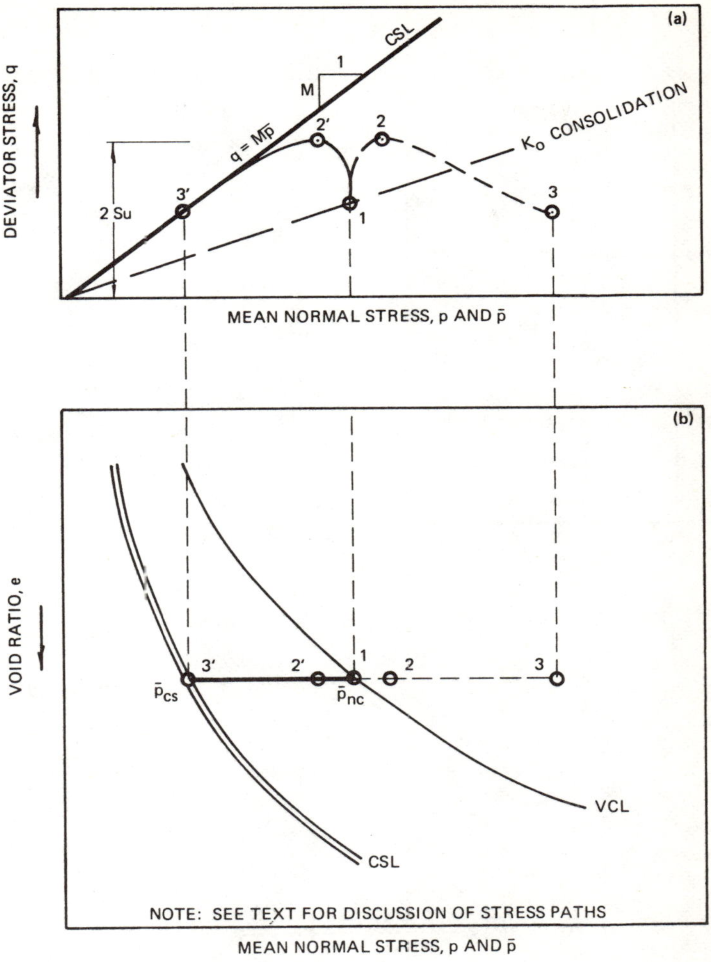

FIG. 14—*Stress paths for stress changes due to pile driving.*

This ratio can be measured directly in the laboratory. However, a tentative relationship between the ratio and the liquid limit of the soil has been proposed by Kirby and Wroth [*43*] and is shown as Fig. 15. It is based on limited test data, mainly from the direct simple shear device, and is therefore considered appropriate for soils that are anisotropically consolidated.

An expression can readily be developed from the relationships described in Fig. 16 to locate the CSL if soils are overconsolidated. That expression is

$$\bar{p}_{cs} = \frac{\bar{p}_{\max}}{(\bar{p}_{\max}/\bar{p}_0)^{C_r'/C_c}} \frac{\bar{p}_{cs}}{\bar{p}_{nc}} \tag{14}$$

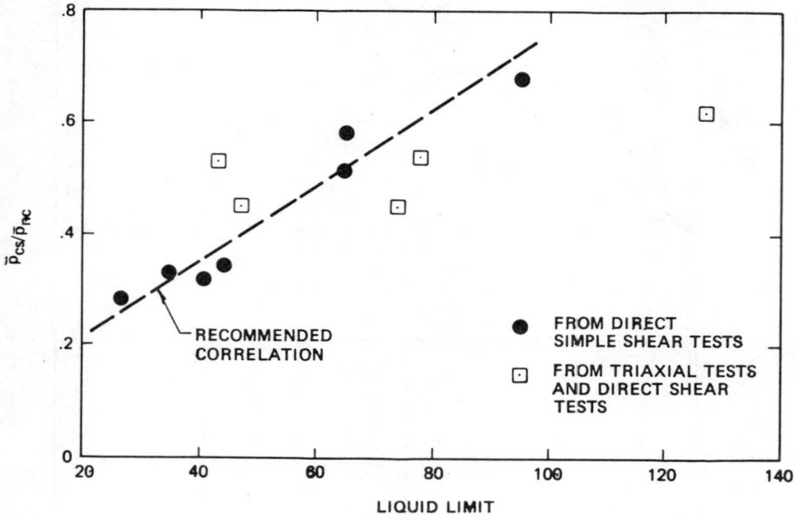

FIG. 15—*Summary of* $\bar{p}_{cs}/\bar{p}_{nc}$ *from laboratory tests on twelve clays.*

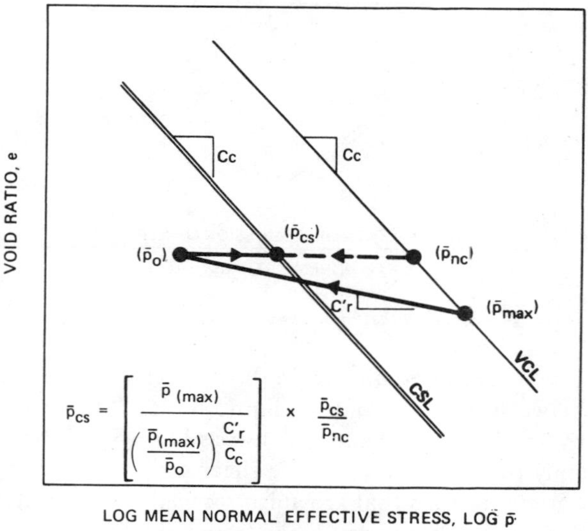

$$\bar{p}_{cs} = \left[\frac{\bar{p}\,(max)}{\left(\frac{\bar{p}_{(max)}}{\bar{p}_o} \right)^{\frac{C'_r}{C_c}}} \right] \times \frac{\bar{p}_{cs}}{\bar{p}_{nc}}$$

FIG. 16—*Basis for derivation of expression for* $\bar{p}_{cs}$ *for driven pile in overconsolidated clay.*

where

$\bar{p}_{max}$ = maximum mean normal effective stress that the soil has been subjected to in its geologic past,
$\bar{p}_0$ = current level of the mean normal effective stress,
$\bar{p}_{nc}$ = mean normal effective stress if the soil were normally consolidated at the current void ratio, and
C_c and C_r' = compression index and the swelling index, respectively, defined in terms of mean normal stress.

It is now appropriate to consider the stress changes due to reconsolidation after pile driving.

Stress Changes Due to Reconsolidation

Magnitude of Stress Changes—The stress changes that occur during reconsolidation after pile driving can be illustrated with the aid of Fig. 17. The yielded region around a driven pile is shown extending to an elastic-plastic boundary at a distance R, measured from the center line of the pile. The elastic region beyond the distance R has been replaced by an equivalent array of elastic springs.

Dissipation of excess pore water pressure generated by pile driving results in volume decrease throughout the yielded region. This volume decrease must be accompanied by an inward movement of the elastic boundary. The inward movement of the boundary represents a decrease in radial total stress occurring from consolidation. No changes in total stress are assumed as occurring in the classic Terzaghi formulation of consolidation; the Terzaghi formulation was utilized by Soderberg [*44*] and others for the pile problem, and therefore, those solutions cannot be relied upon to provide insight into stress changes due to reconsolidation.

Esrig et al [*1*] used the mechanical model of Fig. 17 to provide an indication of the likely stress level at the pile-soil interface after reconsolidation. They concluded that the mean normal effective stress after reconsolidation $\bar{p}$ is likely to be, for normally consolidated soils, about equal to the mean normal effective stress prior to pile driving, while for heavily overconsolidated soils, it is about equal to the mean normal effective stress at the critical state.

In arriving at this conclusion, they assumed, on the basis of the physical reasoning outlined by Kirby and Wroth [*43*], that reconsolidation after pile driving occurs along the critical state line. A closed-form, elastic, Biot type consolidation analysis has been developed by Randolph and Wroth [*45*]. This solution also implies that consolidation occurs along the critical state line.

Miller et al [*46*], uncomfortable with the assumptions of Esrig et al [*1*] and the conclusions of the elastic analysis, produced what may be the first solution to a practical consolidation problem that incorporates the concepts of plasticity. Their finite difference solution permits the introduction of any in-

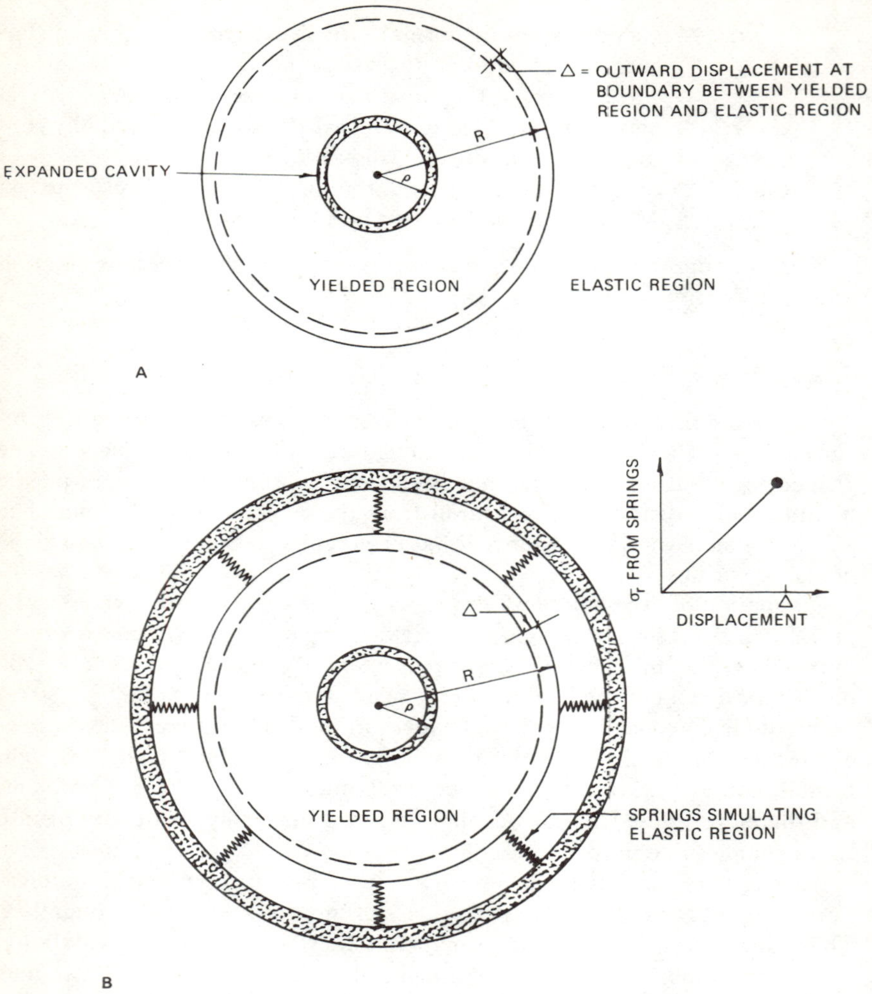

FIG. 17—(A) *Yielded and elastic regions around an expanded cavity in an elastic–plastic medium.* (B) *Yielded region and spring simulation of elastic region around an expanded cavity in an elastic–plastic medium.*

itial pore water pressure distribution and models soil behavior using a soil model developed by Roscoe and Burland [*47*].

Miller et al [*46*] drew two important conclusions from the single case they tested, which approximates reconsolidation after pile driving:

1. Consolidation occurs under radial K_0 conditions. When consolidation is complete, the radial stress $\bar{\sigma}_r$ is the major principal stress and the cir-

cumferential and vertical stresses are equal to $K_0\bar{\sigma}_r$. At no time is consolidation along the critical state line.

2. The mean normal effective stress at the pile-soil interface is smaller than that within a distance of several pile diameters of the interface. Thus, the pile-soil interface is shown to be the surface of least shearing resistance and, therefore, the potential failure surface.

A solution to the consolidation problem that is similar to but more comprehensive than that presented by Miller et al [*46*] has recently been developed by Carter et al [*38*]. Analysis of the limited few cases tested by both of these investigators suggests that, as a first approximation and a temporary expedient until more data become available, it is reasonable to believe that, for normally consolidated soils, the mean normal effective stress *at failure* when the pile is loaded axially is about equal to the mean normal effective stress prior to pile driving. This assumption is consistent with the use of the initial horizontal effective stress in Burland's [*7*] simplified effective stress method. The plasticity solutions have not yet been applied to overconsolidated soils.

Whether the elastic or the plastic solution to consolidation around a driven pile better represents reality is, at present, unclear. Consolidation along the critical state line implies that, on loading a pile to failure, the only excess pore water pressures generated are the consequence of and equal to the increases in mean normal total stress; that is, no changes in mean normal effective stress occur during loading of piles in clay. The Miller et al [*46*] solution implies the generation of excess pore water pressures on pile loading due to applied shear stresses about equal to the difference between $\bar{p}_{nc}$ and $\bar{p}_{cs}$ (see Fig. 15 for the relationship between $\bar{p}_{nc}$ and $\bar{p}_{cs}$) plus those pore water pressures resulting from changes in mean normal total stress. The few data available on pore water pressures due to pile loading [*31,48,49,50*] suggest to the authors that the pore water pressures are low and close in magnitude to the changes in total stress associated with loading. Nevertheless, these field measurements do not provide sufficiently clear data on which to base an absolute conclusion.

The papers to this conference [*23,30*] that discuss residual stresses due to driving a single pile or stresses resulting from driving adjacent piles [*35*] suggest that truth may be found between the elastic and plastic solutions. A series of consolidation lines are shown in Fig. 18 in e–log $\bar{p}$ space. Consolidation along the critical state line is shown by the heavy line marked with the number 2; consolidation along the curve representing K_0 consolidation is marked with the number 1. Between these curves lie an infinite number of curves depicting consolidation at various K values (levels of shear stress). Because of the presence of residual shear stresses at the pile-soil interface, consolidation along a curve between the two boundary curves and closer to the CSL than represented by K_0 consolidation would be understandable. The

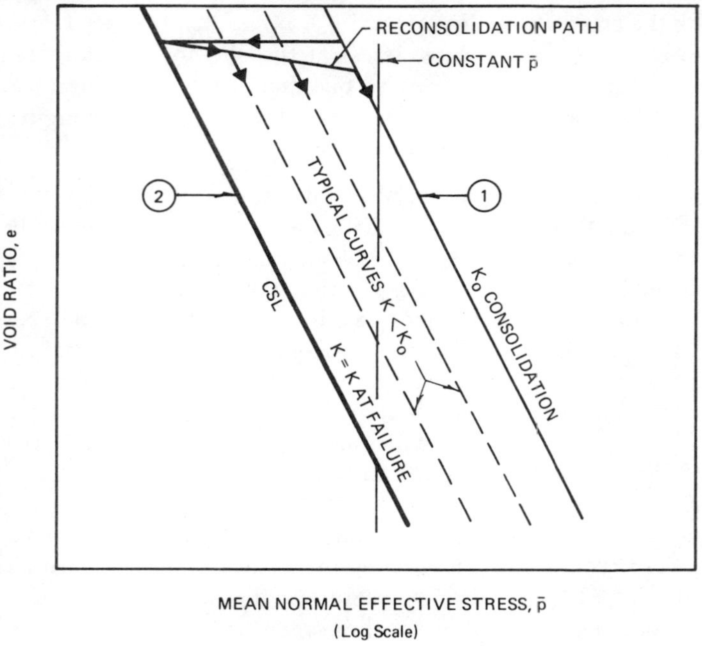

FIG. 18—*Location of consolidation line based on stress ration during consolidation.*

closer to the CSL the final stress state, the smaller the pore pressures generated by pile loading. Such a possibility has not yet been considered analytically, since both the expanding cylinder analyses and the available consolidation analyses have been based on the assumptions that no shear stresses are introduced at the pile–soil interface during pile installation.

It is of interest to note in passing that all analyses of reconsolidation indicate that the radial stress has been increased substantially by pile driving. For example, the Miller et al [*46*] analysis suggests, for some normally consolidated soils, a fourfold increase in $\bar{\sigma}_r$. If this massive increase is real, then it suggests that the quasieffective stress methods that attempt to predict $\bar{\sigma}_r$ as $K_0\bar{\sigma}_v$, thereby assuming no change in the radial stress due to pile driving and loading, provide reasonable predictions of pile capacity for undefined reasons.

Rate of Change of Stress—All solutions for rate of consolidation around a pile that have as yet been published are based on the Soderberg [*44*] formulation. Soderberg assumed that total stresses remain constant during consolidation (the Terzaghi formulation) and that the initial excess pore pressure due to pile driving decays inversely (rather than logarithmically) with distance from the pile surface. Despite these limitations, the rate of consolidation and the shape of the consolidation curve determined by a more complete formulation [45] is represented reasonably well by the Soderberg [44] analysis. The Soderberg consolidation curve is shown in Fig. 19.

The consolidation analyses show that the time required for consolidation is related to the square of the pile radius, which suggests that the distance from the pile within which the soil has been affected by its installation is important. The time required for consolidation is inversely related to the coefficient of consolidation of the soil, which coefficient contains a measure of the elastic soil properties and the soil permeability.

Effect of Pile Loading

Direct measurements of soil response during full-scale tests under axial load reported by Cooke [30] and Lu et al [51] and during model tests reported by Butterfield and Ghosh [41] indicate elastic (linear load-settlement curves and recoverable deformations upon unloading) behavior of piles until, at least, loads equal to one-half the ultimate pile capacity are applied. Similar behavior of model piles embedded in clay and subjected to two-way cyclic loading to stress levels in excess of 85 percent of the ultimate pile capacity was reported for the single stress-controlled test reported by Holmquist and Matlock [52].

The state of stress around a pile resulting from axial loading has been studied analytically by many investigators (for example, [1,36,53,54,55]). These analyses suggest that the axial loading of a pile embedded in an elastic medium produces only small changes in vertical and radial stresses at the

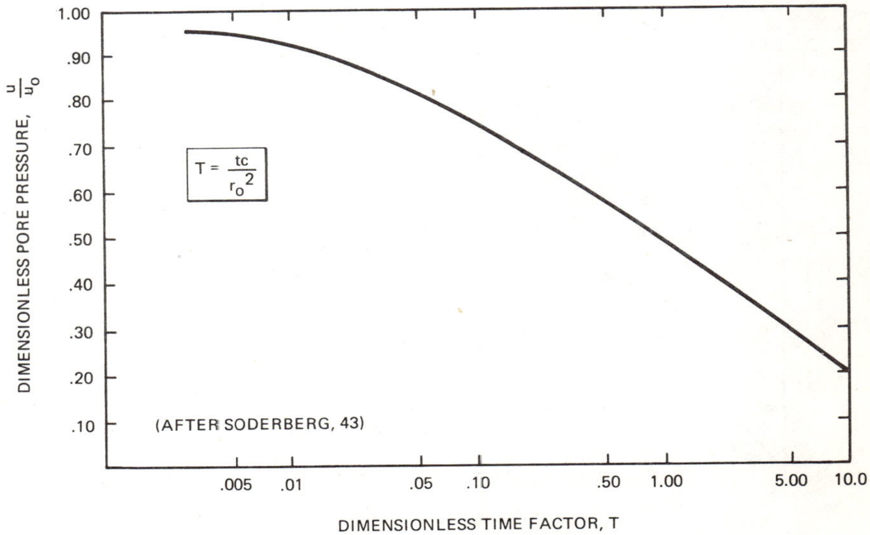

FIG. 19—*Predicted rate of pore pressure dissipation at pile-soil interface.*

pile-soil interface for most of the length of the pile shaft. As a first approximation, the soil elements adjacent to the pile are subjected to a pure shear loading.

Pure shear loading in a soil that behaves as an elastic material can produce no excess pore water pressures. As discussed in above, direct measurement of pore water pressures at the pile-soil interface due to pile loading has suggested only small pore pressures. A possible explanation for these observations may be found in the presence of residual stresses.

Figure 20 depicts in q–$\bar{p}$ stress space the critical state failure envelope and the K_0 consolidation line. Point A in stress space is shown and represents the state of stress at the pile-soil interface at a particular location along the length of the pile if reconsolidation after pile driving has been affected by the residual stress level. A curve through point A in stress space represents the yield envelope; stress levels higher than those bounded by the yield envelope cause plastic deformation, while lower stress levels cause only elastic deformation. Assuming negative skin friction at the point in question along the pile, loading the pile reduces the residual shear stress from its value after reconsolidation to zero. This reduction is represented by the vertical line between A and B, which is within the yield envelope. Such a stress path indicates that the soil behaves elastically, that pure shear loading is applied and that zero shear stress occurs at the K_0 consolidation curve. Continued loading returns the stress level to point A, although now the direction of the shear stress has been reversed. At this point, plastic yield begins and small

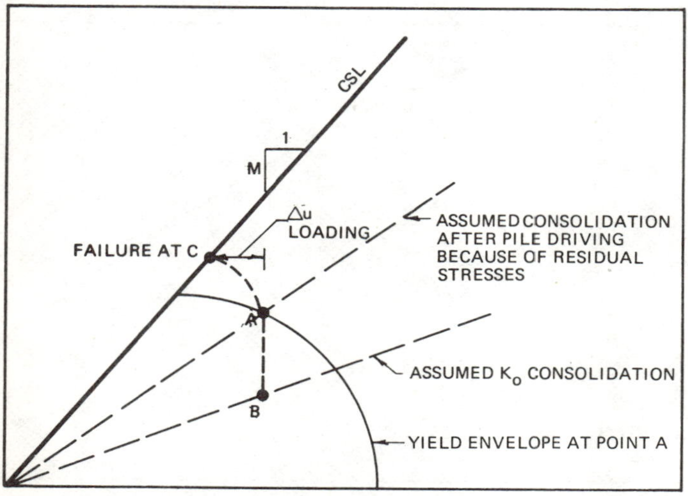

FIG. 20—*Possible effects of residual stresses on stress paths during loading.*

pore water pressures are generated as the stress path continues to failure at point C.

The direct measurements of Cooke and Price [*34*], Cooke [*30*], Butterfield and Ghosh [*42*], and others for piles in clay and those of Hunter and Davisson [*56*] for piles in sand suggest that the residual loads in the pile are 50 percent or more of the ultimate capacity of the tip of the pile and represent mobilization of more than 50 percent of the skin frictional capacity of some points along the shaft. The relationships would be different from those reported to date if the tip capacity were a smaller fraction of the total capacity than provided by London clay or sand. In all cases, however, the residual stresses appear to affect in important ways the distribution of stress along a pile shaft and, perhaps through their effect on the stress system after consolidation, may affect the absolute value of the capacity of the pile.

If residual stresses are as important to pile behavior as suggested in this section, changes in residual stress due to driving adjacent piles in a pile group can be important. Certainly, pile heave caused by driving adjacent piles will change the residual stresses along the pile shaft and will probably alter the load-deflection behavior of the pile. In addition, the stress changes due to pile driving are likely influenced by driving adjacent piles. Therefore, it is to be expected that piles in a pile group will not be uniformly stiff, their behavior is unlikely to be predictable from tests on individual piles, and the proportion of the group load they carry is likely to be dependent on such factors as their position in the group and the sequence of pile driving. Little is known today about group behavior, and much more needs to be learned.

Estimating Pile Capacity

At present, estimates of pile capacity based on the effective stress approach outlined herein still require the use of assumptions of major significance:

1. The final state of stress at the pile–soil interface is not yet predictable. Major steps forward have been made recently and are still to be fully exposed through publication. Approximations summarized below are still necessary.
2. The spacing of the critical state line, which is essential for determination of the capacity of piles in overconsolidated soils, is unique and measureable. It should be noted that the data shown in Fig. 15 are limited, and special tests to determine $\bar{p}_{cs}/\bar{p}_{nc}$ are preferable to the use of the correlation shown in Fig. 15.
3. The effect of possible reductions in effective stress friction angle at the pile–soil interface due to remolding or reorientation of particles during pile driving has not been included in the model to date. This effect may be substantial as suggested by Bozozuk et al [*57*] but may be offset by other uncertainties in the model.

Notwithstanding these assumptions, reasonable estimates of the capacity of piles driven into clay can be obtained. The capacity Q is computed as the sum of the skin frictional capacity Q_s and the tip capacity Q_t:

$$Q = Q_s + Q_t \tag{15}$$

$$Q_s = \pi D \int_0^L q_s \, dz \tag{15a}$$

$$Q_t = 9 S_u A_t \tag{15b}$$

where

L = length of penetration of the pile,
D = pile diameter, and
A_t = area of the pile tip.

In this formulation, the skin frictional resistance q_s, which is the frictional resistance along a vertical failure surface at the pile-soil interface, is given by

$$q_s = \tfrac{1}{2} M \bar{p}_f \cos \bar{\phi} \tag{16}$$

The term $\cos \bar{\phi}$ in Eq 16 is needed to convert from maximum shear stress to shear stress on the vertical plane. The reconsolidation analyses of Miller et al [*45*] and Carter et al [*38*] suggest that, after reconsolidation, the vertical and circumferential stresses are equal, and less than the radial stress. For this condition and using the conventional Mohr-Coulomb failure criterion from Eq 13*a*

$$M = \frac{6 \times \sin \bar{\phi}}{3 - \sin \bar{\phi}}$$

The term $\bar{p}_f$ is the mean normal effective stress at failure at the pile-soil interface. It is assumed as a first approximation that $\bar{p}_f$ has the following values:

1. For normally consolidated clays, $\bar{p}_f$ is assumed equal to the mean normal effective stress in the ground prior to pile driving.
2. For heavily overconsolidated soils that generate (in the laboratory) negative pore water pressures when sheared to the critical state, $\bar{p}_f$ is assumed equal to the mean normal stress after pile driving. This is the mean normal stress at the critical state $\bar{p}_{cs}$ and can be computed from Eq 14.

The reliability of this effective stress approach to the prediction of the axial capacity of driven piles in clay was indicated by Esrig et al [*1*] and is not repeated here.

Acknowledgments

The general effective stress analysis for axial capacity of driven piles was developed under a research contract between Woodward-Clyde Consultants and both Amoco Production Company and Shell Oil Company. The results of that initial study became part of a larger research study that was organized by Amoco and funded by eleven oil companies. The active participation and enthusiasm of B. S. Murphy of Amoco Production Company, who served as Project Administrator, contributed significantly to the success of the work. Without Murphy and R. G. Bea, formerly of Shell Oil Company and now with Woodward-Clyde Consultants, this work would not have been possible.

References

[*1*] Esrig, M. I., Kirby, R. C., Bea, R. G., and Murphy, B. S., *Proceedings*, 1977 Offshore Technology Conference, Vol. 3, 1977, pp. 495-506.

[*2*] Skempton, A. W. *Proceedings*, British Building Research Congress, Vol. 1, 1951, pp. 180-189.

[*3*] Meyerhof, G. G., *Geotechnique*, Vol. 2, No. 4, 1951, pp. 301-332.

[*4*] Vesic, A. S., "Principals of Pile Foundation Design," *Proceedings*, Boston ASCE Lecture Series on Deep Foundations, 1975.

[*5*] Bishop, R. F., Hill, R., and Mott, N. F., *Proceedings*, British Physical Society, Vol. 57, 1945, p. 147.

[*6*] D'Appolonia, D. J., Poulos, H. G., and Ladd, C. C., *Journal of the Soil Mechanics and Foundations Division*, *Proceedings*, American Society of Civil Engineers, Vol. 97, No. SM 10, 1971, pp. 1359-1377.

[*7*] Burland, J. B., *Ground Engineering*, Vol. 6, 1973, pp. 30-42.

[*8*] Terzaghi, K. and Peck, R. B., *Soil Mechanics in Engineering Practice*, Wiley, New York, 1948.

[*9*] Tomlinson, M. J., *Proceedings*, International Conference on Soil Mechanics and Foundation Engineering, Vol. 2, 1957, pp. 66-71.

[*10*] Woodward, R. J., Lundgren, R., and Boitano, J. D., *Proceedings*, 5th International Conference on Soil Mechanics and Foundation Engineering, Vol. 2, 1961, pp. 177-184.

[*11*] Flatte, K. and Selnes, P., *Proceedings*, 9th International Conference on Soil Mechanics and Foundation Engineering, Vol. 2, 1977, pp. 517-522.

[*12*] Tomlinson, M. J., *Proceedings*, Conference on Behavior of Piles, London, England, 1970, pp. 107-114.

[*13*] Zeevaert, L., *Proceedings*, 1st Panamerican Conference on Soil Mechanics and Foundation Engineering, Vol. 3, 1960, pp. 1145-1152.

[*14*] Eide, O., Hutchinson, J. N., and Landva, A., *Proceedings*, 5th International Conference on Soil Mechanics and Foundation Engineering, Vol. 2, 1961, pp. 45-53.

[*15*] Chandler, R. J., *Civil Engineering Public Works Review*, (England), Vol. 63, 1968, pp. 48-51.

[*16*] Meyerhof, G. G., *Journal of the Geotechnical Division*, *Proceedings*, American Society of Civil Engineers, Vol. 102, No. GT3, 1976, pp. 196-228.

[*17*] Vijayvergiya, V. N. and Focht, J. A., *Proceedings*, Offshore Technology Conference, Vol. 2, 1972, pp. 865-874.

[*18*] Seed, H. B. and Reese, L., *Transactions*, American Society of Civil Engineers, Vol. 122, 1955, pp. 731-764.

[*19*] Poulos, H. G. and Davis, E. H., *Elastic Solutions for Soil and Rock Mechanics*, Wiley, New York, 1974.

[*20*] Ellison, R. D., "An Analytical Study of the Mechanics of Single Pile Foundations," Ph.D. thesis, Carnegie-Mellon University, Pittsburgh, Pa., 1968.

[21] Randolph, M. F. and Wroth, C. P., *Journal of the Soil Mechanics and Foundation Division*, *Proceedings*, American Society of Civil Engineers, Vol. 104, No. GT 12, 1978, pp. 1465-1488.

[22] Peck, R. B., Hanson, W. E., and Thornburn, T. H., *Foundation Engineering*, Wiley, New York, 1974.

[23] Holloway, D. M., Audibert, J. M. E., and Dover, A. R., *Proceedings*, Offshore Technology Conference, Vol. 3, 1978, pp. 1915-1924.

[24] Holloway, D. M., Clough, G. W., and Vesic, A. S., *Proceedings*, Offshore Technology Conference, Vol. 4, 1978, pp. 2225-2236.

[25] Casagrande, A., *Proceedings*, 1st International Conference on Soil Mechanics and Foundation Engineering, Vol. 3, 1936, pp. 60-64.

[26] Schmertmann, J., *Transactions*, American Society of Civil Engineers, Vol. 120, 1955, pp. 1201-1233.

[27] Janbu, N., *Proceedings*, 7th International Conference on Soil Mechanics and Foundation Engineering, Vol. 1, 1969, pp. 191-196.

[28] Ladd, C. C., Foott, R., Ishihara, K., Schlosser, F., and Poulos, H. G., *Proceedings*, 9th International Conference on Soil Mechanics and Foundation Engineering, Vol. 2, 1977, pp. 421-494.

[29] Gardner, W. S., "Soil Property Characteristics in Geotechnical Engineering Practice," *Geotechnical/Environmental Bulletin*, Woodward-Clyde Consultants, Vol. 10, No. 2, Winter 1977.

[30] Cooke, R. W., this volume.

[31] Clark, J. I., and Meyerhof, G. G., *Canadian Geotechnical Journal*, Vol. 9, 1972, pp. 351-373.

[32] Massarsch, K. R., "Soil Movements Caused by Pile Driving in Clay," Royal Swedish Academy of Engineering Sciences, Committee on Pile Research, Report No. 51, 1976, p. 261.

[33] Hagerty, D. J., *Some Heave Phenomena Associated with Pile Driving*, Ph.D. thesis, University of Illinois, Urbana, Ill., 1969.

[34] Cooke, R. W. and Price, G., *Proceedings*, 8th International Conference on Soil Mechanics and Foundation Engineering, Vol. 2.1, 1973, pp. 53-60.

[35] Clark, J. I., this volume.

[36] Butterfield, R., and Banerjee, P. K., *Proceedings*, 2nd Southeast Asian Conference on Soil Engineering, Singapore, 1970, pp. 386-394.

[37] Vesic, A. S., *Journal of Soil Mechanics and Foundation Division*, *Proceedings*, American Society of Civil Engineers, Vol. 98, No. SM3, 1972, pp. 265-290.

[38] Carter, J. P., Randolph, M. F., and Wroth, C. P., Cambridge University Research Report CUED/C-SOILS/TR 51, 1978.

[39] Lo, K. Y., and Stermac, A. G., *Proceedings*, 6th International Conference on Soil Mechanics and Foundation Engineering, Vol. 2, 1965, pp. 285-289.

[40] Butterfield, R., and Johnston, I. W., *Proceedings*, 8th International Conference on Soil Mechanics and Foundation Engineering, Vol. 2.1, 1973, pp. 39-43.

[41] Mindlin, R. D., *Journal Applied Physics*, Vol. 7, No. 5, 1936, pp. 195-202.

[42] Butterfield, R. and Ghosh, N., *Proceedings*, 9th International Conference on Soil Mechanics and Foundation Engineering, Vol. 1, 1977, pp. 451-457.

[43] Kirby, R. C. and Wroth, C. P., *Proceedings*, Offshore Technology Conference, Vol. 3, 1977, pp. 483-494.

[44] Soderberg, L. O., *Geotechnique*, Vol. 12, 1962, pp. 217-225.

[45] Randolph, M. F. and Wroth, C. P., Cambridge University Research Report CUED/C-SOILS/TR 50, 1978.

[46] Miller, T. W., Murff, J. D., and Kraft, L. M., *Proceedings*, Offshore Technology Conference, Vol. 4, 1978, pp. 2237-2242.

[47] Roscoe, K. H. and Burland, T. B., "On the Generalized Stress-Strain Behavior of 'Wet' Clay," *Engineering Plasticity*, J. Heyman and F. A. Lecke, Eds., Cambridge University, England.

[48] Airhart, T. P., Hirsch, T. J., and Coyle, H. M., Texas Transportation Institute, Texas A & M, Research Report No. 33-38, 1967.

[*49*] Massarsch, K. R., Broms, B. B., and Sundquist, O., *Proceedings*, Conference on In-Situ Measurements of Soil Properties, American Society of Civil Engineers, 1975, pp. 260-265.
[*50*] Lo, K. Y. and Stermac, A. G., *Canadian Geotechnical Journal*, Vol. 1, No. 2, 1964, pp. 63-80.
[*51*] Lu, T. D., Fischer, J. A., and Miller, B. G., this volume.
[*52*] Holmquist, D. V. and Matlock, H., *Proceedings*, Offshore Technology Conference, Vol. 1, 1976, pp. 553-570.
[*53*] Geddes, J. D., *Geotechnique*, Vol. 16, No. 3, 1966, pp. 231-255.
[*54*] Mattes, N. S. and Poulos, H. G., *Journal of the Soil Mechanics and Foundation Division, Proceedings*, American Society of Civil Engineers, Vol. 95, No. SM1, 1969, pp. 189-207.
[*55*] Mattes, N. S., *Geotechnique*, Vol. 19, No. 1, 1969, pp. 157-159.
[*56*] Hunter, A. H., and Davisson, M. T., *Performance of Deep Foundations, ASTM STP 444*, American Society for Testing and Materials, 1969, pp. 106-177.
[*57*] Bozozuk, M., Keenan, G. H., and Pheeney, P. E., this volume.

M. T. Davisson[1]

Stresses in Piles

REFERENCE: Davisson, M. T., "**Stresses in Piles,**" *Behavior of Deep Foundations, ASTM STP 670*, Raymond Lundgren, Ed., American Society for Testing and Materials, 1979, pp. 64-83.

ABSTRACT: Factors controlling the static, long-term, and dynamic structural strength of pile sections are delineated and discussed for timber, concrete, and steel piles. It is shown that current design stresses for steel and concrete are reasonable, but that timber design stresses equal or exceed long-term constant load ultimate stresses. Reasonable limits to dynamic driving stresses are recommended. Other sources of load and stress in piles are discussed along with pile strength reducing factors; many of these factors are not now considered in design. It is concluded that the normal current pile-to-soil factor of safety of 2.0 is reasonable, and that pile structural factors of safety should always exceed the pile-to-soil factor of safety. It is also concluded that normal factors of safety should not be lowered until the profession attains a higher level of competence in inspection of pile installation.

KEY WORDS: piles, stresses, timber piles, wood piles, concrete piles, steel piles, pile driving stresses, structural design, safety factors, dynamic stresses, inspection, construction problems

The title of this paper implies many things. For practical purposes, pile stresses are important only to the extent that they relate to pile member strength and deformation. Strength of structural members is well understood in other areas of structural engineering. However, a review of structural practice in pile foundations could easily lead to the conclusion that some mysticism surrounds the subject of piles and that structural principles, therefore, do not apply. One of the purposes of this symposium is to draw together in one reference the facts regarding structural strength of piles. To accomplish this, three invited papers have been received: Dismuke [*1*][2] deals with steel, Gamble [*2*] discusses concrete, and Armstrong [*3*] presents a complete review of timber strength.

[1]Professor of Civil Engineering, University of Illinois at Urbana-Champaign, Urbana, Illinois 61801.

[2]The italic numbers in brackets refer to the list of references appended to this paper.

After discussing pile member strength, attention will be given to the various sources of loads and stresses in piles. This will be intermingled with the important implications of each source of load. Finally, conclusions will be offered on the present state of knowledge.

Pile Section Strength

Strength Variables

Pile section strength is a function of several variables:

1. Size.
2. Material.
3. Specimen strength from acceptance test, for example,
   a. f_y from web coupon for H-piles,
   b. f'_c from 28-day cylinder tests for concrete,
   c. tabulation of small clear-wood crushing strengths for species and subspecies of wood material.
4. Statistical treatment of acceptance tests, for example, f'_c.
5. Relationship of acceptance test strength to average section strength.
6. Relationship of acceptance test location to section strength, for example, coupon from web versus average section strength.
7. Reduction for defects (hidden or otherwise), for example, knots in timber.
8. Reduction for section treatment, for example, preservation treatment of wood.
9. Strength increase for rapid loading (that is, strain rates exceeding those in acceptance tests).
10. Strength reduction for long-term creep (that is, strain rates less than those in acceptance tests).
11. Strength reductions for fatigue.
12. ϕ factors (for example, as used in concrete design) covering any items not mentioned above.

The foregoing are best evaluated in light of the nominal factor of safety employed for a pile foundation.

In practice there are two factors of safety involved. The one most commonly referenced relates to slip of the pile relative to the soil without structural failure in the pile. At present, a factor of 2.0 is common, with 2.5 being used where an additional degree of conservatism is desired. It is obvious that the pile itself must possess a structural factor of safety greater than the pile-soil factor of safety; otherwise, piles would fail structurally during load testing, thus obscuring the data in a test wherein the object is to determine the pile-soil factor of safety.

Steel Pile Sections

Steel piles are furnished by size and material according to established ASTM specifications. Strength is verified by coupon tests from the web in the case of H-piles; flange coupons would show lower yield values than web coupons. Dismuke's [*1*] data indicates that a web coupon yield of 115 to 120 percent of guaranteed yield is required in order for the pile section to exhibit the guaranteed yield load in a column stub test. This occurs because of residual loads in the pile caused by differential cooling.

Hidden defects and treatment are not normally considered as a factor in the strength of steel piles. Similarly, the long-term creep strength of steel appears to be at least 90 percent, and perhaps more than 95 percent of the strength from short-term tests. In light of other variables, and the factor of safety normally used, the importance of creep varies from negligible to unimportant.

Increase in strength with strain rate can be an important property when considering pile stresses caused by driving. For efficient driving as well as efficient utilization of pile material, it is desirable to stress the pile to the practical limit during driving. However, this must be tempered by a consideration of low cycle fatigue, because pile driving typically involves of the order of 1000 load repetitions at near-yield stress levels. A review of steel fatigue properties leads to the conclusion that low cycle fatigue is not important in compression. Because steel piles are normally highly stressed in compression relative to tension, fatigue can generally be ignored.

Dismuke recommends a limiting driving stress of $1.4f_y$ to $1.7f_y$; the writer disagrees. It is true that conditions exist where yield increases of 40 percent or more can be obtained on laboratory specimens. However, practical utilization of yield strength increase requires careful consideration of the field conditions and the time delay over which the increased yield can be sustained. For example, at the $1.4f_y$ level, a delay of approximately 0.1 ms is the limit before the material yields; this is too short to be usable in piledriving. Delay times are longer at stress levels closer to the static yield. It is the writer's judgement that a stress level of $1.1f_y$ is the maximum dynamic stress level that is consistently useful in pile driving operations as now conducted.

Dismuke has discussed contact stresses, and has shown that localized yield at contacts is not harmful.

The characteristics of steel pipe were not covered in the papers received for this symposium. Some differences exist relative to H-piles. The comments that follow are based on the writer's investigation and experience. Pipe material is usually purchased under ASTM A252, Grade 2 (f_y = 35 ksi, 241 MPa). This specification is sufficiently loose that pipe rejected for other uses can be marketed for piling. It is common for f_y from coupons to exceed 310 MPa (45 ksi) because of method of manufacture and because of being high-strength material rejected for another use. Consequently, section strength is often much higher than implied by the guaranteed yield.

Two problems that can arise in pipe sections deserve mention. Because of the wide variety in steel marketed under ASTM A252, weldability can be a problem. Unless the steel characteristics are a certainty, the engineer should not assume the acceptability of a weld procedure until checked by test on the actual pipe material to be used. The second problem relates to the quality of welding in spiral-weld pipe. Spiral-weld pipe can be an excellent product, but quality control on the welding is critical. The writer has observed unraveling of such pipe both in handling and driving that is directly and obviously attributable to poor weld seams.

There is an unfortunate misconception among some material suppliers that piling is a noncritical application of steel. On the contrary, piling is subjected to stresses during driving that should exceed any service stresses by a wide margin. It is difficult to think of an application for steel that is subjected to a test as severe as pile driving before or after it goes into service. Consequently, pile fabrication details must be constructed in a reasonable manner if they are to survive the test of driving.

Concrete Pile Strength

Gamble [2] discusses concrete piles in relation to the philosophy and technical provisions of the 1977 American Concrete Institute (ACI) code. Pile size and material are obviously firmly under the control of the engineer, but realistic variations in material strength must be tolerated. Consequently, ACI provides for statistical treatment of cylinder strengths, f'_c. Further a reduction factor of 0.85 on f'_c to relate cylinder strength to average concrete product strength has a long history. Part of the 0.85 factor may account for strength variations with position inside the product, such as a column.

Strength increases with strain rates applicable to pile driving are quite pronounced in concrete (up to 40 percent according to Gamble). This, however, is offset by low cycle fatigue; at 1000 cycles, the reduction is approximately 20 percent. Considering the 0.85 factor applied to f'_c, it may be concluded that a reasonable upper limit to driving compression stress is f'_c. The writer has been using such a limit in his practice with apparent success. It should be appreciated that driving stresses are preferably kept much lower than f'_c if feasible so as not to beg trouble. Driving tension stress is preferably limited to zero (calculated stress equal to the prestress in prestressed piles), but not greater than $6{f_c'}^{1/2}$.

Strength reduction for long-term loading may amount to 15 percent. However, this is normally offset by strength gain within the period of 28 days to 1 year. The normal environment for a pile (buried and wet) is ideal for long time strength gains. Therefore, creep is not likely to be a strength factor in piles unless they are overloaded early in their life and remain so loaded for a long period of time.

Concrete members are not usually penalized for various treatments, as is the case with wood. However, reduction factors (ϕ factors) for (1) hidden defects, (2) relative importance of the member to the stability of a structure, and (3) other factors are well established in the ACI philosophy. A ϕ factor of 0.7 is used for tied columns and would be applicable to most pile configurations.

Gamble correctly advocates the use of moment-thrust interaction diagrams to display the full range of pile section strength. In particular, it should be noted that combinations of tension and moment are very serious loads relative to pile strength. Concrete is usually ignored as a tension member for long-term loads; therefore, tension must be carried solely by steel.

Timber Pile Strength

Armstrong [*3*] has performed a valuable service to the profession by making a thorough presentation of the important factors governing the strength of timber piles. Only a small percentage of the profession is conversant with the important variables in timber strength because of a lack of emphasis in school curricula. The appearance in 1974 of ASTM Method for Establishing Design Stresses for Round Timber Piles (D 2899-74) sparked controversy, because the resulting allowable stresses were recognized by experts to be too high and the whole standard was considered to be in error relative to important strength factors. This atmosphere was the spawning ground for the presentation given herein.

Timber pile size is only partially under the control of the engineer. He may order tip or butt size, but not both, under the current ASTM Specification for Round Timber Piles (D 25-73). Pile diameter can be a controversial topic at the inspection point because of irregular timber shapes. ASTM Method D 25-73 is set up to deal with circumference, which can be more easily measured. Once the engineer selects either tip or butt circumference, he is guaranteed the circumference at the other end according to a table in ASTM Method D 25-73.

The engineer may also specify species such as southern pine. However, he seldom can specify the subspecies (longleaf, shortleaf, etc.). This leads to a major difference with respect to dealing with concrete or steel. The engineer must deal with several subspecies, and they in turn have a range of strengths that must be accounted for. Timber technologists have determined average crushing strengths and standard deviations for each subspecies on small clear green specimens.

Timber technologists have a tradition of dealing with the 5 percent exclusion value. This is determined from the above statistical data and represents a strength wherein 5 percent of samples will fail. Arriving at a definite value for combinations of subspecies is somewhat complicated, but Armstrong recommends 16.5 MPa (2400 psi) for all three leading pile species, oak, pine, and fir. This strength for small clear specimens becomes the basic reference. However, timber is not purchased based on this reference strength, as is the case with concrete and steel. The engineer must recognize timber as a natural material over which he has to exert control in ways other than specifying strength.

Armstrong delineates reduction factors for timber strength due to: (1) knots and spiral or twist, ϕ; (2) duration of loading, β; and (3) treatment, ψ. Another factor relating the statistically higher butt strength to tip strength is ignored because it does not exceed 10 percent and because its practical utility is extremely limited.

Armstrong shows that knots are very important and that knot reduction factors ϕ vary from 0.6 to 0.85. Of great importance is that ASTM Methods D 25-70 and D 25-73 represent a significant departure from ASTM Method D 25-58 in that the quantity of allowable knots was doubled. However, ASTM Method D 2899-74 does not take account of the resulting reduced strength from 0.85 to 0.60.

Armstrong examined duration of loading and found minor disagreement with the values used in ASTM Method D 2899-74. One of the oddities resulting from ASTM Method D 2899-74 is that the average ASTM Method D 2899-74 pile would fail structurally in an ASTM Method of Testing Piles Under Axial Compressive Load (D 1143-74) load test if the entire test load were transferred to the pile tip. Two factors prevent a wholesale epidemic of such failures:

1. Load tests produce positive friction, usually of sufficient amount to prevent such a failure by means of load transfer to the soil.
2. Contractors usually cull the timber supply for a good oversized stick to use as the test pile.

The foregoing statement about potential load test failure emphasizes the importance of duration of loading on timber strength. For a load duration of 10 years, timber fails at a β value of 62.5 percent of the failure stress in reference tests. For longer terms, a β value of 0.56 is commonly used. This property is significantly different from those of concrete and steel; in the writer's experience, the ramifications of this behavior are little known to the profession.

Timber preservative treatment processes have a strength-reducing effect. The allowable techniques for oak and fir lead to a reduction factor ψ of 0.9. However, steam conditioning is allowed for southern pine and results in major strength losses. The important variables are time and temperature. Under current treatment specifications, a ψ value of 0.70 to 0.75 is appropriate. By contrast, ASTM Method D 2899-74 wrongly lists this factor as 0.85.

Armstrong recommended that a minimum factor of safety of 1.2 be used in contrast to the 1.0 used in ASTM Method D 2899-74. Therefore, the ASTM Method D 2899-74 recommended allowables are actually a long term ultimate stress. A comparison of allowable stresses results for treated piles is given in Table 1, which illustrates the distorted results obtained from ASTM Method D 2899-74. Diekmann's [*4*] conclusion regarding the desirability of ASTM Method D 25-58 as a material standard is also amply demonstrated.

Timber is known for its ability to withstand transient loads, such as pile driving. This must be tempered somewhat by low cycle fatigue properties. Armstrong suggests a β value of 1.2 for driving stresses. Based on Armstrong's study, the writer's [*5*] study of driving limitations involved a β value of approximately 1.5, which may be somewhat unconservative.

Driving Stresses

In almost all cases, the highest stress levels that will occur in a pile happen during driving. Thus, driving is a serious test of the integrity of the pile section and, when properly observed, its strength. High driving stresses are necessary to cause pile penetration sufficient to develop static pile loads that equate to reasonable static stresses under service loads. The introduction of the wave equation analysis of pile driving has made it possible to dissect the problem of driving stresses versus static load stresses. Acceptance of the wave equation analysis during the past decade has been somewhat of a revolution within the pile foundation art. Of the 32 papers submitted to this symposium, 13 mention wave analysis favorably with 1 of the 13 disliking some of its applications. None of the authors proposed or used an energy formula; it is hoped that such formulas have been purged from practice.

Davisson [*5*] used the wave equation analysis to study driving stresses and pile drivability for purposes of defining the maximum readily attainable static pile loads and stresses at a pile–soil safety factor of 2 that can be developed at the driving stress levels cited previously using current hammer-

TABLE 1—*Allowable stresses results for treated piles.*

Pile	Oak and Fir, psi (MPa)		Southern Pine, psi (MPa)	
	D 2899 (1)	Armstrong (1.2)	D 2899 (1)	Armstrong (1.2)
D 25-58, A, B		925 (6.4)		775 (5.3)
D 25-58, C		800 (5.5)		675 (4.7)
D 25-70, D 25-73	1100–1250 (7.6–8.6)	650 (4.5)	1200 (8.3)	550 (3.8)

cushion combinations. This was done on the assumption of no soil "setup" or "freeze," as is called the process of increasing pile skin friction strength with time. There are many soil deposits for which freeze does not exist; in rare cases, relaxation occurs. Thus, the Davisson study represents an important bound to the problem of developing pile load capacity.

Other soil deposits are well known for prodigious amounts of the process of freeze, for example, New Orleans. The proponents of high allowable pile stresses invariably reference load test results from sites where high soil freeze exists; silence reigns regarding load tests from sites without freeze.

The generally attainable static design stresses derived by Davisson are:

steel	83 MPa (12 000 psi)
concrete	11 MPa (1 600 psi)
timber	8.3 MPa (1 200 psi)
	[Friction 5.5 MPa (800 psi) point bearing]

The importance of these stress levels as a bound is that the engineer always has control over working pile loads up to these stress levels, providing he follows the driving recommendations that go with them. By contrast, higher allowable stresses as proposed, for example, by the steel industry ($0.5f_y$) are 1.5 to 2.0 times the above value and can consistently be developed only with the aid of soil freeze. The engineer does not have control over soil freeze; he can only observe it. If steel stresses of $0.5f_y$ are to be used, prior local experience or load tests during design are necessary if the job that goes to bid is to be a certainty rather than a speculation. In summary, the bounds presented by Davisson represent that which can be earned by the engineer's effort. Load capacity beyond these values must usually rely on soil freeze.

The paper by Thompson and Thompson [*6*] bears heavily on this subject; both praise and criticism of their efforts are offered herein. The authors are commended for making and reporting systematic observations of field behavior, and for recognizing and showing that developed pile load capacity bears a relationship to driving stresses.

Thompson and Thompson's figure 5 has been redrawn as Fig. 1 to illustrate the importance of their data as viewed by the writer. The figure shows pile stress at static load test failure versus peak pile head stress during driving. Thompson and Thompson suggest a correlation of static stress equal to 1.2 times the dynamic driving stress; obviously, the dynamic stress must be limited to $0.8f_y$ for practical reasons and because a static ultimate load beyond the pile yield load is of no value.

Also on Fig. 1 is a line at static stress equal to 0.8 dynamic stress, which represents approximately the writer's bound for piles; this is two-thirds of the value proposed by Thompson and Thompson. Two horizontal lines have been drawn at ultimate static load stress levels of 166 to 193 MPa (24 000 and 28 000 psi). This represents a zone that the writer has said

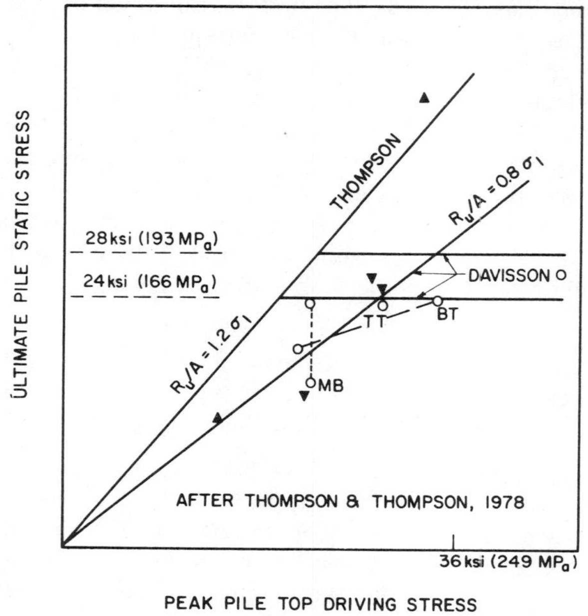

FIG. 1—*Peak pile top driving stress.*

can be obtained consistently only with the best of driving techniques in the absence of soil freeze. Three piles omitted by Thompson and Thompson as examples of relaxation, and perhaps the one case of damage they also omitted, can be reinterpreted as described above. The writer has added several other data points for sites at which soil freeze does not exist that fall close to the writer's aforementioned bounds. Many more data points can be added to illustrate the point. Thus, the writer offers that the conclusion of Thompson and Thompson represent an optimistic view of the situation and that much of their data involves soil freeze. Further, their correlation is not recommended for design unless the risks are understood.

Another conclusion by Thompson and Thompson that involves risk is the recommendation of a pile head driving stress of $0.8f_y$. This fails to address the real nature of the problem and would lead to pile tip damage in many point-bearing pile applications, because peak driving stresses usually occur at the tip because of reflections. Although the margin between $0.8f_y$ and $1.0f_y$ covers many practical cases, the margin is insufficient in others.

Present allowable stresses for concrete piles (in excess of 2 000 psi, 13.8 MPa) exceed the limitations from driving, 11 MPa (1 600 psi) determined for no-freeze soil conditions. It is expected, therefore, that problems with broken piles would develop for highly stressed designs. Gerwick and Brauner [7] confirm this by relating their experience. They strongly advocate more pile head cushioning, which reduces the damage problem in tension but may also make it impossible to stress the pile high enough to develop the desired compression load. The requirements for cushioning and pile load capacity are incompatible beyond certain limits as indicated by Davisson [*5*]. Gerwick and Brauner correctly state the solution for such problems—increase the prestress from 5.5 to 8.3 MPa (800 to 1 200 psi). Unfortunately, this also reduces pile load, but this can be resolved by raising f'_c.

Another current indication of driving stress problems is given in a recent *Engineering News Record* [*8*] article about a high-capacity timber pile project in Nevada. The design load is 623 kN (70 tons). This is one of the projects referred to by Diekmann. The fir piles have steel tips and are being banded with steel at 3-m (10-ft) intervals, presumably to prevent splitting. It is also noted by the writer that the soil conditions giving rise to this pile selection provide a large measure of freeze.

Sources of Stresses in Piles

Normal dead and live load pile stresses are always accounted for by designers. It is generally recognized that error in assignment of loads can occur, which is part of the reason for the pile-to-soil factor of safety of 2.0 to 2.5. These factors of safety also must account for below average pile-to-soil capacities and pile structural capacities. In this regard, the system used by ACI provides more insight into the problem, as follows

$$1.4D_L + 1.7L_L \leq \phi \times \text{nominal strength}$$

Simply put, 140 percent of design dead load (D_L) plus 170 percent of design live load (L_L) should always be less than nominal member strength reduced by a ϕ factor to account for the likely lower bound to strength. Thus, it can be seen that increases in load or reductions in strength are equally effective in consuming the margin indicated by the factor of safety used in design.

Several sources of load increase, stress increase, or strength reduction may be applicable on any given project and may or may not have been design considerations. These are briefly described in the following paragraphs.

Negative Skin Friction

Bozozuk et al [*9*] and Lacy [*10*] describe case histories wherein negative skin friction was considered during design. The mechanism of negative friction can have serious design consequences. Positive friction observed in load tests must not only be disregarded as support, it must be reversed and considered as a load. This means that maximum pile service load occurs at the bottom of the subsiding soil layers. Under such conditions, structural strength requirements may be higher along the embedded portion of the pile than at the butt. In cases where negative friction was neglected or underestimated embarrassing failures have occurred. The incidence of failure is higher for timber than for other pile materials, and is interpreted by the writer to be expected based on the low to negligible structural factor of safety employed in timber relative to concrete and steel.

Negative skin friction failures have occurred long after construction. In one case, failure was due to placement of fill adjacent to an existing structure 20 years or more after construction. In another, failure occurred 9.5 years after construction, with no aggravating circumstances. The piles were timber in both cases.

Group Behavior

One of the unsolved problems in pile foundations is how interior piles in a group transfer their load to the soil. It is easy to reason that pile butt loads transfer toward the tip more so for interior piles than exterior or isolated piles, such as test piles. One set of measurements clearly shows this trend [*11*].

Tip Load Analysis

Designers often assume sufficient knowledge exists about transfer of pile load to the soil that they can calculate the location of the critical or most highly stressed portion of the pile. Thus, deeper portions of the pile carry less load and do not need to be as strong as at the butt. Analyses of this type are heavily promoted by suppliers of tapered piles. Other designers prefer to bound the problem by using pile sections with sufficient strength throughout to handle all load combinations regardless of the position along the pile at which the loads occur.

Serious defects exist in our knowledge of load transfer, because residual loads in piles caused by driving are not accounted for. The result is an underestimate of tip load. The influence of residual loads was made known to the profession a decade ago in another ASTM symposium [*12*]. In spite of this, of nine papers submitted to this symposium wherein recognition of residual loads was important, five failed to mention or show awareness of them. The practical significance of this is that actual pile tip loads are higher than normally considered. When coupled with negative skin friction and pile group behavior, it is clear that many designers are unwittingly unconservative in selecting the critical pile section. Residual loads do not exist in cast-in-place mandrel-driven shells.

Earthquake

Earthquake, hurricane, or other natural forces may or may not be taken into account in design. Ductile pile sections capable of large flexural rotations and having large residual strengths are preferred to resist earthquake.

Corrosion

Designers may account for minor corrosion by discounting the cross-sectional area. This effectively raises pile stresses under a given load. Unanticipated corrosion obviously has the same effect.

Error in Loads

There is always uncertainty in the live loads that may be experienced. This is a source of stress usually considered to be covered by a factor of safety.

Damage

Stresses are effectively raised by pile damage, even though the actual mechanism is a reduction of pile strength. Undetected damage may result from encountering obstructions or an adjacent pile. More commonly, damage occurs at the tip and consists of buckling of steel, spalling of concrete, and brooming of timber.

Sweeps and Bends

Wu and Fox [*13*] describe case histories with crooked piles. It is obvious that increased stresses are produced that usually are not considered in design.

Mislocation

Piles out of position but within location tolerance can lead to increased stresses. For one four-pile cluster, the writer found that the increase in load could be as much as 24 percent with the normal 0.076-m (3-in.) tolerance. This is usually ignored in design. Field layout errors can have the same influence.

Differential Settlement and Unequal Bearing

This is a natural occurrence that varies only in magnitude. It is seldom considered in design.

Adjacent Driving and Heave

Clark [*14*] describes a very complete case history involving pile heave and shaft separation. In this case, the shaft was unreinforced concrete and, therefore, vulnerable to separation. Designers seldom consider stresses and movements caused by driving of adjacent piles.

Slope Instability

Wang et al [*15*] describe bending loads on piles caught within an unstable slope. This problem surfaces from time to time and is experienced with caissons also. Such loading has been cause for replacement of entire foundations.

Construction Activity

Piles may be laterally loaded by accidental collison with construction equipment. They may also be subject to lateral loads from adjacent surcharge loads or become part of an unstable slope due to adjacent trenching activities. Free-standing batter piles are stressed by their own weight and may be stressed further by pulling operations to fit within the pile cap.

Obviously, many other situations occur during construction that can be added to the foregoing. Some of these situations may occur during future construction activities.

Inspection

Inspectors can obviously be a cause of pile damage due to overdriving, or a cause of undercapacity by leniency regarding specification requirements. The writer regards inspection of pile installation as a major variable in achieving in-place pile strength. A complete view of the subject is given by Davisson [*16*].

The foregoing list of topics may not be complete but should be sufficient to illustrate that the factor of safety used for piling involves a margin that is always eroded in practice. Thus, a factor of safety of 2.0 implies a 100 percent margin over the design loads, but one or more of the foregoing topics invariably will be the cause of actual pile capacity being less than indicated by the limited number of topics considered in design. The margin of safety must cover damage and undercapacity as well as overload.

Summary and Conclusions

Dismuke, Gamble, and Armstrong have made it possible to collect into one reference the pertinent features of pile section strength for steel, concrete, and timber. It has been shown that timber pile strength is much less than currently advertised and that pile foundation failures confirm this defect.

Efficient use of pile material requires that it be stressed near the limits during driving. The writer has proposed limits of $1.1f_y$ for steel, f_c' and the prestress value for concrete, and 1.2 to 1.5 times the static crushing strength for timber.

A review of other sources of pile stresses reveals serious defects in our knowledge of residual loads and pile load transfer. A better understanding of pile load transfer to the soil within groups rates a high priority in future investigations; negative skin friction is considered to be part of this effort. Effective stress analyses of the complete history of a pile group from before driving to service conditions appears to be the best hope for improving knowledge. Meanwhile, designers are cautioned about being overly certain about the location of the critical pile section.

Pile inspection is an area urgently in need of educational efforts. Many of the problems with pile foundations are rooted in construction defects. Knowledge of this area must be brought up to the level of other portions of the art before significant changes can be made in design practice, particularly with regard to allowable stresses.

A discussion of several common causes of pile load increases, pile stress increases, and pile capacity reductions reveals that most are not accounted for in design. These events are covered by the margin of safety, which in practice is always eroded during and after construction. Hence, the writer concludes that our current normal practice of using a pile-to-soil factor of safety of 2.0 and a pile structural factor of safety greater than 2.0 should be continued until a better balance in our knowledge of deep foundations exists, particularly with respect to load transfer and inspection of pile installation.

References

[*1*] Dismuke, T. D., "Behavior of Steel Bearing Piles During Installation and Service," this volume.

[*2*] Gamble, W. L., "Capacity of Reinforced and Prestressed Concrete Pile Sections," this volume.

[*3*] Armstrong, R. M., "Structural Properties of Timber Piles," this volume.

[*4*] Diekmann, E. F., "Timber Piles in Standards, Codes and Practice," this volume.

[*5*] Davisson, M. T., "Pile Load Capacity," *Proceedings*, Design Construction and Performance of Deep Foundations, American Society of Civil Engineers, University of California, Berkeley, Feb. 1975.

[6] Thompson, C. D. and Thompson, D. E., "The Influence of Driving Stresses on the Development of High Pile Capacities," this volume.
[7] Gerwick, B. C. and Brauner, H. A., "Design of Concrete Piles for Static and Dynamic Loads," this volume.
[8] *Engineering News Record*, 8 June 1978, p. 12.
[9] Bozozuk, M., Keenan, G. H., and Pheeney, P. E., "Load Testing Instrumented Steel Test Piles in Compressible Silty Soil," this volume.
[10] Lacy, H. S., "Load Testing of Instrumented 225-Foot-Long Prestressed Concrete Piles," this volume.
[11] York, D. L., Miller, V. G., and Nabil, F. I., "Long-Term Load Transfer in End Bearing Pipe Piles," Transportation Research Record No. 517, 1974, pp. 48-60.
[12] Hunter, A. H. and Davisson, M. T. in *Performance of Deep Foundations, ASTM STP 444*, American Society for Testing and Materials, 1968, pp. 106-117.
[13] Wu, A. H., and Fox, R. R., "Capacity of Axially Loaded Bent Piles in a Bearing Stratum Overlain by a Thick Layer of Soft Clay," this volume.
[14] Clark, J. I., "The Failure During Construction and Subsequent Rehabilitation and Performance of a Displacement Caisson Foundation," this volume.
[15] Wang, M. C., Wu, A. H., and Scheessele, D. J., "Stress and Deformation in Single Piles Due to Lateral Movement of Surrounding Soils," this volume.
[16] Davisson, M. T., "Inspection of Pile Driving Operations," Technical Report M-22, Department of the Army, CERL, Champaign, Ill. July 1972.

DISCUSSION

W. A. Norum [1] (*written discussion*)—Guidance being offered by Professor Davisson to engineers on the strength properties and use of timber piles needs some response. My comments on this subject are offered in light of the purpose of the symposium expressed by Davisson: "One of the purposes of this symposium is to draw together in one reference the facts regarding structural strength of piles."

First, it should be immediately recognized that Davisson's comments on the subject of timber piles is almost totally dependent upon information and judgments expressed in Armstrong's paper.[2] Interestingly, he has seen fit to completely ignore the information in Diekmann's paper[3] which contains much relative data and discussion that is not in total agreement with Armstrong's conclusions. Differing conclusions to Armstrong's have also been reached by myself. Details of these are included in my comments on Armstrong's paper.

Second, there are specific comments made by Professor Davisson that need response. They are enumerated below along with my response:

1. In his comments about how engineers are to specify timber piles in accordance with ASTM Specification for Round Timber Piles (D 25-73), he

[1]National Forest Products Association, Mountain View, Calif. 94040.
[2]See p. 118.
[3]See p. 264.

implies that engineers cannot exercise as much control over selection of timber pile sizes as was done by him previously under ASTM Method D 25-58. The fact is engineers have much greater control referencing ASTM Method D 25-73 in their specifications. The tables of pile sizes conveniently list the sizes that can be made available. The table of sizes provide guidance on how trees grow. Engineers can select any size desired but within the limits of nature's control.

2. Davisson refers to the southern pine subspecies as an example of further problems in specifying timber piles. He fails to point out that the other pile species do not have subspecies like southern pine, longleaf, slash, loblolly, and shortleaf, nor does he point out that subspecies per se do not reflect differences in pile strengths. It is density of timber that is a direct measure of strength among subspecies and, fortunately, some visual guidance is possible on jobsite by measuring ring count.

3. The author's comment about timber having low cycle fatigue properties is misleading. According to the U. S. Forest Products Laboratory *Wood Handbook*, fatigue need not be a design consideration for wood construction until repetitions of design stress or near design stress are expected to be more than 100 000 cycles during the normal life of a structure.

4. Regarding the subject of pile selection for load testing, the author's statements are again misleading. He implies that all test piles are selected by contractors who will purposefully select the strongest piles for pile load testing and mislead the engineer. It's been my experience that test piles are selected by soil or structural engineers, at least this is the practice on the West Coast. Engineers do not usually give that option to contractors.

5. As for comments made by the author about the reported 70-ton design load and soil conditions at the Nevada state highway project, he needs to be reminded that the choice was made after extensive soil investigations by both the Nevada State Highway Department and the Federal Highway Administration. It was found that the Douglas fir timber piles had the capacity to support the 70-ton design load and that they were found more economical than the concrete and steel pipe alternates.

6. Statements made about failures due to negative friction are questionable. The author says, "The incidence of failure is higher for timber than for other pile materials and is interpreted by the writer to be expected based on the low to negligible structural factor of safety employed in timber relative to concrete and steel." The author should, in my opinion, substantiate his claim that there are more timber pile failures due to negative friction. I have no such information, and I doubt that any good statistical information is available. The author's conclusion is strictly speculative.

7. Statements made about pile tip loading being higher than normally considered overlooks situations discussed by the author relative to soil freeze. If the author really believes that all loads are eventually carried

to the tip, it's obviously impossible to justify the design loads on piles embedded in the Louisiana mud.

8. In his Summary and Conclusions, the author says, "It has been shown that timber pile strength is much less than currently advertised and that pile foundation failures confirm this defect." My response is a question—Where is the evidence? There was no such evidence contained in the paper.

Some very important and practical benefits of timber piles have been overlooked by the author. He fails to point out that most all factors that influence the strength of timber piles are evident at the jobsite. These factors include knots, ring count, and grain pattern. The same cannot be said for steel; and as it effects concrete, test results cannot be made immediately available. Concerns over the effects of preservative treatment are mollified by knowledge that only southern pine is steam conditioned prior to treatment, so it is easy to identify which piles would be effected. Also, the author fails to point out that driving stresses are not important in all situations. Importance is dependent upon soil conditions and level of design load. Where soil conditions are difficult, preboring is always a possibility. In those cases where driving is tough, banding timber piles may be a good solution. It has been proven that banding timber piles reduces the amount of fiber separation experienced during hard driving. Banding makes it possible to experience higher blow counts without failure.

As pointed out in my response to Armstrong's paper, timber has a multivalued "factor of safety" that will probably range from 1.12 to 2.74 when assuming there will be creep occurring in timber due to the effects of load and time that will probably not occur because the levels of design stresses being recommended are relatively low. This "factor of safety" also considers that the piles will contain knots having the maximum effect on strength and that all piles will consist of wood having near minimum wood density that is found in trees of the species specified. Since procedures of ASTM Establishing Design Stresses for Round Timber Piles D 2899 call for design stresses to represent the weaker material, then about 95 percent of the piles will theoretically have a greater unit stress than is recommended by ASTM Method D 2899. Add this element of conservatism to the fact timber pile sizes specified by an engineer are minimum sizes, and since timber piles are nature's product, most piles will be oversize. The effect oversize piles has on a factor of safety is very significant. For example, the average pile could support 4.6 more tons than is calculated by the designer when the average pile tip is 24 in. in circumference rather than 22 in. minimum as specified. A value of 1 250 psi design stress in compression parallel to grain is used in the example. With all factors considered, including conservative design loads and tip bearing that may never be experienced, the "real factor of safety" for most projects will, undoubtedly, be considerably higher.

Detailed comments on quality of timber piles and recommended design stresses are included with my written comments on Armstrong's paper.

M. T. Davisson (author's closure)—Replies to the various comments made by Mr. Norum will be given in approximately the order that they appear in his discussion.

The writer found that Armstrong's paper (see footnote 2) was based on a review of all the readily available and important data regarding timber pile strength. Hence, it was the primary source for the writer's paper. Norum's criticism of Armstrong's paper is thoroughly answered by Armstrong in his closure.

It is true that the writer ignored part of Diekman's paper (see footnote 3), but only because of the errors it contained. For Mr. Norum's enlightenment the errors will be described.

Diekmann states: "The calculations for both Southern Pine and Red Oak have been made using the lowest characteristics of the group. Averages, weighted or otherwise, which may conceal poor material do not seem appropriate for piling where each piece may be structurally critical." He then applies the rule and calculates a design stress of 7.5 MPa (1092 psi) for pile tips of steam conditioned southern pine, by using the ASTM Method D 2889-74 formula. This value is lower than the 8.3 MPa (1 200 psi) promoted by the timber industry, because of Diekmann's application of *his rule* of using the lowest species. While Diekmann is in error when referring to the 7.5 MPa (1 092 psi) as the ASTM Method D 2899-74 value, it is the value used as the basis for his subsequent calculations and conclusions.

Upon comparing this long term design stress of 7.5 MPa (1 092 psi) to the test results on steamed southern pine tips reported by Wilkinson, Diekmann concluded: "Steam-conditioned southern pine tip bearing piles should be designed for two thirds ASTM Method D 2899-74 stresses." This conclusion results in a recommended design stress of 5 MPa (728 psi), (two-thirds of 1 092).

In making his evaluation Diekmann compared the long term design stress of 7.5 MPa (1 092 psi) to short-term test results, hence ignoring the influence of load duration, a well documented fundamental property of timber. Applying the load duration factor of 1.52 (which was included in the ASTM Method D 2899-74 formula) to correct for his error, results in a recommended design stress of 3.3 MPa (479 psi) (728/1.52). Hence if Diekmann's review is taken at face value, his recommendations are even lower than those proposed by Armstrong.

Both Armstrong and Diekmann draw attention to the knot provisions of ASTM Method D 25-73, with Diekmann drawing the conclusion that ASTM Method D 25-58 provisions are preferable to ASTM Method D 25-73. Hence the writer's comments are supported by Diekmann as well as Armstrong.

Norum's numbered comments are answered in the same sequence.

1. Norum disputes the writer's statement that pile size under ASTM Method D 25-73 is only partially under the engineer's control. A comparison of ASTM Methods D 25-58 and D 25-73 illustrates the writer's point. It is true as Norum states, that the product actually furnished is still a tree grown according to mother nature's specifications.

2. Norum's statement that "other pile species do not have subspecies like southern pine" is patently in error. Reference to ASTM Establishing Clear-Wood Strength Values (D 2555) and Armstrong's Table 1 provides more than enough evidence.

3. Norum's statements regarding fatigue illustrate his failure to understand the problem of driving stresses. The FPL Wood Handbook refers to *design stress* in saying fatigue is not a problem unless 100 000 cyles or more occur. Driving stresses are close to *ultimate stress.* Therefore, design stresses of 20 to 35 percent of ultimate stress do not present a fatigue problem, but repeated driving stresses at ultimate stress levels are clearly in the range of low cycle fatigue.

Failure to consider fatigue when assessing installation problems can lead to costly delays and remedial requirements during the construction stage. While steel banding may reduce the amount of fiber separation during driving, it does not necessarily eliminate all fiber damage experienced during installation. The influence of the fiber damage experienced during driving on the long term strength has not been addressed by either the timber industry or timber technologists. The writer feels that it could only accelerate the strength reduction–time relationships experienced with undamaged fibers. This is of particular concern at the high stress levels the timber industry is promoting.

4. Norum's statements regarding selection of test piles are in error. Engineers may select a pile location for testing, but the contractor generally selects the particular piece of wood that will be driven in that location.

5. Norum states that 70 ton timber piles in Nevada were selected after extensive investigations by the Nevada State Highway Department and the Federal Highway Administration, and implies that this is authority for others to follow. The writer accepts the statement and rejects the implication. Bureaucracies and lack of error do not correlate.

6. Norum doubts the author's statements about negative skin friction and asks for proof. It is not coincidental that 60 percent of the examples of negative skin friction failure cited in the MIT Symposium on Downdrag

of Piles[4] were timber. There are numerous unpublished failures that become suppressed by legal entanglements.[5,6] The failures are fact, not speculation as stated by Norum.

7. Norum dislikes the writer's statements regarding stress transfer to the tip of the pile. He follows up by attributing a statement to the writer, "believes that all loads are eventually carried to the tip," that did not appear in the paper. Therefore, it is impossible to respond.

With respect to piles in the Louisiana mud, it is important to note that the New Orleans building code wisely limits timber pile loads to 25 tons.

8. Norum asks for evidence for the writer's statements that timber pile strength is less than currently advertised, and that pile foundation failures confirm this defect. The errors in ASTM Method D 2899-74 result in promotion of the average 10 year constant load failure stress as a design working stress.[7] As a result, there is no factor of safety as is common with other piles. Timber pile failures are documented (see footnotes 4 through 6).

Norum follows his numbered list with a discussion of factor of safety, which by his admission is as low as 1.12. Further, he assures the readers that only 5 percent of the piles will have a factor of safety less than 1.12. This situation is unacceptable to both the structural and geotechnical engineering professions, especially in light of Armstrong's closure which shows that Norum's factors of safety are optimistic.

Finally, the dangers of engineers relying on the timber industry for establishment of useful design stresses is illustrated[8] by structural failures of timber columns. In the subject case the following have been claimed.

1. The failures are a result of misrepresentation of strength by the timber industry.
2. The timber industry claims that an agency that accepts and uses the timber industry's design values assumes responsibility for them.

It is clearly unacceptable for both the structural and geotechnical professions to be influenced by the claims of timber industry, when the timber industry will deny responsibility when put to the test. The establishment of design stress levels is best left to those who ultimately are responsible for the results.

[4]*Proceedings of a Symposium on Downdrag of Piles,* J. E. Garlanger and T. W. Lambe, Eds., Research Report No. 73-56, Department of Civil Engineering, Massachusetts Institute of Technology, Cambridge, Nov. 1973.

[5]Continental Grain Company versus Fegles-Power Service, U. S. District Court, St. Paul, 1972.

[6]*Engineering News Record,* 8 Oct. 1959, p. 26.

[7]ASTM Subcommittee D07.07 on Round Timbers, Report of Task Group by R. M. Wolfe, Oct. 1978.

[8]*Engineering News Record,* 2 Feb. 1978, p. 15.

F. M. Fuller[1]

State-of-the-Art Pile Design Practice—Current and Proposed as Reflected in Building Codes

REFERENCE: Fuller, F. M., **"State-of-the-Art Pile Design Practice—Current and Proposed as Reflected in Building Codes,"** *Behavior of Deep Foundations, ASTM STP 670,* Raymond Lundgren, Ed., American Society for Testing and Materials, 1979, pp. 84–117.

ABSTRACT: Building codes generally specify maximum allowable compressive stresses for basic pile materials that control the structural design of piles. Sometimes these prescriptive type provisions are qualified by performance type provisions that permit such allowable stresses to be exceeded under controlled conditions. Combined prescriptive and performance regulations governing pile foundation design and installation is recommended because totally prescriptive or totally performance type provisions are impractical under the present state of the art. A relatively high structural factor of safety must be reflected in allowable stresses for pile material to provide for the many uncertainties currently existing in pile foundation design and installation and the several conditions under which the actual structural factor of safety can be less than assumed. Until we have more precise answers to all of our problems in designing and installing satisfactory pile foundations, any move to reduce the structural factors of safety below current levels would be ill advised. An examination of the current allowable stresses and proposed increases indicates that under today's state of the art, a practical limit has been reached or in some cases exceeded and proposed increases are generally unwarranted or undesirable.

KEY WORDS: allowable stresses, building codes, concrete piles, piles, pile design, steel piles, timber piles

The topic for session four of this symposium is design practice—current and proposed—including consideration of standards and codes. It was not intended that this session embrace the entire subject of present and proposed pile design practice, as attested to by the fact that only four papers were selected for this session. The primary objective of this session is to focus attention on those provisions of major model building codes that relate to or affect deep foundation design. The major model building codes

[1]Assistant vice president, Raymond International Builders Inc., Houston, Texas.

have been singled out because they are often considered as more reflective of current practice than other codes due to the continuing development process they undergo. Of the two major subdivisions of deep foundations—piles and caissons—most of the codes cover pile foundations, but very few have specific requirements governing caissons or large-diameter drilled shafts. This paper will be limited to pile foundations only.

Pile foundation design considerations can be classified into two major areas—geotechnical and structural. The geotechnical aspects of pile foundation design in terms of specific requirements are generally not covered in building codes except for minor provisions governing such things as pile spacing, group action, negative friction, and possibly load testing. Most of the codes limit their provisions to those governing the structural design of the pile foundation, with its related provisions covering installation and construction. This paper will be further limited to pile foundation design practice relating to the structural design of the piles.

In this regard, the most important single factor is the design stress—the maximum allowable compressive stress (or in some cases the maximum allowable load) that can be applied to the pile material resulting from externally applied axial loads. This one number—allowable stress—often establishes the required pile size and often dictates the type of pile to be used. For these reasons, this one number has substantial commercial ramifications and therefore is the one item in building codes that receives the most attention from suppliers or manufacturers of various pile materials or producers of special pile types. There is a continuous competitive drive to push these allowable stresses (or loads) as high as possible.

One of the purposes of this symposium is to examine the current high allowable stress levels and the trend toward even higher allowable stresses for various basic pile materials. The basis and validity of some of these current stresses or proposed increases will be examined.

Building Code Formats, Objectives, and Development

There can be two basic types of building codes—the performance type and the prescriptive or specification type. Performance-type codes specify the end result, not the means for obtaining it. Prescriptive type codes set forth in detail the step-by-step procedures covering either design or construction or both.

The functions and use of structures as a whole can be defined in terms of performance. Many of the materials, construction systems and assemblies, methods, components, etc. used in a structure can also be defined in terms of performance. Most of these items are manufactured and installed or are constructed in place so that the end results are inspectable and the performance can be observed and measured. The materials involved have known and consistent properties.

These structural items are generally above ground and involve portions of the frame, the shell, and the interior of the structure. It should be noted that these superstructure components are generally not designed for or subjected to stresses higher than those resulting from the service loads, which are readily determined.

In contrast, for pile foundation design and construction, it is difficult to adequately translate the requirements into performance criteria. Each pile foundation, regardless of pile type, is custom made; thus, each pile foundation is unique. As a matter of fact, each individual pile is unique.

Furthermore, the installation and performance of pile foundations always involve and are dependent upon a separate material—the soil or rock—the properties of which are not completely known and can be infinitely variable and can change with time. And finally, the structural component, the pile, is generally subjected to stresses during installation that could be considerably higher than those resulting from service loads acting on the pile, which in turn are not always readily determined. Therefore, a true performance-type building code for pile foundations would be most difficult to evaluate and enforce.

For example, a code could merely require that the structure be adequately supported by the piles. Such a provision would be impossible to administer, would result in the use of a wide range of safety factors, and would not protect the owner. The term "adequate support" would be a judgment factor or an opinion and may not reflect actual long-term performance.

This raises the question as to the purpose of a building code. Generally, in a broad sense, building codes are considered to exist to secure the life, safety, health, and general welfare of the public. However, building codes in some areas of construction must go beyond that broad mandate. One of the major model building codes [*1*][2] includes in its purpose the safeguarding of property by regulating and controlling the design, construction and quality of materials. Another model code [*2*] provides for securing the general welfare through structural stability. A third model code [*3*] considers insuring the public welfare through structural strength. As Olson [*4*] points out, not only must the building code guard against catastrophic failure of the structure, but it must also protect owners from financial loss resulting from distress of the structure that affects the owner's and adjacent property owners' equity as well as the owner's maintenance costs. Distress would include unsightly conditions resulting from poor performance, short or long-term.

The concept of a purely performance type code, where responsibility for compliance rests solely with the engineer, has weaknesses, especially in the area of protection of property owners. If an engineer were not fully qual-

[2]The italic numbers in brackets refer to the list of referencs appended to this paper.

ified in the design and installation of pile foundations, he could be successfully promoting cheaper, unsatisfactory foundations or, on the other hand, designing ultraconservative foundations of excessive cost. Furthermore, enforcement of such a type code would require considerable specialized expertise in all local building departments, which is generally not available.

A totally prescriptive type code for pile foundations is also impractical. The state of the art for the design and installation of pile foundations has not yet advanced sufficiently to permit a detailed step-by-step procedure that will guarantee end results. There is still a considerable amount of judgment or art in the proper design and installation of pile foundations. Furthermore, on a practical note, the design procedures and analyses are too complicated and involved to be written into a building code. However, even if this could be accomplished, the enforcement problems would be formidable.

It would appear, therefore, that as far as pile foundations are concerned, building codes should be written as a combined prescriptive and performance code with the prescriptive provision based on conservative accepted engineering practice, in order to prevent unqualified designers from exceeding such practice. Provisions could be included for waiving such prescriptive requirements if a qualified engineer in charge can adequately substantiate his design and the subsequent pile installation.

There are four so-called model building codes in the United States: the Uniform Building Code [UBC] [*1*], the Standard Building Code (SBC) [*2*], the Basic Building Code (BBC) [*3*], and the National Building Code (NBC) [*5*]. These codes are adopted by various jurisdictions, such as states, counties, and cities, either as written or as modified locally. In addition to these codes, some of the states and major cities develop their own building codes.

Of the four model codes, three of them—Basic, Uniform, and Standard—are promulgated by an organization comprised of a substantial headquarters staff and a large membership made up of building officials, industry, trade associations, etc. These organizations have standing committees and handle code revisions on a formalized basis involving publication of proposed revisions, public hearings, recommendations from technical groups and a membership ballot limited to building officials only. In contrast, the National Building Code is promulgated by the American Insurance Association and is written without the benefit of a broad input. For this code, the trade associations are most influential. Technical societies and trade associations influence the development of other building codes but their recommendations and proposals are often subject to a review process and challenge by opponents.

Technical societies actively engaged in proposing and submitting recommendations on code changes include the Structural Engineers Associations of California and Washington and the Soils and Foundation Engineers Association (California). Other technical societies such as the American Society for Testing and Materials (ASTM) and the American Concrete Institute (ACI) influence code development through their various publications. They, along with the American Society of Civil Engineers (ASCE), do not participate as an organization in building code development; local ASCE groups may get involved. Thus, except for the Uniform Building Code and locally written codes, structural and geotechnical engineering organizations do not take an active role in the development of building codes. ASTM, through its standards, is especially influential in formulating some building code provisions. Such standards are readily adopted because they are considered to be consensus standards reflecting the current state of the art. Therefore, extreme care should be taken as to what is issued under the ASTM label, and ASTM should avoid promulgating design standards.

On the other hand, trade associations are actively engaged in this field. These include the American Iron and Steel Institute (AISI), the Portland Cement Association (PCA), the National Forest Products Association (NFPA) the American Wood Preservers Institute (AWPI), and many more. Those involved in the process of developing building codes do need the valuable input offered by trade associations. However, in matters as highly specialized and involved as the design and installation of pile foundations, technical societies should participate more actively in code development to balance the commercial influences.

Allowable Stresses in Building Codes

It would appear that the most practical approach for specifying the provisions governing the structural design of piles in building codes is the working stress method. However, such stresses established in the code can be based upon a strength-design analysis as was done for the current provisions governing allowable stresses for concrete piles. In order to convert strength-design to working stress, appropriate load reduction factors and load factors (factor of safety) must be selected for the conditions and material involved. Such allowable stresses should reflect an accidental eccentricity factor, as do allowable concrete stresses.

A slightly different reduction factor approach has been suggested by Rempe [*6*] and Gamble [*7*]. Different load reduction factors would be applied depending upon the type of pile, potential defects, potential loss of structural integrity with time, uncertainties involved in the installation process and other factors. These would include driveability of the pile (which would involve the pile size, length, and material, required ultimate capacity, and soil conditions.) Although intriguing, this approach does present difficulties in adoption and application. It would require different allowable stress levels, depending upon the type of pile, its method of installation, the subsoil conditions involved, and other factors. Obviously such an approach would be objected to strenuously by the proponents of those pile types penalized by this method.

Rempe [6] calls attention to a common defect in pile design practice and the establishment of working stresses in building codes, which is a failure to account for pile driveability. Driveability includes sufficient stiffness to transmit driving forces large enough to overcome the necessary soil resistance and to develop the required pile capacity. Driveability also involves the pile having sufficient strength to withstand the driving forces without damage. For many pile installations, driveability is a very important limitation and could be reflected in code provisions by either incorporating driveability limits into allowable stress provisions or requiring a check on driveability as part of a substantiation for a pile design. Load tests are essential in checking driveability but not totally conclusive where soil conditions are variable. Driveability could be checked by a wave equation analysis as a supplement to load testing. Another defect in the establishment of working stresses (for steel and timber piles) is the lack of consideration of accidental eccentricity that is likely to occur.

There are two principal factors of safety to be considered in pile foundation design and installation—the factor of safety for the pile-soil system and the structural safety factor for the piles alone. The former involves many unknowns, such as exact soil conditions at each pile location, effects of installation methods, load transfer rates, and time effects. The latter involves not only the basic material and loads, but also installation and time effects.

Because of the unknowns, current practice generally requires a factor of safety of 2 against pile bearing capacity failure. There are some practitioners today who are advocating reducing this pile-soil system safety factor down to 1.5 or less to justify reducing the structural safety factor down to 2 or less. It would appear that proponents of such a move are overly confident of their ability to predict or determine the ultimate capacity of each pile as installed.

Under the present state of the art, considering the uncertainties involved in our knowledge of soil conditions at each pile location, the actual loading on each pile, installation and time effects, and several other indeterminate factors, it would seem prudent to retain a reasonable factor of safety of 2 for the pile-soil system.

In this case, the structural factor of safety should be somewhat greater than 2 to provide for adquate testing of the pile-soil system and to preclude premature structural failure. Rempe [6] points out the actual structural factor of safety can be less than the assumed or design value for several reasons such as:

1. Use of substandard material.

2. Damage to pile during installation (including bending, rupture, deformation, reduction in cross section).
3. Unequal tip bearing on rock or boulder.
4. Errors in calculating design loads.
5. Errors in determining critical section.
6. Failure to adequately account for driving stresses.
7. Failure to account for loads due to pile misalignment or mislocation.
8. Failure to account for negative friction loads.
9. Failure to account for residual stresses.
10. Deterioration of pile material with time.

Rempe concludes, "Considering the several conditions under which the actual structural factors of safety can be less than the nominal factor, existing code provisions corresponding to a structural factor of safety of 2 or less should be examined critically." The above list of possible causes for reductions in the assumed structural factor of safety would require that this safety factor be greater than 2.

Any reduction in the structural factor of safety, including those resulting from increased design stresses, affects the other factors of safety of the pile foundation. Until we have more precise answers to all of our problems in designing and installing satisfactory pile foundations, such a move as reducing the factors of safety would be unwise.

In connection with allowable working stresses in general, it should be noted that code provisions of the specification type tend to become those on which the design is based, regardless of pile type, driving requirements, or soil conditions. The designer is under commercial pressures to use the maximum allowable stresses for pile materials, which often results in the lowest apparent first cost. Therefore, what is specified in the code should in the general case assure the owner of a satisfactory foundation on a long-term basis.

In order to discuss specifics on pile structural design practice, piling can be conveniently considered according to the basic material involved—concrete, steel, or timber. For each material, the present structural design criteria in codes as well as what has been proposed or implemented in the recent past will be examined.

Concrete Piles

Concrete piles are of two basic types—precast and cast in place. Precast can be either conventionally reinforced or prestressed. The actual structural design of precast concrete piles is not covered in most building codes. Such details are found in various standards and recommended practices such as the ACI Committee 543 report [*8*], and the comparable (PCI) Prestressed Concrete Institute report [*9*] covering prestressed piles only. In addition, standards are developed by various state highway departments, the Joint PCI-AASHTO Committee and some pile producers. These standards are generally adequate and up to date for most applications of precast concrete foundation piles. In connection with the complete structural design of precast piles, it should be noted that Gerwick and Brauner [*10*] make no recommendations for changing the present design methods except to increase the amount of spiral reinforcement, especially for piles designed to resist high bending forces or seismic loading.

Insofar as building codes are concerned, the only structural desigin parameter normally included is an allowable working stress, which in turn determines the required cross-sectional area for a given load or the maximum design load that can be used for a given size pile. Table 1 after Rempe [*6*] shows the allowable compressive stresses for precast concrete piles as found in some building codes. The allowable stresses are expressed in terms of a percentage of the 28-day concrete strength, and thus, the actual maximum design stresses vary with the concrete strengths. In the case of prestressed concrete piles, the maximum allowable stress is also controlled by the level of effective prestress in the pile.

Except for the allowable stress shown for conventionally reinforced piles for Los Angeles and the allowable stresses included in the New York City code, these allowable stresses all result from a strength design analysis made by the Portland Cement Association (PCA). The broad acceptance of the PCA report [*11*] accounts for the uniformity of allowable stresses.

These stress levels have been in use now for more than 8 years, and in general, the experience has been quite satisfactory. Occasionally, installation problems have been encountered for normal foundation piles when design loads were excessively high. As loads increase, heavier hammers, harder driving, and deeper pile penetration are required. This subjects the pile to higher or more prolonged driving stresses, which could produce cracking or other damage if proper installation methods are not used or if the pile is not adequately designed and constructed. As Gamble [*7*] points out, under the high stress rates imposed during driving, the com-

TABLE 1—*Allowable stresses for precast concrete piles.*[a]

Code/Standard	Reinforced	Prestressed
UBC, NBC, SBC	$0.33f'_c$	$0.33f'_c - 0.27f'_{pe}$
BBC	Provisions subject to interpretation	
Los Angeles	$0.225f'_c$	$0.33f'_c - 0.27f'_{pe}$
New Orleans	$0.33f'_c$	$0.33f'_c - 0.27f'_{pe}$
New York City	$0.25f'_c$ to $0.36f'_c$	$0.25f'_c$ to $0.36f'_c$
PCA	$0.33f'_c$	$0.33f'_c - 0.27f'_{pe}$
ACI	$0.33f'_c$	$0.33f'_c - 0.27f'_{pe}$

[a]After Rempe [6].

pressive strength of the concrete is considerably higher than the slow loading strength, but the beneficial effects of having a high stress rate are offset by the low-cycle fatigue effects. Therefore, the higher strengths at the higher stress rates cannot be counted on under prolonged pile driving.

For the structural design of reinforced and prestressed concrete piles, especially those subjected to combined axial loading and bending, Gamble [*7*] proposes using strength design methods according to ACI 318-77 [*12*], and the use of interaction diagrams. Anderson and Moustafa [*13*] proposed a similar approach. The use of ACI column design combined with interaction diagrams is common practice for unsupported concrete piles or piles subjected to combined loading.

The allowable loads and thus the design stresses resulting from the ACI 318 approach are generally of the same order of magnitude as those used in current practice for fully embedded concrete piles. The example cited by Gamble [*7*] results in an allowable load somewhat higher than that which would result from the allowable stresses shown in Table 1 for the size pile used.

It should be noted that the allowable compressive stresses in building codes apply only to fully embedded and laterally supported foundation piles. In most building codes, for piles that extend through air or water or through soils incapable of providing lateral support, the piles are to be designed in accordance with current practice for columns, except that specified allowable stresses for piles are not to be exceeded. In the pile foundation sections of building codes, the allowable stresses due to bending or combined loading are not normally addressed.

Table 2, also from Rempe [*6*], shows the allowable compressive stresses for cast-in-place concrete piles currently in representative building codes. The allowable stresses in the Uniform, National and Standard Building Codes are based upon the ACI [*8*] and PCA [*9*] recommendations, which in turn result from a strength design analysis. For these three codes, the allowable stress of up to 40 percent of the 28-day concrete strength applies only to concrete confined in a non-load-bearing steel shell. This value of

TABLE 2—*Allowable stresses for cast-in-place concrete piles.*[a]

Code/Standard	Uncased	Cased (Shell)	Cased (Pipe)
UBC, NBC, SBC	$0.33f'_c$	$0.33f'_c$ to $0.40f'_c$	$0.33f'_c$
BBC		Provisions subject to interpretation	
Chicago	...	$0.40f'_c$	$0.40f'_c$
Los Angeles	$0.225f'_c$	$0.225f'_c$	$0.25f'_c$
New Orleans	$0.225f'_c$	$0.25f'_c$	$0.275f'_c$
New York City	...	$0.25f'_c$ to $0.36f'_c$	$0.25f'_c$ to $0.36f'_c$
PCA	$0.33f'_c$	$0.33f'_c$ to $0.40f'_c$	$0.33f'_c$
ACI	$0.33f'_c$	$0.33f'_c$ to $0.40f'_c$	$0.33f'_c$

[a]After Rempe [*6*].

$0.40f'_c$ was obtained from the PCA report [*11*], as shown in Fig. 1. The actual allowable stress can be determined based upon the thickness of the steel shell, the yield strength of the steel, the diameter of the shell, and the 28-day concrete strength. To simplify matters, an upper limit of 40 percent was selected, based upon using a minimum 14 gage, 1.9 mm (0.07 in.), 207 MPa (30 ksi) steel shell, of maximum 410-mm (16-in.) nominal diameter and a maximum 35 MPa (5000 psi) concrete for computational purposes.

The allowable stresses in Tables 1 and 2, which follow the ACI [*8*] and PCA [*9*] recommendations reflect a 5 percent accidental eccentricity factor. These design stress levels have been in satisfactory use for more than 8

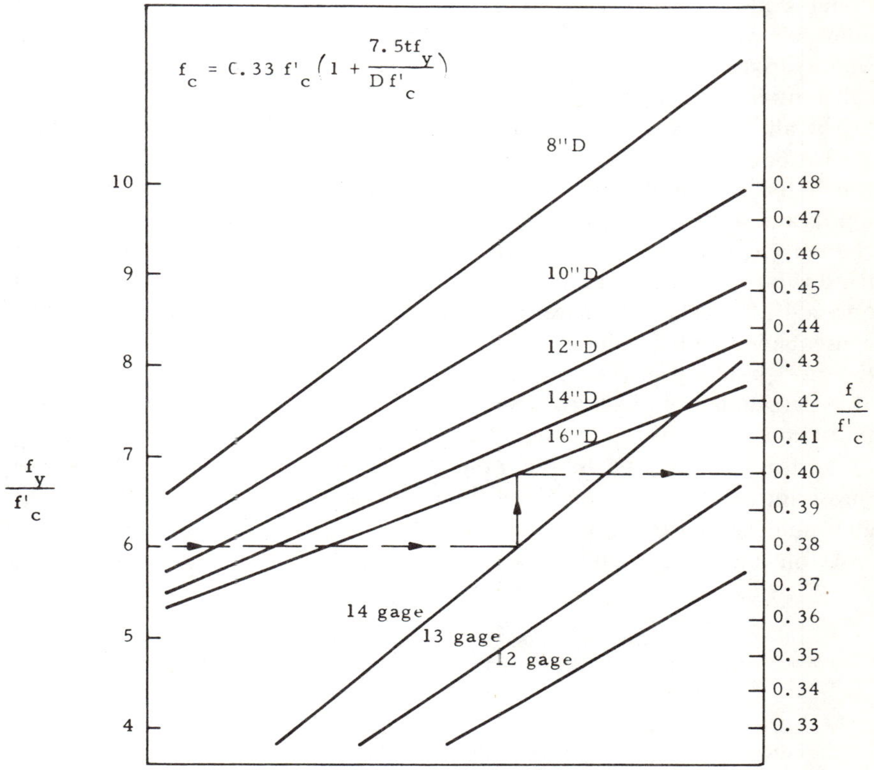

f_c = allowable concrete stress
f'_c = 28-day concrete strength
t = steel shell thickness
f_y = steel yield strength (shell)
D = steel shell diameter

FIG. 1—*Allowable stress for concrete confined in non-load-bearing steel shell (from PCA Report, June 1971).*

years; no structural problems have been reported. There are no proposals to raise the allowable stress for either precast or cast-in-place concrete piles above those currently in the Uniform, Standard, and National Building Codes.

Steel Piles

In Table 3, Rempe [*6*] indicates the range in allowable stresses for steel piles in some representative codes and standards. Except for the National Building Code, the New Orleans Code, and the AISI manual on Pile Foundations [*14*], the allowable maximum stress is uniformly around 83 MPa (12 000 psi). As mentioned previously, provisions incorporated in the National Building Code are strongly influenced by trade associations. The allowable values shown for the New Orleans Code represent a condition where driveability is not a factor; the New Orleans soils are noted for substantial soil setup or freeze.

The allowable stress of 87 MPa (12 600 psi) for UBC, SBC, and the New York Code (NYC) is based upon 35 percent of the specified yield value, which for computational purposes is limited to 248 MPa (36 ksi). This limitation was originally proposed by the steel industry in 1967 when changing the concept for specifying allowable stresses for steel piles from a fixed value of 83 MPa (12 000 psi) to a percentage of yield. This resulted in only a 4 MPa (600 psi) increase, which at the time was considered of little consequence and no objections were voiced. However, in the recent past, the steel industry has proposed code changes either increasing the allowable percentage of yield from 35 to 50 or removing the 248 MPa (36 ksi) limitation or both. Except for New Orleans and the National Building Code, this was resolved by retaining a basic stress of 87 MPa (12 600 psi) (prescriptive code) and including in the code provisions for using higher allowable stresses if justifiable (performance code).

As an allowable compressive stress for steel piles, the steel industry is

TABLE 3—*Allowable stresses for steel piles.*[a]

Code/Standard	Pipe	H-Piles
UBC	$0.35f_y$ max, 12.6 ksi	$0.35f_y$ max, 12.6 ksi
NBC	$0.50f_y$	. . .
SBC	12.6 ksi	12.6 ksi
BBC	Provisions subject to interpretation	
Chicago	12.0 ksi	12.0 ksi
Los Angeles	12.0 ksi	12.0 ksi
New Orleans	$0.50f_y$ max, 25 ksi	$0.50f_y$ max, 25 ksi
New York City	$0.35f_y$ max, 12.6 ksi	$0.35f_y$ max, 12.6 ksi
AISI	$0.50f_y$	$0.50f_y$

[a]After Rempe [*6*].

currently actively promoting the use of 50 percent of yield regardless of yield values [*15*]. This promotional effort is being directed to engineers designing pile foundations and to owners paying for them, the inducement being to reduce foundation costs. Pile load testing is proposed as a means of proving a satisfactory design and installation, but no guarantees are offered or responsibility accepted by the material suppliers that 50 percent of yield will work.

Under such conditions, load testing a limited number of piles is not sufficient to insure that all piles are being installed to the required ultimate capacity. Dismuke [*15*] cautions against putting too much reliance on pile load test results. He states, "The pile test is not as effective in this (verifying or establishing the allowable load) as the reliance indicates because of many factors." Also he says, "The pile load test can be a valuable tool for control and design purposes, but its limitations should be recognized." Furthermore, such proof testing should be done far enough in advance of any final design or contract award to avoid any substantial commitments before the design is proven to be feasible.

It should be noted that steel pile stiffness or driveability does not increase with increasing yield strength. Pile stiffness is a function of the cross-sectional area, the pile length, and the Young's modulus, which is the same for 248, 345, or 414 MPa (36, 50, or 60 ksi) steel. But with increasing design stresses, the required cross-sectional areas decrease, and generally, required pile lengths could increase. Both of these trends decrease the pile's stiffness or driveability.

In general, Davisson [*16*] shows and experience has indicated that the maximum steel stress level of about 83 MPa (12 000 psi) for average driving conditions is reasonable and appropriate. Dismuke [*15*] contends that current "experience with the steel pile stress level of $0.5f_y$ (172 MPa for 345 MPa steel) (25 000 psi for 50 ksi steel) is considered to be sufficient for inclusion (in codes) as an allowable stress." He does not say who considers this experience to be sufficient nor does the experience cited by Dismuke support this broad conclusion. Davisson [*16*] has presented a convincing case for retaining a basic stress level of 83 MPa (12 000 psi) for steel piles.

Numerous tests have shown that often the stresses resulting from an ultimate load are far below the yield strength. For example, Lu et al [*17*] report a steel H-pile having an ultimate capacity of 1245 kN (140 tons) resulting in an ultimate stress of 124 MPa (18 000 psi), which is half the yield strength. Thompson and Thompson [*18*] shows that of six H-piles tested, only one reached the yield strength at the failure load. For all others the ultimate stress was less than the yield strength—one as low as 87 MPa (12 600 psi). Thatcher Engineering Corporation [*19*] reports load test results on a steel H-pile for which the flanges started to buckle at a load of 2002 kN (225 tons), or 80 percent of yield. The pile failed at a load of 2491 kN (280 tons) for which the ultimate stress was equal to the yield value. Williams [*20*] reports that friction H-piles failed according to the acceptance criteria at stress levels as low as 78 MPa (11 350 psi), which is 32 percent of the yield value. In this same test program, steel H-piles bearing on rock failed at stress levels as low as 181 MPa (26 200 psi), or 73 percent of the yield strength.

Dismuke [*15*] states in support of using an allowable stress of $0.50f_y$ that numerous load tests have been conducted to the yield point of steel piles and presents a selected summary of such tests. According to Dismuke's [*15*] tabulated data of the six load tests summarized, four were not tested to the specified yield strength (Table 4). Of the remaining two tests, one involved a pipe pile filled with 35 MPa (5000 psi) concrete; thus, the stress in the steel at full load is not known. For the remaining test, the pile failed by flange buckling at the maximum test load of 2135 kN (240 tons), as indicated in the AISI report [*21*] (test no. 12). Thus, in addition to presenting too small a data base to be convincing, Dismuke [*15*] in his tabulated test results has not offered support for the use of $0.50f_y$ as a working stress for steel piles.

Dismuke [*15*] refers to the AISI report [*21*] on load tests on steel piles as further experience proving that a design stress of 50 percent of yield is satisfactory. An analysis of those data as reported by AISI shows the following:

1. Of the 28 tests, three were on concrete-filled pipe piles and thus not applicable to the question of allowable stresses for steel piles.
2. Of the remaining 25 tests, 14 (56 percent) were not tested to the minimum specified yield value. Of these 14, 6 indicated failure by buckling.
3. Of the remaining 11 tests of H-piles, 3 failed by buckling or yielding at a stress level at or below the yield value, and 3 resulted in a bearing capacity failure at about the yield value.

TABLE 4—*Selected summary of high capacity steel pile tests.*[a]

Test No.	Minimum Specified Yield Strength, psi (MPa)	Maximum Applied Steel Stress, psi (MPa)	Remarks
1	33 000 (228)	38 866 (268)	Pile failed by flange buckling
2	45 000 (310)	...	Concrete-filled pipe pile
3	33 000 (228)	31 687 (218)	
4	50 000 (345)	41 311 (285)	Pile cross-sectional area increased by 16 percent with instrumentation protective plates, etc.; pile buckled at maximum stress level
5	36 000 (248)	35 128 (242)	
6	35 000 (241)	33 710 (232)	

[a]Extension of Table 1 from Dismuke [*15*].

Therefore, of the 28 tests, only 5 (18 percent) could be considered as supporting an allowable stress of $0.50f_y$ for steel piles, which is hardly conclusive. This, in conjunction with the tests cited from references [*17-20*] inclusive indicates that 50 percent of yield is not a reasonable basic allowable stress.

Generally, the pile-soil capacity is less than the structural capacity based on the yield strength of the steel. Therefore, to accept 50 percent of yield as a basic allowable stress for steel piles would require reducing the acceptable factor of safety for the pile-soil system to something less than 2. This would appear to be undesirable, considering the many unknowns associated with pile foundation design and installation as previously discussed.

It is advisable that allowable design stresses reflect at least a 5 percent eccentricity factor and the structural factor of safety be somewhat greater than 2. An allowable stress of $0.35f_y$ at an eccentricity of 5 percent results in a structural factor of safety of 2.0, which is minimal. An allowable stress of $0.50f_y$ combined with a 5 percent eccentricity results in a safety factor of 1.4, which is considered unsatisfactory.

Timber Piles

Diekmann [*22*] shows that over the years, allowable stresses for timber piles have about tripled. Timber piles are a product of nature, and during the many years that timber piles have been used there have not been any great changes in the growth processes that influence strength properties. Norum [*23*] states that improvements associated with timber piles are limited to the selection of trees and processing by the manufacturer including debarking (mechanical peeling), trimming and treating with preservatives. Because of the substantial increase in allowable stresses for timber piles, and because these are not manmade products resulting from improved technology, the trend for increasing these stresses should receive careful attention.

Table 5 from Rempe [*6*] shows some typical allowable compressive stresses for timber piles currently found in various building codes. These are design stresses for treated piles; for untreated piles, allowable stresses are considered to be 10 to 15 percent higher. In general, there is a uniformity of about 8274 kPa (1200 psi) for southern pine and Douglas fir piles, 7584 kPa (1100 psi) for red oak, and about 5516 to 5861 kPa (800 to 850 psi) for the weaker woods. For the Basic and National Codes, allowable stresses for timber piles are determined directly from ASTM Method for Establishing Design Stresses for Round Timber Piles (D 2899). This standard was developed by the ASTM Committee D07 on Wood and first issued in 1970. Thus allowable stresses for timber piles in both the BBC and NBC vary according to several factors and can range from about 8274 kPa (1200 psi) to more than 11 032 kPa (1600 psi).

TABLE 5—*Allowable stresses (psi) for timber piles.*[a]

Code/Standard	Southern Pine	Red Oak	Douglas Fir
UBC	1200	1100	1200
SBC	1200	1100	1250
BBC, NBC		Varies per ASTM D 2899	
Baltimore	500	500	500
Atlanta	1200	1200	1200
New Orleans	1200	1100	1250
New York City	1200	1200	1200
NFPA	1200	1100	1250
ASTM D 2899	Varies	Varies	Varies

[a]After Rempe [*6*].

It is not surprising to see a basic uniformity in most of the current codes (Table 5), because such provisions were predicated on ASTM Method D 2899, which because of its ASTM label, was accepted without question as a consensus standard.

ASTM Method D 2899 contains the formula

$$C = (S - 1.645\text{SD})/1.88 \tag{1}$$

where

C = working stress in compression parallel to grain at tip of green untreated pile,

S = average crushing strength of a small clear test specimen obtained from ASTM Establishing Clear Wood Strength Values. (D 2555), and

SD = standard deviation corresponding to *S* also from ASTM Method D 2555.

Subtracting 1.645SD from *S* converts the average ultimate strength of a small clear specimen to a 5 percent exclusion limit; statistically, 5 percent of the piles will have an ultimate strength less than that computed.

The 1.88 in the denominator is intended to convert from the ultimate strength value of a small clear specimen to the ultimate strength at the tip of a full-size pile (accounting for size and growth characteristics) and to convert that strength to the ultimate strength under normal load duration. Normal load duration is defined in ASTM Method for Establishing Structural Grades and Related Allowable Properties for Visually Graded Lumber (D 245) as the application of the full design load (design stress) for ten years either continuously or cumulatively.

Figure 2 taken from ASTM Methods D 2555 and D 245 shows the relation of strength of timber with duration of load. With the duration of the standard strength test as an index of 100, the strength under 10-year

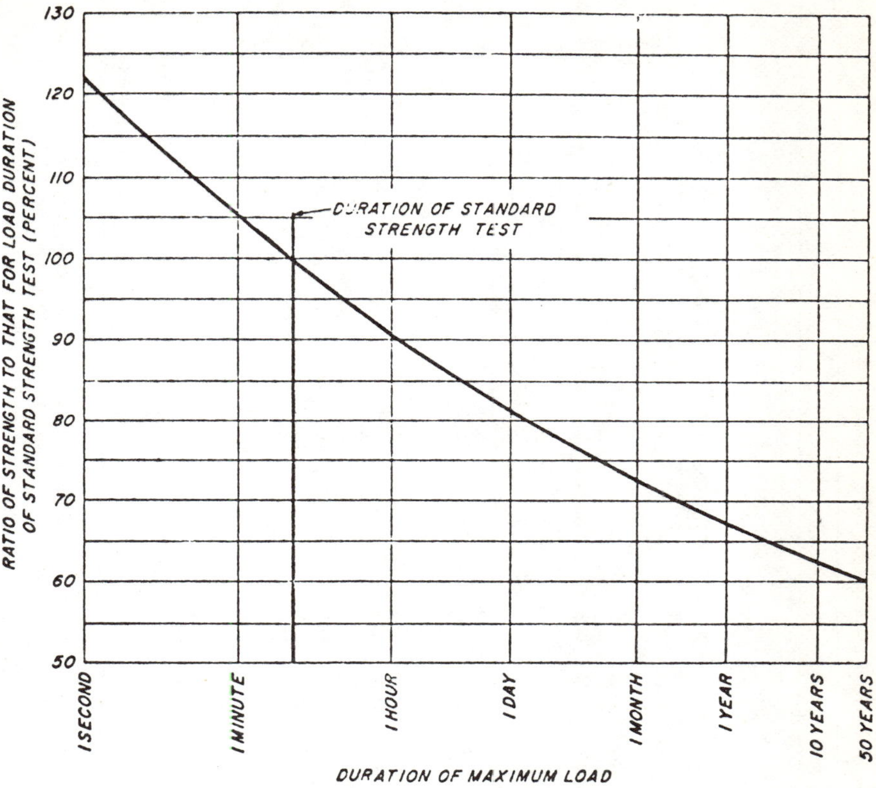

FIG. 2—*Relation of strength of timber to duration of load (from ASTM Method D 2555-78).*

loading is about 62 percent of that determined by the standard test. The actual load duration reduction factor used in ASTM Method D 2899 formula is only 66 percent according to Norum [*23*], which from Fig. 2 represents a continuous or cumulative loading of about 1.5 to 2 years.

ASTM Method D 2555 lists average ultimte values of the compression strength parallel to the grain (*S*) and the standard deviation (SD) for various species and subspecies of timer, all of which have different values for *S* and SD; Table 6 is an example of such listings. According to ASTM Method D 2899 the value of S to be used in formula 1 is an average weighted value for the species or subspecies (or group) involved. Diekmann [*22*] has suggested using the lowest strength of the group. He states, "Averages, weighted or otherwise, which may conceal poor materials do not seem appropriate for piling where each piece may be structurally critical."

For treated piles, the allowable working stress at the tip of a green untreated timber pile obtained from formula 1 is to be adjusted according to

TABLE 6—*Clear wood strength values unadjusted for end use and measures of variations for commercial species of wood in the unseasoned condition (method A).*[a,b]

Species or Region or Both	Modulus of Rupture[c]			Modulus of Elasticity[d]			Compression Parallel to Grain, crushing strength, max			Shear Strength			Compression, Perpendicular to Grain;[e] Fiber Stress at Proportional Limit		Specific Gravity		
	Average, psi	Variability Index	Standard Deviation, psi	Average, 1000 psi	Variability Index	Standard Deviation, 1000 psi	Average, psi	Variability Index	Standard Deviation, psi	Average, psi	Variability Index	Standard Deviation, psi	Average, psi	Standard Deviation, psi	Average, psi	Variability Index	Standard Deviation
Douglas fir[f]																	
Coast	7665	1.05	1317	1560	1.05	315	3784	1.05	734	904	1.03	131	382	107	0.45	...	0.057
Interior west	7713	1.03	1322	1513	1.04	324	3872	1.04	799	936	1.02	137	418	117	0.46	...	0.058
Interior north	7438	1.04	1163	1409	1.04	274	3469	1.04	602	947	1.03	126	356	100	0.45	...	0.049
Interior south	6784	1.01	908	1162	1.00	200	3113	1.01	489	953	1.00	153	337	94	0.43	...	0.045
White fir	5854	1.01	949	1161	1.02	249	2902	1.02	528	756	1.01	78	282	79	0.37	...	0.045
California red fir	5809	1.01	885	1170	1.01	267	2758	1.01	459	767	1.00	146	334	94	0.36	...	0.043
Grand fir	5839	1.03	680	1250	1.03	164	2939	1.04	363	739	1.04	97	272	76	0.35	...	0.043
Pacific silver fir	6410	1.07	1296	1420	1.05	255	3142	1.06	591	746	1.05	114	225	63	0.39	...	0.058
Noble fir	6169	1.07	966	1380	1.08	310	3013	1.08	561	802	1.04	136	274	77	0.37	...	0.043
Western hemlock	6637	1.03	1088	1307	1.02	258	3364	1.03	615	864	1.02	105	282	79	0.42	...	0.053
Western larch	7652	1.04	1001	1458	1.02	249	3756	1.04	564	869	1.03	85	399	112	0.48	...	0.048
Black cottonwood	4890	1.00	951	1083	1.00	197	2200	1.00	360	612	1.00	92	165	46	0.31	...	0.034
Southern pine																	
Loblolly	7300	1.08	1199	1402	1.08	321	3511	1.09	612	863	1.05	112	389	109	0.47	1.06	0.057
Longleaf	8538	1.07	1305	1586	1.07	295	4321	1.07	707	1041	1.05	120	479	134	0.54	1.05	0.069
Shortleaf	7435	1.04	1167	1388	1.04	268	3527	1.05	564	905	1.05	125	353	99	0.47	1.05	0.052
Slash	8692	1.09	1127	1532	1.08	295	3823	1.07	547	964	1.05	128	529	148	0.54	1.09	0.062

[a] From ASTM Method D 2555-78.

[b] All digits retained in the averages and standard deviations through the units position to permit further computation with minimum round-off error (specific gravity excepted). For tension parallel and perpendicular to grain and modulus of rigidity, see 4.3.

[c] Modulus of rupture values are applicable to material 2 in. (51 mm) in depth.

[d] Modulus of elasticity values are applicable at a ratio of shear span to depth of 14.

[e] All maximum crushing strength perpendicular to grain values are based on standard test data only.

[f] The regional description of Douglas firs is that given on pp. 54–55 of U.S. Forest Service Research Paper FPL 27, "Western Wood Density Survey Report No. 1."

the conditioning process prior to treatment. The adjustment factor in ASTM Method D 2899 for steam conditioning is 0.85. Armstrong [*24*] indicates that the reduction factor for steam conditioning should be about 0.75 rather than 0.85. Wilkinson [*25*] agrees it should not have been 0.85. Diekman [*22*] has suggested using a factor of 0.56.

ASTM Method D 2899 also contains a provision that permits increasing the allowable tip stress determined by the formula if the critical section is above the tip of the pile. The rate of allowable stress increase is 0.2 percent for each foot of length from pile tip to critical section. Thus, if critical section is 15.2 m (50 ft) above the tip, the allowable stress as determined by ASTM Method D 2899 formula is increased by 10 percent.

Because of its natural taper, the design load on a timber pile would depend not only on the allowable unit stress used but also on the cross-sectional area of the assumed critical section. The combination of increased allowable unit stress from tip to butt and the normal increase in cross-sectional area from tip to butt, could result in dangerously high design loads, especially when full driving stresses are felt at the pile tip. Furthermore, if with time, the critical section drops to a lower depth than assumed, the pile could be overstressed. Those building codes using ASTM Method D 2899 as a basis for computing or tabulating allowable stresses for timber piles generally include this provision for increasing the allowable stress from tip to butt. Thus unsafe design loads could result from such codes.

In the final analysis, the formula in ASTM Method D 2899 actually solves for the ultimate stress of the pile material under the 1.5 to 2 years duration of loading and the other reduction factors used. Diekmann [*22*] states that ASTM Method D 2899 should be looked upon as a materials standard and compares its use and end results to ASTM specification for Structural Steel (A 36-37), which gives the yield strength of the steel and not working stresses. ASTM Method D 2899 does state that no formal safety factor is included in the formula for determining the allowable stress in compression parallel to the grain.

ASTM Method D 2899 is tied in with the ASTM, Specifications for Round Timber Piles, (D 25) also developed by the ASTM Wood Committee. The 1970 revision of ASTM Method D 25-58 substantially liberalized the knot limitations for timber piles. Other requirements, such as minimum tip and butt dimensions, were reduced. The important fact, however, is that the current version of ASTM Method D 25 (1973) permits larger single knots and more knots for a given length of pile than allowed in the 1958 version for even class C piles. Armstrong [*24*] notes that the pile test specimens used in the laboratory tests from which the knot reduction factor reflected in the ASTM Method D 2899 formula was developed generally met the knot limitations of ASTM Method D 25-58 for class A or B piles. Since the stress levels from ASTM Method D 2899 are being applied to piles produced in accordance with the current (1973) version of

ASTM Method D 25 which permits more and larger knots, the knot reduction factor reflected in the ASTM Method D 2899 formula is totally inadequate. Diekmann [*22*] has recommended that the knot limitations in ASTM Method D 25-58 are preferable to those in ASTM Method D 25-73.

Inasmuch as ASTM Method D 2899 does contain many shortcomings and results in the determination of a type of ultimate stress rather than working stress and inasmuch as these stresses are being applied as design stresses to piles being produced under standards lower than those of the test specimens on which ASTM Method D 2899 was based, it would appear that the current stress levels for timber piles in building codes predicated on ASTM Method D 2899 are too high.

For average driving conditions, Armstrong [*24*] recommends allowable compressive stresses ranging from 3792 to 4482 kPa (550 to 650 psi) for treated piles satisfying the knot requirement of ASTM Method D 25-73. If, D 25-58 were used, these stresses according to Armstrong could be increased by about 40 percent. Armstrong shows that such stress levels are much more realistic than those currently found in building codes.

Diekmann [*22*] points out that ASTM Method D 2899 does not address the capacity (bearing) of a timber pile nor its driveability. He does mention the higher impact strength of timber, but unfortunately this beneficial property is offset by a type of low-cycle fatigue comparable to that for concrete which was brought out by Gamble [*7*]. For timber piles, the application of relatively few hammer blows at high tip penetration resistance tends to quickly break down the internal structure of the wood and separate the fibers ("brooming").

Current and Future Trends

Recently, some of the model codes, in addition to the prescriptive provisions for allowable stresses, have included performance-type provisions, permitting the basic specified allowable compressive stresses to be exceeded if adequately substantiated. Such substantiation includes the requirement for a comprehensive subsoil investigation and report prepared by an engineer knowledgeable in this area, which report would include recommended design loads for the type or types of piles considered and the necessary installation procedures to achieve the required pile capacity with an adequate factor of safety. Under such provisions, load testing of piles is mandatory, and the installation of piles must be done under the supervision of the engineer knowledgeable in this field who can certify that the piles as installed do in fact meet the design requirements. This appears to be a reasonable approach to provide the means for the exercise of sound engineering judgment based upon sufficient data and solid experience to permit the use of higher design values where they are justified.

As a basic philosophy, we should proceed with caution in permitting further blanket increases in allowable stresses and loads on piles or reductions in safety factors regardless of the material involved. As Dismuke [*15*] states, the probability of failures occurring becomes greater as pile stresses and loads increase.

Increased stresses lead to the use of smaller piles or greater loads or both. The use of greater loads leads to the use of fewer piles in a foundation making the structural and bearing capacity of each pile more critical. As loads increase, so will job delays, problems, claims and litigation costs. The higher the loads on piles, the greater the degree of problems.

Rempe [*6*] rightly reminds us that as pile design stresses increase, the engineer is faced with certain problems which heretofore have largely been avoided by the use of a generous factor of safety against pile structural failure. We should add that the engineer's problems also become problems for the contractor and the owner. For work controlled by building codes, these problems also involve the building official and his department.

Conclusions

1. For the design and installation of pile foundations, building codes should contain both prescriptive and performance provisions.
2. Prescriptive provisions should be based on conservative accepted and proven engineering practice for general conditions.
3. Performance provisions should be based on the development of adequate subsurface data, the conduct of meaningful tests, and with the design and installation under the direct supervision of an engineer knowledgeable in this field.
4. A factor of safety of 2 should be retained for the pile-soil system.
5. The structural factor of safety should be greater than 2, the degree depending on the application of appropriate load reduction factors for the type pile and soil conditions.
6. Prescribed allowable compressive stresses should reflect some consideration of driveability or installation limitations for different pile types, materials, and soil conditions.
7. Current allowable stress levels for conventional steel and concrete piles as reflected in the Uniform and Standard Building Codes appear to be reasonable for general conditions.
8. Current allowable stresses for timber piles in model building codes appear to be too high, considering the ultimate long-term strength of the material involved.
9. The development of allowable stresses based on the application of strength design methods and the use of suitable load reduction factors reflecting such things as pile type, method of installation, and reliability of structural integrity should be given consideration.
10. Technical societies and engineering organizations having expertise in this field should participate more actively in building code development.

References

[*1*] *Uniform Building Code,* International Conference of Building Officials, Whittier, Calif., 1976.

[*2*] *Standard Building Code,* Southern Building Code Congress International, Inc. Birmingham, Ala., 1976.

[*3*] *Basic Building Code,* Building Officials and Code Administrators International, Inc. Chicago, 1978.

[*4*] Olson, R. E., "Provisions of Building Codes and Standards Relating to Foundations," Report prepared for National Bureau of Standards, July 1973.

[*5*] *National Building Code,* American Insurance Association, New York, 1976

[*6*] Rempe, D. M., "Building Code Requirements for Maximum Design Stresses in Piles," this volume.

[*7*] Gamble, W. L., "Capacity of Reinforced and Prestressed Concrete Pile Sections," this volume.

[*8*] "Recommendations for Design, Manufacture and Installation of Concrete Piles," ACI 543 *ACI Manual of Concrete Practice, Part 3,* American Concrete Institute, Detroit, Mich., 1974.

[*9*] "Recommended Practice for Design, Manufacture and Installation of Prestressed Concrete Piling," *Journal of the Prestressed Concrete Institute,* Vol. 22, No. 2 March–April 1977.

[*10*] Gerwick, B. C., Jr. and Brauner, H. A., "Design of High Performance Prestressed Concrete Piles for Dynamic Loading," this volume.

[*11*] *Report on Allowable Stresses in Concrete Piles,* Portland Cement Association, Skokie, Ill., 1971.

[*12*] *Building Code Requirements for Reinforced Concrete,* ACI 318, American Concrete Institute, Detroit, Mich., 1977.

[*13*] Anderson, A. R., and Moustafa, S. E., *Journal of the American Concrete Institute,* Vol. 67, No. 8 Aug. 1970.

[*14*] *Pile Foundations,* American Iron and Steel Institute, New York, 1973.

[*15*] Dismuke, T. D., "Influence of Codes and Standards on Design of Pile Foundations," this volume.

[*16*] Davisson, M. T., "Pile Load Capacity" *Proceedings,* Seminar on Design, Construction and Performance of Deep Foundations, University of California, Berkeley, Calif., 1975.

[*17*] Lu, T. D., Fischer, J. A., and Miller, D. G., "Static and Cyclic Axial Load Tests on a Fully Instrumented Pile," this volume.

[*18*] Thompson, C. D. and Thompson, D. E., "The Influence of Driving Stresses on the Development of High Pile Capacities," this volume.

[*19*] Thatcher Engineering Corporation, Report to Inland Steel Company on Loading of 12BP53 Bearing Pile to Failure, Dec. 1970.

[*20*] Williams, J. A., "Report on Test Pile Program Conducted by Kansas and Missouri State Highway Departments," *Bulletin 279,* Highway Research Board, Washington, D.C., 1960.

[*21*] *Steel Pile Load Test Data,* American Iron and Steel Institute, Washington, D.C., 1975.

[*22*] Diekmann, E. F., "Timber Piles in Standards, Codes and Practice, this volume.

[*23*] Norum, W. A., Discussion on "Timber Piles in Standards, Codes and Practice" by E. F. Diekmann, this volume.

[*24*] Armstrong, R. M., "Structural Properties of Timber Piles," this volume.

[*25*] Wilkinson, T. L., Discussion on "Timber Piles in Standards, Codes, and Practice," by E. F. Diekmann, this volume.

DISCUSSION

W. A. Norum[1] (*written discussion*)—The author has detailed much information on codes that should prove helpful. It is to be hoped that his paper will prove to be a stimulus for engineers and encourage them to get more involved in code writing activities. They will add greatly to the value of building codes and code writing organizations, such as Building Officials and Code Administrators International (BOCA), International Conference of Building Officials (ICBO), and Southern Building Code Congress International (SBCCI). These organizations write and publish the model codes that greatly influence the content of local codes and, consequently, all building construction. It is estimated that the model codes directly control 80 percent of building construction in the United States. As stated by the author, these codes are titled Basic Building Codes (BOCA), Uniform Building Code (ICBO), and Standard Building Code (SBCCI).

On the subject of the role of codes in pile foundation design and construction, I agree that codes should be written as a combined prescriptive and performance regulation, with the performance regulation allowing freedom in design based on substantiation by a qualified engineer. All model codes are written in this manner, and the reason is well known to both soils and structural engineers. We just don't know how to write a code to recognize all the variables that can be experienced with pile foundations, such as soils, driving equipment, personnel qualifications, and pile materials. We do know something about all of these variables, and possibly some engineers know just about all there is to know about all of them. The important question is, What can be written in performance terms rather than be written in the form of specifications? This is the challenge of the professional engineers and building officials. In my opinion, we should always improve codes by writing more and more in performance terms whenever possible. Getting rid of specified load limits on piles would be my first objective. All model codes fortunately have been rid of that archaic method of design, but some local codes continue. Engineers should be given more credit for knowledge. To shackle them by such a restriction is an insult. Performance criteria should be adopted that are based on soils investigation, pile strength properties, interaction of piles with soils, and technique of installation. Proof loading or test loading may be necessary as part of the standards for special situations to justify the design.

I take exception to the author's remarks directed at the role of trade

[1] District manager, National Forest Products Association, Mountain View, Calif. 94040.

associations in code and standard writing activities and his specific criticisms of ASTM Establishing Design Stresses For Round Timber Piles (D 2899). Indirectly, the author is taking issue with ASTM procedures and in the specific case made of ASTM Method D 2899; he in reality is taking issue with the federal government. Membership on ASTM committees are selected so as to balance the consumer, producer, and general interest groups, which leaves little room for control by trade associations. In the case of ASTM Method D 2899 or any other ASTM standard on wood products, there is an additional strong influence exercised by the public interest through representatives of the U.S. Department of Agriculture, Forest Service, Forest Products Laboratory (FPL). Representatives of the FPL are recognized authorities in wood technology, so there is no doubt that proposed standards have little chance for advancement to standard by ASTM without their acceptance. More important, however, is the question, Does the standard stand up satisfactorily to a critical review? The answer will come from ASTM Committee D07.07 membership who are examining the criticisms expressed in papers presented for this symposium by Armstrong and Diekmann and in my discussions of their papers.

Fuller has also made some critical comments directed to round timber piles that need response. His specific comments and my response to each are set forth below:

1. The author said, "The actual load duration reduction factor used in the ASTM Method D 2899 formula is only 66 percent according to Norum,[2] which represents a continuous or cumulative loading of about 1.5 years."

As a matter of factor, I said "it includes a factor for duration of load so that now it relates to normal loading conditions, which means 10 years of accumulated time at maximum stress conditions."

It is obvious that a conflict exists between these two statements. Apparently, the author neglected to take certain facts into consideration when drawing a conclusion that the ASTM Method D 2899 formula derives stresses for short-term loading only. Probably he overlooked a need to compensate for a difference in the effect time has on the results of full-size pile tests, which averaged 20 min. Instead, he referred only to the 2 min test period applicable to small clear specimens when interpreting the meaning of the 0.66 or 1/1.52 coefficient explained in my discussion of Diekmann's paper.

2. The author said, "In the final analysis the formula in ASTM Method D 2899 actually solves for the ultimate stress of the pile material under the 1.5 years duration of loading and the other reduction factors used."

[2] Norum, W. A., Discussion on "Timber Piles in Standards, Codes, and Practice," by E. F. Diekmann, this volume.

As previously explained, the 1.5 years should read 10 years of accumulated time at maximum stress conditions. This is not the only exception I take, however, with the author's comment. He introduces the thought that stress values derived under procedures of ASTM Method D 2899 represent ultimate values. This is wrong! The conventional engineering use of the word "ultimate" refers to breaking values, which is obviously not the case under ASTM Method D 2899. If the values derived under ASTM Method D 2899 represent ultimate values, then the pile tip parallel to grain values for Douglas fir would be about 2960 psi. This figure is the average crushing value experienced by FPL in their test program. On the other hand, the value of 1250 psi is being recommended by ASTM Method D 2899.

3. The author said, "ASTM Method D 2899 does state that no formal safety factor is included in the formula for determining the allowable stress in compression parallel to the grain."

It is a misleading statement as presented by the author. His statement implies that the derived stresses are crushing values, which is not correct. A factor of safety is inherent in the procedure used for deriving stresses for round timber piles similar to procedures used satisfactorily for deriving stresses for lumber (ASTM Method D 245). It is a multivalued factor of safety that is accountable by use of mathematical techniques that predict probability of risk.

4. The author said, "The important fact, however, is that the current version of ASTM Method D 25 (1973) permits twice as many knots for a given length of pile than allowed in the 1958 version for Class A and B piles."

It is a misleading statement as presented by the author. It implies that the growing characteristics of trees will adjust to a change in ASTM Method D 25. The fact is that man can only select trees for acceptance under ASTM Method D 25 and that the growing characteristics of trees prevents the occurrence of a radical change in knot patterns. Changing the standard to allow the same size and number of knots to occur along a 6-in. pile length that previously was limited to a 1-ft length does not actually double the amount and size of knots. Please refer to my discussion of Armstrong's paper for more details and reasoning in support of my comments. Fuller's conclusion based on the premise that the allowable knot sizes and numbers in ASTM Method D 25-73 are not adequately considered in ASTM Method D 2899 is wrong. The standard reflects these knot limits, as explained in my discussion of Armstrong's paper.

5. The author said, "He does mention the higher impact strength of timber but unfortunately this beneficial property is offset by low-cycle fatigue effects just as stated by Gamble[3] for concrete."

The facts are that the U.S. Forest Products Laboratory *Wood Handbook*

[3]Gamble, W. L., "Capacity of Reinforced and Prestressed Concrete Pile Sections," this volume.

reveals that fatigue need not be a design consideration for wood construction until repetitions of design stress or near design stress are expected to be more than 100 000 cycles during the normal life of a structure. It is apparent to me that this physical property of timber does not justify being classified "low-cycle fatigue".

Conclusions reached by Fuller, at least as they relate to timber piles, are very questionable. Much of his rationale is based on papers presented by Armstrong and Davisson, both of whom have arrived at conclusions related to timber piles that are judgmatical and without documentation. The several areas of disagreement with these authors are covered in my discussions of their papers.

F. M. Fuller (*author's closure*)—The discusser has made a valuable contribution to the general subject of allowable stresses for timber piles as determined by ASTM Method D 2899 by bringing out for further discussions some very important considerations to which the writer will respond. A more comprehensive discussion is essential for an understanding of these controversial issues.

Design versus Ultimate Stress

The discusser emphatically denies that the stress values derived under procedures of ASTM Method D 2899 result in ultimte values. The proper terminology for such stresses is "failure stress at the 5 percent exclusion value under normal load duration." This is a type of "ultimate" stress and regardless of what this stress value is called, it is not a working or design stress, although identified as such in ASTM Method D 2899.

The development of a true design stress for 10-year loading at the tip of an untreated timber pile, starting with the short-term crushing strength of a small clear specimen is illustrated in Fig. 3. To obtain a design stress for treated piles, an additional strength-reduction factor must be applied to design stress in Fig. 3, depending upon the type of conditioning used. To obtain a long-term design stress (for more than 10-year loading) the load duration reduction factor must be increased.

As Armstrong[4] states, the basic reference strength or 5 percent exclusion strength for small clear timber specimen ($S - 1.645$SD) (or product of the first two terms in Fig. 3) can be considered as comparable in nature to the design strength f'_c used in concrete design under ACI 318 Building Code Requirements for Reinforced Concrete. To this basic reference strength, whether for timber or concrete, are applied various appropriate strength reduction factors and an appropriate factor of safety (load factor) to arrive at a design or working stress.

[4] Armstrong, R. M., "Structural Properties of Timber Piles," this volume.

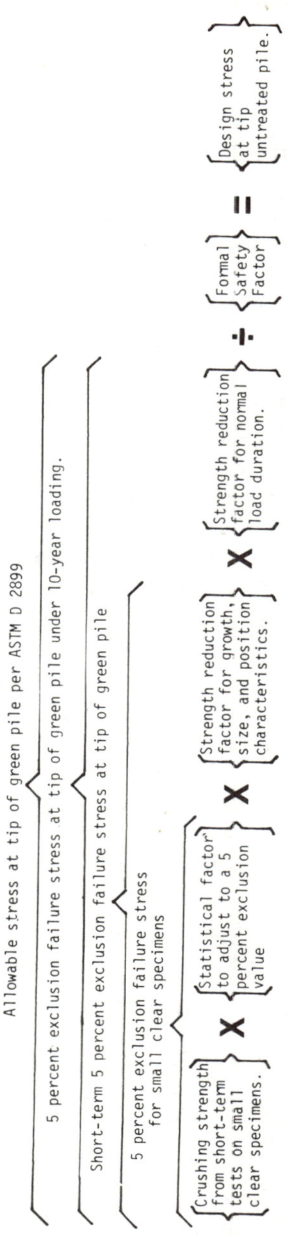

FIG. 3—*Development of a true design stress for 10-year loading at the tip of an untreated timber pile, starting with the short-term crushing strength of a small clear specimen.*

A recent study by Randolph[5] indicates that the 5 percent exclusion failure stress for treated southern pine piling under normal load duration (10 years) is about 5.5 MPa (800 psi). When the ASTM Method D 2899 recommended factor of safety of 1.25 is applied to this "failure stress," the resulting allowable design or working stress is about 4.4 MPa (640 psi). This is of the same order of magnitude as that recommended by Armstrong (see footnote 4) and is far less than the so-called design or working stresses computed by ASTM Method D 2899.

In defense of his stand that the application of procedures in ASTM Method D 2899 does not result in an "ultimate" stress, the discusser compares the average crushing value of 20.4 MPa (2960 psi) resulting from Forest Products Laboratory (FPL) tests on treated Douglas fir pile tips as reported by Wilkinson[6], with an allowable design stress value of 8.6 MPa (1250 psi) which the discusser claims is derived from ASTM Method D 2899 for treated Douglas fir piles.

The value 20.4 MPa (2960 psi) from the FPL tests is the average short-term crushing strength, whereas the 8.6 MPa (1250 psi) from ASTM Method D 2899 is a 5 percent exclusion stress under reportedly 10-year (normal) load duration as shown in Fig. 3. Because of the difference in load duration and statistical considerations, the two values cannot be compared directly. In order to make a valid comparison, the reportedly 10-year 5 percent exclusion strength, 8.6 MPa (1250 psi), must be converted to a short-term average strength.

The value 8.6 MPa (1250 psi) is converted to a short-term strength by applying as follows the load duration factor used in the ASTM Method D 2899 formula 1 [Norum (see footnote 2)]

$$1250 \times 1.52 = 1900 \text{ psi} \qquad 5\% \text{ exclusion, short-term}$$

For compression parallel to the grain, the standard deviation SD according to ASTM Method D 2555 is approximately 18 percent of the average crushing strength S. Hence, the 5 percent exclusion value is approximately 70 percent of the average strength:

$$(S - 1.645 \times 0.18\, S) = 0.70S$$

Therefore, the 5 percent exclusion short-term value 13.1 MPa (1900 psi) can be adjusted to an average short-term crushing strength as follows:

$$1900/0.70 = 2714 \text{ psi} \qquad \text{average short-term}$$

[5] Randolph, M. W., "Application of Monte-Carlo Method to Strength of Timber Piles," Special Problem, Civil Engineering Department, University of Illinois, Urbana, Ill., April 1979.

[6] Wilkinson, T. L., "Strength Evaluation of Round Timber Piles," U.S. Department of Agriculture Forest Service Research Paper FPL 101, Forest Products Laboratory, Madison, Wis., Dec. 1968.

Thus, the only valid comparison between the results of the FPL tests (see footnote 6) on treated Douglas fir pile tips and the "allowable" stress from ASTM Method D 2899 is 20.4 MPa (2960 psi) versus 18.7 MPa (2714 psi). This comparison indicates a difference of only about 8 percent. However, for the FPL tests only 15 Douglas fir pile tips were tested, which is too small a data base to make an accurate comparison.

A better comparison can be made between the ASTM Method D 2899 design stress 8.6 MPa (1250 psi) and results from the Oregon tests[7] on untreated Douglas fir piles, for which many more samples were tested giving a much broader data base. Armstrong (see footnote 4) shows (his Fig. 8) that the short-term 5 percent exclusion crushing strength for the 114 untreated Douglas fir pile tips from the Oregon State University tests was 13.6 MPa (1975 psi). The 8.6 MPa (1250 psi) stress from ASTM Method D 2899 represents a 5 percent exclusion value for treated piles under reportedly normal load duration. Therefore, to compare the results of the Oregon tests with the allowable stress 8.6 MPa (1250 psi) from ASTM Method D 2899, that allowable stress must be converted from "normal" load duration to short-term loading and from a treated to untreated condition.

The allowable stress 8.6 MPa (1250 psi) is converted from normal load duration to short-term loading by applying as follows the load duration factor which according to Norum (see footnote 2) was used in deriving formula 1:

$$1250 \times 1.52 = 1900 \text{ psi} \qquad 5\% \text{ exclusion, short-term, treated}$$

To convert the resulting short-term 5 percent exclusion value for treated piles, 13.1 MPa (1900 psi), to that for untreated piles, the strength reduction factor for the Bolton process per ASTM Method D 2899 is applied as follows:

$$1900/0.90 = 2111 \text{ psi} \qquad 5\% \text{ exclusion, short-term, untreated}$$

The resulting stress, 14.6 MPa (2111 psi), is basically comparable to the stress 13.6 MPa (1975 psi) from the Oregon tests; both are short-term 5 percent exclusion failure stresses for untreated piles. It should be noted that using the enlarged data base from the Oregon tests as a comparison, the stresses resulting from ASTM Method D 2899 are about 7 percent higher.

It is quite evident in analyzing both the FPL and the Oregon tests that the ultimate stress levels obtained in both cases are of the same order of

[7] Peterson, J., "Final Report—WWPI Pile Tests," Report on testing project funded by the Western Wood Preservers Institute, Civil Engineering Department, Oregon State University.

magnitude as the so-called design stress derived by ASTM Method D 2899 when the proper adjustment factors are applied; there is no wide difference, as claimed by the discusser. In developing and supporting his arguments, the discusser drew only upon the FPL test and totally ignored the results of the extensive testing done at both Oregon State University as reported by Peterson (see footnote 7) and Mississippi State College (Forest Products Utilization Laboratory) as reported by Thompson[8]. Wilkinson[9] points out that while the FPL test program was one of those used in developing ASTM Method D 2899, it was not used as heavily as other studies.

Considering the above, especially Fig. 3, the only conclusion that can be reached is that the procedures in ASTM Method D 2899 result in a type of "ultimate" stress and not a working or design stress.

Load Duration Factor

The discusser objects to the writer's statement that the 66 percent load duration reduction factor, used according to Norum (see footnote 2) in deriving formula 1 from ASTM Method D 2899, relates to a continuous or cummulative loading of about 1.5 years and claims that the factor used represents normal 10-year loading. To justify the 66 percent factor, he also claims that the duration of tests on full-size piles averaged 20 min.

The strength ratio versus load duration curve in Fig. 2 (from ASTM Methods D 2899 and D 245) was developed based upon results of bending tests for which the test duration under load ranged from 5 to 10 min; the index line of 100 is plotted at approximately 5 min. Although the curve in Fig. 2 is supported by studies in bending, ASTM Method 245 suggests that the same relationship of strength versus load duration may be used for other allowable stresses (including compression parallel to the grain). An examination of the curve in Fig. 2 shows that the duration of load corresponding to the 0.66 factor used in the ASTM Method D 2899 formula is about 1.5 to 2 years as stated. The general definition in wood terminology of "normal load duration" is a duration of 10 years under the maximum service load stresses either continuous or cumulative. Long-term loading is considered as loading in exces of 10-years (factor = 0.90 × normal load duration factor).

If the tests on full pile sections averaged 20 min, as claimed by the discusser, this would have the effect of moving the 100 percent index line (Fig. 2) to the right, thus raising the curve and justifying the use of a 0.66 factor for 10-year loading.

[8]Thompson, W. S., "Results of Strength Tests on Piling Sections," Report submitted to American Wood Preservers Institute by the Forest Products Utilization Laboratory, State College, Miss.

[9]Wilkinson, T. L., Discussion on "Timber Piles in Standards, Codes, and Practice," by E. F. Diekmann, this volume.

However, there is no indication in the test data upon which ASTM Method D 2899 was based that the test loads were applied to full pile sections for a duration of 20 min. These data resulted from tests at the University of Wisconsin (Forest Products Laboratory) Oregon State University (see footnotes 6 and 7) and Mississippi State College (see footnote 8). Although actual load durations were not reported, information reported on the strain to failure and the rate of loading can be used to calculate the load duration. Armstrong and Davisson[10] show that such calculations lead to the following load test durations on full pile sections:

FPL tests	1.5-5 min
Oregon tests	2.2 min
Mississippi tests	2.1-5.1 min

These load durations are far less than the 20 min claimed by the discusser, and it is obvious that the index line as currently plotted in Fig. 2 at 5 min duration is compatible with the actual load duration of tests on full pile sections. Thus, the load duration factor of 0.66 used in the ASTM Method D 2899 formula does in fact relate to a load duration of from 1.5 to 2 years as stated by the writer and does not represent 10-year loading as claimed by the discusser. The recognized load duration factor is 0.625 (5/8) for normal load duration (10 years) and is 0.5625 (9/16) for long-term loading as indicated in Fig. 2 and as shown by Diekmann,[11] Armstrong (see footnote 4), Wood,[12] Gurfinkel,[13] and others.

Factor of Safety

The discusser claims that a factor of safety is inherent in the procedures used for deriving "long-term" allowable stresses for timber piles in accordance with formula 1 from ASTM Method D 2899.

Norum (see footnote 2) shows the basic derivation of formula 1 for design compressive stress parallel to the grain from ASTM Method D 2899. The basis for the ASTM Method D 2899 formula is also illustrated in Fig. 3. It should be noted that the adjustment factors are intended to convert the crushing strength of small clear specimens to a comparable 5 percent exclusion crushing strength for full-size pile tips under normal load duration. Section 13.1 of ASTM Method D 2899 recommends a safety factor of 1.25 for compression parallel to the grain "if a formal factor of safety is

[10] Armstrong, R. M. and Davisson, M. T., "Review of Some Factors Controlling Allowable Loads on Timber Piles" unpublished critique and analysis of test data on which ASTM Method D 2899-74 was based, Urbana, Ill., July 1976.

[11] Diekmann, E. F., "Timber Piles in Standards, Codes and Practice," this volume.

[12] Wood, L. W., *Journal of the Structural Division*, American Society of Civil Engineers, Vol. 84, No. ST7, Nov. 1958.

[13] Gurfinkel, G., *Wood Engineering*, Southern Forest Products Association, New Orleans, La., 1973.

considered to be required," but no formal factors of safety are included in the formulas used to calculate working stresses.

Armstrong (see footnote 4) recommends a formal factor of safety for compression parallel to the grain of 1.2 based upon a load duration factor of 0.60. This is comparable to the factor of safety of 1.25 recommended in ASTM Method D 2899 combined with a 10-year load duration factor of 0.625. If the load duration factor of 0.66 currently in ASTM Method D 2899 is retained, the comparable required factor of safety would be 1.32.

These factors of safety are considerably lower than those used for structural design of other pile types; generally, the structural factor of safety is about 2.2. However, a formal factor of safety of 1.25 (for a load duration factor of 0.625) is considered satisfactory for timber piles because of the informal safety factor inherent in the load duration factor. However, if the pile is subjected to the full design stress either continuously or cumulatively over the load duration period reflected in the reduction factor used, this informal safety factor disappears and only the formal safety factor (if used) is left. As the actual load duration period under full design stress continues to increase, any formal factor of safety used continues to be eroded. As pointed out in the case of the ASTM Method D 2899 formula, the load duration factor actually used reflects only a 1.5 to 2 year total load duration. After the expiration of this full-load duration period, there is no safety factor left.

Knot Limitations

The discusser claims that the revision to ASTM Method D 25-58 resulting in the current specification ASTM Method D 25-73 did not increase the amount and size of knots permitted in timber piles and contends that the allowable knot sizes and sum of knot sizes permitted in ASTM Method D 25-73 are adequately considered in ASTM Method D 2899.

Table 1 shows a direct comparison of the knot limitation requirements in both the 1958 and 1973 versions of ASTM Method D 25. It will be noted that

1. Specific limitations for maximum-size single knot and maximum sum of knots in a given length of pile were included in the 1958 version for piles of various qualities or lengths. These specific maximum sizes could not be exceeded regardless of the pile diameter.

2. The 1973 version does not contain any such specific limitations. Thus, the permitted knot size or sum of knot sizes in a given length of pile can increase with the increasing pile diameter.

3. In the 1973 version, the maximum allowable knot size and sum of knot sizes in a given pile length for all piles are equivalent to or greater than those for class C piles in the 1958 version.

TABLE 1—*Comparison of knot limitations between ASTM Methods D 25-58 and D 25-73.*

Knot Property	Knot Limitations[a] 1958		1973	
Maximum-size single knot	Class A and B piles < 50 ft	$\frac{1}{3}D$ max 4 in.	All piles	$\frac{1}{2}D$
	Class A and B piles > 50 ft			
	$\frac{3}{4}L$ from butt	$\frac{1}{3}D$ max 4 in.		
	$\frac{1}{4}L$ from tip	$\frac{1}{2}D$ max 5 in.		
	Class C piles	$\frac{1}{2}D$ max 5 in.		
Maximum sum of diameters all knots per length indicated	Class A and B piles < 50 ft	$\frac{2}{3}D$ max 8 in. in 12 in. length	All piles	$\frac{1}{3}C = D$ in 6 in. length
	Class A and B piles > 50 ft			
	$\frac{3}{4}L$ from butt	$\frac{2}{3}D$ max 8 in. in 12 in. length		
	$\frac{1}{4}L$ from tip	D max 10 in. in 12 in. length		
	Class C piles	D max 10 in. in 12 in. length		
Clusters of knots	Class A and B piles	not permitted	All piles	permitted up to single knot size $= \frac{1}{2}D$
	Class C piles	not permitted		

Conversion factors
1 ft = 0.30 m
1 in. = 2.54 cm
[a] D = pile diameter, C = pile circumference, L = pile length.

4. In the 1958 version, the maximum allowable sum of knot diameters is based on any 12-in. length of pile; for the 1973 version, it is based on any 6-in. length of pile. The possibility of having an increased sum of knots exists.
5. In the 1958 version, cluster knots were prohibited for all piles, including class C piles.
6. In the 1973 version, cluster knots are permitted for all piles up to the maximum size allowed for single knots.

It is quite obvious that when ASTM Method D 25-58 was revised in 1970, the knot restrictions were substantially liberalized. As noted above, the quality of piles permitted in the current specification, as far as knots are concerned, is that which would be classified as class C piles or poorer in the 1958 specification. Class C piles were identified as suitable for use in foundations that will always be completely submerged or for cofferdams, falsework (temporary construction), or light construction. Thus, in today's market for timber piles, all piles, regardless of what type structure they are to support or what loads they are to carry, could be of class C quality or poorer as far as knots are concerned.

Formula 1 from ASTM Method D 2899 was derived principally from results of tests on timber piles conducted at the University of Wisconsin

(Forest Products Laboratory), Oregon State University, and Mississippi State College. Armstrong (see footnote 4) shows that of the total piles tested under these three programs, approximately 60 percent were of class A or B quality as far as knots are concerned. Thus, the majority of piles used in tests on which ASTM Method D 2899 was based were of higher quality than those allowed in today's market and to which ASTM Method D 2899 is being applied. For this reason, and because the strength of a pile is reduced as the size and number of knots increase, the allowable stresses determined by ASTM Method D 2899 do not adequately reflect the poorer quality of piles permitted under the current specification ASTM Method D 25-73.

Low-Cycle Fatigue

The discusser objects to the writer's use of the term "low-cycle fatigue" in discussing the physical driving limitations of timber piles, and discusses "fatigue strengths" on the basis of more than 100 000 repetitions of the design stress or near design stress during the normal life of the structure. For a life of 20 to 30 years, this is equivalent to about 1 cycle for each 10 to 16 min. The "fatigue strength" to which the writer refers, relates to dynamic driving stresses, which are considerably higher than design stresses and are applied at the rate of about 60 or more cycles per minute.

A timber pile will not stand up under sustained hard driving. At high point resistance, it does not take many hammer blows to break down the cellular structure of the wood and leave broomed fibers. Norum[14] recognizes this by stating "It has been proven that banding timber piles reduces the amount of fiber separation experienced during hard driving." This phenomenon of brooming exists whether it is called "low-cycle fatigue" or something else.

Conclusions

Conclusions regarding the items of controversy can be summarized as follows:

1. The procedures of ASTM Method D 2899 do not result in a design or working stress as indicated in the standard.

2. The load duration factor in the ASTM Method D 2899 formula 1 represents a load duration of 1.5 to 2 years and is not justified; it should be at least 0.625 (1.60) for normal 10-year loading.

3. The ASTM Method D 2899 formula 1 does not contain a factor of safety (load factor) in accordance with conventional timber design.

[14] Norum, W. A., Discussion on "Stress in Piles," by M. T. Davisson, this volume.

4. The revision to ASTM Method D 25-58 resulting in the current ASTM Method D 25-73 did substantially liberalize the knot limitations for timber piles.

5. ASTM Method D 2899 does not adequately reflect the size and extent of knots permitted under the current pile material specification, ASTM Method D 25-73.

6. Timber piles are subject to a type of "low-cycle fatigue" resulting in the breakdown of wood structure (brooming) under repetitively high impact stresses often occurring during pile driving.

R. M. Armstrong[1]

Structural Properties of Timber Piles

REFERENCE: Armstrong, R. M., "**Structural Properties of Timber Piles,**" *Behavior of Deep Foundations, ASTM STP 670,* Raymond Lundgren, Ed., American Society for Testing and Materials, 1979, pp. 118-152.

ABSTRACT: Results of test programs with short pile sections and small clear specimens are reviewed to evaluate the influence of imperfections and preservative treatment on the strength properties of timber piles. Reduction factors developed from this review, available information on natural wood strength variations, and available information on the influence of load duration are used to evaluate the structural capacity of embedded timber piles under axial loading. Indicated allowable axial unit stresses for treated timber piles range from 3.8 MPa (550 psi) to 6.4 MPa (925 psi), depending on the extent of knots and treatment conditioning that are permitted in current (1978) timber pile standards.

KEY WORDS: piles, timber piles, wood, strength, mechanical properties, structural design, safety factor, standards, dynamic loads, static loads, evaluation

Nomenclature

A Cross-sectional area
D Pile diameter
P Design load
f_s Factor of safety
K_L Largest knot diameter
K_s Sum of knot diameters in 1 ft of pile length
s_{cr} Average green small clear crushing strength
s_d Standard deviation of green small clear crushing strength
s_c' Basic reference strength
s_u' Ultimate unit axial stress for untreated short pile sections loaded to failure in approximately 2 min
s_u Ultimate unit axial stress for timber piles

[1] Raymond Technical Facilities, Inc., Houston, Tex. 77027; formerly graduate teaching and research assistant, Department of Civil Engineering, University of Illinois at Urbana-Champaign.

s_a Allowable working stress on axially loaded timber piles
β Load duration factor
ψ Strength-reduction factor for treatment
ϕ Strength-reduction factor for imperfections
ϕ_L Strength-reduction factor for imperfections based on the largest knot
ϕ_s Strength-reduction factor for imperfections based on the sum of knots

A fundamental design requirement for all pile foundations is that the piles be structurally capable of resisting, with an appropriate factor of safety, the forces they will experience during installation and under service loading. In order to satisfy this requirement, the designer must be familiar with the structural properties of the various materials used in piling. Structural properties and design procedures for concrete and steel piles are discussed in other papers presented at this symposium [*1-3*].[2] This paper will discuss timber piles.

A complete and detailed review of all factors influencing the engineering properties of wood is beyond the scope of this paper and would be repetitious, because such information is readily available [*4-9*]. Evaluation of the structural capacity of a timber member generally starts with a strength property of green clear wood measured in short-term laboratory tests on small specimens. This strength property, adjusted for the large natural variability, is used as a basic reference strength. The basic reference strength must then be adjusted to account for any influences of defects and imperfections, load duration, treatment, and in-use moisture content. Finally, allowable design stresses are developed by applying an appropriate factor of safety.

In the following sections, the approach outlined above is followed to evaluate the axial structural capacity of embedded timber piles. The term "embedded" is intended to imply that the soil provides sufficient lateral support to allow the pile to be treated as a short column.

Basic Reference Strength

The general reference strength used in timber design for compression parallel to grain is the crushing strength of small specimens of green wood cut from a tree so that the grain runs parallel to the direction of loading and that the specimens are of clear wood (free of defects or imperfections). Specimens have a cross-sectional area of 6.5 to 26 cm^2 (1 to 2 $in.^2$) and are 10 to 20 cm (4 to 8 in.) long. The tests are performed at a strain rate of approximately (0.003 in./in./min), which results in a test duration of approximately 1 to 2 min. The unit stress at failure of the specimens is

[2] The italic numbers in brackets refer to the list of references appended to this paper.

referred to as the green small clear crushing strength. Methods for performing such tests are given in the ASTM Methods of Testing Small Clear Specimens of Timber (D 143-52).

The green small clear crushing strength varies considerably, not only between different species, but also between trees of the same species. To further compound the issue, strengths of specimens from different locations in the same tree will vary. ASTM Method for Establishing Clear-Wood Strength Values (D 2555-76) contains a tabulation of the average green clear wood strength and standard deviations for various species. Three of the more frequently used timber species for piling are Douglas fir, red oak, and southern pine. Average green small clear wood crushing strengths parallel to grain and standard deviations for these species are presented in Table 1. The standard deviation provides a measure of the variability of strength values within the species.

Because of the large variability of strength within a given species and

TABLE 1—*Green clear wood crushing strength parallel to grain (after ASTM Method D 2555-76).*

	Crushing Strength, psi		
Species	Average, s_{cr}	Standard Deviation, s_d	Basic Reference Strength,[a] s_c'
Douglas fir			
Coast	3784	734	2577
Interior west	3872	799	2558
Interior north	3469	602	2479
Interior south	3113	489	2309
Oak, red			
Black	3470	625	2442
Cherrybark	4620	832	3251
Northern red	3440	618	2422
Southern red	3030	545	2133
Laurel	3170	571	2231
Pin	3680	662	2591
Scarlet	4090	736	2879
Water	3740	673	2633
Willow	3000	540	2112
Southern pine			
Loblolly	3511	612	2504
Longleaf	4321	707	3158
Shortleaf	3527	564	2599
Slash	3823	547	2923
Pitch	2950	531	2077
Pond	3660	659	2576
Spruce	2835	580	1881
Sand	3440	619	2422
Virginia	3420	616	2407

Conversion factor—1000 psi = 6.9 MPa.
[a] Based on Eq 1.

within a given tree, the average strength cannot be used directly as the basic reference strength. ASTM Methods for Establishing Structural Grades and Related Allowable Properties for Visually Graded Lumber (D 245-74) recommends a statistical basis for selecting the basic reference strength. Under ASTM Method D 245-74, the probability is 1 in 20 that a random individual strength test will be below the specified basic reference strength. Using this approach, the basic reference strength is related to the average small clear crushing strength and standard deviation by Eq 1. The basic reference strength denoted herein by the symbol s_c', is generally referred to as the 5 percent exclusion strength in most literature on wood.

$$s_c' = s_{cr} - 1.645 s_d \tag{1}$$

where

s_c' = basic reference strength,
s_{cr} = average green small clear crushing strength, and
s_d = standard deviation of green small clear crushing strength

Values for the basic reference strength, based on Eq 1, are tabulated in the last column of Table 1 for the various subspecies of Douglas fir, red oak, and southern pine.

The piles provided on a particular job will generally consist of more than one subspecies. In such cases, the statistical evaluation of the basic reference strength becomes slightly more complex than Eq 1, because it is dependent on the combined strength frequency distribution of all subspecies present. ASTM Method D 2555-76 outlines methods for evaluating the 5 percent exclusion limit for specie combinations if the volume of the various species are known. The use of these methods for specie combinations in timber piling is somewhat specious since neither the designer or the contractor are generally qualified to distinguish between the various subspecies.

The writer recommends that a design value of 17.2 MPa (2400 psi) be used for the basic reference strength s_c' for Douglas fir, red oak, and southern pine. This value has been selected on the basis of the data contained in ASTM Method D 2555-76 on the standing timber volume for the species grown in the United States, which indicates that pitch and spruce combined comprise less than 3 percent of the total southern pine volume and that interior south Douglas fir comprises less than 4 percent of the total Douglas fir volume. The southern red, laurel, and willow subspecies of red oak have been excluded, because they do not appear to be frequently used for piling from what little information the writer has been able to obtain from oak piling suppliers. The value of 17.2 MPa (2400 psi) for the basic reference strength s_c' will be used throughout the remainder of this paper.

The basic reference strength s_c' might be thought of as the ultimate axial unit stress under a load duration of 1 to 2 min for an untreated short timber pile section containing no defects. Since real piles are generally treated, contain imperfections, and have load durations of other than 1 to 2 min, s_c' is only an index to ultimate strength of a timber pile. It is comparable in nature to the design strength f_c' used in concrete design under American Concrete Institute (ACI) 318.

Influence of Imperfections on the Strength of Timber Piles

General

Timber piles contain imperfections such as spiral grain, knots, checks, shakes, and splits. Such imperfections are natural growth characteristics of trees and are present to some degree in all timber piles. Because of these imperfections, the ultimate axial unit stress is less than s_c' for timber piles. The *Wood Handbook* [*4*] and ASTM Method D 245-74 both indicate that shakes, checks, and splits have little or no effect on the strength properties in axial compression. Hence, the imperfections exerting the greatest influence on the axial compression strength of timber piles are associated with knots and spiral grain. Since it would be impractical to require that timber piles be entirely free of knots and spiral grain, the only alternative is to limit knots and spiral grain to a degree that is commercially acceptable and to use a strength reduction factor compatible with the extent of imperfections permitted.

Knot Limitations

Limitations on knots are generally specified in terms of the largest knot diameter K_L, the sum of knot diameters in 1 ft of pile length K_s, and the pile diameter at the point where the knots occur D. Using these symbols, knot limitations can be expressed in terms of either a specified value of K_L and K_s, by the dimensionless ratios K_L/D and K_s/D, or both.

One of the most frequently used specifications for timber piles is the ASTM Specification for Round Timber Piles (D 25). This standard was originally issued in 1915 but has gone through many revisions. The most recent edition is ASTM Method D 25-73, which replaced ASTM Method D 25-70. With respect to knot limitations, ASTM Methods D 25-70 and D 25-73 are identical. ASTM Method D 25-70 replaced ASTM Method D 25-58 in 1970. The knot limitations in ASTM Methods D 25-70 and 73 are significantly different from ASTM Method D 25-58, as illustrated in Table 2. ASTM Method D 25-70 removed all limitations on the knot dimensions K_L and K_s, increased the permissible largest knot ratio K_L/D for all piles to the level previously permitted only in Class C piles, and increased the permissible summation of knots ratio K_s/D for all piles to a level twice that previously permitted in Class C.

TABLE 2—*Comparison of knot limitations for ASTM Method D 25-58 and ASTM Method D 25-70.*

Knot Limitation	ASTM D 25-58[a] Class A and B Piles	ASTM D 25-58[a] Class C Piles	ASTM D 25-70 and D 25-73
K_L	4 in.	5 in.	no limitation
K_s	8 in.	10 in.	no limitation
K_L/D	⅓	½	½
K_s/D	⅔	1	2.09

Conversion factor—1 in. = 2.54 cm.
[a] Knot limitations quoted for class A and B piles are for lengths less than 50 ft.

The major revisions of ASTM Method D 25 that surfaced in the 1970 edition are reportedly based on the results of test programs with full-size pile sections [*10,11*]. The test programs on which the revisions are reportedly based were conducted at the U.S. Forest Products Laboratory [*12*], Mississippi State University [*13,14*], and Oregon State University [*15*]. In two of these test programs [*12,13*], the crushing strength of both small clear wood and short pile sections are reported, along with the maximum knot diameters and the maximum sum of knot diameters in 1 ft.

The largest knot sizes present in the Mississippi State University test program [*13*] is illustrated in Fig. 1, where the frequency distribution of the largest knot ratio K_L/D is shown for the 150 pile sections that had companion small clear specimens. Also shown on Fig. 1 are the upper limits for the ratio K_L/D permitted by ASTM Method D 25-58 and ASTM

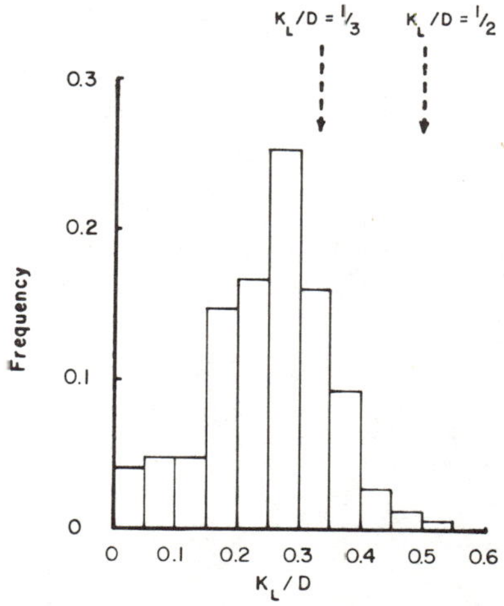

FIG. 1—*Knot size distribution for Mississippi State University tests* [13].

Methods D 25-70 and 73. Based on the largest knot ratio K_L/D limitations, approximately 78 percent of the test piles met the ASTM Method D 25-58 Class A and Class B requirements, and only one pile failed to meet the ASTM Method D 25-58 Class C and ASTM Methods D 25-70 and 73 requirements.

The frequency distribution of the sum of knots ratio K_s/D for the 150 pile sections with companion small clears from the Mississippi State University program is presented in Fig. 2. Also shown on Fig. 2 are the upper limits of the sum of knots ratio for ASTM Method D 25-58 and ASTM Method D 25-70 and 73. Based on the sum of knots ratio limitations, approximately 60 percent of test piles met the ASTM Method D 25-58 Class A and B requirements, and only 4 percent failed to meet the ASTM Method D 25-58 requirement for Class C piles. Note that all of the piles meet the ASTM Methods D 25-70 and 73 requirements for the sum of knots ratio.

The U.S. Forest Products Laboratory test program [*12*] consisted of 15 piles for each of the three major species (Douglas fir, southern pine, and red oak), or a total of 45 piles. With respect to the largest knot ratio, 30 of the piles met the requirements for ASTM Method D 25-58 Class A and B piles, and only two of the piles failed to meet the requirements for ASTM Method D 25-58 Class C and ASTM Methods D 25-70 and 73. Similarly considering only the sum of knots ratio K_s/D; 13 of the piles meet the Class A and Class B requirements, and 38 piles met the requirements for Class C piles under ASTM Method D 25-58. While seven of the piles exceeded the limitation for the sum of knots ratio of ASTM Method D 25-58 Class C piles, all 45 piles meet the knot requirements for ASTM Methods D 25-70 and 73.

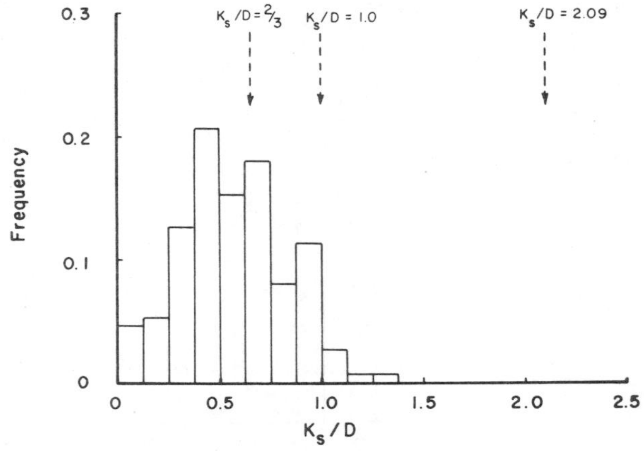

FIG. 2—*Knot size distribution for Mississippi State University tests* [13].

In the Oregon State University test program [*15*], tests were not performed on small clear specimens, and only the sum of knot diameters in 1 ft were reported. Data on 268 pile sections 3 ft long were reported. The frequency distribution of the sum of knots ratio K_s/D, is presented in Fig. 3 along with the ASTM Methods D 25-58 and D 25-70 and 73 limitations on this ratio. Approximately 55 percent of the samples met the ASTM Method D 25-58 Class A and B requirement for the sum of knots ratio, and only 5 percent failed to meet the ASTM Method D 25-58 Class C requirement. All pile sections met the ASTM Methods D 25-70 and 73 limitation for the sum of knots.

The preceeding review of the knot limitations and the knot data for the test programs on which the 1970 revision of ASTM Method D 25 was based indicate (*a*) the 1970 revision results in a major relaxation in knot limitations, and (*b*) the sum of knots ratio permitted under ASTM Methods D 25-70 and 73 represents an extrapolation to approximately twice the upper boundary of the available data. Since the strength of a timber pile is reduced as the size and number of knots increase, the allowable stress used with ASTM Method D 25-70 and 73 piles should be less than the allowable stresses used with ASTM Method D 25-58 piles. The influence of knots on strength are discussed in the following section.

Strength Reducing Effects of Knots

The strength reduction due to the presence of knots is denoted herein by ϕ, which is defined as the ratio of the unit crushing stress of a pile containing knots to the unit crushing stress of clear wood. Where the strength reduction is based on the largest knot ratio K_L/D, the ϕ value is denoted

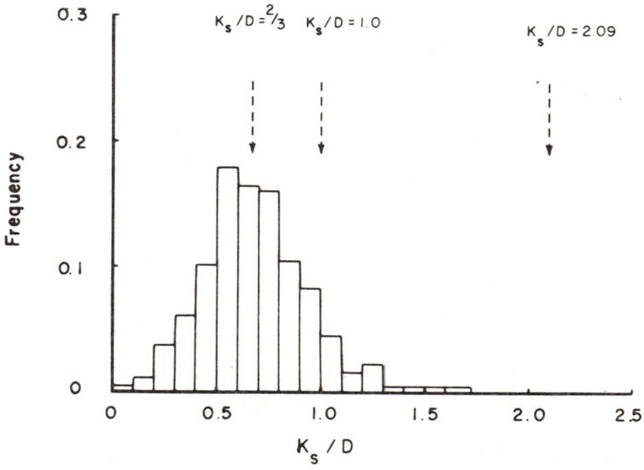

FIG. 3—*Knot size distribution for Oregon State University tests* [15].

by ϕ_L, and where the strength reduction is based on the sum of knots ratio K_s/D, the ϕ value is denoted by ϕ_s.

ASTM Method D 245-74 gives equations for evaluating the strength reduction for sawn lumber, which are functions of both the largest knot diameter and the width of the member. Using the pile diameter as the width of the member in these equations, with the largest knot ratio limitations of ASTM Method D 25-58 and ASTM Methods D 25-70 and 73 (Table 2), relationships for the strength reduction factor ϕ_L were developed. These relationships are presented in Fig. 4. The ASTM Method D 245-74 relationships indicate ϕ_L values of 0.55 to 0.7 for ASTM Method D 25-58 Class A and B piles, and values of 0.35 to 0.55 for ASTM Method D 25-70 piles and ASTM Method D 25-58 Class C piles. It appears that an increase of the largest knot ratio K_L/D from ⅓ to ½ results in a 20 to 40 percent reduction in pile strength.

ASTM Method D 245-74 does not contain equations for the strength-reducing effect of the sum of knots ratio K_s/D. However, it does restrict the sum of knot diameters in any 6 in. of length to twice the largest permitted knot diameter. Since the permitted sum of knots under ASTM Method D 25-58 and ASTM Methods D 25-70 and 73 are less than the ASTM Method D 245-74 limitation, it appears that the largest knot ratio controls the value of the strength-reducing factor ϕ.

The *Wood Handbook* [4] states that knots have less effect on the strength of round timber piles than in sawn lumber, but it does not give any numerical values for this suggested increased efficiency of timber piles over sawn lumber. While the above statement indicates the strength reduction factors based on ASTM Method 245 are conservative, use of smaller reductions

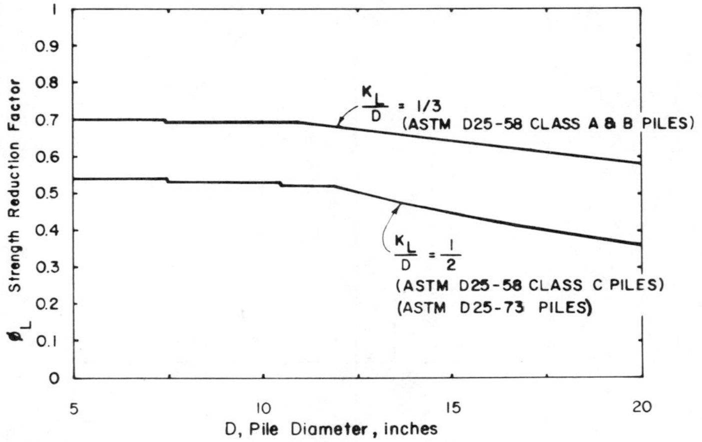

FIG. 4—*ASTM Method D 245-74 strength reduction for knots.*

require an evaluation of strength data on round timber piles. Available data are reviewed in the following paragraphs.

Data from two of the pile test programs previously mentioned [*12,13*] can be used to evaluate the influence of knots on the ultimate strength of timber piles, since both clear wood and short pile specimens were tested. Relationships between the average ϕ values and the average knot ratios from these two test programs are indicated by the data points in Fig. 5 and 6. The vertical dashed lines on these figures indicate the permissible knots under ASTM Methods D 25-58 and D 25-70 and 73. Based on these data points, it appears the influence of the largest knot ratio and the sum of knots ratio on strength can be approximated by

$$\phi_L = 1.0 - 0.45(K_L/D) \tag{2}$$

$$\phi_s = 1.0 - 0.20(K_s/D) \tag{3}$$

The controlling φ value can be determined by examining the coefficients in Eqs 2 and 3. When the sum of knots ratio K_s/D is less than 2.25 times the largest knot ratio K_L/D, the ϕ value is controlled by the largest knot and is given by Eq 2. When the sum of knots ratio is greater than 2.25 times the largest knot ratio, the ϕ value is controlled by the sum of knots and is given by Eq 3. Hence, based on the permissible knot ratios (Table 2), the ϕ value is controlled by the largest knot for ASTM Method D 25-58

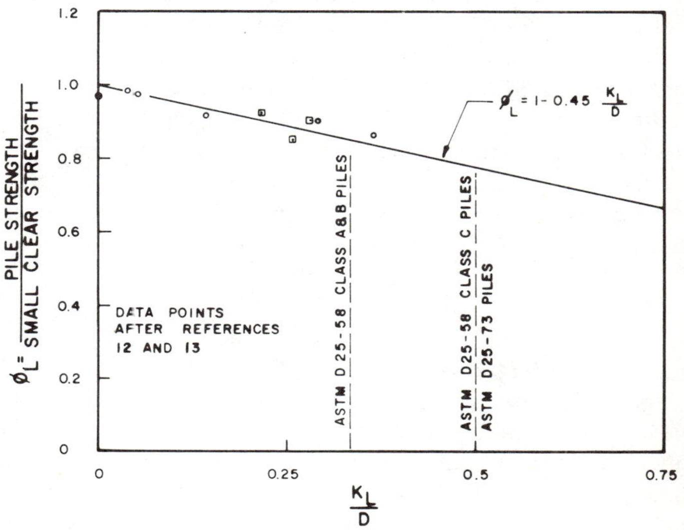

FIG. 5—*Influence of largest knot on pile strength.*

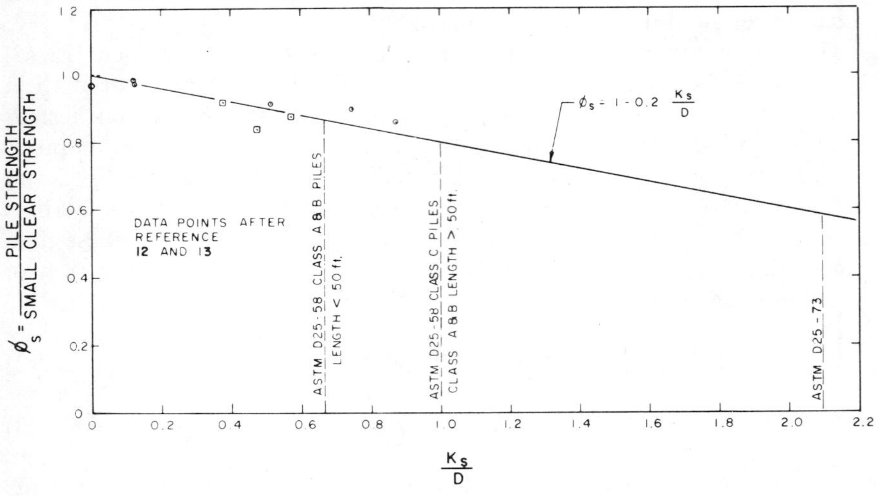

FIG. 6—*Influence of sum of knots on pile strength.*

piles, while the ϕ value for ASTM Methods D 25-70 and 73 piles is controlled by the sum of knots.

While the information available on the strength-reducing effect of knots is somewhat limited, Figs. 4 through 6 all indicate that the strength of a timber pile decreases as the size and number of knots increase. Appropriate ϕ values must be selected so that they are compatible with the size and number of knots permitted. The writer recommends the use of the ϕ values in Table 3, which have been selected on the basis of Eqs 2 and 3, and the knots permitted in ASTM Methods D 25-58 and D 25-70 and 73 (Table 2).

The ultimate unit axial stress for an untreated short timber pile section, loaded to failure at a strain rate of approximately 0.003 in./in./min, is denoted herein by s_u'. Using the strength-reduction factor ϕ and the basic reference strength s_c', the ultimate unit axial stress s_u' for an untreated short timber pile section can be expressed in the form

$$s_u' = \phi s_c' \tag{4}$$

Since the basic reference strength s_c' in Eq 4 is based on a probability of 20 to 1 against failure, the ultimate strength denoted by the symbol s_u' also represents a probability of 20 to 1 against failure. Hence approximately 5 percent of a random sample of pile sections tested at a strain rate of approximately 0.003 in./in./min should fail at a unit axial stress less than the value predicted by Eq 4. Using the writer's recommended ϕ values (Table 3) and the recommended s_c' value of 16.6 MPa (2400 psi), Eq 4 gives the values of s_u' tabulated in Table 4.

TABLE 3—*Recommended ϕ values for ASTM Method D 25-58 and ASTM Method D 25-73 piles.*

Pile Quality Standard	ϕ
ASTM Method D 25-58	
Class A and B piles with lengths of 50 ft or less	0.85
Class A and B piles with lengths greater than 50 ft	0.75
Class C piles	0.75
ASTM Method D 25-73	
piles	0.60

TABLE 4—*Ultimate unit axial stress for untreated timber piles in short-term tests, s'_u, psi.*

Pile Quality Standard	s'_u [b]
ASTM Method D 25-58 class A and B piles[a]	2050
ASTM Method D 25-58 class C piles	1800
ASTM Method D 25-73 piles	1450

Conversion factor—1000 psi = 6.9 MPa.
[a] Assumes pile lengths of 15.2 m (50 ft) or less.
[b] Based on s'_c value of 16.6 MPa (2400 psi) and ϕ value in Table 3.

Distributions of crushing strengths for 150 untreated southern pine pile sections from the Mississippi State University test program [*13*] and 114 untreated Douglas fir pile sections from the Oregon State University test program [*15*] are presented in Figs. 7 and 8, respectively. The actual 5 percent exclusion limit of 14.6 MPa (2120 psi) in Fig. 7 is in close agreement with the value of 14.1 MPa (2050 psi) given in Table 4 for class A and B piles, because approximately 78 percent of the piles in this test program met the ASTM Method D 25-58 Class A and B requirements for the largest knot ratio (Fig. 1). This close agreement is to be expected, since the Mississippi State University tests [*13*] were used in developing Eqs 2 and 3.

The actual 5 percent exclusion limit for the Oregon State University tests on Douglas fir [*15*] is 13.6 MPa (1975 psi), as indicated in Fig. 8, which is between the values of 14.1 MPa (2050 psi) and 12.4 MPa (1800 psi) given in Table 4 for ASTM Method D 25-58 piles. This intermediate position is to be expected, because the Oregon State University test program consisted of approximately 55 percent ASTM Method D 25-58 Class A and B piles, 40 percent ASTM Method D 25-58 Class C piles, and 5 percent that failed to meet all ASTM Method D 25-58 requirements (Fig. 3). Because small clear wood specimens were not tested in the Oregon State University test program, it was not used in developing Eqs 2 and 3. Hence, this test program represents an independent test of the values recommended in Table 4.

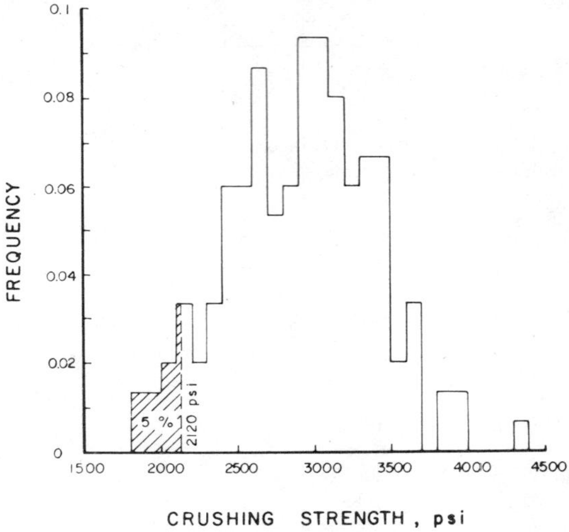

FIG. 7—*Pile section crushing strength for untreated southern pine* (*after* [13]).

Effects of Spiral Grain on Pile Strength

ASTM Method D 25-58 and ASTM Method D 25-70 and 73 both permit piles to have a 180-deg twist of grain over a length of 20 ft. The *Wood Handbook* [*4*] indicates the strength reduction in compression is less than 1 percent for grain slopes of 1 in 10 or less. Hence, the twist of grain permitted in ASTM Method D 25 does not result in a serious strength reduction. Localized grain distortion around knots may significantly influence the strength. This influence was inherently included in the preceeding data on knots.

Influence of Load Duration on the Strength of Timber Piles

Under service loading conditions, piles are subjected to sustained stress levels for time durations that are considerably longer than the duration of approximately 2 min experienced in standard short-term strength tests on small clear specimens or short piles sections. By contrast, stress waves caused by hammer impact have durations considerably less than 2 min. The strength of wood is highly dependent on the duration of loading [*4,6,7,9,12,16-24*], and any attempt to evaluate the structural adequacy of a timber member must include an adjustment for load duration.

The influence of load duration and wood strength can be accounted for by the use of a reduction factor, denoted herein by β, defined as the ratio of strength for a given load duration to the strength measured during standard tests of short duration. The general relationship between β and

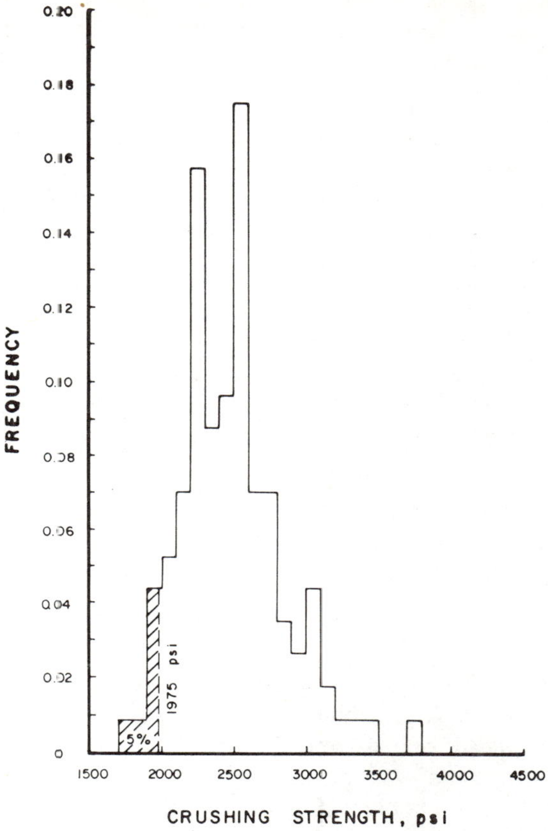

FIG. 8—*Pile section crushing strength for untreated Douglas fir* (*after* [15]).

load duration recommended by ASTM Method D 2555-76 is presented in Fig. 9.

Figure 9 indicates a β value of approximately 0.625 for a 10-year duration, which is commonly referred to as normal load duration in wood design. For long-term permanent loading, a β-value of approximately 0.5625 is generally used [*7,21,23,24*]. Since the relationship in Fig. 9 is only approximate, the use of a β value of 0.6 appears appropriate for long-term loading conditions.

Under very short load durations such as the blow of a pile hammer, the value of β is larger than 1.0. Additional information on the strength of wood under very short stress durations has been reported by Wilkinson [*12*] and Keeton [*20*]. Wilkinson reports impact tests at rise times to failure of approximately 0.25 ms, which resulted in β values ranging from 1.9 to 3.2. Keeton reports test data on Douglas fir indicating β values of

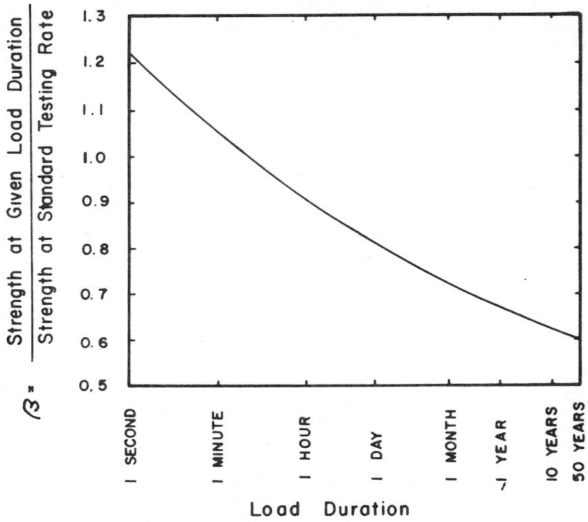

FIG. 9—*Influence of load duration on strength (after ASTM Method D 2555-76).*

1.3 to 1.5 for loading rates compatible to stress wave rise times experienced during pile driving.

At working stress levels, the possibility of fatigue failure due to repeated loading is not generally a problem in wood [*16*]. Peak stress levels experienced by a pile during driving are generally much higher than static working stress levels and in many instances may approach the yield strength of the material. At such high stress levels, the potential for the fatigue failure increases. In addition, defects such as knots, checks, shakes, and splits have a more adverse effect on the fatigue strength of wood than in static loading conditions [*4,16*]. Because the number of blows a pile experiences during driving can range from a hundred to a thousand or more, the writer would caution against using a β value greater than 1.2 when considering driving stresses in timber piles.

During construction, designated or selected piles are frequently load tested to twice the design load for 1 to 2 days duration. Based on Fig. 9, a β value of approximately 0.8 appears appropriate for normal load test durations.

Influence of Treatment on Strength of Timber Piles

In most cases, timber piles are treated to protect them from the destructive action of fungi, insects, or marine borers. Processes associated with the treatment can lead to a reduction in the strength of wood [*4,6,8*]. The primary strength reduction appears to be controlled by the conditioning process used prior to pressure treatment [*12*]. These conditioning processes are necessary to achieve the required preservative penetration and retention.

The principle conditioning processes used are air seasoning, the Boulton process, and steaming. It is possible that piles could be kiln dried prior to treatment, but at present this is not a general practice. The strength loss associated with these conditioning processes can be accounted for by the use of a reduction factor denoted herein by ψ, defined as the ratio of the strength of a treated specimen to the strength of an untreated specimen.

Of the currently used conditioning processes, only air seasoning does not appear to cause an appreciable strength loss [*12*]. Hence for air seasoning, a ψ value of 1.0 is recommended. Available data on the strength loss caused by the Boulton process [*12,15,25*] indicate the appropriate ψ value for this process is approximately 0.9. Information on the influence of kiln drying on pile strength is limited. What information is available [*12,26*] indicates the loss may range from 0 to 15 percent and is a function of drying temperature and duration. Until more information becomes available, a ψ value of 0.9 is recommended for kiln drying.

Wilkinson [*12*] and Thompson [*13,26*] present data on steam-conditioned southern pine piles. In addition, extensive information on the strength reduction caused by steaming was gathered during the ASTM Pole Research Program [*25*]. The strength reduction caused by steaming is dependent on both the steam temperature and the time duration of steaming. The available data indicate the strength loss at a temperature of 245°F is approximately 1.35 percent per hour of steaming. For example, 20 h of steaming at 245°F would result in a strength loss of approximately 27 percent.

The strength-reduction factor for the treatment, like the strength reduction for knots, must be selected so that it is compatible with the specified conditioning process. The current treatment standard most frequently used in specifications [*27*] specifies the permissable treatment processes for various species, the maximum permissable temperature, and the maximum duration of the temperature.

The writer recommends ψ values of 0.75 for southern pine and 0.9 for Douglas fir and oak for use with the American Wood Preservers Association (AWPA) C3-77 standard [*27*]. This recommendation is based on the assumption that Douglas fir and oak are conditioned by the Boulton process and southern pine is conditioned by steaming. When restraints more specific and restrictive than the AWPA C3-77 standard are used, ψ values can be selected on the basis of the recommendations in the preceeding paragraphs.

Design Stresses for Axially Loaded Timber Piles

The ultimate unit axial stress for timber piles, denoted herein by symbol

s_u, can be expressed by Eq 5 using the strength-reduction factors defined in the preceeding sections.

$$s_u = \beta\psi\phi s_c' \tag{5}$$

where

s_u = ultimate unit axial stress,
s_c' = basic reference strength,
ϕ = strength-reduction factor for imperfections,
ψ = strength-reduction factor for treatment, and
β = load duration factor.

Ultimate capacity for a given case can be calculated from Eq 5 by using values of ϕ, ψ, and β compatible with the permitted imperfections, permitted treatment, and load duration. For instance, Eq 5 indicates that Douglas fir piles treated in accordance with AWPA C3-77 ($\psi = 0.9$) meeting ASTM Method D 25-58 class B knot limitations ($\phi = 0.85$) have a long term ($\beta = 0.6$) ultimate unit stress of 7.6 MPa (1100 psi). During driving ($\beta = 1.2$) such piles, tip damage can be anticipated at stress levels greater than approximately 15.0 MPa (2200 psi).

The structural design of piling has historically been based on working stress formulas. The ultimate unit axial stress given by Eq 5 can be reduced to an allowable working stress s_a by dividing with an appropriate factor of safety f_s. While the term "factor of safety" is used herein, the quantity denoted by the symbol f_s is compatible to the load factor used in strength design methods.

$$s_a = \frac{s_u}{f_s} = \frac{\beta\psi\phi}{f_s} s_c' \tag{6}$$

The factor of safety f_s that should be used with Eq 6 is dependent on the reliability with which the loads can be determined. Since the reliability with which loads can be predicted is dependent on the nature of the loads, the factor of safety f_s should also be dependent on load duration.

The writer believes that a value of approximately 1.2 is appropriate for f_s under long-term durations where a β value of 0.6 was recommended. It is also the writer's opinion that allowable stresses derived on the basis of Eq 6 using $\beta = 0.6$ and $f_s = 1.2$ will lead to reasonable design stresses at other load durations. Using this recommendation, the factor of safety at other load durations can be estimated by

$$f_s = \frac{1.2}{0.6}\beta = 2\beta \tag{7}$$

Combining Eq 6 and 7,

$$s_a = 0.5\psi\phi s_c' \quad (8)$$

Allowable working stresses based on Eq 8 and the writer's recommended values for ψ, ϕ, and s_c' are presented in Table 5. The allowable working stresses of 4.7 to 6.4 MPa (675 to 925 psi) recommended for treated ASTM Method D 25-58 piles compare reasonably well with values of 3.5 to 6.9 MPa (500 to 1000 psi) commonly specified in pre-1970 building codes. The American Institute of Timber Construction [*23*] recommends allowable unit stresses of 4.1 to 6.2 MPa (600 to 900 psi) for green posts and timbers under long-term loading. These values are in close agreement with the values of 5.0 to 7.1 MPa (725 to 1025 psi) in Table 5 for untreated piles.

Discussion

In the preceeding sections, the influence of natural variability, imperfections, load duration, and treatment on the strength of a timber pile section were discussed. Available data were reviewed and reduction factors were recommended to account for the various variables influencing timber pile strength. Allowable design stresses based on the recommended reduction factors are summarized in Table 5.

In 1970 ASTM Committee D07 on Wood published ASTM Method for Establishing Design Stresses for Round Timber Piles (D 2899). This standard was intended for use with piles meeting the requirements of ASTM Method D 25-70 and 73. Working stresses for axial loading derived on

TABLE 5—*Recommended maximum*[a] *allowable axial unit stress for timber piles.*

	Allowable Axial Unit Stress, s_a, psi			
	Douglas Fir and Red Oak		Southern Pine	
Pile Quality Standard	Treated[b]	Untreated	Treated[b]	Untreated
ASTM Method D 25-58 class A and B[c]	925	1025	775	1025
ASTM Method D 25-58 class C	800	900	675	900
ASTM Method D 25-73	650	725	550	725

Conversion factor—1000 psi = 6.9 MPa.

[a] Assumes the pile tip is selected as the critical section, where a factor of safety f_s of 1.2 appears appropriate with a load duration factor of 0.6. Also assumes s_c' = 16.6 MPa (2400 psi)

[b] Assumes treatment to AWPA Standard C3-77.

[c] Assumes pile lengths of 15.2 m (50 ft) or less. For class A and B piles of lengths greater than 15.2 m (50 ft), use recommended values for class C piles.

the basis of ASTM Method D 2899-74 for treated timber piles are shown in Table 6. The ASTM Method D 2899-74 values in Table 6 are 1.7 to 2.2 times the writer's recommended values (Table 5) for piles meeting ASTM Methods D 25-70 and 73. It is the writer's opinion, based on the data presented in the preceeding sections, that the working stress formulas for timber piles presented in ASTM Method D 2899-74 are perilous and should not be used. The deficient ASTM Method D 2899-74 equations are the result of improper scaling for the influence of imperfections and treatment conditioning and the failure to include an appropriate factor of safety.

The ultimate unit axial stresses developed from Eq 5 and allowable working stresses developed from Eq 8 are directly applicable to pile tips. In some cases, the clear wood strength at other positions along the pile may be greater. For instance in the work of Wilkinson [*12*], small clear wood strength at the butt of Douglas fir and southern pine was approximately 10 percent greater than the tip small clears, and the crushing strength of short pile sections was approximately 20 percent greater at the butt than the tip. This higher butt strength for pile sections was caused by the presence of fewer knots in addition to the higher clear wood strength. While some designers may want to consider the variation of strength along the pile when the critical section is considered at some point other than the tip, the 10 to 20 percent higher butt strength is generally negligible in comparison to the increased cross-sectional area at the butt due to the natural taper of timber piles.

The topic of critical section under axial loading is frequently considered when piles are of nonuniform cross section. If piles transmit an appreciable percentage of their load through side friction, the critical section will be located at some point other than the tip. The location of the critical section

TABLE 6—*ASTM Method D 2899-74*[a] *working stresses for treated*[b] *round timber piles* [24].

Species	ASTM Method D 2899-74 Working Stress for Compression Parallel to Grain, psi
Douglas fir[c]	1250
Southern pine[d]	1200
Red oak[e]	1100

Conversion factor—1000 psi = 6.9 MPa.
[a] Assumes piles conforming to ASTM D 25-73.
[b] Assumes treatment in accordance with AWPA Standard C3.
[c] Assumes Pacific Coast Douglas fir.
[d] Assumes longleaf, slash, loblolly, and shortleaf pines.
[e] Assumes northern and southern red oaks.

is dependent on complex soil-pile interactions, which are beyond the scope of this paper. In any case, the selection of the critical section for design should be tempered by a great deal of engineering judgement.

This paper has been restricted to the consideration of axially loaded embedded timber piles. The stresses derived on the basis of the formulas are not applicable to cases where adequate lateral support does not exist. In many instances, pile heads are subjected to moments and shear forces in addition to axial forces. Where piles have unsupported sections or are subjected to combined loading, the structural design of a timber pile can be based on column design methods used with structural sawn lumber [*7,23*]. Imperfections such as checks, shakes, and splits may be significant when dealing with shear and flexural stress conditions.

The actual unit stresses that can be developed are generally less than the values given by Eq 8, because factors other than structural adequacy under service loading conditions control the required cross-sectional dimensions of a timber pile. Generally the required minimum cross-sectional dimensions will be controlled by driveability [*28*] or soil properties. Load test requirements of standards such as ASTM Method D 1143-74 or building codes which require testing to twice the design load can also control the design stress.

For the loading condition of a pile load test of 1 to 2 days duration (β equal to approximately 0.8), Eq 7 indicates a factor of safety or load factor of 1.6. Hence, it would not generally be possible to load test timber piles to twice the design load based on Eq 8 unless the test piles are specially selected to be relatively free of knots, have a tip diameter approximately 12 percent greater than the minimum specified diameter, or more than 40 percent of the test load is transferred to the soil through side friction during the test.

Conclusions

The preceeding review of factors influencing the structural properties of axially loaded timber piles provided the background for the following conclusions and recommendations:

1. Timber piles must be structurally capable of resisting, with an appropriate factor of safety, the forces they will experience during installation and under service loading.
2. The strength properties from short-term tests on clear wood specimens must be adjusted for the influence of natural strength variations, imperfections, load duration, and treatment.
3. The 1970 revision of ASTM Method D 25 included a major relaxation in knot limitations, which represent an extrapolation to approximately twice the upper boundary of the data on which the revision was based.

4. The strength reduction caused by knots appears to scale with the dimensionless knot ratios K_L/D and K_s/D, and can be approximated by Eqs 2 and 3.

5. The strength-reduction factor for knots should be compatible with the knot limitation specified. Recommended strength-reduction factors are given in Table 3.

6. The strength of timber piles is dependent on the duration of load, with the long-term static strength being only approximately 60 percent of the strength measured in short duration standard tests.

7. For evaluating pile driving stress levels, a load duration factor of 1.2 may be used.

8. Treatment of piles may cause a strength reduction. The strength-reduction factor should be selected to be compatible with the permitted conditioning. Recommended strength reduction factors are 0.9 for Douglas fir and red oak and 0.75 for southern pine for current (1978) treatment standards.

9. Recommended allowable stresses for axially loaded timber piles are given in Table 5. The recommended values agree well with pre-1970 building codes.

10. Use of ASTM Method D 2899-74 for determining design stresses on timber piles, results in calculated allowable working stresses, which may be 1.6 to 1.85 times the long-term ultimate axial stress, when the critical section is near the tip and the piles provided have knots near the ASTM Method D 25-73 limits. Hence, ASTM Method D 2899-74 should be withdrawn until appropriate revisions are made.

11. The minimum cross-sectional dimensions of a pile may be controlled by either pile driveability, soil properties, or structural requirements. To develop a safe and adequate design, all three variables must be checked.

References

[1] Gamble, W. L., "Capacity of Reinforced and Prestressed Concrete Pile Sections," this volume.

[2] Gerwick, B. C., Jr., and Brauner, H. A., "Design of Concrete Piles for Static and Dynamic Loads," this volume.

[3] Dismuke, T. D., "Behavior of Steel Bearing Piles During Installation and Service," this volume.

[4] U.S. Forest Products Laboratory, *Wood Handbook: Wood as an Engineering Material*, U.S. Department of Agriculture Handbook No. 72, 1974.

[5] Brown, H. P., Panshin, A. J., and Forsaith, C. C., *Textbook of Wood Technology*, Vol. 1, McGraw-Hill, New York, 1949.

[6] Brown, H. P., Panshin, A. J., and Forsaith, C. C., *Textbook of Wood Technology*, Vol. 2, McGraw-Hill, New York, 1952.

[7] Gurfinkel, G., *Wood Engineering*, Southern Forest Products Association, 1973.

[8] Hunt, G. M. and Garratt, G. A., *Wood Preservation*, 2nd ed., McGraw-Hill, New York, 1953.

[9] Markwardt, L. J., "Wood as an Engineering Material," Edgar Marburg Lecture, Presented Before the Forty-sixth Annual Meeting of the American Society for Testing Materials, *ASTM Proceedings*, Vol. 43, 1943.

[10] Dwyer, W. W., Chairman ASTM Subcommittee D07.07, letter to John G. Shope, Secretary ASTM Committee D07.00, Mar. 29, 1972.
[11] Douglas, A. H., Jr., "A-S-Ti-Mber Piling," *AE Concepts in Wood Design*, No. 4, American Wood Preservers Institute, McLean, Virginia, July–August 1975.
[12] Wilkinson, T. L., "Strength Evaluation of Round Timber Piles," Research Paper FPL 101, Forest Products Laboratory, Madison, Wisconsin, Dec. 1968.
[13] Thompson, W. S., "Results of Strength Tests on Piling Sections," Report submitted to American Wood Preserver's Institute by the Forest Products Utilization Laboratory, State College, Miss.
[14] Thompson, W. S., "Factors Affecting the Variation in Compressive Strength of Southern Pine Piling," *American Wood-Preserver's Association Proceedings*, Vol. 65, 1969.
[15] Peterson, J., "Final Report-WWPI Pile Test," Report on testing project funded by the Western Wood Preservers Institute, Civil Engineering Department, Oregon State University.
[16] Horner, A. C., Lewis, W. C., Ruble, E. J., and Wood, L. W., *Journal of the Structural Division*, American Society of Civil Engineers, Vol. 83, No. ST5, Sept. 1957.
[17] Wood, L. W., "The Factor of Safety in Design of Timber Structures," *Journal of the Structural Division*, American Society of Civil Engineers, Vol. 84, No. ST 7, Nov., 1958.
[18] Sugiyama, H., "On the Effect of the Loading Time on the Strength Properties of Wood: A Review on Japanese Research," *Wood Science and Technology*, Vol. 1, 1967.
[19] Sunley, J. G., "Changes in Basic Assumptions in UK Timber Design Codes," Building Research Establishment, Princes Risborough Laboratory, Princes Risborough, Aylesbury, Buckinghamshire, 1974.
[20] Keeton, J. R., "Dynamic Properties of Small Clear Specimens of Structural Grade Timber," Technical Report R573, California Naval Civil Engineering Laboratory, 1968.
[21] Boyd, J. D., "The Strength of Australian Pole Timbers—II: Principles in the Derivation of Design Stresses for Poles," Division of Forest Products Technological Paper No. 22, Commonwealth Scientific and Industrial Research Organization, Melbourne, Australia, 1962.
[22] Committee of the Waterways Division on Timber Piles and Construction Timbers, "Timber Piles and Construction Timbers," Manual of Engineering Practice, No. 17, American Society of Civil Engineers, 1939.
[23] American Institute of Timber Construction, *Timber Construction Manual*, 2nd ed., Wiley, New York, 1974.
[24] "National Design Specification for Wood Construction," Recommended Practice for Structural Design, National Forest Products Association, Washington, D.C., 1977.
[25] Wood, L. W., Erickson, E. C. O., and Dohr, A. W., "Strength and Related Properties of Wood Poles," *Final Report of the ASTM Wood Pole Research Program, ASTM STP 295*, American Society for Testing and Materials, Sept. 1960.
[26] Thompson, W. S., "Effect of Steaming and Kiln Drying On the Properties of Southern Pine Poles," *Forest Products Journal*, Vol. 19, No. 1, Jan. 1969.
[27] American Wood Preservers Association Book of Standards, Standard C3-77, "Piles-Preservative Treatment by Pressure Processes," American Wood Preservers Association, Washington, D.C., 1977.
[28] Davisson, M. T., "Pile Load Capacity," *Proceedings*, Design Construction and Performance of Deep Foundations, American Society of Civil Engineers, University of California, Berkeley, Aug. 1975.

DISCUSSION

W. A. Norum[1] *(written discussion)*—This paper contains an excellent analysis of the strength properties of timber piles, but unfortunately,

[1] District manager, National Forest Products Association, Mountain View, Calif. 94040.

several conclusions have been based on erroneous and misleading assumptions. The paper is an academic review of test data and technical information without enough attention given to the fact that timber is nature-made, in contrast to products manufactured under man's control, such as steel or concrete. This failure to appreciate the difference has resulted in improper extrapolation of test data. The growth characteristics of trees must be taken into consideration. Fortunately, this subject has been studied by wood technologists over the centuries. Their work has provided us with a wealth of knowledge about the patterns of nature that must be considered when attempting to make accurate strength predictions of timber. Much of the data collected by authorities on strength properties is published in ASTM Method D 2555, and information on how to apply it to sawn timber is published in ASTM Method D 245. Reports published by the U.S. Department of Agriculture, Forest Products Laboratory, are a source of much more information used to support current practices for deriving recommended design stresses for timber. Information contained in ASTM Method D 2899 is an expansion of this knowledge to round timber piles. Armstrong questions the validity of these practices and recommends much lower design stresses for round timber piles than what is derived by use of ASTM Method D 2899. Armstrong's recommendations could unnecessarily lead to the need for larger piling or more timber piling to accomplish the same purpose. This, of course, means greater cost for timber pile foundations and fewer used because competitive piling could be favored economically. It is my opinion, however, that Armstrong's recommendations are ultraconservative and should not be followed. Reasons for my belief are detailed in the following paragraphs.

Man-Made Products versus Nature's Products

Well controlled man-made products should have little variability among all manufactured pieces; consequently, its strength properties can be predicted with great confidence. Theoretically, any factor of safety used by engineers in such cases would only reflect their concern about man maintaining good controls over manufacturing and his concern about being able to properly identify the products at the job site. Adding to this, he may worry about the unknowns that could effect the use of all products, such as its misuse or improper handling experienced during construction and unforeseen loading or environmental conditions to which the material will be exposed. In other words, his "factor of safety" becomes a "factor of ignorance."

Timber, on the other hand, has strength properties controlled by nature and subject to natural laws. Strength of clear wood has a normal variability among trees or among pieces of lumber from the same tree. It is not subject to man's control or, conversely, lack of control, so possibly the engineer's "factor of ignorance" does not need to be as large, that is, providing the recommended stresses are based on the lower level of stress values found in a tree of each species. In a general sense, this is done in ASTM Method D 2899. According to L. W. Wood, author of "Factor of Safety in Design of Timber Structures," a general representative value of standard deviation for clear wood specimens is approximately 16 percent; for example, a 1000 psi mean value varies between 1160 and 840 for 68 percent of the pieces. Considering only clear wood, any recommended design stress for timber piles that takes into account the standard deviation and the data contained in ASTM Method D 2555 will probably represent a value lower than that found in timber piles supplied to a job site. As a matter of fact, the first step in deriving either lumber or timber pile stresses is to reduce average clear wood strength values to a level that is exceeded by 95 percent of clear wood specimens. This near minimum strength value is then further reduced to include a factor of safety and to account for various wood characteristics. This means that the engineer will receive timber having a factor of safety based on near-minimum values, with a high probability that he will actually receive much stronger timber. In Armstrong's paper, there is no recognition given to this fact, which is extremely important to the intelligent use of recommended design stresses for timber.

Factor of Safety In Nature's Product

The differences between man-made products and those made by nature calls for a different approach for deriving recommended design stresses. With nature-grown materials, there is need for greater dependence on a statistically sound data base and use of mathematical techniques based on the theory of probability. On the other hand, test information on man-made products can be directly applied to recommended design stresses with dependence on quality control and identification. Since most engineers have become more accustomed to man-made products, at least in the use of "safety factors," there is confusion created when they try to relate these experiences to timber. With timber, a "safety factor" is initially built into the system by deriving stresses based on low-line quality. The net result is that an engineer can theoretically expect 95 percent of the material to be greater in unit strength than the value used in his design when he accepts design values derived by use of ASTM Method D 2899. Also, the factor of safety will be multivalued because each piece of timber will have a different strength value. Using an example to illustrate the idea, let's take Douglas fir (coast) having a clear wood average short-term duration of load value of 3784 psi and a standard deviation of 734 psi in compression parallel to grain (ASTM Method D 2555). Assuming round timber piles have the same unit strength as clear wood specimens (the difference will be discussed later), this means that the engineer will probably receive material having short-term unit strength values ranging from (3784 − 734) 3050 psi to (3784 + 734) 4518 psi. If the material shipped was representative of the Douglas fir growing in the coastal area of the Pacific Northwest, the shipment would relate to the normal distribution curve, which means that 68 percent of the material will have a value between 3050 psi and 4518 psi. Procedures in ASTM Method D 2899 specify that recommended stresses are to represent the weaker material, except that the lower 5 percent is to be excluded from consideration. Following this procedure, then, the short-term recommended design value is calculated to be 3784 − (1.645 × 734) = 2577 psi. Piles supplied to a job site will probably have greater unit strengths in 95 percent of the pieces, and 68 percent of the pieces will have unit strengths that exceed the recommended design value by 473 to 1941 psi (4518 − 2577 and 3050 − 2577). A calculated factor of safety is obviously multivalued and will range from 1.18 to 1.75. When all other factors are considered, such as long-term duration of load and the effect of knots on strength, the factors of safety appear to remain the same. At least it is in the case discussed by E. F. Diekmann, *Timber Piles in Standards, Codes and Practice.* He shows a variation from 1.12 to 1.88 (corrected to consider duration of load) with an average of 1.58. This comparison suggests that the theoretical approach used to calculate the factor of safety, which includes the statistical data base and the normal distribution curve developed by the U. S. Forest Products Laboratory, relates to real conditions examined by Diekmann, who calculated the factor of safety using test results from crushed timber pile tips.

The concept for deriving recommended design stresses expressed by ASTM Method D 2899 should be accepted with little reservation where the designer is involved in projects requiring several timber piles or where piles are installed to share the load. In cases, however, where few piles are installed with little possibility of load sharing, an even more conservative design value may be advisable. Engineers should consider the fact that procedures of ASTM Method D 2899 recognize there may be 5 percent of the piles having a lower unit strength than the recommended design value by taking appropriate risk. He has the option of (1) adding further controls at the job site so as to select the obviously stronger piles for application where only single piles or relatively few supporting piles are used or (2) reducing the design stresses. If he takes the latter option, ASTM Method D 2899 provides guidance. An additional factor of 1.25 for compression parallel to grain and an additional factor of 1.30 for bending is recommended. Interestingly, Armstrong's recommended design stresses are only about 66 percent of what is derived by applying this additional factor of safety. For example, Armstrong recommends a design stress for treated Douglas fir of only 650 psi (ASTM Method D 25-70), while ASTM Method

D 2899 indicates it should be 1233 psi, which reduces to 986 psi by applying the 1.25 reduction factor.

In Lyman Wood's paper "Factor of Safety in Design of Timber Structures," he identifies many other considerations when computing a factor of safety and inspects them using techniques based on probability. He concludes:

> A few major and many minor factors affect safety in timber design. A near-minimum margin of safety can be readily estimated by applying near-minimum values for the major factors. For a more thorough consideration, however, the various factors must be considered as multivalued, and a multivalued factor of safety computed from them.
>
> The designing engineer has the right and the responsibility to use his own judgment in evaluating the factors that affect the safety of his design. This paper is presented as an aid in forming this judgment. The manner in which safety is affected by use as well as strength and how safety is related to the working stress level are indicated. Thus, through better understanding and logical application of the factors affecting the safety of structural lumber, progress is made toward the goal of safe and economical timber design.

This paper is recommended reading for engineers concerned with the subject of factor of safety in timber.

Effect of Knots on Strength

A great deal of concern was expressed by Mr. Armstrong over the effect knots have on the strength properties of timber piles, and he devoted much of his paper to this subject, making it a major issue in support of lower design stresses in the belief ASTM Method D 25-70 has more liberal knot limitations than does ASTM Method D 25-58. This has prompted my intensive research and evaluation of all known and available test information in response. Results of my study bring me to a different conclusion for reasons stated in the following paragraphs.

Before revealing the detailed results of my study, there is need to again detail important differences between a nature-made and a man-made building material, because it appears Armstrong has failed to make that distinction. For example, the growth characteristics of conifer trees establishes a ring pattern of branches formed in annual growth layers that leave vertical clearances between branches in the growing tree. This pattern of growth normally causes the trunk of the tree to be free of knots between rings of knots formed by the layered branches. When comparing differences in knot limitations between ASTM Methods D 25-58 and D 25-70, the author did not recognize this, but instead made a mathematical extrapolation as if timber piles were man-made. In ASTM Method D 25-58, the sum of sizes of all knots relates to a 1-ft rather than a 6-in. timber pile length as specified in ASTM Method D 25-70. This was interpreted by the author to have "increased the permissible summation of knots ratio K_s/D for all piles to a level twice that previously permitted in class C". This isn't true for reasons explained. On the other hand, it would probably be true if it were a man-made product, especially if timber piles could be produced cheaper by man, adding more knots to the length of piles. The reason for the revision, as I recall, was to prevent the occasional rejection of perfectly sound piling because of a random knot occurring away from the normal circle of knots. The author also fails to recognize the importance of the cellulous and fiber structure of timber when he compares the knot limitations for sawn timber specified in ASTM Method D 245 with round timber piles. They simply cannot be compared, because round timber piles retain the inherent fiber pattern of a growing tree, which, on the other hand, becomes disrupted by sawing a tree into lumber. The process of severing wood fiber that flows around knots as is done by sawing trees into lumber does cause a weakening in the area of the knot. Maintaining the existing wood fiber around knots obviously helps maintain the strength found inherent in trees. Some recognition by the author is given this fact by his reference to the *Wood Handbook,* which states that knots have less effect on the strength of round timber piles than in sawn lumber but, unfortunately, he proceeded to make a direct comparison between ASTM Methods D 245 (lumber) and D 25 (piles).

What should really matter to design engineers and to pile producers is the degree to which knots effect strength of round timber piles. In my effort to arrive at an answer, I reviewed the pile reports referenced by Armstrong and, in addition, two pole test reports. These reports are: (1) *Strength and Related Properties of Wood Poles* published by ASTM, September 1960, and (2) "Pole Strength Tests" published in the *Journal of the Forest Products Research Society,* April, 1952. The latter publication is a report on testing done by the Bell Telephone Laboratories. These two timber pole test programs were comprehensive and are of considerable value to our interest in timber piles, since they both involve the tree, even though the specification details are not identical. Each report contains information on the nature and location of the pole rupture as it relates to knots. It is reported that the results of testing full-length poles, which were tested in cantilever bending by use of test equipment that gripped the poles at a specified ground line, indicate that knots are "a relatively unimportant factor in the strength of poles for conventional uses." In the ASTM test, it is reported, "The poles showed about the same strength as standard test specimens of clear wood. A few poles in the ASTM program indicated, however, that larger knots near the groundline reduce the strength". The test program done by Bell Telephone Laboratory included testing pole top segments taken from the broken full-length poles. All pole top segments were 13 ft or greater in length. Results of this study is of special interest, since it relates more directly to our concern over knots located in pile tips. It was reported that "less than three percent of the pole breaks and less than nine percent of the pole top breaks were at maximum knot locations."

The report goes on to say, however, that "many of the breaks occurred at single knots and groups of knots having diameters or sums of diameters less than the permitted maxima." The conclusion reached is, "While the maximum knots permitted in a pole generally have little effect on pole strength, because of their location, knots in general may have a decided effect. In these tests relatively small knots, occurring both singly and in groups, were associated with the break of 33⅓ percent of the poles and 70 percent of the pole tops." The question remains, How significant are knots to the strength of piles? On the one hand, the ASTM reports no significant effect on the ultimate strength of poles breaking near the ground line, and Bell Lab's reports that knots were associated with the breaks occurring in a significant number of cases, even though the knots located near the break were not the maximum-size knots found on the poles. The answer may be found in the Bell Lab's comparison made between strength of poles (full size) and strength of pole tops. They report that "the strength of the pole tops to be about 85 percent of the strength of the poles from which the tops were cut. Considering the fact that the density of a pole generally decreases from the butt to the top and the fact that the top of a Southern pine pole usually contains more and larger knots than the lower sections, to say nothing of the possibility of non-visible injury to the pole top during the breaking test of the pole, the strength of the tops was remarkably high." In other words, the effect of knots located near the ground line of poles (approximately 5 to 10 ft from butt) on the strength of poles in bending does not appear significant, and where breaking occurs near the tip end, the maximum stress in bending is found to be 15 percent lower. The reduction is probably effected by both the knots and lower density existing in tip sections.

I believe the same conclusion can be reached from the results of compression tests on pile tips reported by T. C. Wilkinson, U. S. Forest Products Laboratory, Madison, Wis., and W. S. Thompson, Forest Products Utilization Laboratory, State College, Miss. (both reports are referenced by Armstrong). Wilkinson reports that red oak, Douglas fir, and southern pine pile tips were demonstrated to have 85, 91, and 90 percent, respectively, of the compression strength found in small clear specimens taken from the full-size pile tip test specimens. Since the small-size clear specimens should have possessed a specific gravity approximately equal to that of the full-size pile tip, it can be assumed that the major difference in strength from the clear specimens relates to the effect of knots. In Part 2 of Wilkinson's report, further testing is done with southern pine. Results of this supplemental work demonstrated that southern pine tips taken from green, kiln dried, and kiln dried-treated piles had 93 to 96 percent of the crushing strength, tested parallel to grain, of small clear specimens. This suggests that the process of steam conditioning southern pine piles, which was done to these piles reported in Part 1, affects the strength of the wood located in the area of knots. It appears that where steam conditioning of southern pine piles is not practiced, the effect of knots on strength of pile tips varies from 4 to 7 percent, and where steam conditioning is practiced, it may cause a strength reduction of 10 percent. The tip strength of red oak appears to be more adversely effected by knots (15 percent) than any other species. This may be at least partially explained by the fact that only 5 of the 15 red oak piles conformed with the knot limitations specified in ASTM Method D 25-58.

As part of the discussion on knots and their effect on strength, Armstrong makes a point that knot limitations in ASTM Method D 25-70 have been made more liberal. The author's conclusion on this point is, "Since the strength of a timber pile is reduced as the size and number of knots increase, the allowable stress used with ASTM Method D 25-70 piles should be less than the allowable stresses used with ASTM Method D 25-58 piles." This conclusion seems very rational when considered separately from the author's intent, but when considering it in light of his intent to use it as an argument against pile stresses derived in accordance with procedures of ASTM Method D 2899, it is wrong. Wilkinson reports that all tested pile segments broke through a cross section confined to a 6-in. length of pile. This is evidence that the revision to ASTM Method D 25 allowing knots to be increased along a 6-in. segment rather than 1-ft segment does not in fact liberalize the standard. Armstrong's conclusion also presumes that procedures for deriving stresses under ASTM Method D 2899 were based on tests of piles conforming to ASTM Method D 25-58. The author has himself proven that the tested piles did not conform. Data taken from test reports by Mississippi State University, U.S. Forest Products Laboratory, and Oregon State University, which the author includes in his paper, proves that many piles had knots larger and greater in number than allowed by the standard. Knot limitations of ASTM Method D 25-58 were violated in every test program. The violations were most significant among the southern pine and red oak piles supplied for testing, but even several Douglas fir piles were found nonconforming. It was not surprising to find Douglas fir piles closer to conformance, because the growth characteristics of this species of trees is more favorable. I suspect that all piles furnished for testing were representative of what could be furnished from the forest rather than having been especially selected to influence test results. This is a conclusion that I have reached along with a conclusion that the piles tested were more representative of ASTM Method D 25-70 than ASTM Method D 25-58.

Other Influences on Strength of Timber Piles

In addition to knots and specific gravity, the author discusses other influencing factors to which I will only briefly respond because they are covered fairly well by Armstrong and Diekmann. I believe Armstrong has properly dismissed spiral grain from much discussion. This then leaves load duration and preservative treatment effects on strength for consideration.

Regarding the question of load duration on the strength of timber piles, I believe Armstrong to be more conservative than is justified. Procedures of ASTM Method D 2899 relates data taken from short-term loading conditions (±2 min) to "normal duration of load," which is 10 years of continuous loading at maximum recommended stress. The factor used to convert basic strength data contained in ASTM Method D 2555 to full size piles is 1/1.52, or 0.658. This value was recommended by representatives of the U.S. Forest Products Laboratory as appropriate for application to foundation piles. There is good reason to believe it to be sufficiently conservative, especially in light of Diekmann's comments on this subject. Diekmann reported on studies that indicate there may be no time effects on lumber being stressed below 2600 psi. Proposed pile design stresses are 1250 to 1500 psi.

The author's discussion of the influence of treatment on strength of timber piles is forthright and leaves me with no cause to differ with his conclusions. Procedures established by ASTM Method D 2899 are adequate for considering the effect of preservative treatment on strength except for steam conditioning. Probably there should be greater reductions applied in the case of steam conditioning, which by the way, only applies to southern pine. Only this species is steam conditioned prior to preservative treatment. Armstrong recommends a reduction factor of 0.75, which I believe is appropriate. Diekmann has recommended the more conservative factor of 0.56. Apparently, he arrived at this figure by relating it to the test results of southern pine tips, which he computed to have an average factor of safety of only 1.67. His recommended factor of 0.56 would cause an increase in this factor of safety to 2.98. It would also increase the average factor of safety to 4.43 for the butt sections of steam-conditioned southern pine piles, which I believe is too conservative.

Conclusions

1. Strength properties of nature-made round timber piles can be predicted by use of the techniques specified in ASTM Method D 2899.

2. Reduction factors specified in ASTM Method D 2899 are reasonable to use for deriving recommended design stresses, except that a reduction factor of 0.75 may be more appropriate than 0.85 to recognize the effects steam conditioning has on the crushing strength of timber pile tips.

3. Knot limitations specified in ASTM Method D 25-70 are more representative of the southern pine and red oak piles that were tested for strength than are specifications in ASTM Method D 25-58. Growth characteristics

of all trees control the pattern of knots, so liberalized standards do not necessarily mean piles will be produced with more or larger knots. Growth characteristics of Douglas fir trees allows for production of round timber piles with less concern over knot limitations than for the other named species.

4. Design stresses derived by use of ASTM Method D 2899 reflects the near-minimum stress value of available round timber piles conforming to the quality specifications of ASTM Method D 25. Approximately 95 percent of the available piles will have a greater value.

5. Factors of safety will be multivalued. Even where load duration is considered, factors of safety for pile tips will probably range from 1.12 to 1.88 for Douglas fir and 1.94 to 2.41 for red oak. Southern pine that is kiln dried prior to preservative treatment will probably have a factor of safety range from 1.91 to 2.74 and when steam conditioned 1.43 to 2.28, providing the design stresses are derived as recommended in conclusion 2 above. The average factor of safety for pile butts among these species will probably range from 1.62 to 2.37. It is very doubtful that a duration of load effect will ever be experienced at the level of design stresses recommended, so the factors of safety calculated by Diekmann are probably closer to reality. He shows pile tips ranging from 1.83 to 3.39 when the value for steam conditioned Southern pine is adjusted to reflect the recommendations in conclusion 2 above.

6. An additional factor of safety may be appropriate where the load is not shared among several foundation piles. An additional factor of 1.25 for compression parallel to grain and an additional factor of 1.30 for bending is recommended in the National Design Specification for Wood Construction published by the National Forest Products Association. These additional factors of safety also appear in ASTM Method D 2899.

R. M. Armstrong (*author's closure*)—Norum has raised several points that require further comment. His discussion concerns five issues: (1) the approach used by the writer in deriving recommended design stresses, (2) the influence of treatment on strength, (3) the influence of load duration on strength, (4) the influence of knots on strength, and (5) the selection of an appropriate factor of safety. With regard to the first point, the discusser states that the writer questions the validity of current practices for deriving recommended design stresses for timber. The writer strongly questions the validity of allowable design stresses derived on the basis of ASTM Method D 2899-74 equations. However, this does not mean that the writer questions current practices for deriving recommended design stresses for timber in general. The approach used by the writer was stated in the second paragraph of the text and is *exactly* the current practice for derivation of design stresses in timber outlined in ASTM Method D 245 and other references cited by the author [*4, 7, 9, 17, 19, 21, 23*].

The approach used in deriving the ASTM Method D 2899-74 design stresses is outlined in Sections 13, 14, and 16 of that standard, and deviates from current practice for such derivations since a formal factor of safety is not included. This deviation can not by itself account for the ASTM Method D 2899-74 design stresses being 1.7 to 2.2 times the writer's recommendations, because the formal factor of safety normally included is only of the order of 1.2 to 1.3 for compression parallel to grain. The remaining difference, 1.4 to 1.8, results from variations in values used for the strength reductions for imperfections, load duration, and treatment.

With regard to the second issue, influence of treatment on strength, Norum agrees ASTM Method D 2899-74 provisions on steaming are inadequate and that the writer's recommended strength reduction factor of 0.75 is more appropriate. Norum points out that Diekmann[2] recommends a value of 0.56 for the treatment strength-reduction factor, then discards Diekmann's value as being too conservative without considering the full implications of Diekmann's recommendation.

Diekmann (see footnote 2) reached his recommendation of 0.56 through a subjective comparison of the test results on steam-conditioned southern pine pile sections reported by Wilkinson (Ref *12* of paper) with design stresses calculated by using ASTM Method D 2899-74 equations. Diekmann felt the calculated stresses were too high when compared to the actual test failure stresses reported by Wilkinson (Ref *12* of paper), and a reduction to two-thirds of ASTM Method D 2899-74 stresses was required to reach a safe design value for treated southern pine. His subsequent recommendations of 0.56 ($0.66 \times 0.85 = 0.56$) for the steam treatment reduction factor is based on the dubious assumption that the treatment factor was the only deficiency in ASTM Method D 2899-74. Diekmann failed to point out that Wilkinson's test specimens were steamed at 245°F for only 15 h, while current treatment standards (Ref *27* of paper) allow 20 h of steaming at 245°F. Hence, Diekmann's recommended reduction for steaming would have been even lower (approximately 0.51) if adjusted to be compatible with the currently permissible treatment of 20 h. Diekmann's recommended reduction to two-thirds of ASTM Method D 2899-74 stresses is a lumped factor for deficiencies in Method D 2899-74, of which treatment is only one of many.

Norum in his comments on the third issue, influence of load duration, expresses the belief that the load duration factor of 0.6 used by the writer for long-term loading conditions is more conservative than justified. The load duration factor used for long term loading is 0.5625 (Refs *7,21,23,* and *24* of paper and footnote 2). The writer rounded this value to 0.6, because the use of four significant figures is unwarranted. For 10-year loading, a value of 0.625 is normally used. Norum[3] states the value included in ASTM Method D 2899-74 equations was 0.658. The writer has

[2] Diekmann, E. F., "Timber Piles in Standards, Codes and Practice," this volume.

[3] Norum, W. A., Discussion on "Timber Piles in Standards, Codes and Practice," by E. F. Diekmann, this volume.

not been able to find the discusser's factor of 0.658 for 10-year loading in any of the common references on load duration (Refs *4,10,16–18* of paper) or in references on current timber design practice (Refs *4,7,19,21–24* of paper). Although by itself minor, this exemplifies the many deviations of ASTM Method D 2899-74 from current practice in the derivation of allowable stresses for timber.

Norum also makes reference to the work of Madsen[4] on load duration by reference to Diekmann's comments (see footnote 2). Madsen's work was on the bending strength of dry wood with a moisture content of approximately 8 percent. Sugiyama (Ref *18* of paper) and Cousins[5] both indicate that the influence of the load duration effect is less in dry wood than wet wood. Furthermore, the 17.8 MPa (2600 psi) stress level quoted by the discusser is a modulus rupture value for dry wood and is not directly comparable to design stresses for compression parallel to grain in green piles.

The work of Madsen does not justify a change in current design procedures in wood, although it does help to understand why the formal factor of safety included in wood is only of the order of 1.2 to 1.3. As stated by Madsen:

> In the past the development of allowable stresses has included a specified factor of safety on the 5th percentile of 1.3. This factor is low compared to other materials, and the factor for duration of load (9/16) has in reality provided an extra factor of safety.

Madsen's statement implies that it is the gross ratio of the specified factor of safety to the load duration factor, 1.3/0.5625 = 2.3, that is important and that this ratio is comparable to the factor of safety for other materials. In this light, it is of interest to note that ASTM Method D 245 does not provide separate factors of safety and load duration factors, but provides one factor that includes the adjustment for normal duration of load and a factor of safety. This factor is 2.1 to 2.3 for bending and 1.9 to 2.1 for compression parallel to grain. The writer used a value of 2.0 (Eq 7) for compression parallel to grain.

The fourth objection of the discusser concerns the influence of knots on pile strength. Data available on the influence of knots on the compressive strength of piles are limited, as was stated in the text. The writer presented Fig 5 based on ASTM Method D 245, which is for sawn lumber, since it is one source of information and represents a lower bound for the imperfections reduction factor. The writer's recommended strength reductions for imperfections in Table 3 are based on Figs. 6 and 7, which contain the available data on pile sections tested in compression parallel to grain. Hence, the writer has attempted to account for the difference between

[4] Madsen, Borg, *Forest Products Journal*, Vol. 23, No. 2, Feb. 1973.
[5] Cousins, W. J., *Wood Science and Technology*, Vol. 8, 1974.

sawn lumber and round timber piles, to the extent the available data permit.

All the references reviewed by the writer (Refs *12–15* of paper) containing compressive tests on pile sections reported the sum of knot diameters per foot of pile length. In order to compare the knot limitations with the available data, the writer chose to transform the ASTM Method D 25-73 sum of knots limitation per 6 in. to a mathematical equivalent per foot, rather than distort the data. In support of the ASTM Methods D 25-70 and 73 increase in the sum of knots ratio, the discusser states; "Wilkinson reports that all tested pile segments broke through a cross section confined to a 6-in. length of pile." This statement is not contained in Wilkinson's report on the Forest Products Laboratory tests (Ref *12* of paper).

Norum states that the writer's comments on ASTM Method D 25 and the strength-reducing effect of knots "prompted his intensive research and evaluation of all known test information in response." The only data the discusser offers on compression tests are from Wilkinson's report (Ref *12* of paper). The strength-reduction factors of 0.85, 0.91, and 0.90 he quotes, respectively, for red oak, Douglas fir, and southern pine tips are average values and represent three of the data points shown on the writer's Figs. 5 and 6. The discusser then reports the strength-reduction factors ranged from 0.93 to 0.96 for the green, kiln dried, and treated kiln-dried southern pine piles from Part 2 of Wilkinson's report (Ref *12* of paper).

The discusser compared the average strength ratios from Parts 1 and 2 of Wilkinson's report and concluded the difference is the result of treatment. The writer disagrees with this conclusion. The piles in Part 2 of Wilkinson's report were specially selected to be relatively free of knots. Hence, the difference can be attributed to knots as easily as treatment.

The other information on the influence of knots referenced by the discusser[6] (also Ref *25* of paper) was from cantilever bending tests on poles which result in maximum moments near the ground line where fewer knots are generally present. When knots were present, both references (Ref *25* of paper and footnote 6) indicate they had a significant influence. The writer's paper was restricted to the consideration of axially loaded timber piles. Hence, a detailed review of the influence of natural imperfections on bending strength was not presented. The strength reductions for bending would be different from compression, and imperfections such as checks, shakes, and splits may be significant. Strength-reduction factors for bending, like compression, should be scaled to the extent of knots permitted. It is interesting that Diekmann (see footnote 2) also concluded that the knot limits of ASTM Method D 25-58 are preferable to the Method D 25-73 limits.

The last issue on which comment is required is the discusser's view of

[6]Eggleston, R. C., *Journal of the Forest Products Research Society*, April 1952.

the factor of safety. With other materials, such as concrete and steel, the designer specifies the minimum basic reference strength used in design. Strength reduction factors and load factors are applied to the basic reference strength, not the average strength. To meet these minimum strengths the average strength must always exceed the specified basic reference strength. In the ACI 318-77 "Building Code Requirements for Reinforced Concrete," a statistical basis is used to establish the average strength required to assure attainment of the specified basic reference strength f'_c used in design.

The design philosophy used in wood as outlined in ASTM Method D 245 and other references (Refs *4,7,8,19,21,* and *23* of paper) is the same as other materials in that the basic reference strength used is a statistically selected near-minimum value referred to as the 5 percent exclusion limit. Strength reductions and a factor of safety are applied to this basic reference strength. Hence, the factor of safety as defined by the writer is consistent with current timber design procedures. Its small value is justified by the use of the strength reduction factors to account for variables that could cause the pile strength to be less than the basic reference strength and the inclusion of a load duration factor.

The discusser appears to think the large natural variability is a positive rather than negative feature of wood. He then proceeds to play a numbers game by using this variability to calculate an illusory factor of safety. Both the discusser and Diekmann (see footnote 2), in a similar numbers game, used data from Wilkinson (Ref *12* of paper) to support their views. However, Wilkinson's tests included only 15 piles from a given species, which is hardly a sufficient volume of data to support their conclusions.

Norum's numbers game falls apart when tested against large data sets (Refs *13* and *15* of paper). For instance, the short-term design stress for untreated coast Douglas fir, as determined by ASTM Method D 2899-74, is approximately 2100 psi. If the minimum possible factor of safety is 1.12 as stated in the discusser's conclusions, all Douglas fir piles tested under short-term conditions should have a failure stress greater than approximately 2350 psi. Figure 8 clearly shows that Norum's conclusions are in error, since almost 40 percent of this sample of over 100 piles failed at a stress level less than 2350 psi. Evidently mother nature doesn't use the same "natural laws" as were used by Norum in computing the factor of safety.

In summary, the writer finds Norum's conclusions unsupported by his data or factual discussion. In fact, all data referenced by the discusser tend to support the writer's recommendations, rather than refute them.

M. Bozozuk,[1] *G. H. Keenan,*[2] *and P. E. Pheeney*[3]

Analysis of Load Tests on Instrumented Steel Test Piles in Compressible Silty Soil

REFERENCE: Bozozuk, M., Keenan, G. H., and Pheeney, P. E., **"Analysis of Load Tests on Instrumented Steel Test Piles in Compressible Silty Soil,"** *Behavior of Deep Foundations, ASTM STP 670,* Raymond Lundgren, Ed., American Society for Testing and Materials, 1979, pp. 153–180.

ABSTRACT: Axial compression and tension tests were performed on two end-bearing piles (one pipe and one H-pile) and two friction piles (one pipe and one H-pile) 3 months after they were driven through a granular test fill into compressible clayey silts. In analyzing the behavior of the piles, skin friction equations were developed based upon effective stresses in the soil, the soil-pile friction factor and the Poisson's ratio of the pile. These equations satisfactorily estimated the distribution of negative skin friction loads mobilized on the piles before they were test loaded and the skin friction mobilized in the axial compression and tension pile tests at failure. The equations were also applied to some well-documented pile tests, and showed that the skin friction that can be mobilized to resist axial compression loading can be estimated from pullout tests. This paper discusses the pile test procedures and the observed relative movements between the pile tips and the surrounding soil at various loads applied to the piles and describes a laboratory test method for measuring the coefficient of friction between the pile and the soil.

KEY WORDS: test fill, pile tests, axial compression, axial tension or pullout, negative skin friction, positive skin friction, effective stress, coefficient of soil-pile friction, Poisson's ratio, relative pile tip settlement

A new bridge over the Saint John River at Fredericton, New Brunswick, will be constructed on concrete piers supported by steel end-bearing piles driven through compressible silts into a dense gravel formation. The south

[1]Senior research officer, Geotechnical Section, Division of Building Research, National Research Council of Canada, Ottawa, Ontario K1A OR6, Canada.

[2]Head, Geotechnical Section, New Brunswick Department of Transportation, New Brunswick, Canada.

[3]Geotechnical engineer, Geotechnical Section, New Brunswick Department of Transportation, New Brunswick, Canada.

shore is being redeveloped and requires large volumes of fill for new roads and an interchange with the bridge. It was expected that the high approach embankments required at the south end of the bridge would cause significant settlements and generate large negative skin friction loads in the piles.

To obtain design parameters for the foundations of the proposed bridge, the New Brunswick Department of Transportation, in cooperation with the Division of Building Research, National Research Council of Canada, embarked on an extensive full-scale field testing program. During the winter of 1977, an 11-m-high granular test embankment with a surface area of 25 by 40 m, was constructed and instrumented on the south bank of the river. In July 1977, seven steel instrumented test piles, consisting of two end-bearing and five friction piles, were driven through the test fill to study the development of negative skin friction loads. In October 1977, the two end-bearing and two friction piles were test loaded to failure in axial compression and pullout by means of a 500-ton hydraulic jack and a specially constructed steel reaction frame.

Description of the Fill and Subsoils

The 11-m-high granular test fill was constructed on the south shore of the river in line with the proposed bridge. The first stage was constructed in water to elevation +3 m. This stage provided the working base for the installation of most of the field instrumentation. The second stage, constructed from 12 March to 7 April 1977, brought the surface elevation to +9 m. Numerous in situ density tests performed during construction indicated an average density of 1825 kg/m^3.

Dredged granular fill was placed around the test fill from 26 August to 19 October for the redevelopment of the south shore of the river. This dredged fill was placed over a very large area and imposed a relatively uniform distribution of vertical load on the subsoils (Fig. 1). The average density of this fill was 1800 kg/m^3.

The subsoils were divided into six distinct soil formations. The test fill and dredged fill were placed directly upon layer A, a heterogeneous deposit of soft organic silt, sand, pebbles, gravel, wood, etc. Although it varied considerably in thickness, it was about 4 m thick beneath the center of the test fill. It was not possible to obtain undisturbed samples of this material for testing, but it was assumed to have a density of 1680 kg/m^3.

Layer B was encountered from elevation −6 to −12.5 m. This formation was a gray layered clayey silt with a water content varying from 23 to 38 percent, a liquid limit of 29 percent, and plasticity index of 9 percent. The average grain size distribution was 40 percent clay and 60 percent silt. Its average wet density was 1840 kg/m^3. The maximum *in-situ* vane shear strengths varied from 0.3 to 0.5 kg/cm^2.

Layer C extended from elevation −12.5 to −18 m. This was a gray-

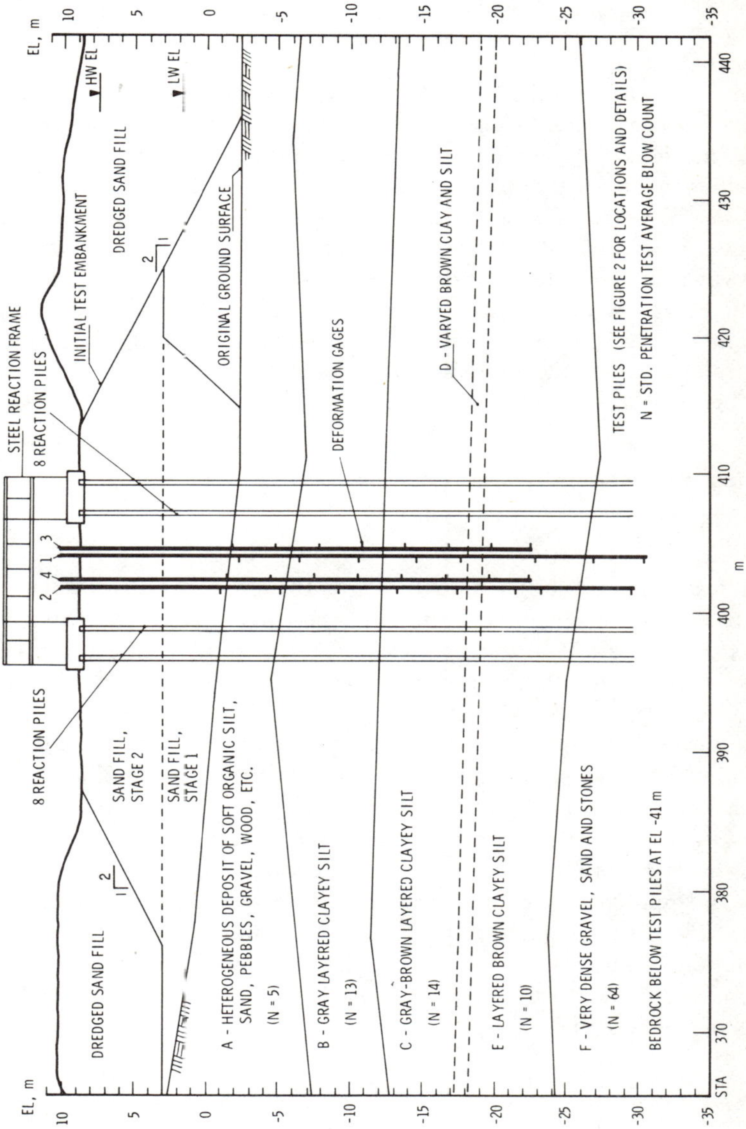

FIG. 1—*Section through test embankment showing test piles, dredged sand fill, and subsoil formations.*

brown layered clayey silt with properties similar to those of layer B. The water content varied from 29 to 37 percent; the average liquid limit was 34 percent. The plasticity index was 12 percent, the grain size distribution was 45 percent clay and 55 percent silt. The average wet density was 1840 kg/m^3. The maximum in-situ vane shear strengths varied from 0.6 to 0.9 kg/cm^2.

Layer D was a 1-m-thick varved brown clay and silt.

Layer E was a layered brown clayey silt that extended from elevation −19 to −26 m below the center of the test fill. The water content varied from 35 to 45 percent, the average liquid limit was 40 percent and plasticity index was 20 percent. The average grain size distribution was 64 percent clay and 36 percent silt. The wet density was 1840 kg/m^3. The maximum *in-situ* vane shear strength varied from 0.9 to 1.2 kg/cm^2.

Layer F, a very dense gravel with sand and stones, was encountered at elevation −26 m. It was not possible to obtain undisturbed samples from this formation. Borings showed that it extended to an elevation −41 m, where bedrock was encountered.

Field Instrumentation

Test Fill

The test fill was designed to measure the consolidation properties of the subsoils and to verify the stability of the embankment constructed with various side slopes. Before construction started, settlement platforms were placed on the river bed to provide a complete settlement record from the start of construction. After stage one was completed, settlement gages and vibrating-wire piezometers were installed to measure the settlements and excess pore water pressures when the test fill was completed. These observations were also required when the test piles were installed 5 to 13 m (16 to 43 ft) from this instrumentation and test loaded.

The instrumentation of the test fill was completed with reference benchmarks and reference Geonor hydraulic piezometers installed outside of the construction area. Gloetzl earth pressure cells were installed to check the distribution of vertical and horizontal stresses, and slope indicator casings were installed to measure horizontal soil movements under and beyond the toe of the test fill.

Test Piles

The locations of the test piles are shown on the plan on Fig. 2. The two end-bearing piles (nos. 1 and 2) and the two friction piles (nos. 3 and 4) were driven along two lines 3.048 m apart. Piles 1 and 4 were HP 304.8 mm 132 kg/m (12 in. by 89 lb/ft); piles 2 and 3 were closed-end steel pipes

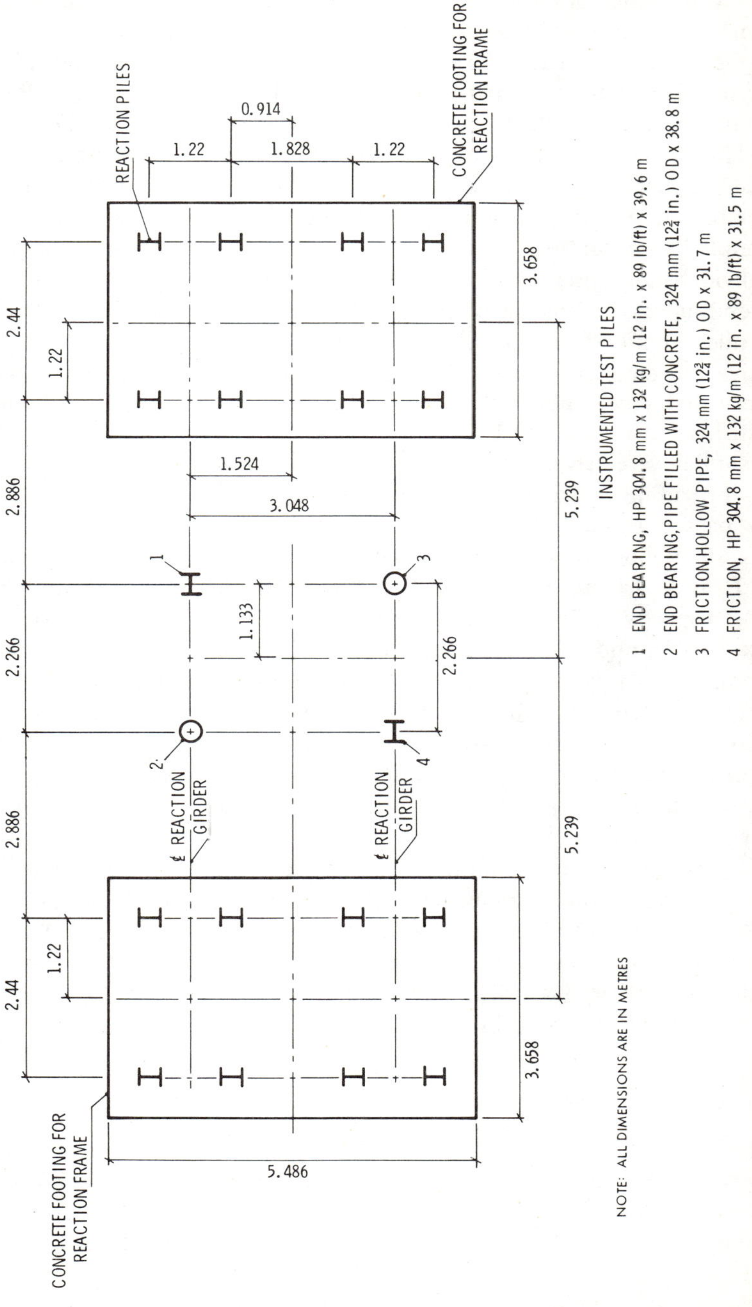

FIG. 2—*Plan of test installation.*

324 mm outside diameter by 7.92 mm wall (12¾ in. outside diameter by 0.312 in. wall). The center-to-center spacing between the piles was seven pile diameters (2.266 m) along each line.

Each pile was instrumented with eight deformation gages *[1]*[4] or telltales. They consisted of two concentric steel pipes with the outer casing welded continuously to the outside of the pipe or to the inner web of the steel H-piles. The inner pipe formed the stress-free reference rod for the measurements. The upper 11 m of each pile was coated with asphalt to prevent the development of skin friction in the granular fill. Therefore the first gage extended to a depth of 11 m. The remaining uncoated length of pile was subdivided into equal lengths for the remaining seven gages, with one gage extending to the bottom of the pile. Steel saddles were mounted on the test piles about 25 mm (1 in.) above each deformation gage so that the relative movements between the pipes could be measured with a micrometer or dial gage. These measurements were made to within 0.0025 mm (0.0001 in.).

The test piles were driven with a diesel pile driver using a 1360-kg (3000-lb) hammer, providing a driving energy of 4011 kg-m (29 000 ft-lb). When the piles passed through the granular fill, the driving resistance for all piles was about 66 blows/metre (20 blows/ft). Piles 3 and 4 terminated in layer E at a driving resistance of 187 and 102 blows/m (57 and 31 blows/ft), respectively. Piles 1 and 2 terminated in layer F at a driving resistance of 476 and 836 blows/m (145 and 255 blows/ft), respectively.

When piles are driven into the soils in the Fredericton area, the soil is usually squeezed out as a fluid along the pile shafts. This is due to the collapse of the soil structure and because the liquid limit of the soil is equal to or greater than the natural water content. Under these conditions, it is improbable that substantial residual stresses could exist in the piles after they were driven.

The piles were installed during the period 14 to 21 July; on 7 September, test pile no. 2 was filled with concrete to strengthen it for load testing. This concrete had a 28-day strength of 359 kg/cm^2 (5100 psi) and a 171-mm (6¾-in.) slump, which varied significantly from that specified. At 41 days, its modulus of elasticity was measured to be 2.74×10^5 kg/cm^2 (3.9×10^6 psi). This pile was test loaded in compression on 21 October.

Load Testing Equipment

In order to load the piles to failure in axial compression and pullout, a twin girder reaction frame was designed, constructed, and positioned over the piles as shown schematically in Fig. 1. Each steel girder, 10.46 m long by 0.61 m wide and 2.51 m deep, was positioned directly over each

[4]The italic numbers in brackets refer to the list of references appended to this paper.

row of test piles as shown in Fig. 2. They were supported at the ends by steel towers anchored to heavy reinforced concrete footings, which were in turn anchored to eight HP 304.8 mm by 79 kg/m (12 in. by 53 lb/ft) steel reaction piles driven to end bearing in the dense gravel formation. The girders were cross braced for added stiffness. The closest center-to-center distance between the reaction and test piles was nine pile diameters.

Each test pile was fitted with a 76.2-mm (3-in.) thick steel bearing plate to provide a rigid, full-bearing surface for a 500-ton calibrated load cell and a 500-ton hydraulic jack. Steel reaction plates 50.8 mm (2 in.) thick were also welded to the top and bottom flanges of each girder directly above the test piles to distribute the reaction loads exerted during the test.

To carry out the axial compression tests, the load cell was first carefully centered over the test pile. A 25.4-mm-thick flat steel plate was then placed over it and under the hydraulic jack, which was carefully centered over the load cell. A second 76.2-mm-thick flat steel plate was clamped above the jack to the underside of the reaction girder. This provided a sturdy reaction plate for the jack to push against (Fig. 3).

To monitor the deformation response of the test pile to the applied loads, the micrometer saddles were replaced by dial gage mounts, which supported the dial gages and electric linear variable displacement transducers (LVDT). This double measuring system produced a continuous automatic recording of all the deformation gage movements as well as mechanical measurements while the pile was loaded. Two additional dial gages were fastened to each side of the pile head; each reacted against a steel beam reference bridge. They provided an independent measure of the vertical movement of the pile head relative to the surrounding ground and showed any tilting of the pile head during loading.

To perform the pullout tests, the load cell and the hydraulic jack were carefully positioned on top of the reaction girder directly over the test pile following the same procedure as for the compression tests. An upper pullout yoke was positioned over the jack and a bottom yoke securely bolted to the head of the test pile. The two were connected together with four 44.5-mm outside diameter steel rods, which thus allowed the jack load to pull upon the pile.

Effect of Changes in Pore Water Pressures on Negative Skin Friction

Field observations on all the test piles showed that as excess pore pressures dissipated and the fill settled, they compressed axially due to negative skin friction, but when the pore pressures increased, the piles rebounded. This behavior is illustrated by the observations for test pile no. 4 (Fig. 4).

The axial compressions on test pile no. 4 started immediately after it was installed in July and increased with time, reflecting the buildup of negative skin friction load as the pore pressures dissipated and the fill

FIG. 3—*Setup for applying axial compression loads showing load cell, 500-ton-capacity hydraulic jack, and measuring gages.*

settled. In late August, a 7460-kW (10 000 hp) dredge started pumping 1.15×10^6 m^3 of saturated clean sand and gravel around the test fill (Fig. 1) for the redevelopment of the south shore of the Saint John River. The *in-situ* pore water pressures increased considerably and almost reached their previous maximum values measured when the test fill was constructed. During this buildup of excess pore pressure, the pile compressions reduced markedly, falling to 25 percent of their previous maximum value when the excess pore pressures were at their highest on 7 October. After this date, the excess pore water pressures started dissipating and the axial compressions in the pile increased again.

From 19 to 28 October, when the excess pore pressures were still very

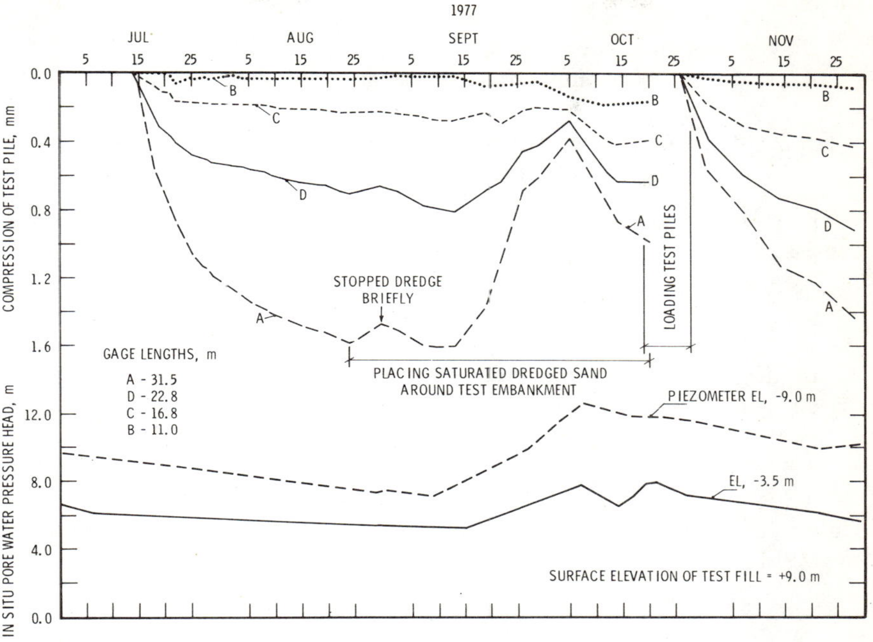

FIG. 4—*Variations in compression of pile no. 4 caused by changes in negative skin friction load due to changes in pore water pressures.*

high, the test piles were subjected to axial compression and pullout loads. At the completion of these tests, the piles continued to compress axially as the negative skin friction loads developed with the dissipation of the excess pore water pressures.

These observations along with the work of many recent investigators [*2-7*] show that negative skin friction is affected by changes in pore water pressures and consequently is related to the effective stresses in the soil around the pile.

Skin Friction Load on a Single Pile

Skin friction on the pile is directly proportional to the effective normal pressure of the soil [*8*]. Consider a pile driven into a compressible soil and neither it nor the soil is loaded; that is, there are no relative movements between the pile and the soil. Then the potential skin friction that could be mobilized would be given by

$$P_0 = \sigma_h' \tan \delta' \tag{1}$$

where

P_0 = potential skin friction,
σ_h' = total horizontal effective normal soil pressure acting on the pile, and
$\tan \delta'$ = coefficient of friction between the pile and the soil, where δ' is the angle of friction between the soil and the pile.

The total horizontal effective normal soil pressure acting on the pile will consist of

$$\sigma_h' = \sigma_{h1}' + \Delta\sigma_{h2}' + \Delta\sigma_{h3}' \tag{2}$$

σ_{h1}' is the in-situ horizontal effective stress in the soil before the pile is driven. It is related to the vertical effective stress by

$$\sigma_{h1}' = K_0 \, \sigma_v' \tag{3}$$

Unless K_0 is measured, it can be approximated for normally consolidated soils [*9*] from

$$K_0 = 1 - \sin \phi' \tag{4}$$

where ϕ' = coefficient of internal friction of the soil. $\Delta\sigma_{h2}'$ is due to any surcharge placed on the surface of the ground around the pile. For large uniformly loaded areas, it can be calculated from $K_0 \, \Delta\sigma_v'$, where $\Delta\sigma_v'$ is the applied effective vertical uniform surcharge pressure. For surcharges such as embankments, it can be calculated using normal engineering methods. $\Delta\sigma_h{'}_3$ is developed when the pile is driven into the ground. For an open-ended pipe that is cleaned as it is driven and for H-piles, the soil displaced will be negligible and this term can be neglected. For all other driven piles, the soil displaced will equal the volume of the pile. The fact that pore pressures are generated when the pile is being driven shows that horizontal stresses develop during driving.

Published literature tends to show that this increase in horizontal stress dissipates quickly with time for a single pile. Chandler [*4*] stated that the lateral stresses around bored piles in London clay gradually returned to the *in-situ* conditions. Yang [*10*] reported that when steel piles were driven in sands and organic silts and redriven about 5 days later, the blow count was often reduced by as much as 50 percent due to soil relaxation around the pile. Furthermore, Baguelin and Jezequel [*11*] measured the increase in horizontal stresses on an instrumented 1 m square closed-end large displacement steel test pile driven 4 m into soft silty clay. They reported that after 1 month, the horizontal effective stresses had reduced to the original in-situ values. Consequently, for long-term conditions, $\Delta\sigma_{h3}'$ can be neglected for single piles driven in weak compressible silty clays. It should, however, be considered for large groups of closely spaced piles.

The coefficient of friction between the pile and the soil can be measured in a laboratory test or estimated from

$$\tan \delta' = M \tan \phi' \tag{5}$$

where ϕ' is obtained from undrained triaxial strength tests with pore pressure measurements, and the ratio M [*2,12*] has the following range for steel and concrete piles

$$0.5 \leq M \leq 0.8$$

Assuming a circular pile of circumference (or perimeter) C installed to a depth L in a uniform soil deposit with the groundwater table at the surface, the axial load due to skin friction mobilized along the shaft will be given by

$$P_P = C \int_L^0 K_0 \gamma_s' \tan \delta' \, z \, dz \tag{6}$$

where γ_s' is the submerged unit weight of the soil. Let

$$\beta = K_0 \gamma_s' \tan \delta' \tag{7}$$

where β is the soil-pile friction factor with units of kilograms per metre cubed.

Introducing β into Eq 6 and simplifying,

$$P_P = \tfrac{1}{2} C \beta L^2 \tag{8}$$

To calculate the skin friction resistance from any depth D to L, Eq 8 becomes

$$P_P = \tfrac{1}{2} C \beta (L^2 - D^2) \tag{9}$$

Similarly, negative skin friction is given by

$$P_N = \tfrac{1}{2} C \beta (L^2 - D^2) \tag{10}$$

When the pile is loaded rapidly to failure in axial compression or tension, the skin friction resistance mobilized in the compression test is usually much greater than in the tension test [*13*]. In the rapid compression test, the vertical stress in the soil around the pile and the horizontal normal contact pressure increases. The pile also expands under the applied axial load, increasing the contact pressure and the skin friction resistance along the shaft of the pile, thus raising the axial compression load at failure. In the tension test, the reverse takes place; the horizontal pressure between the shaft of the pile and the soil decreases and the pullout load is reduced.

The incremental increase or decrease in axial load due to skin friction mobilized along the shaft of the pile at failure was found to be related to Poisson's ratio ν of the pile material. Hence, for axial compression loads due to positive skin friction, Eq 9 was modified to

$$P_P = \tfrac{1}{2}(1 + \nu)C\beta(L^2 - D^2) \tag{11}$$

Similarly, the skin friction for axial tensile loads applied to the pile is given by

$$P_T = \tfrac{1}{2}(1 - \nu)C\beta(L^2 - D^2) \tag{12}$$

These equations can be used to estimate skin friction loads on piles driven into layered soil formations.

Pile Load Tests

Test Procedure

The test piles were subjected to a constant rate of loading, applied in increments of 10 tons every 10 min. During the test, each load was maintained for 9 min, allowing 1 min to increase the load to the next level. Readings were made on all the deformation gages (dials, LVDTs) and settlement dial gages on the pile head at 0, 3, 6, and 9 min. It was therefore possible to observe the load transferred to the soil by positive skin friction as the applied loads increased until the pile failed. The sequence of pile testing was 3, 4, 2, and 1; that is, the friction piles were loaded before the end-bearing piles. Thus, the loading equipment was checked out before the higher-capacity end-bearing piles were tested.

The compression loads were removed from the friction and end-bearing piles in five and ten increments, respectively, whereas the tension loads were removed in about half the number. Engineering level surveys were performed during the loading tests to obtain a secondary set of pile head movements to check those made with the dial gage and steel beam reference bridge.

Axial Compression Tests

Before the piles were tested, they were compressed by negative skin friction loads. The measured axial deformations (settlement) due to the "prestress" from negative skin friction was used as the starting point of the test curves for the four piles in Fig. 5. Upon loading, the initial pile settlements reflected this "prestress" until it was exceeded by the applied axial load, after which the piles performed normally until failure occurred.

The friction piles nos. 3 and 4 performed about the same. Pile head settlements were small until the ultimate load was reached, when the piles "plunged" into the ground. The bearing capacity of the pipe pile no. 3 was 159 tonnes (175 tons) and that of the H-pile, no. 4, 141 tonnes (155 tons). The difference was attributed to the greater end-bearing load developed by the pipe pile.

The end-bearing piles had similar load settlement curves, except that the H-pile (no. 1), settled more than the concrete-filled pipe pile (no. 2) during the test for the same loads. Using Davisson's failure criteria [*14*], the ultimate bearing capacity of the pipe pile was 372 tonnes (410 tons), which was greater than the 350 tonnes (385 tons) obtained for the deeper (by 1 m) H-pile.

Not only did the H-pile heads settle more than the pipe pile heads but, from the elastic compression measurements, the H-pile tips also settled more as will be shown later. The same observations applied to the friction piles.

During unloading all the piles easily supported the initial estimated shaft friction load overnight (Fig. 5). With the exception of pile 4, which settled slightly, the other three piles showed a slight rebound. When the loads were reduced to about 200 tonnes (220 tons) on piles 1 and 2, the rebound after 1 day was even greater. These observations show that the test piles easily supported axial compression loads smaller than the estimated maximum skin friction load.

When the piles were completely unloaded, they rebounded to within 0.3 mm (0.01 in.) of the lengths they had before the loads were applied. This indicates that the only significant residual stresses at the beginning and end of the tests were due to negative skin friction.

Axial Tension Loads

The pullout load-deformation behavior of the test piles is shown in Fig. 6. For friction piles 3 and 4, the curves are similar to the axial compression curves (Fig. 5), except that the failure loads were considerably less. The curves for bearing piles 1 and 2 resemble those obtained for the friction piles. It was observed from the deformation gage measurements that as soon as the pile tips started to move upward relative to the soil, the pile started slipping out of the ground at constant load. The ultimate pullout failure loads in order of magnitude were 59, 68, 127, and 150 tonnes (66, 76, 140, and 165 tons) for test piles 4, 3, 1, and 2 respectively. Note that

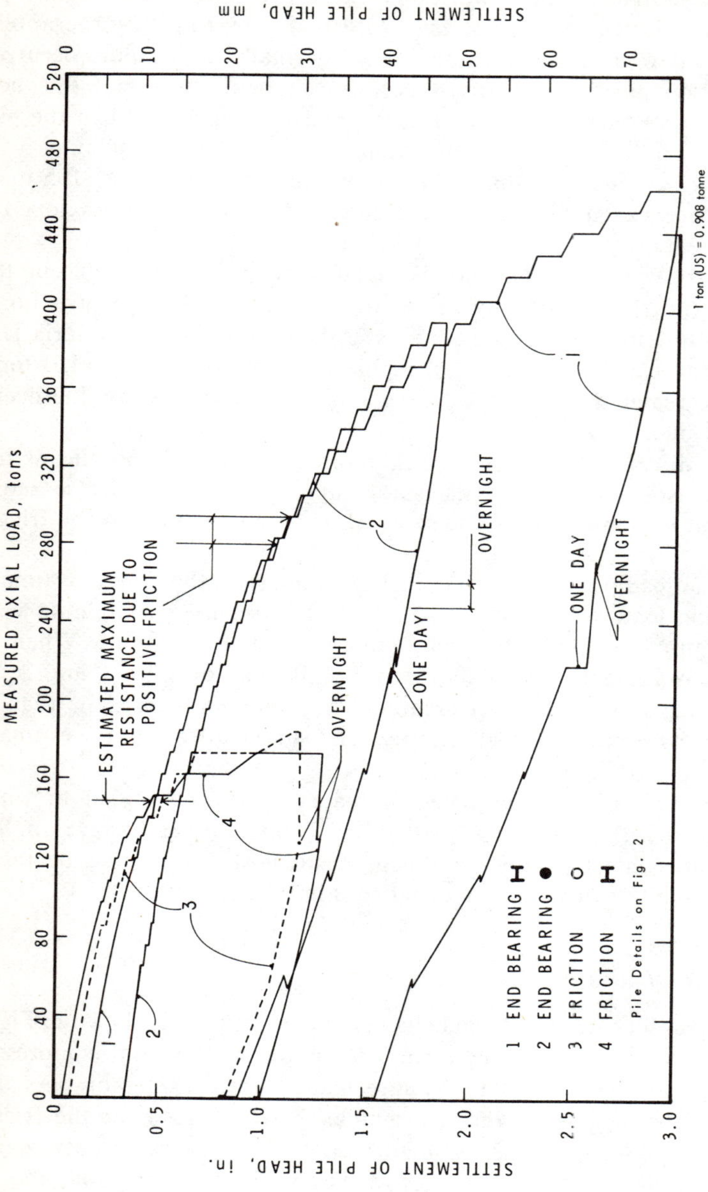

FIG. 5—*Axial compression pile load tests, Fredericton Bridge.*

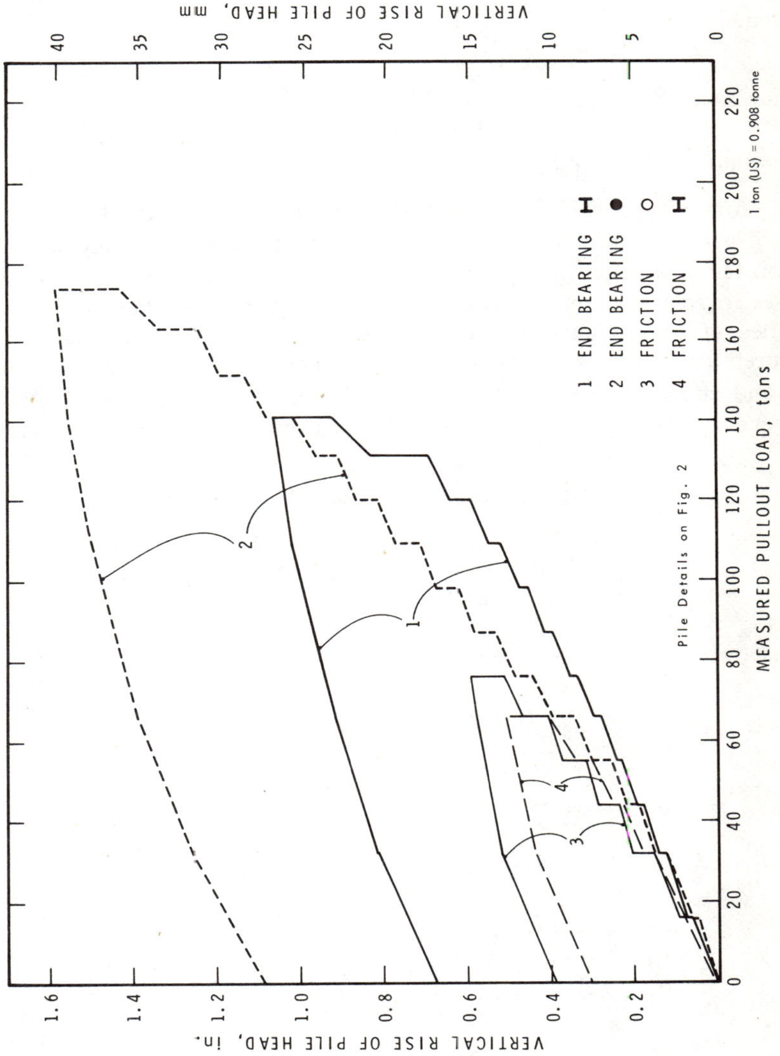

FIG. 6—*Axial tension pile load tests, Fredericton Bridge.*

the pullout loads for the H piles were smaller than for the corresponding pipe piles.

Because the piles were prestressed with negative skin friction loads, the elongation of the piles would be somewhat greater in the pullout tests. However, the magnitude of the failure loads given above would not be affected.

Friction Load Estimates for Test Piles

Determination of Soil Parameters

To estimate the skin friction loads that can be mobilized along the shaft of a pile subject to axial tension or compression loads, the soil-pile friction factor β given in Eq 7 must be evaluated. The horizontal effective stresses are related to the vertical effective stress by Eq 3. From the measured densities of the various soil formations and the *in-situ* pore pressures measured at the time the piles were test loaded, the distribution of vertical total and effective stresses shown in Fig. 7 were obtained. The coefficient of internal friction of the soil ϕ' was measured at five elevations within

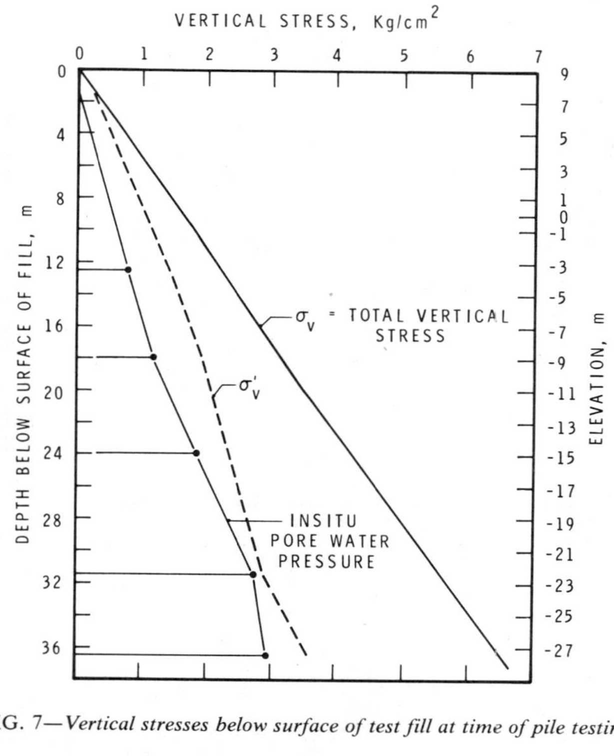

FIG. 7—*Vertical stresses below surface of test fill at time of pile testing.*

soil formations B, C, and E (Fig. 1) by performing isotropically consolidated undrained triaxial shear strength tests with pore pressure measurements. The range of ϕ' was 30 to 32 deg, with an average of 31 deg. Using Eq 4, $K_0 = 0.48$ was obtained, which was assumed to apply to the whole soil profile except for layer F.

A soil-pile friction test apparatus was designed and constructed to evaluate the coefficient of friction tan δ' between the steel pile and the surrounding soil. The design was based on the fact that a driven pile disturbs the soil and that the stress transfer between the pile and the soil must be through the disturbed soil. Consequently, the test apparatus measures the coefficient of friction between remolded and reconsolidated soil and a steel cylindrical model pile in terms of effective stress.

A schematic diagram of the apparatus is shown in Fig. 8. It consists of a heavy bronze cubical chamber with outside dimensions of 152 mm on each side. The three interior vertical walls and the front cover are fitted with expandable rubber diaphragms, which are used to consolidate the soil against the model pile. Drainage occurs through the top and bottom of the cell. The steel cylinder or model pile (51 mm outside diameter by 102 mm long) is fitted with three pore pressure sensors. Relative movements between the pile and the soil are obtained by applying a torque through the torque transducer mounted on the shaft connected to the top of the cylinder.

In operation, the cell containing the pile is filled with soil remolded at a water content exceeding its liquid limit. A consolidation pressure σ_h is applied to the soil by inflating the diaphragms with water under pressure. When consolidation is complete, the steel cylinder is rotated through the torque transducer at the rate of 2 deg per hour. The torque and the applied consolidation pressure corrected for the excess pore pressure define an effective stress soil-pile friction test curve shown in Fig. 9. At the end of the test, the soil is consolidated to a higher pressure, the direction of rotation reversed, and another test curve obtained. This procedure is repeated for four or five consolidation pressures until the friction envelope is defined (Fig. 9).

The soil-pile friction angle for the clayey silt at the test site was measured to be $\delta' = 24$ deg, and tan $\delta' = 0.445$. The ratio M for this silty clay on a normal steel surface is

$$M = \frac{\tan \delta'}{\tan \phi'} = 0.74$$

Axial Tension and Compression Loads

With this information and assigning a Poisson's ratio $\nu = 0.29$ for the steel piles, the skin friction resistance to pullout was determined from

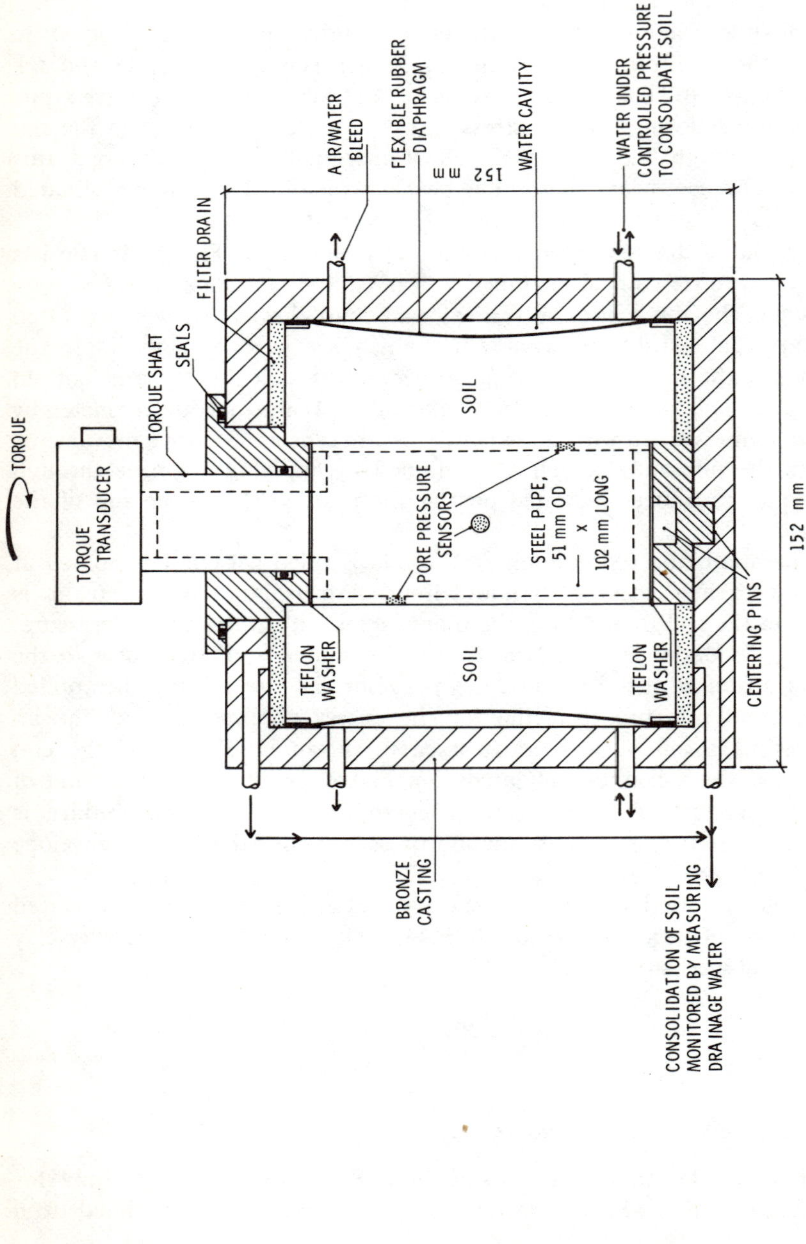

FIG. 8—*Soil-pile friction apparatus.*

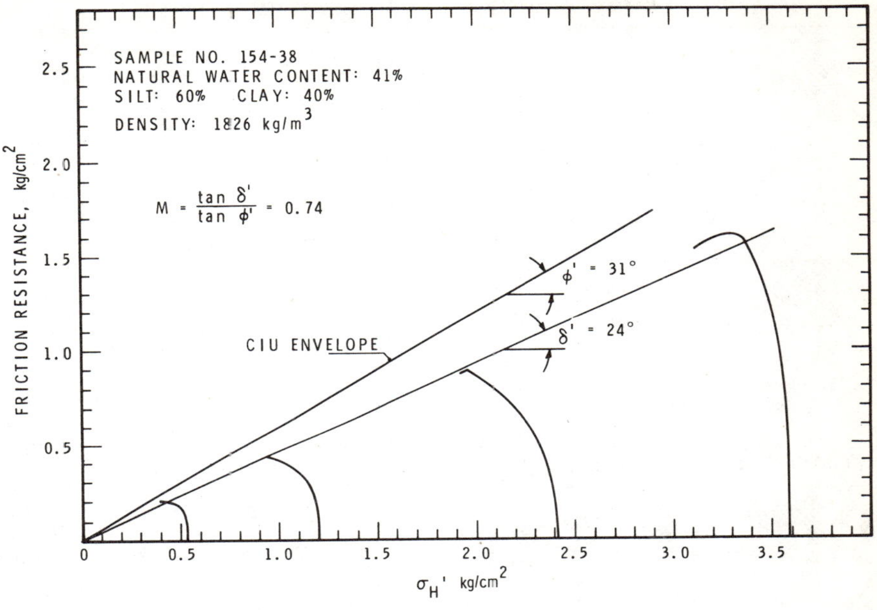

FIG. 9—*Soil-pile friction tests.*

Eq 12. The contact perimeter C for the pipe piles was corrected for the additional area of the deformation gages welded to the outside of the pile. For the H-piles, C was assumed to be a square section equal to the flange width. The asphalt coating in the upper part of the piles was assumed to be 50 percent effective in reducing friction. The weights of the piles were not considered.

The estimated pullout loads are compared with the measured values in Table 1. The comparisons are very good for the friction piles. It was not possible to predict the friction resistance for the end-bearing piles in the dense gravel, as no samples were tested from this layer. The 50 tons entered in the table for pile 2 was taken as the difference between the measured load and the value estimated for the clayey silt formation and allowing for the weight of the concrete filled pile. It was then possible to calculate the pullout resistance for pile no. 1, which overestimated the measured load by about 20 tons or 14 percent.

The axial compression loads estimated with Eq 11 are compared with the field test values in Table 2 and in Fig. 5. The point resistance for the friction piles was estimated from Meyerhof [*15*]. The comparisons are again very good for the friction piles.

The point resistance for the end-bearing piles was estimated from Standard Penetration Tests after Vesić [*23*]. The estimate of skin friction load

TABLE 1—*Comparison of estimated and measured axial pullout loads.*

Pile No.	Pile Type	Depth of Penetration, m	Estimated Side Friction, tons: Clayey Silt	Estimated Side Friction, tons: Dense Gravel	Estimated Side Friction, tons: Sum	Measured Load at Failure, tons
3	pipe	31.7	80	. . .	80	76
4	H	31.5	80	. . .	80	66
2	pipe	38.8	105	50	155	165
1	H	39.6	105	55	160	140

Conversion factor—1 ton = 0.908 metric tonne.

TABLE 2—*Comparison of estimated and measured axial compression loads.*

Pile No.	Pile Type	Depth of Penetration, m	Estimated load, tons: Side Friction: Clayey Silt	Estimated load, tons: Side Friction: Dense Gravel	Estimated load, tons: Point Resistance	Estimated load, tons: Total	Measured Load at Failure, tons
3	pipe	31.7	150	. . .	20	170	175
4	H	31.5	150	. . .	5	155	155
2	pipe	38.8	190	90	140[a]	420	410
1	H	39.6	195	100	30[a]	325	385
			195	100	85[b]	380	385

Conversion factor—1 ton = 0.908 metric tonne.

[a] Estimated from Vesić (1977) using Standard Penetration Test and actual area of tip of pile.

[b] Estimated from Vesić (1977) assuming effective end area equals 50 percent of box enclosing tip of H pile.

in the dense gravel, however, was based on the pullout resistance of 50 tons (pile 2, Table 1). The comparison between the estimated and the measured ultimate bearing capacity was again quite satisfactory.

If the estimated maximum friction resistance is realistic, then the tip of the pile loaded in axial compression should move down relative to the soil just enough to mobilize the full skin friction on the whole length of the pile. Selecting load test values that were close to the estimated maximum friction loads the corresponding vertical pile movements were plotted as curve 2 for each pile in Fig. 10. In each case, the tip movements were about equal to or less than 4 mm; the amount specified by Davisson [*14*] and Vesić [*7*] required to mobilize full skin friction. At the smaller loads (curve 1), the pile tips showed very little movement. At loads close to the ultimate failure load (curve 3), the piles moved bodily downward but remained essentially parallel to curve 2, indicating that no additional friction resistance was mobilized. The pile movements due to negative skin friction loads up to the time of the field tests are indicated on the figure as curve N.

Downdrag Loads

Since the negative skin friction loads were known from the pile compressions measured prior to the field tests, it was possible to compare the

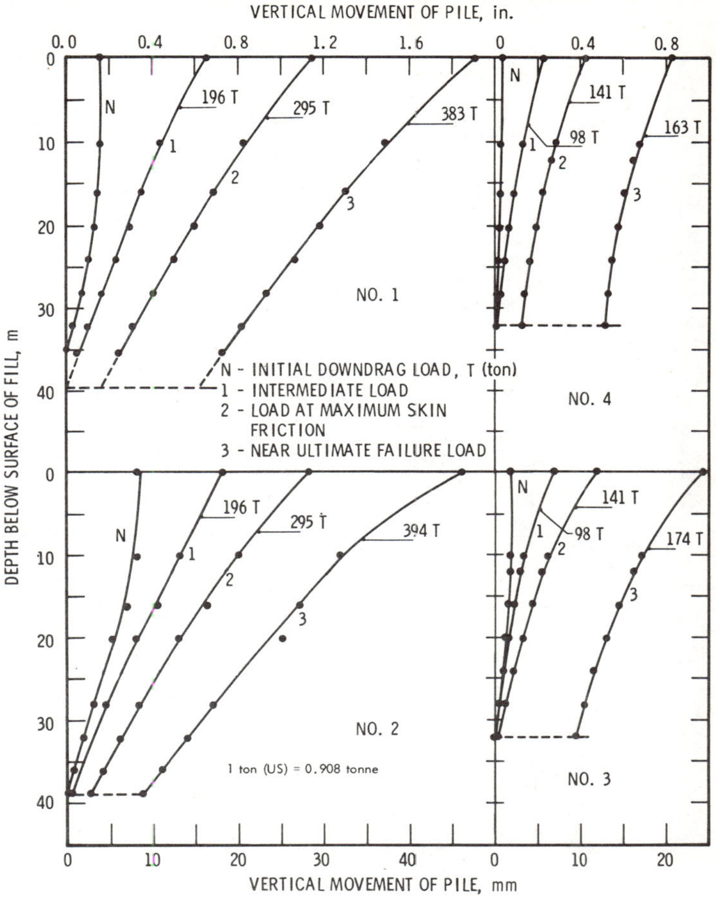

FIG. 10—*Effect of axial compression loads on pile movements.*

distribution of these loads along the piles with those estimated with Eqs 9 and 10 for the effective stress conditions that existed in the soil at that time. The downdrag loads in the upper part of the friction piles 3 and 4 would be resisted by positive skin friction developed at the lower end. The downdrag loads on the end-bearing piles 1 and 2 would develop to layer C (Fig. 1), as no settlements were measured in this layer. It was assumed that these piles (1 and 2) would be unloaded by positive skin friction to the bottom of layer E (Fig. 1). The results are compared in Fig. 11.

The field measurements (curves A) showed that the friction piles had from 60 to 70 tons of downdrag load, whereas the end-bearing piles had from 70 to 100 tons. It is expected that these values will increase when all the excess pore pressures have dissipated.

The distribution of downdrag loads was estimated assuming the asphalt

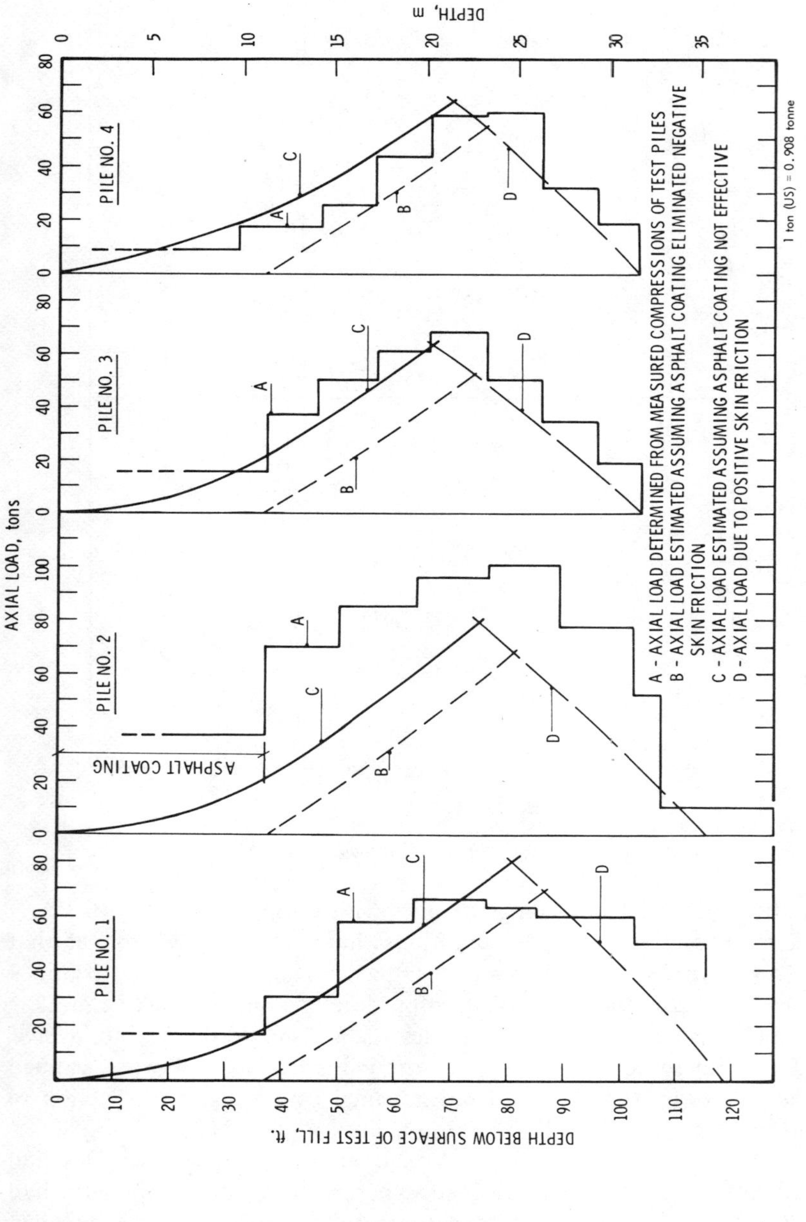

FIG. 11—Comparison of estimated and observed negative skin friction load distributions in test piles.

coating eliminated skin friction (curves B) and assuming the coating did not exist, that is, was not effective (curves C). The unloading due to positive skin friction is given by curves D. End bearing was not considered, but it appears that some end bearing was mobilized. The figure shows a reasonably good correlation between the field and estimated load distribution for all piles except no. 2. This pile was filled with poor-quality, very wet concrete and it is certain that the field loads were overestimated from the measured pile deformations. The figure also shows that the asphalt coating did not eliminate negative skin friction.

Case Records

The results of the pile test program on the Fredericton test piles indicate that pullout tests may be used to estimate the skin friction loads for axial compression. When different piles are used, the calculations would be based on Eqs 11 and 12. If, however, both tests are performed on the same pile as was done in Fredericton, the soil-pile friction factor β does not change. Consequently the axial compression and axial tension friction loads on the pile are related through the Poisson's ratio of the pile material

$$P_P = \frac{(1 + \nu)P_T}{(1 - \nu)} \tag{13}$$

From various handbooks [*16–18*], Poisson's ratio for various pile materials are

Cold rolled steel	$\nu = 0.29$
High strength concrete	$\nu = 0.24$
Timber loaded parallel to grain	$\nu = 0.37$

Friels [*19*] conducted a number of these tests in Tennessee on various lengths of HP 10 × 42 piles driven into alluvial silt and sandy silt. The measured pullout and axial compression failure loads are summarized in Table 3. The estimated positive skin friction loads for three of the H-piles (nos. 46, 48, and 49) is about the same as the axial compression failure load, indicating no mobilization of point resistance or that the piles were not loaded to ultimate failure. The estimated value for pile 47 exceeds the load test value by about 16 percent.

D'Appolonia [*20*] conducted push and pull tests on concrete-filled pipe, of 12¾ in. outside diameter, and 14BP89 steel piles instrumented with telltale gages at two different sites along the southern shore of Lake Michigan. The test results for four of the piles are given in Table 3.

At site A, two pipe piles were driven through fill and stiff silty clay to

TABLE 3—*Comparing skin friction compression loads determined from pullout tests with measured compression failure loads.*

Pile No.	Pile Details: Description	Pile Details: Penetration Depth, ft	Soil Profile from Ground Surface	Measured Ultimate Pullout Load, tons	Estimated Load, tons: Positive Friction	Estimated Load, tons: Point Resistance	Measured Failure Load in Axial Compression, tons
			Friels [19]				
46	steel, HP 10 × 42	39.5	29 ft silt	23	40	...	38
47	steel, HP 10 × 42	55.2	9 ft silty sand	68	120	...	103
48	steel, HP 10 × 42	45.5	5 ft sandy clayey silt	52	95	...	90
49	steel, HP 10 × 42	27	8 ft silty sand	18	30	...	27
			D'Appolonia [20]				
A-1	12¾ in. OD pipe and concrete	94	32 ft slag and sand fill; 53 ft stiff silty clay	149	270	...	360
A-3	12¾ in. OD pipe and concrete	80–93	"hardpan"	140	255	...	350
B-4	steel 14BP89	111.9–128.9	35 ft sand; 7 ft lake clay	160	290	...	310
B-5	steel 14BP89	44.6	65 ft lake till; glacial till	175	320	...	300
			Fellenius and Samson [13]				
9	precast concrete H800, 30 cm OD	55	marine clay	51.4	70	10	90
			Mansur and Hunter [21]				
2	16-in. pipe	52.8	Relatively dense; medium to	100	180	...	195
4	16-in. concrete	40.2	fine sands, with some	85	140	...	170
7	steel 14BP73	52.1	layers of silt and clay	70	130	...	190
8	timber (class A)	38.6		30	65	...	80

Conversion factors—
1 ft = 0.305 m, and
1 ton = 0.908 metric tonne.

about the same depth into "hardpan." Pile A-1 was driven as a normal pile, but pile A-3 was cased to 24.4 m (80 ft). The difference between the estimated positive friction loads and the measured compression load at failure gave about the same end-bearing resistance for both piles showing consistency in the field tests and estimates. Furthermore, the estimated 270-ton friction load for A-1 coincided with the sharp break in the field loading curve (Fig. 12).

At site B, the two steel H piles were driven through sand, lake clay, lake till, and glacial till. Pile B-5 terminated in the lake till, and the estimated friction load exceeded the ultimate failure load by 7 percent. Pile B-4 was driven through a casing extending to the glacial till. The estimated friction load was less than the compression failure load by 6 percent.

Fellenius and Samson [*13*] performed a pullout load test on an 18.6-m (61-ft) long precast concrete H 800 pile driven in sensitive marine clay in Quebec. An axial compression test was performed on another similar pile (no. 9, Table 3), 180 m away, driven to a depth of 16.8 m (55 ft). The estimated combined friction and end-bearing load of 80 tons was 11 percent below the measured test value but coincided with the minimum postpeak value of 80 tons shown in Fig. 12.

Mansur and Hunter [*21*] performed numerous pullout and axial compression pile load tests on concrete, steel, and timber piles driven into the alluvial sediments along the Arkansas River below Pine Bluff. The alluvial sediments were composed of relatively dense medium to fine sands with seams or layers of silt and clay.

The four test piles reported in Table 3 were subjected to several cycles of pullout and axial compression loads. The pullout load used to estimate the positive friction load was determined from the last load cycle of each test and consequently differ from the values reported by the authors. The estimated positive friction loads for the 40.6-cm (16-in.) pipe (no. 2), 40.6-cm (16-in.) concrete (no. 4), 14BP73 steel (no. 7), and timber (no. 8) piles were, respectively, 180, 140, 130, and 65 tons, which were less than the measured failure loads reported in Table 3. These friction loads generally coincided with the start of the rapid change in slope of the load settlement curves shown on Figs. 12C–F, indicating that the estimates reasonably defined the actual skin friction loads mobilized on the steel, concrete, and timber piles.

Considering the above cases, it appears that the skin friction resistance that is mobilized to carry axial compression loads can be estimated from the axial pullout failure load and the Poisson's ratio of the pile material. Vesić [*22*], however, reports results where pullout and compression skin friction loads are about the same. Consequently, the proposed relation requires further study.

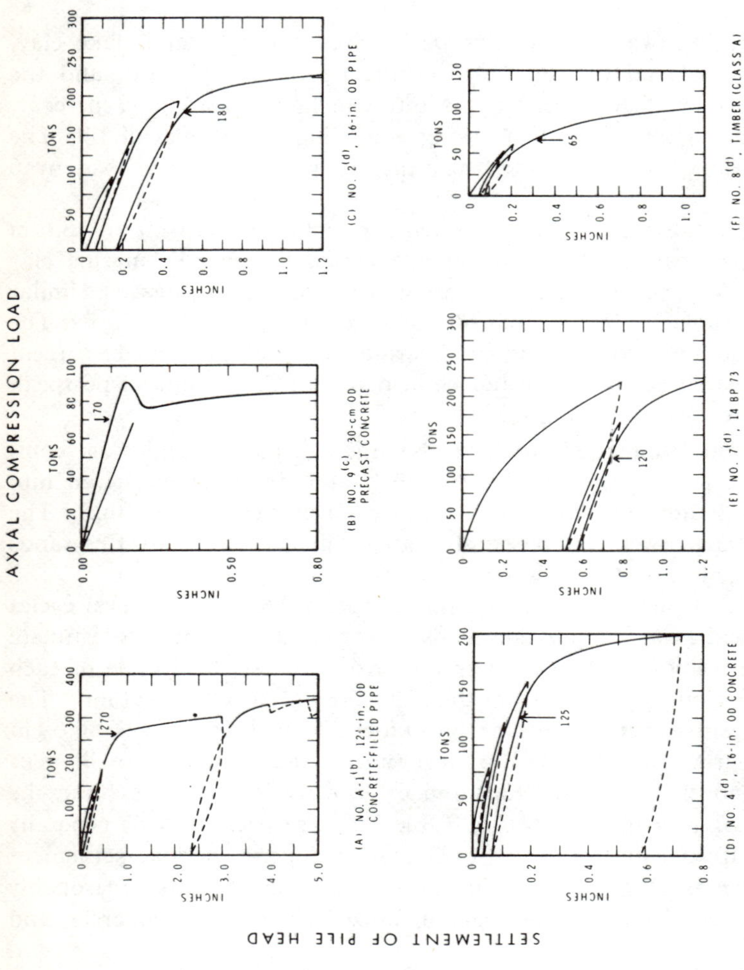

1 TON (US) = 0.908 tonne
1 INCH = 25.4 mm

FIG. 12—*Estimated skin friction loads compared with measured load-settlement behavior of piles;* (b) *D'Appolonia* [20], (c) *Fellenius and Samson* [13], *and* (d) *Mansur and Hunter* [21].

Summary

Two friction and two end-bearing pipe and steel H-piles driven through a granular test fill constructed on compressible clayey silts were field loaded to failure in axial compression and tension three months after they were installed. During this time, the instrumented piles compressed axially due to negative skin friction, then rebounded as the excess *in-situ* pore water pressure increased when additional fill was placed around the test embankment. This phenomenon was attributed to changes in effective stresses around the pile.

Relating skin friction to the horizontal effective stresses acting on the pile, equations were developed to estimate negative skin friction and skin friction due to axial compression and tension loads on single piles. The friction loads that can be mobilized under rapid loading conditions in compression or tension are related to the Poisson's ratio of the pile material. A soil-pile friction apparatus was described that was used to measure the coefficient of friction $\tan \delta' = M \tan \phi'$.

Conclusions

1. Effective stress equations were developed to estimate the magnitude and distribution of negative skin friction and the skin friction mobilized on test piles loaded rapidly to failure in axial tension and compression.

2. The coefficient of friction $\tan \delta'$ between the steel pile and the silty soil measured on remolded soil with the soil-pile friction apparatus was 0.445.

3. The soil-pile friction factor β can be evaluated from the properties of the soils or from field pullout load tests on piles.

4. An analysis of the pile tests and other published data indicates that positive friction mobilized to resist axial compression loads may be related to the skin friction mobilized in the axial tension test through the Poisson's ratio of the pile material.

5. The pile tip movement required to mobilize the maximum skin friction to resist axial compression loads is about 4 mm.

6. The asphalt coating applied to the upper part of the test piles did not eliminate skin friction.

Acknowledgments

This project was financed by the New Brunswick Department of Transportation, but the study was a cooperative effort with the Division of Building Research of the National Research Council of Canada. The authors are indebted to W. J. Oudemans, who initiated the project, and to D. Smith, who designed the test facilities for loading the piles. The assistance of the technical staff from both departments is gratefully acknowledged. This paper is published with the approval of G. D. Reeleder,

Deputy Minister of the New Brunswick Department of Transportation, and of C. B. Crawford, Director of the Division of Building Research.

References

[1] Bozozuk, M. and Jarrett, P. M., "Instrumentation for Negative Skin Friction Studies on Long Piles in Marine Clay on the Autoroute du Québec," presented at the Workshop Meeting of the International Bridge, Tunnel and Turnpike Association, Montreal, Nov. 1966, (reprint available—NRC 10046).

[2] Bozozuk, M., *Canadian Geotechnical Journal,* Vol. 9, No. 2, 1972, pp. 127-136.

[3] Burland, J. B., *Ground Engineering,* Vol. 6, No. 3, 1973, pp. 30-42.

[4] Chandler, R. J. *Civil Engineering and Public Works Review,* Vol. 63, No. 738, 1968, pp. 48-51.

[5] Esrig, M. I., Kirby, R. C., and Bea, R. G., "Initial Development of a General Effective Stress Method for the Prediction of Axial Capacity for Driven Piles in Clay," presented at 9th OTC (Offshore Technology Conference) Houston, Texas, 2-5 May 1977.

[6] Parry, R. H. G. and Swain, C. W., *Ground Engineering,* Vol. 10, No. 3, 1977, pp. 24-26.

[7] Vesić, A. S., "Design of Pile Foundations," National Cooperative Highway Research Program Synthesis of Highway Practice No. 42, Transportation Research Board, National Research Council, Washington, D.C., 1977.

[8] Bakholdin, B. V. and Bol'shakov, N. M., "Investigation of the State of Stress of Clays During Pile Driving," Scientific-Research Institute of Bases and Underground Structures, translated from *Osnovaniya, Fundamentij i Mekhanika Gruntov,* No. 5, 1973, pp. 7-9.

[9] Jaky, J., *Proceedings,* Second International Conference on Soil Mechanics and Foundation Engineering, Vol. 1, 1948, pp. 103-107.

[10] Yang, N. C., *Journal of the Soil Mechanics and Foundations Division, Proceedings,* American Society of Civil Engineers, Vol. 96, No. SM2, 1970, pp. 395-409.

[11] Baguelin, F. and Jezequel, J. F., "Étude Expérimentale de Fondations Profondes Rigides Sollicitées Horizontalement," Laboratoire Central des Ponts et Chaussées, Département des Sols, Paris, Rapport No. 71-B-687, mai 1971.

[12] Potyondy, J. G., *Géotechnique,* Vol. 11, No. 4, 1961, pp. 339-353.

[13] Fellenius, B. H. and Samson, L., *Canadian Geotechnical Journal,* Vol. 13, No. 2, 1976, pp. 139-160.

[14] Davisson, M. T., "High Capacity Piles," *Proceedings,* American Society of Civil Engineers, Lecture Series, Innovations in Foundation Construction, Illinois Section, 1972, also in *Foundation Engineering,* 2nd ed. Peck, Hanson, and Thornburn, Eds., Wiley, New York, 1974.

[15] Meyerhof, G. G., *Géotechnique,* Vol. 3, No. 7, 1953, pp. 267-282.

[16] Baumeister, T., Avallone, E. A., and Baumeister III, T., Eds., *Standard Handbook for Mechanical Engineers,* 8th ed., McGraw-Hill, New York, 1978.

[17] Urquhart, L. C., Ed., *Civil Engineering Handbook,* 4th ed., McGraw-Hill, New York, 1959.

[18] *Wood Structures, A Design Guide and Commentary,* Compiled by Task Committee on Status-of-the-Art: Wood, Committee on Wood, ASCE Structural Division, American Society of Civil Engineers, New York, 1975.

[19] Friels, D. R., "Axial Load and Uplift Resistance of Steel H-Piles," Piletips, Piling and Foundation News, Associated Pile and Fitting Corporation, July-Aug. 1977, pp. 5-7.

[20] D'Appolonia, E., "Load Transfer-Bearing Capacity for Single Piles and Pile Clusters," Lecture presented to the Illinois Section, Soil Mechanics and Foundations Division, American Society of Civil Engineers, April 1968.

[21] Mansur, C. I. and Hunter, A. H., *Journal of the Soil Mechanics and Foundations Division, Proceedings,* American Society of Civil Engineers, Vol. 96, No. SM5, 1970, pp. 1545-1582.

[22] Vesić, A. S., *Journal of the Soil Mechanics and Foundations Division, Proceedings,* American Society of Civil Engineers, Vol. 96, No. SM2, 1970, pp. 561-584.

[23] Vesić, A. S., "Design of Pile Foundations," National Cooperative Highway Research Program Synthesis of Highway Practice No. 42, Transportation Research Board, National Research Council, Washington, D. C., 1977.

G. S. Brierley,[1] *D. E. Thompson,*[2] *and C. W. Eller*[1]

Interpreting End-Bearing Pile Load Test Results

REFERENCE: Brierley, G. S., Thompson, D. E., and Eller, C. W., **"Interpreting End-Bearing Pile Load Test Results,"** *Behavior of Deep Foundations, ASTM STP 670,* Raymond Lundgren, Ed., American Society for Testing and Materials, 1979, pp. 181–198.

ABSTRACT: Interpretation of pile tip settlement behavior obtained from rod telltales on fourteen full-scale test piles at five different sites indicates that the applied load can be divided into end-bearing and skin friction components. The end-bearing component is used to obtain stiffness information about the bearing stratum. An equation for estimating pile settlement based on information accumulated during this study is given, and a procedure is proposed to estimate settlement from downdrag forces. Information about bearing strata obtained from the load test results can be accumulated and used to assist with design of future end-bearing pile foundations.

KEY WORDS: foundations, pile foundations, end-bearing piles, field testing, pile driving, pile load testing, downdrag, pile settlement

The settlement behavior of end-bearing piles is frequently assumed equal to the settlement behavior of similarly driven test piles at the same site. Frequently, however, test piles are supported during the relatively short duration of the test by positive skin friction in the surrounding soil. With time, positive skin friction may reduce to zero or even reverse in direction to augment the applied structural loads.

To compensate for this potentially serious change in pile support, it is desirable to use a technique for conducting and interpreting pile load tests that reveals the mechanism of support of the test pile and provides information about the behavior of the assumed end-bearing stratum. A relatively simple method for accomplishing this objective utilizing a rod telltale to measure movement at the tip of the pile is described herein. The interpretive procedure allows test pile support to be divided into end-bearing and skin friction components and provides stiffness data about the end-bearing stratum that may be useful for other sites. Data used in developing this

[1] Manager and staff engineer, respectively, Haley and Aldrich of New York, Rochester, N.Y. 14604.

[2] Principal and senior vice president, Haley and Aldrich, Inc., Cambridge, Mass. 02142.

technique was obtained from 14 full-scale pile load tests at five different sites.

Conducting the Load Test

The procedure for conducting the load tests is generally in accordance with ASTM Test for Load-Settlement Relationship for Individual Vertical Piles Under Static Axial Load [(D 1143-69) presently Testing Piles Under Axial Compressive Load (D 1143-74)]. All of the devices for measuring deflection at the top of the pile, including three dial gages, piano wire and mirror, and an engineer's level, are employed as described by ASTM. The load is applied by jacking against an overhead reaction (usually concrete blocks) with a hydraulic jack, which is monitored by an accurately calibrated hydraulic gage. A swivel head and electrical load cell are also used to provide a better degree of axial load application and a more accurate load measurement.

In addition to the standard measuring devices, a rod telltale is used to measure settlement at the tip of the pile. The telltale consists of a 0.65 cm (0.25 in.) diameter solid steel rod inside a 1.25 cm (0.5 in.) diameter steel pipe that is used to isolate the rod from the pile. A 90-deg bend at the top brings the rod out the side of the pile, where a dial gage is used to measure its vertical movement. This rod telltale system worked very well for all of the case histories.

Loads are applied in increments of various percentages of the assumed design load of the pile to a level equal to twice the design load. Loads at the design load and at twice the design load are held constant for at least 24 h, with intermediate loads held constant for 4 h. The pile is unloaded in four equal increments, which are each held constant for 4 h. In some cases, the pile is reloaded to the capacity of the loading system or to failure.

Case Histories

Five of the fourteen case histories used to develop this procedure are described herein. These five case histories are representative and are specifically selected to illustrate the most interesting behavioral patterns, such as deflection of the bearing stratum, loss of positive skin friction and failure of a pile during testing.

Basic data for each of the five case histories is presented in Figs. 1 through 5. Shown on the left side of each figure are the soil profile as revealed by a nearby test boring, the driven position of the pile, and the driving record. Shown on the right side of each figure are the measured settlements at the top and tip of the piles. All of these piles are 35.5 cm (14 in.) outside diameter A36 steel pipes driven closed end to bearing and filled with concrete. Twenty-eight-day concrete strength was generally specified at 316

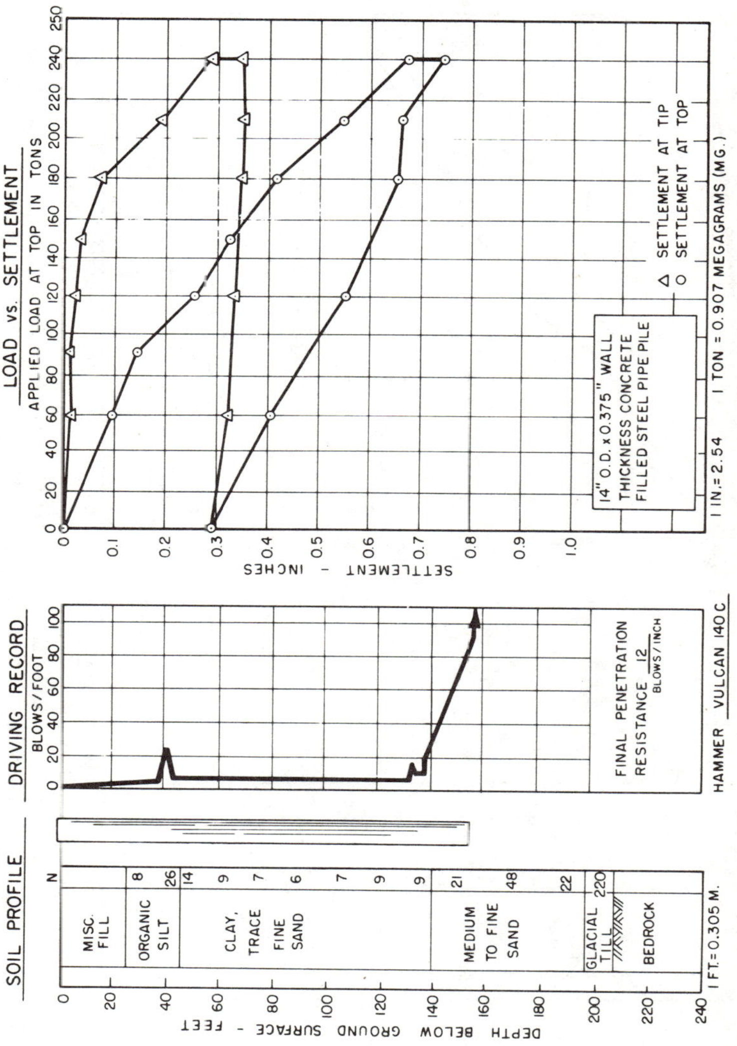

FIG. 1—*Case history I (Site A, test pile no. 1).*

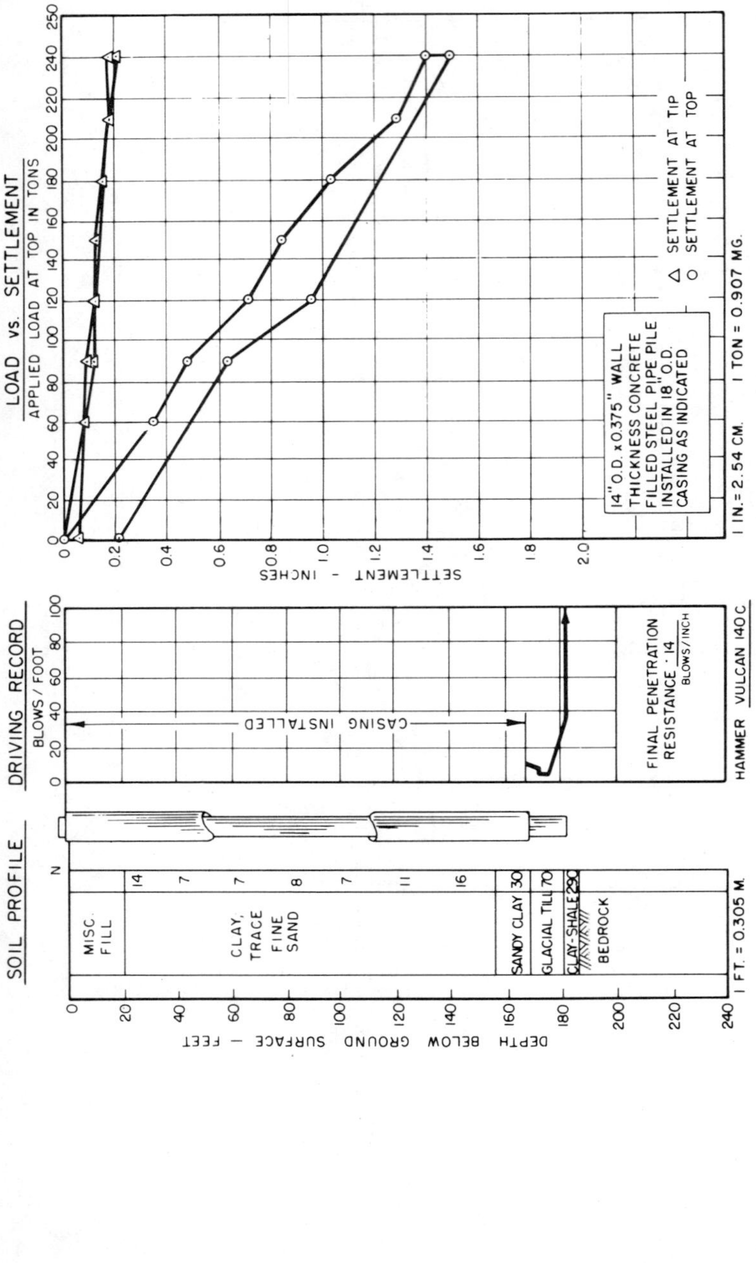

FIG. 2—*Case history II (Site A, test pile no. 2).*

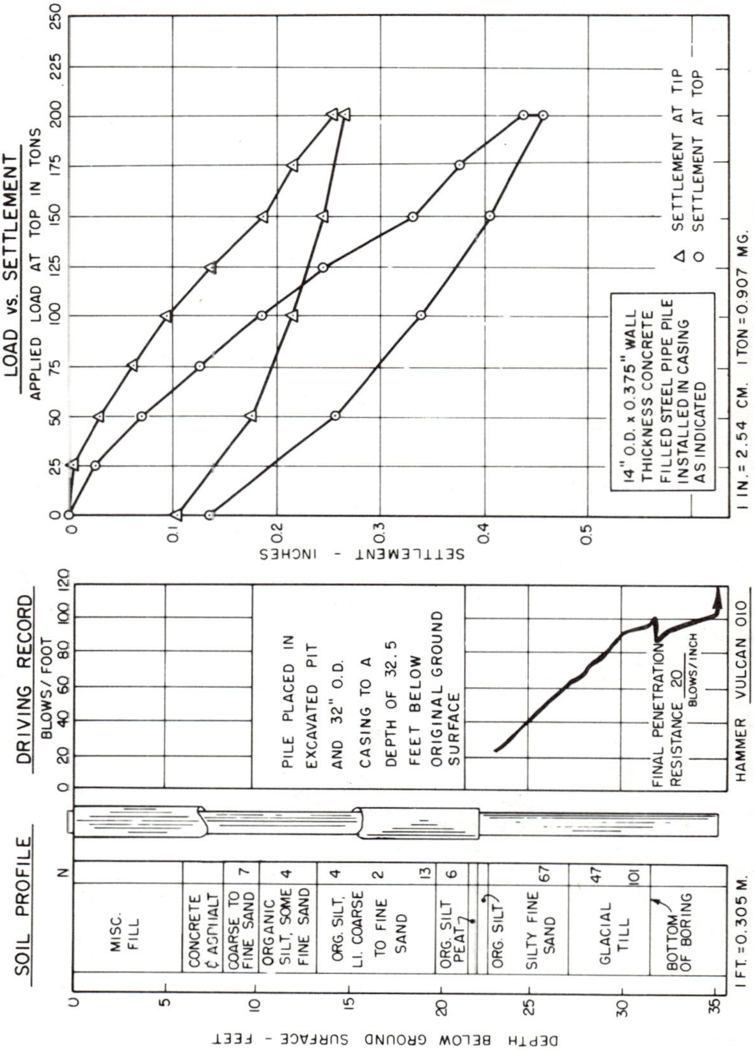

FIG. 3—*Case history III (Site B, test pile no. 1).*

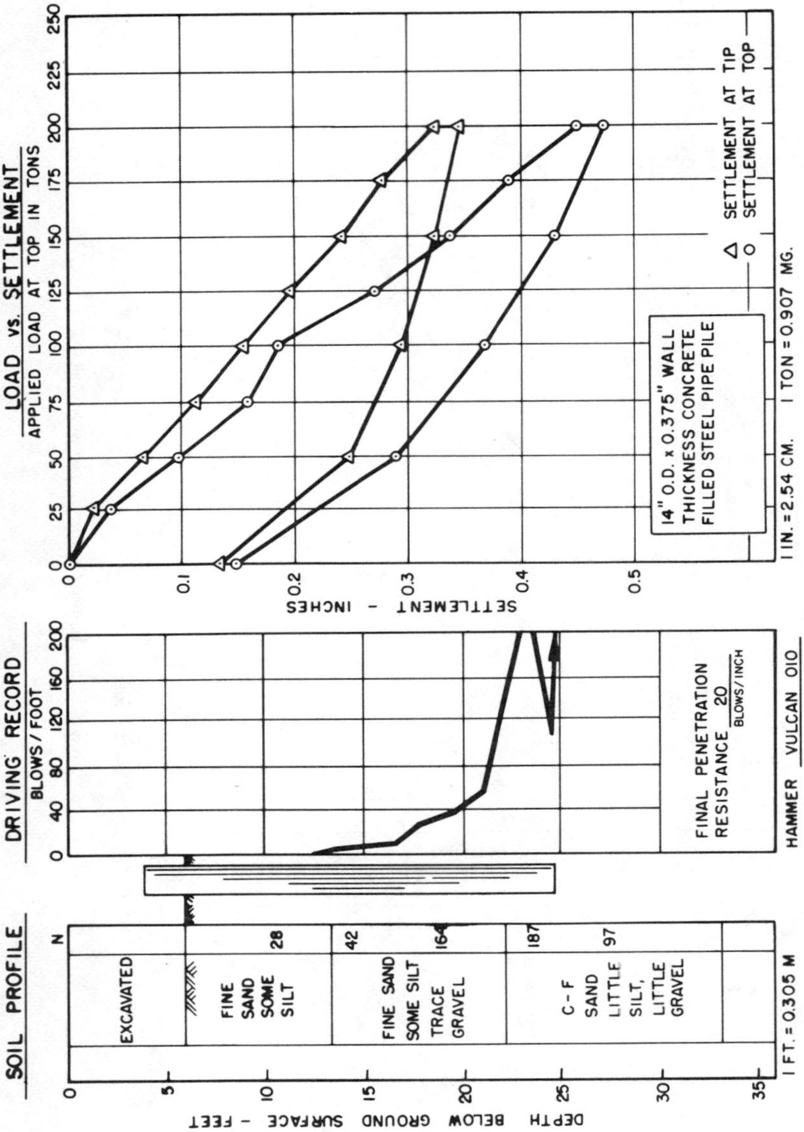

FIG. 4—*Case history IV (Site B, test pile no. 2).*

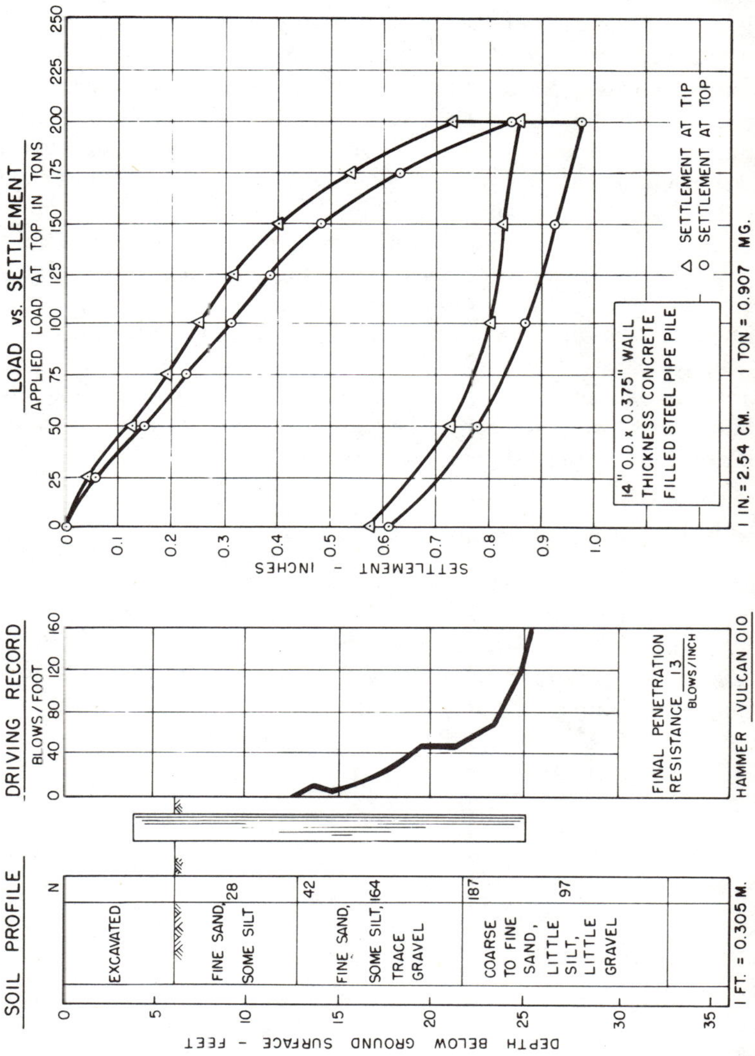

FIG. 5—*Case history V (Site B, test pile no. 3).*

kg/cm² (4500 psi) and was accomplished in the test pile by high early strength cement. Design loads for the piles were 90 or 110 Mg (100 or 120 tons).

Case histories 1 and 2 were driven at Site A, and case histories 3 through 5 at Site B. Subsurface conditions at Site A consist of approximately 46 m (150 ft) of fill, organic material, and soft clay overlying compact sand, glacial till, and bedrock. Subsurface conditions at Site B are more variable as shown in Figs. 3 and 4, but the piles are all bearing in glacial till or very dense sand.

Test pile no. 1 at both sites was conventionally driven and tested except that at Site B, test pile no. 1 was placed in a partially penetrating casing to reduce skin friction. Test pile no. 2 at Site A was placed in a deep casing for the purpose of nearly eliminating skin friction and driven to refusal on bedrock. Test pile nos. 2 and 3 at Site B were production piles, which were tested during production driving to establish redriving criteria for piles that experienced heave during driving of adjacent piles.

Settlements at the tops of the piles for the five case histories varied from approximately 1.25 to 3.75 cm (0.5 to 1.5 in.) at maximum load. Settlements at the tips varied from approximately 0.65 to 2.25 cm (0.25 to 0.9 in.). Note that it was necessary to change the scale of settlement in Figs. 1 through 5 in order to show all the data conveniently.

Proposed Interpretive Technique

The proposed interpretive technique is based on the diagram shown in Fig. 6, which is a plot of applied load at the top of the pile versus settlement at the tip of the pile. This curve is characteristic of a pile that is driven through some thickness of compressible soil to a firm bearing stratum. Below is a description of the numbered segments of the curve in Fig. 6.

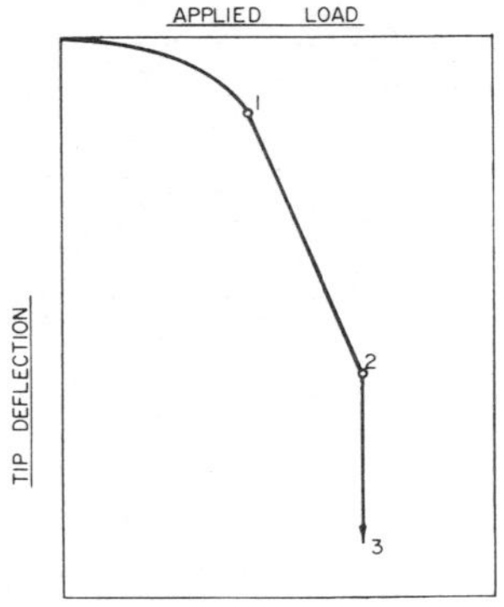

FIG. 6—*Idealized pile tip behavior.*

Segment 0-1—The applied load is being supported by skin friction along the pile, and the tip is deflecting according to some unknown load application.

Segment 1-2—Skin friction has maximized and the entire magnitude of additional applied load is transmitted directly to the pile tip. For most strata capable of supporting end-bearing piles, such as dense sand, glacial till, or bedrock, this portion of the curve will be nearly linear.

Segment 2-3—The pile has experienced a bearing capacity failure.

It should be noted that point 1 in Fig. 6 does not signify a failure but is simply a break in the curve caused by maximization of the skin friction. Pile tip movement at this point is usually very small and settlement at the top of the pile is usually well within tolerable limits.

Values of tip deflection used to plot a curve as shown in Fig. 6 must be based on effective stress distribution to the bearing stratum. The duration of load application may vary with different bearing strata, but it should be sufficiently long to allow dissipation of excess pore water pressures beneath the pile tip. The time duration should also be chosen in order to standardize curves for different tests and to minimize discrepancies due to creep movement. Use of several different load durations in this investigation resulted in very little difference in bearing strata behavior. One hour was finally selected as a reasonable time limit.

Two assumptions are inherent in the use of this interpretive technique:

1. The amount of skin friction support for the pile is limited by some maximum value, and this value of skin friction is substantially mobilized with pile movement considerably less than that required to cause a bearing capacity failure.

2. The bearing stratum will deflect linearly until the applied load approaches its ultimate bearing capacity.

Neither of these assumptions have been demonstrated as generally valid by this study although they appear valid based on the investigated case histories. If these assumptions are not valid for a particular test pile, it will be difficult to interpret pile behavior with this technique and additional, more sophisticated instrumentation such as strain gages will be required. It is also possible in some cases that the skin friction force may be so great that it cannot be overcome prior to a structural failure of the top of the pile.

If the skin friction force can be overcome with small deformation and the bearing strata does deflect linearly with increased load, then the straight-line segment 1-2 in Fig. 6 is directly related to the load-carrying capacity of the bearing strata. A line parallel to segment 1-2 through the origin will give the load supported by the tip for any value of tip deflection. This line through the origin is referred to as the "tip capacity line," as shown in Fig. 7.

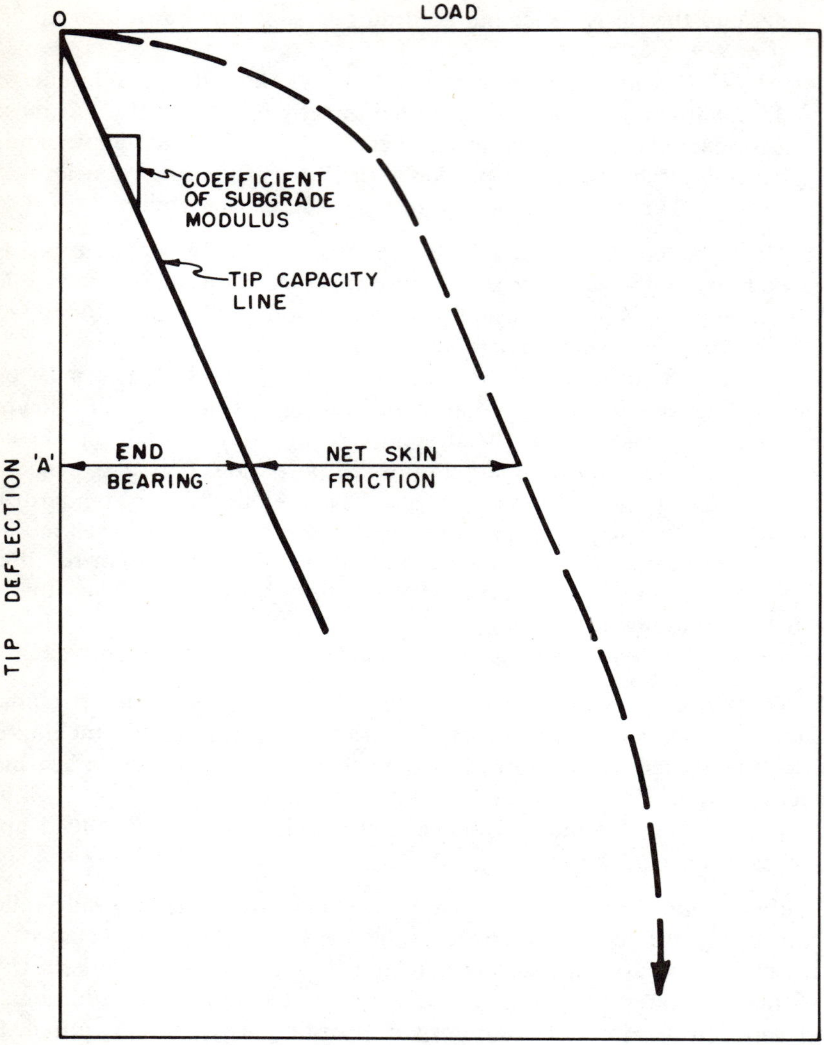

FIG. 7—*Pile tip capacity.*

Figure 7 is similar to a diagram used by Van Weele[3] to interpret pile load test results. Van Weele, however, obtains tip deflections through an indirect procedure, which is critically dependent on assumptions about pile properties and about soil-pile load transfer along the axis of the pile which cannot be easily verified during the test. Reference to Van Weele's paper for

[3]Van Weele, A. F., *Proceedings*, Fourth International Conference on Soil Mechanics and Foundation Engineering, Vol. 2, London, 1957 pp. 76–80.

a complete description of his procedure may be helpful for understanding the technique described herein.

Another diagram that is helpful in analyzing pile behavior is a comparison of the actual and the theoretical elastic deformations of the pile. Actual elastic deformations of the pile at any load can be computed by subtracting the tip deflection from the top deflection. Theoretical elastic deformations for an assumed end-bearing pile can be calculated using the simple formula

$$\text{elastic deformation} = \frac{PL}{AE} \tag{1}$$

where

P = the applied load,
L = length of the pile,
A = cross-sectional area of the pile, and
E = elastic modulus of the pile.

A friction pile with uniform stress distribution to the soil throughout the length of the pile will have an elastic deformation exactly one-half of the above. Elastic deformations for this study were calculated using the moduli

$$E_{\text{steel}} = 2.0 \times 10^6 \text{ kg/cm}^2 \text{ } (29 \times 10^6 \text{ psi})$$

$$E_{\text{concrete}} = 0.25 \times 10^6 \text{ kg/cm}^2 \text{ } (3.5 \times 10^6 \text{ psi})$$

Load Test Interpretations

Elastic deformations and tip settlements versus applied load at the top of the pile are plotted for the five case histories in Figs. 8 through 12. Case history 1 (Fig. 8) is for a conventionally driven and tested pile. As mentioned previously, the pile was driven through 46 m of compressible soils to bearing in dense sand and was acceptable as a bearing pile in accordance with local building code requirements.

The measure tip deflection, plotted according to the proposed method, is very similar to the assumed ideal behavior shown in Fig. 6. The higher test load increments show distinct straight line behavior, and the tip capacity line is constructed as described previously. The tip capacity line suggests that a considerable portion of the applied load for this load test was being supported by skin friction. In fact, the tip capacity line indicates that even though the applied load was carried to twice the assumed design load, the tip of the pile was not loaded to the level of the design load. This distribution of support in the test pile is also indicated by the computed elastic deformations in Fig. 8, which are less than the theoretical elastic deformations for a friction pile.

Case history 2 is at the same site as case history 1, but in this instance, the pile was placed inside a casing and driven through a few feet of soil to

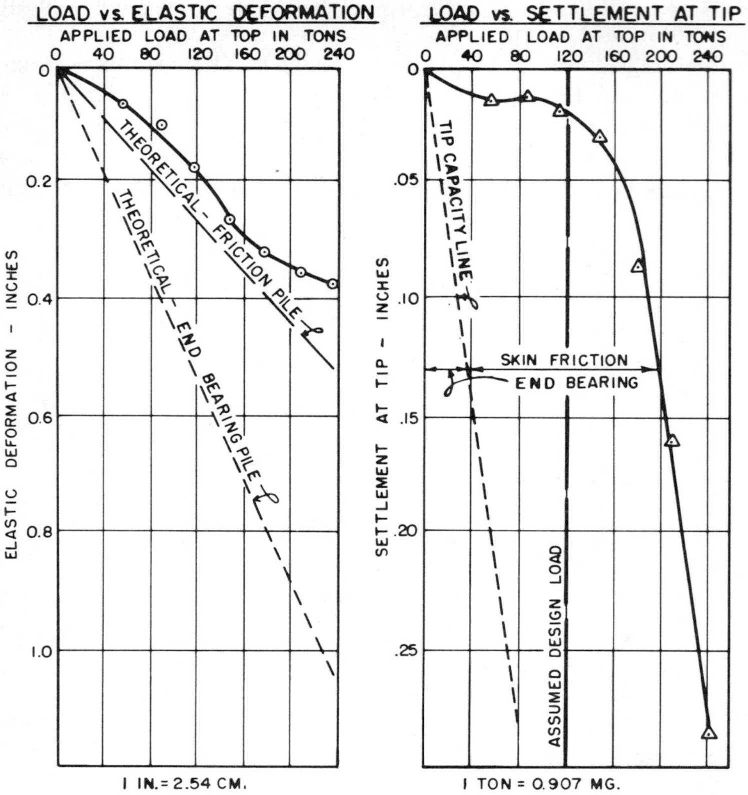

FIG. 8—*Case history I (Site A, test pile no. 1)—data interpretation.*

bearing on bedrock. Skin friction resistance for this pile can be safely assumed as negligible.

Interpreting the load test results as suggested herein clearly reveals end-bearing behavior for this pile. As shown in Fig. 9, the tip settlement plots as a straight line through the origin and the computed elastic deformations are nearly equal to the calculated theoretical elastic deformation for an end-bearing pile. It should be noted that the tip capacity line in Fig. 9 is less steep than in Fig. 8, indicating a stiffer end-bearing stratum, which is, in fact, the case—bedrock versus dense sand. More discussion about the slope of the tip capacity line is given later in this paper.

Case histories 3 through 5 are all at the same site, Site B, as mentioned previously. Case history 3 is for a conventionally driven and tested test pile. Interpretation of the load test results as shown in Fig. 10, clearly suggests end-bearing behavior for this pile, which is reasonable, since the pile is relatively short and was driven inside a partially penetrating

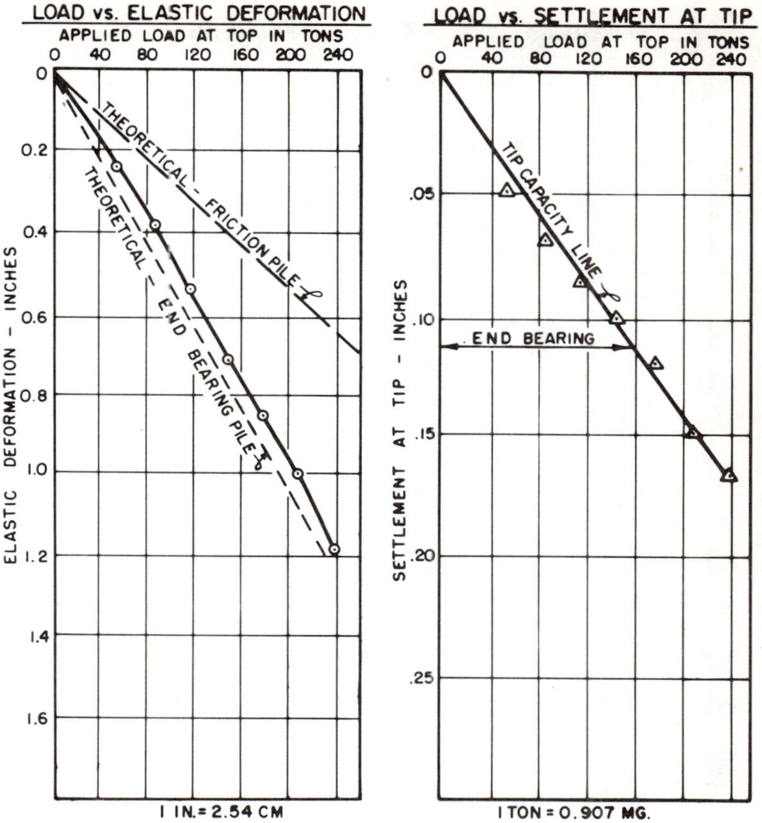

FIG. 9—*Case history II (Site A, test pile no. 2)—data interpretation.*

outside casing. In this instance, the tip capacity line indicates that the tip of the pile was subjected to a load equal to 1.6 times the assumed design load without a bearing capacity failure.

During construction at Site B, many piles experienced heave during driving of adjacent piles. Upon redriving of the heaved piles, it was noted that as much as 1 to 2 m of driving was necessary to obtain acceptable penetration resistance even though the original amount of heave was only a few centimeters. Since this amount of redriving required expensive splicing of piles, it was decided to delay work and to test two production piles that had only been redriven the amount of the measured heave despite the redriving penetration resistance.

The results of the first of these redriven test piles is shown in Fig. 11. A comparison of tip settlement for Case histories 3 and 4 reveals that the redriven pile experienced less skin friction than the original test pile but

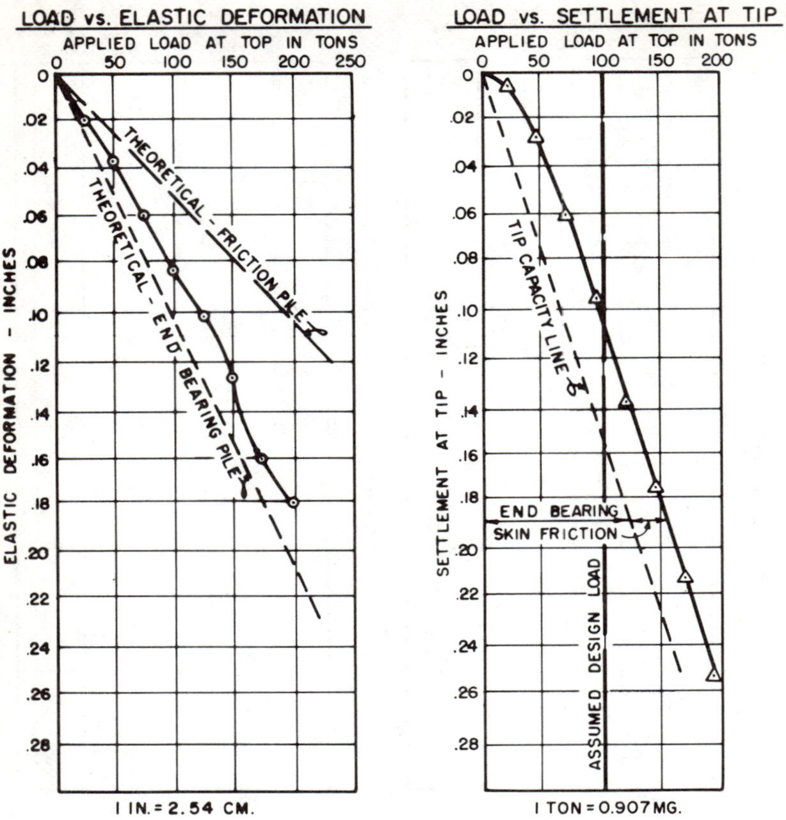

FIG. 10—*Case history III (Site B, test pile no. 1)—data interpretation.*

that the tip capacity lines for both cases were nearly parallel. This demonstrates that the end-bearing stratum for these two piles was equally competent and that simply reseating heaved production piles at this site was a valid criterion. It can be assumed that the driving of numerous production piles induced high excess pore pressures in the surrounding soil and caused a reduction in skin friction resistance. Reduced skin friction allows more driving energy to be transmitted to the tip of the pile and resulted in pile penetration through competent end-bearing material. In fact, the tip capacity line for Case history 4 shows that the production pile was capable of supporting nearly twice the assumed design load without a bearing capacity failure.

Case history 5 is the second redriven test pile at Site B, and its interpreted behavior is summarized in Fig. 12. This pile shows even less skin friction than the previous two piles and a slightly steeper but acceptable

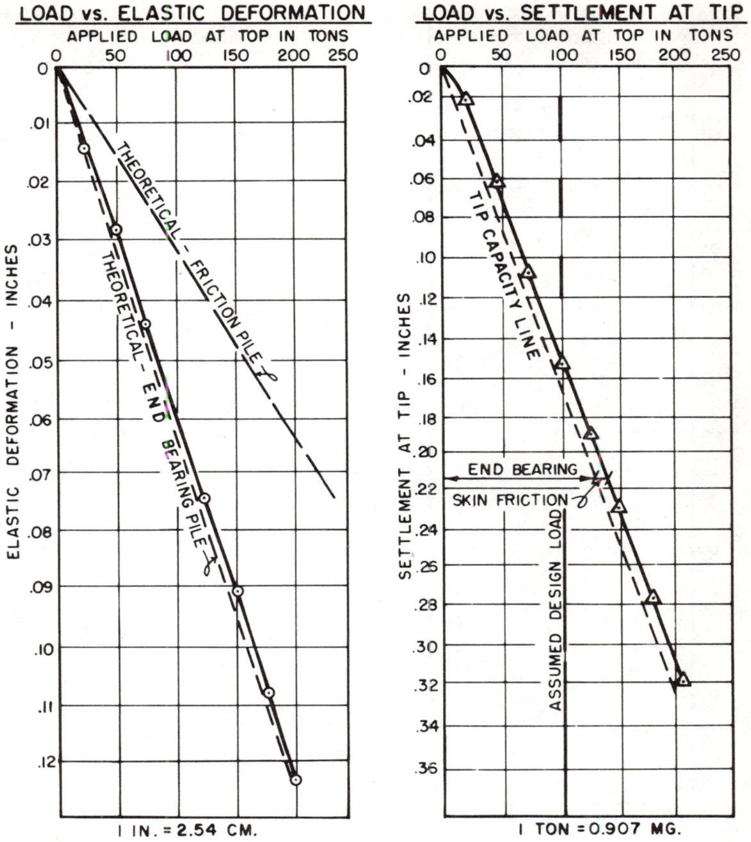

FIG. 11—*Case history IV (Site B, test pile no. 2)—data interpretation.*

slope for the tip capacity line. The almost complete lack of skin friction resistance for this pile results in a severe test of the bearing stratum, and it appears that a bearing capacity failure occurred at a load of approximately 136 Mg (150 tons) or 1.5 times the assumed design load. Despite the bearing capacity failure, the pile fulfilled the local building code requirements for acceptable end-bearing piles.

Computing Pile Settlement

A simple formula for computing settlement of assumed end-bearing piles based on the results of this study is

$$\text{pile settlement} = \frac{PL}{AE} + \frac{P}{AT} + \text{creep} \tag{2}$$

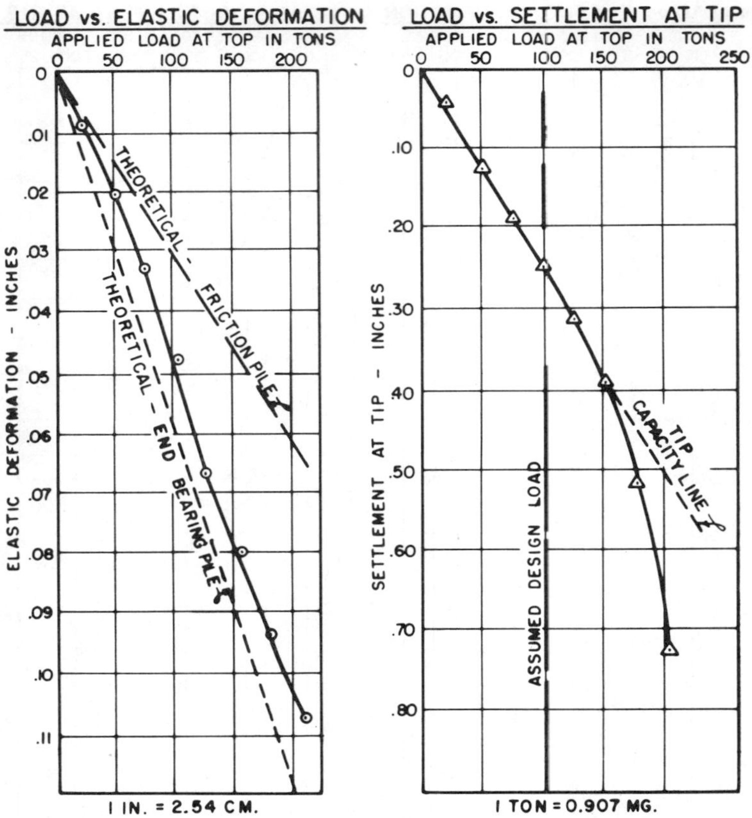

FIG. 12—*Case history V (Site B, test pile no. 3)—data interpretation.*

where

PL/AE = the elastic deformation of the pile as given in Eq 1, and
T = the coefficient of subgrade modulus of the bearing stratum

If a pile is assumed to act as an end-bearing member and positive skin friction is not present, it is reasonable to include in the settlement calculation the entire magnitude of elastic deformation of the pile caused by the design load. In fact, the elastic deformation of the pile would be the lower limiting value of the total settlement. It is then necessary to add the amount of bearing stratum deflection to the pile deformation for an estimate of total settlement. In Eq 2, the "elastic" deformation of the bearing stratum is computed using a coefficient of subgrade modulus T obtained from the tip capacity line for the test pile as shown in Fig. 7. It is possible that additional time-dependent creep deformation of the bearing stratum will occur, and allowance for this should also be included in the settlement calculations. Typical values of subgrade moduli as indicated by this study are given in Table 1. It is emphasized that these values are based on limited information and that the coefficient of subgrade modulus is dependent on the size of the loaded area and the installation procedures for the pile. The coefficients should not be extrapolated to other foundation types such as caissons or footings.

For a 25 m (82 ft) long, 35.5 cm (14 in) diameter concrete filled steel pipe pile driven to bearing in glacial till and subjected to a 90-Mg (100-ton) design load without the benefit of positive skin friction, the settlement calculations would be as follows:

elastic deformation of pile	$= PL/AE$	$\cong$ 1.0 cm
"elastic" deformation of bearing stratum	$= P/AT$	$\cong$ 0.7 cm
"creep" deformation of bearing stratum	= Estimate	$\cong$ 0.2 cm
	total	1.9 cm

It is noted that Eq 2 also provides a method of estimating the settlement that might be induced by downdrag. Instead of the applied structural load P, a load of varying intensity as a function of the soil profile could be input into the pile and the resulting elastic deformation of the pile calculated. Based on the assumed downdrag load distribution, a total downdrag force could then be obtained and its effect on the bearing stratum computed in the same manner as for the applied structural load.

Summary, Conclusions, and Recommendations

A method of interpreting the results of pile load tests on assumed end-bearing piles has been presented and discussed. All load tests were conducted in accordance with ASTM or local building code standards with the addition of a rod telltale to measure deflection of the pile tip. Five case histories were presented to demonstrate the validity and the practical application of the proposed interpretive procedure.

In general, the interpretive technique involves careful examination of pile tip deflection. If skin friction forces can be overcome with small de-

TABLE 1—*Typical values of subgrade moduli.*

Bearing Stratum	Coefficient of Subgrade Modulus	
	kg/cm^3	$lb/in.^3$
Dense sand	85	3 000
Glacial till	140	5 000
Bedrock	420	15 000

formation and the bearing stratum deflects linearly with increased load, it is possible to divide the applied load into skin friction and end-bearing components and to obtain information about the adequacy of the bearing stratum to support the assumed design load.

Based on known pile properties and information obtained from the load test, it is possible to accurately estimate the settlement of pile foundations. With time, it appears that information about bearing strata behavior can be accumulated and used to improve end-bearing pile design procedures. It also appears likely that slight variation of design procedures for structural loads can be used to estimate settlement resulting from downdrag.

The proposed interpretive procedure has many advantages that seem to warrant its additional use and evaluation. Some of the more important advantages are:

1. The procedure is relatively inexpensive and easy to implement.
2. The interpretive technique does not depend on a detailed knowledge of pile properties or of soil-pile stress transfer throughout the length of the pile.
3. The interpretive technique provides information about the assumed bearing strata directly and its ability to support the imposed structural load.

One disadvantage of the proposed technique is the difficulty of installing telltales in certain pile types such as H- or wood piles. The advantages of this approach, however, warrants the additional cost of welding a pipe to an H-pile or routing a trough in a wood pile to allow insertion of the telltale.

It is felt that the results of this study strongly suggest that rod telltales to the pile tip are useful for interpreting pile load tests and that the approach deserves additional study to evaluate its validity better. In particular, strain gages on the piles could be used for an independent check of load transfer to the bearing stratum and long-term measurements of actual building settlements could be obtained for comparison with computed design values.

J. R. Cheeks[1]

Analytical Methods to Predict Pile Capacities

REFERENCE: Cheeks, J. R., **"Analytical Methods to Predict Pile Capacities,"** *Behavior of Deep Foundations, ASTM STP 670,* Raymond Lundgren, Ed., American Society for Testing and Materials, 1979, pp. 199–208.

ABSTRACT: A comprehensive test pile program was conducted for a large manufacturing facility. During this program, several pile types, including timber, precast concrete, concrete-filled pipe, and shells were driven and load tested to failure. Prior to load testing, the ultimate pile capacity was predicted by three static and two dynamic techniques. After testing, the predicted and measured capacities were compared. The wave equation and the Dutch cone penetration soundings resulted in the most consistent predictions.

KEY WORDS: bearing capacity, cyclic loads, failure, forecasting, piles, pile driving, Dutch cone, wave equation, standard penetration test, skin friction, geotechnical engineering

Cost and time factors generally limit opportunities to conduct pile load tests to failure. Opportunities to load test a variety of pile types on one site are even more rare. This paper describes a unique test program, which included tests to failure of several pile types on a single site and compares various predictive techniques employed during the test program.

Prior to load testing, ultimate pile capacities were predicted by a variety of static and dynamic techniques. It was neither possible nor practical to utilize all available predictive techniques. However, the analytical methods selected for this program are representative and include techniques that are both commonly used by geotechnical engineers in their daily practice as well as techniques still in the research and development stages. The predicted pile capacities are compared to load test results, and the accuracy of each predictive technique is discussed.

[1] Vice president, Stokley, Cheeks and Associates, Lexington, Ky. 40587.

General Site and Soil Conditions

The pile test site is an island in the St. Johns River near Jacksonville, Fla. (Fig. 1). The entire area is underlain by Mesozoic and Cenozoic sediments typical of the Lower Atlantic Coastal Plain. The island is divided into north and south sides by a shallow estuary known as Back River. The north side of the site originally consisted of several small islands, interconnected by tidal sloughs. The south side was originally a tidal marsh.

For years, the site has been used for spoil placement during channel deepening and maintenance dredging of the St. Johns River. Most of the original surface features are now covered by these dredged deposits, which range in thickness from 0 to 8 m (0 to 25 ft). These fills are undifferentiated and range from fine sands containing limerock fragments to very soft clayey silts and silty clays.

The generalized profile (Fig. 2) illustrates the soil stratification together with the standard penetration resistance (SPT) and the Dutch cone penetration resistance (DCPT) at each test pile location.

Test Pile Program

The purpose of the test pile program was to determine the optimum pile types and maximum allowable pile capacities to use for foundation design of a major manufacturing facility. A number of pile types were driven and load-

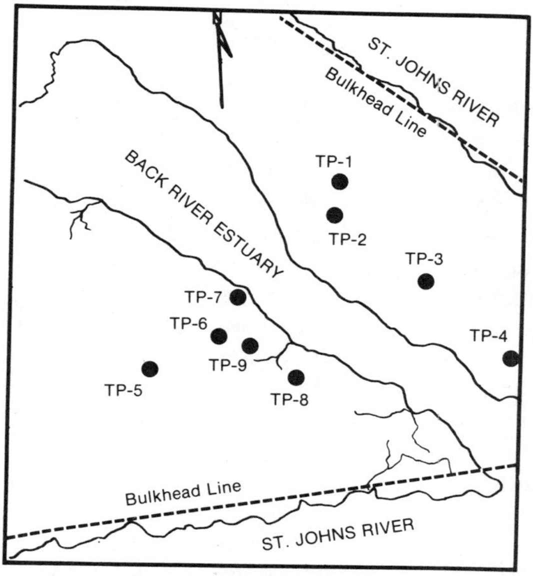

FIG. 1—*Test pile location plan.*

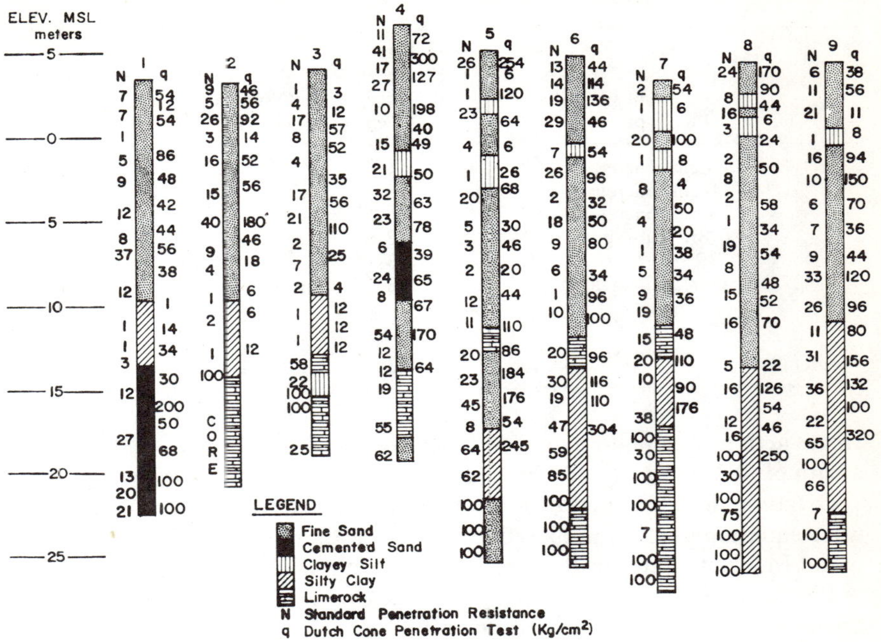

FIG. 2—*Generalized profile.*

tested during this program. Included were 45.7 cm (18 in.), 35.6 cm (14 in.), and 25.4 cm (10 in.) square prestressed concrete piles; 000 BR and 0 BR tip step-taper piles; 27.3-cm (10.75-in.) outside diameter pipe piles with a 0.64-cm (0.25-in.) wall thickness; and timber piles.

To obtain load test data for a range of soil conditions at the site, piles were driven at the nine locations shown on the test pile location plan (Fig. 1). Soil test borings and Dutch cone soundings were made at each test location. A total of 23 test piles were driven at these nine locations. Of these, ten were statically load tested in compression only, and six in both compression and tension. The untested piles were driven primarily to provide comparisons of length and driving behavior for the different pile types. Table 1 summarizes the distribution of piles by type and indicates the number of load tests for each type of pile.

Driving Criteria

Two hammers were used during this program. For most piles, a 65C hammer was used. This hammer is a differential steam hammer with a 2944.5-kg (6500-lb) ram and a rated energy of 26,430 N·m (19 500 ft·lb), and it was used to install all of the piles except the 45.7-cm concrete piles. The 5/0, a

TABLE 1—*Summary of pile types and load tests.*

Pile Type	Total Number of Piles Driven	Number of Compression Tests Only	Number of Compression and Tension Tests
25.4 cm concrete	1	1	0
35.6 cm concrete	3	1	1
45.7 cm concrete	6	2	2
000 BR step-taper	2	1	1
0 BR step-taper	3	2	0
27.3 cm outside diameter pipe	5	1	2
timber	3	2	0
total	23	10	6

single-acting steam hammer, which develops 59 299 N·m (43 750 ft·lb) with a ram of 7827.5 kg (17 500 lb), was used for these larger concrete piles.

Dynamic driving criteria for the test piles were established based upon a wave equation analysis prepared by Raymond Concrete Pile Division of Raymond International. For this analysis, pile lengths of 18.3 m (60 ft) and 27.5 m (90 ft) were assumed for each pile type. One-half the total resistance was assumed to be at the tip and the other half in a triangular distribution of frictional resistance along the shaft.

In planning the test pile program, a desired design pile capacity for each pile type was established based upon local experience. The driving criteria, utilizing a factor of safety of 2, was then determined using the wave equation solutions. In the field, every effort was made to cease driving as soon as this criterion was satisfied. Table 2 summarizes the final driving resistance, length, and ultimate capacity, as indicated by the wave equation solution for each pile.

Load Test Procedures

Load tests were performed, with minor exception, according to ASTM Testing Piles Under Axial Compressive Load (D 1143), and the Jacksonville Building Code. The load was applied with a calibrated 272.2 Mg (300-ton) hydraulic jack. In most cases, eight step-taper tension piles provided the reaction for the compression loads. For the tension tests, the reaction force was provided through steel beams on timber mats supported on the ground.

Deflections were measured by two systems. Dial measuring instruments, accurate to 0.00254 cm (0.001 in.), were mounted on a reference beam, and a small wire scale was mounted on the pile.

The piles were loaded to 200 percent of the estimated design capacity in 25 percent increments. Rebound cycles were made after the 50, 100, and 200

TABLE 2—*Summary of test pile data.*

Pile Number	Pile Type	Pile Length, m	Final Driving Resistance, Blows/cm	Predicted Capacities by Wave Equation, Mg	Load Test Results, Mg
2	timber	19.2	1.8	82	91
3	25.4 cm PSC	21.3	1.1	...	73
4	45.7 cm PSC	20.4	2.3	301	...
5	000 BR	17.3	1.2	102	127
6	pipe	36.9	3.1	105	...
8	35.6 cm PSC	18.6	4.7	205	236
11	000 BR	18.6	2.7	208	163
11A	0 BR	20.7	2.0	...	181
12	pipe	17.1	3.6	122	95
13	45.7 cm PSC w/casing	18.9	3.5	344	209
14	45.7 cm PSC w/casing	24.7	1.2	228	236+
16	45.7 cm PSC w/casing	19.8	1.4	246	272+
17A	timber	12.8	2.0	74	127
18	35.6 cm PSC	17.1	2.6	154	...
19	45.7 cm PSC	14.9	1.1	178	154
20	pipe	22.6	2.4	100	163
20A	pipe	15.2	2.4	100	63
21	0 BR	16.2	1.5	...	...
22	45.7 cm PSC	23.5	0.2	218	...
23	timber	15.2	2.2	83	...
24	35.6 cm PSC	15.8	4.6	148	145
26	pipe	24.7	4.7	127	...
27	0 BR	15.2	1.8	181	163

percent loads. The 200 percent load was maintained for a minimum of 24 h and until the settlement rate was less than 0.15 mm (0.006 in.) per day. If the piles continued to move at greater rates after 36 to 48 hs, then the load was reduced until the movement criterion was satisfied.

After the pile was rebounded from either the 200 percent load or the reduced load, the load was reapplied and then increased in small increments until the ultimate bearing capacity of the pile was exceeded or the capacity of the jack and reaction system was reached. The ultimate bearing capacity of the pile was defined for this test program as the load that caused a gross pile deflection of 2.54 cm (1.0 in.).

Predictive Techniques

Static Methods

Prior to the pile load tests, three static analyses were utilized to estimate pile capacities. Both tension and compression capacities were estimated with these static methods. Initially, approximate lengths were estimated for bid quantities using the static analysis described in *Soil Mechanics, Foundations and Earth Structures,* NAVFAC DM-7.

The DM-7 analysis applies conventional soil mechanics theories of bearing capacity and frictional resistance for a given pile geometry and soil stratification. The frictional component for piles embedded in sand is computed by estimating the lateral stress on the pile. The point resistance is computed by estimating the bearing capacity at the pile tip.

The second method used during this project utilizes the standard penetration test (SPT) *N* value. The method was developed by L. C. Nottingham and R. H. Renfro and has been completely documented in Research Bulletin 121-B, State of Florida, Department of Transportation. This analytical tool was prepared for use by personnel of the Department of Transportation to estimate pile capacities for highway structures, and the authors were uncertain about the potential uses. Therefore, the technique has large safety factors, and conservative results should be anticipated.

The third static technique is presently in the research and development stages at the University of Florida under the direction of Professor John Schmertmann. In this method, the Dutch cone penetration test (DCPT) values are utilized to estimate the pile capacity. A recent doctoral dissertation by Larry Nottingham at the University of Florida describes this developing technique. His study concentrated on small-diameter model piles. During this test pile program, University of Florida personnel performed the DCPT at each test pile location and used the analytical procedures developed for the model piles to estimate the ultimate capacities of the test piles.

Dynamic Methods

During the test pile installation, two dynamic techniques were utilized to predict ultimate pile capacities. Only compression loads were predicted with these methods.

As mentioned above, the dynamic driving criteria were established by applying the wave equation analysis to the desired capacity for each pile type. Actual final driving resistances, however, generally differed from the established criteria; therefore, the final driving resistance was used to estimate the ultimate pile capacities.

Dynamic measurements using the methods developed by Professor G. G. Goble were made on six test piles, three of which were statically load tested. Near the end of driving for each of these piles, transducers were attached near the pile top. These transducers provided continuous electronic signals giving strain and acceleration of the pile top. The signals were recorded on tape and later processed using the Case project processing system, from which the ultimate pile capacities were predicted.

Effectiveness of Predictive Techniques

The effectiveness of the five predictive techniques is illustrated in Figs. 3-7. The vertical axis in each case depicts the measured ultimate pile capacities, and the horizontal axis shows the predicted values. The diagonal line represents the group of points where Q predicted equals Q measured.

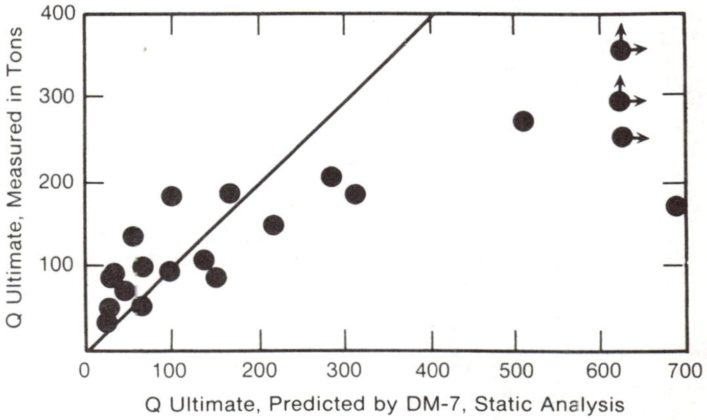

FIG. 3—*Capacities using DM-7 analysis (• = actual pile capacity greater than plotted). Note that prediction by DM-7 refers to Nav Fac DM-7, March 1971, Chapter 13. Soil parameters were selected to each boring location on the basis of soil type and standard penetration resistance (1 ton = 0.9 Mg).*

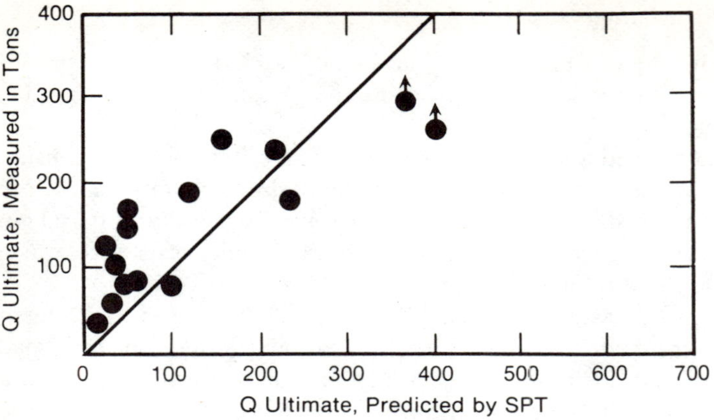

FIG. 4—*Measured versus predicted pile capacities using standard penetration tests (⬆ = actual pile capacity greater than plotted). Analysis preformed by Larry Nottingham in accordance with Research Bulletin 121-B, State of Florida, Department of Transportation. Standard penetration test* N-*values were used for pile predictions (1 ton = 0.9 Mg).*

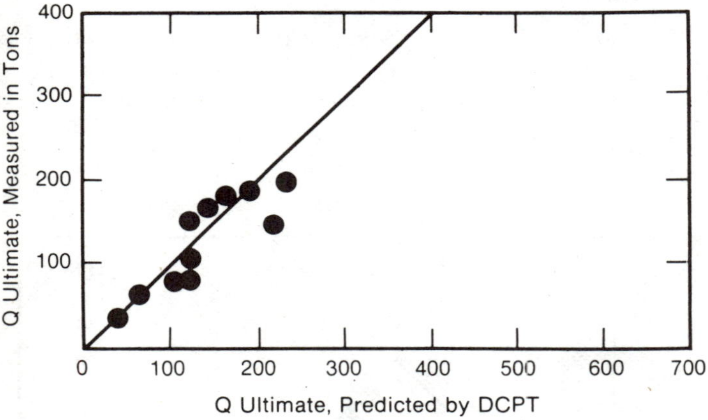

FIG. 5—*Measured versus predicted pile capacities using Dutch cord penetration resistances. Predictions by Larry Nottingham and John Schmertmann in accordance with their model pile methods (1 ton = 0.9 Mg).*

Static Methods

The DM-7 analysis provided a very erratic and poor indication of pile capacities. The six tension test predictions were reasonably accurate, indicating that the erratic nature of the technique is possibly derived from the point resistance component. For ultimate capacities in excess of 90.7 Mg (100 tons), the analysis seems to overpredict the capacities by a factor of 2 or more.

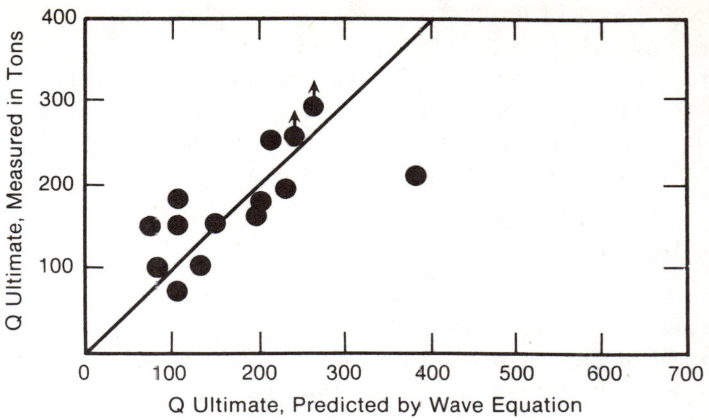

FIG. 6—*Measured versus predicted pile capacities using the wave equation (↑ = actual pile capacity greater than plotted). Wave equation solution prepared by Raymond Concrete Pile Division for this project. The analysis assumed one-half of the total capacity at the tip and one-half of the total capacity as friction. The data was initially generated to establish pile driving criteria, and later used to derive predicted ultimate capacities from actual final driving resistances (1 ton = 0.9 Mg).*

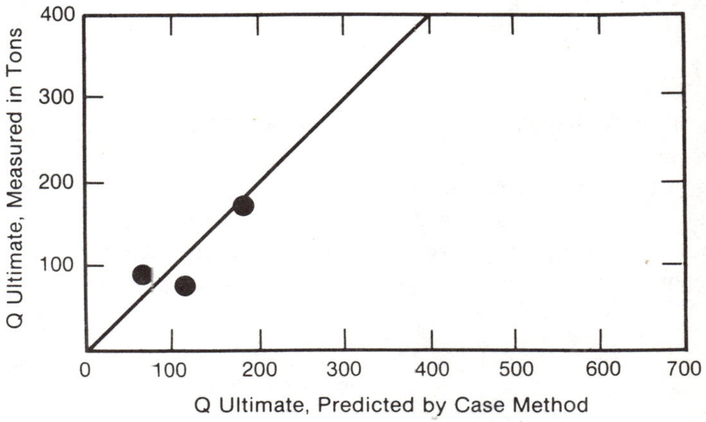

FIG. 7—*Measured versus predicted pile capacities using the case method. Predictions performed by Professor G. G. Gable. (1 ton = 0.9 Mg).*

The SPT static analysis also provided scattered results, as indicated on Fig. 4. However, this technique appears to be conservative in its approach, since all but four pile capacities were underpredicted. As discussed above, the developers of this analysis had so intended. Therefore, this technique can probably be used to provide a preliminary indication of the ultimate pile capacities.

The DCPT method resulted in very accurate pile capacity predictions. However, this technique is still in the research and development stages. Therefore, the use of the technique for routine applications will require complete documentation, instruction on its use and sufficient local experience. After obtaining this data and experience, this technique should prove to be a valuable means of predicting static pile behavior.

Dynamic Methods

The wave equation analysis provided a consistent indication of the ultimate pile capacities. Several of the predictions by this method were within ±10 percent of the test value. Since complete documentation of the wave equation exists and is readily available, the wave equation can be used on a rather widespread basis. It is an excellent technique for the practicing engineer to use in establishing a dynamic pile driving criteria.

The Case method was only used on three load-tested piles. The correlations for these piles were very good; however, this technique is also in the development stages, and much of the data now available had been limited to pipe piles. Therefore, more data, documentation, and local experience with the technique will be required before it can be routinely used with some degree of confidence.

Acknowledgments

The author wishes to thank E. W. Lingo, Chief Engineer with Law Engineering Testing Company in Jacksonville, Florida, for the opportunity to work on this very interesting project. His help and guidance were most valuable. The author also wishes to thank John Schmertmann, Larry Nottingham, G. G. Goble, and Raymond International for their cooperation and assistance with the pile capacity predictions that each performed.

J. I. Clark[1]

Failure During Construction and Subsequent Rehabilitation and Performance of a Dynamically Cast-in-Place Concrete Pile Foundation

REFERENCE: Clark, J. I., "**Failure During Construction and Subsequent Rehabilitation and Performance of a Dynamically Cast-in-Place Concrete Pile Foundation,**" *Behavior of Deep Foundations, ASTM STP 670,* Raymond Lundgren, Ed., American Society for Testing and Materials, 1979, pp. 209-230.

ABSTRACT: During construction of a dynamically cast-in-place concrete pile foundation for a heavily loaded library tower at the University of Calgary, extensive pile heave was recorded. Consultants were engaged to assess the effect of pile heave on the pile capacity. A down-hole sonic tool was used to assess several piles where damage was suspected, and the results indicated severe damage to the pile shafts. Load tests confirmed the results when piles failed at a fraction of the required load. Further assessment led to the conclusion that 200 of the total of 292 piles were not capable of supporting the design load.

The contractor elected to rehabilitate the damaged piles by adopting a pressure grouting technique recommended by the consultants. Load tests confirmed the viability of the grouting method, and it was implemented for all damaged piles. During the rehabilitation, extensive fracturing of the soil was revealed at various depths over the entire building site. The piles were successfully rehabilitated, and the structure was instrumented to record settlement during and after construction. Settlement records have been maintained for seven years since construction.

KEY WORDS: piles, dynamically cast-in-place expanded base concrete piles, pile heave, cohesive soils, sonic tests, pile load tests, grouting, pile rehabilitation, settlement observations

The University of Calgary library tower was designed for a total of 21 stories with a full basement, extending approximately 3.7 m (12 ft) below

[1] Vice President, Technical Services, R. M. Hardy and Associates, Ltd., Calgary, Alberta, Canada T2P 2W5.

grade. The building is located within the university complex nearby other buildings founded in similar soil conditions. The configuration of the building required that the foundations carry relatively high loads. The design involved a central core, which would carry a total load of approximately 203 MN (45 540 kips) and a number of perimeter foundations with loads varying from 5 782 to 12 455 kN (1 300 to 2 800 kips). The foundation layout and design loads are shown in Fig. 1. The building was subsequently built to 14 stories, but the change was not related to foundation problems. The building design allows for an additional seven stories if required in the future.

The designers selected dynamically cast-in-place expanded base concrete piles (Franki type) to carry the building loads. A total of 292 piles was required. The design submitted by the contractor comprised 132 piles of 580-mm (20-in.) diameter with a driven length of 10.7 m (35 ft) to support the central core and a total of 160 piles of 610-mm (24-in.) diameter in groups of four to eight with a driven length of 1.2 m (40 ft) to support the columns. The design load for the core piles was approximately 1535 kN (345 kips) per pile and for the perimeter piles ranged from 1446 kN (325 kips) to 1557 kN (350 kips) per pile. The difference in length was due to a deeper cutoff of the core piles. The spacing of all piles was 2.5 diameters center to center. A one-story link connecting the new tower with the existing library was founded on drilled cast-in-place concrete piles.

The piles were constructed by driving a steel casing with a gravel plug in the casing. When the casing was driven to the required depth, it was attached to the leads of the driver, and the gravel plug was forced out by the drop hammer. Successive charges of 0.14 m^3 (5 ft^3) of zero slump concrete were expelled from the casing to form a bulb at the base. The specified minimum strength of the concrete was 35 MPa (5000 psi). The 508-mm (20-in.) diameter piles had a base of 0.42 to 0.56 m^3 (15 to 20 ft^3), and the 610-mm (24-in.) diameter piles had a base of 0.42 to 0.70 m^3 (15 to 25 ft^3). A steel reinforcing cage consisting of six No. 8 reinforcing bars was placed inside the casing and the pile shaft was formed by ramming 0.14-m^3 (5-ft^3) charges of concrete from the casing as it was withdrawn in increments of about 0.3 m (1 ft).

During construction, the inspector observed that a substantial amount of heaving of previously driven piles was occurring as adjacent or nearby piles were driven. Construction of the piles was approximately 80 percent completed at that time. Foundation engineering consultants were engaged to assess the problem, and recommendations were presented for preboring and for a staggered driving sequence for the remainder of the piles. However, the contractor elected to follow the same procedures to complete the project that had been used for the previously driven piles. Detailed records of heave for the remainder of the piles were obtained, and an investigation of the potential problems that might be associated with heave was initiated.

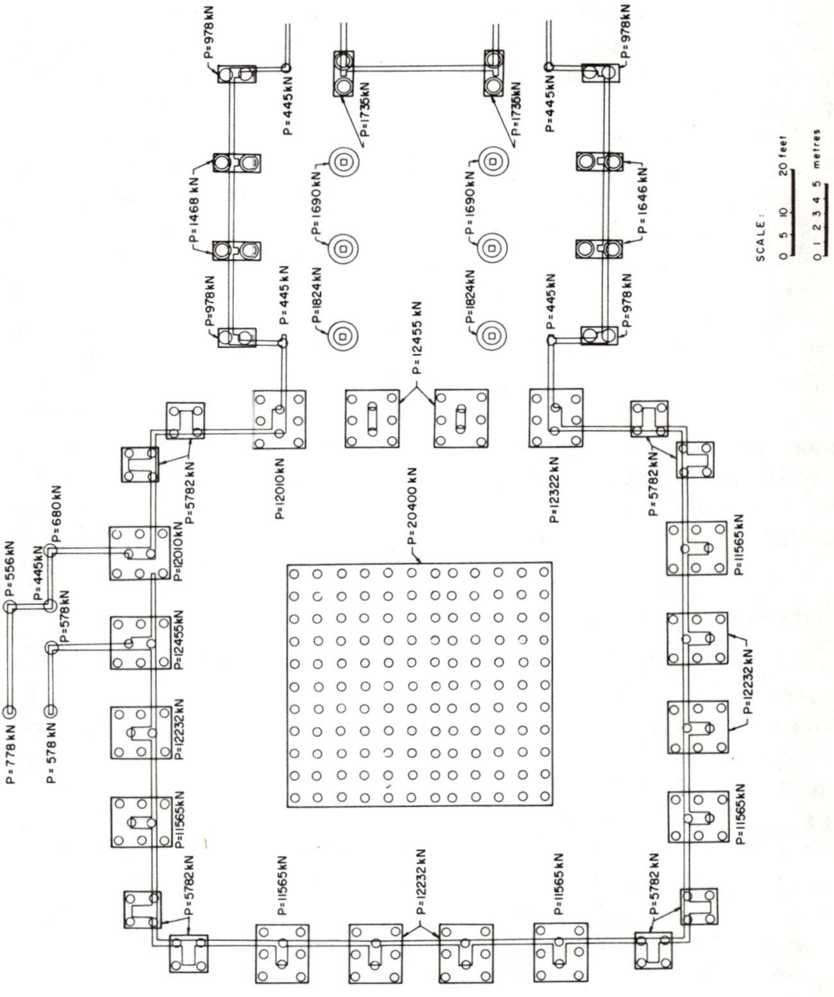

FIG. 1—*Foundation layout and design loads.*

Site Geology

The site is underlain by sandstone and shale of early Tertiary age, which lies at a depth of approximately 36.6 m (120 ft). The bedrock is covered by a layer of preglacial gravel about 4.6 m (15 ft) deep, which is overlain by Pleistocene deposits of glacial and lacustrine origin. Glaciers invaded the area at least twice and left the bedrock and preglacial gravel covered with glacial drift (till). During the second glacier advance, the eastern ice blocked the Bow Valley, and the area was inundated by a large lake, called Lake Calgary.

The boundary between lake deposits and till is transitional, as the upper portion of the till was reworked by the lake waters. Lacustrine deposits, which form the upper part of the profile, typically range from about 12.2 to 24.4 m (40 to 80 ft) in thickness. They are made up of clayey silt, clay, and silty sand with poorly defined strata. The sediments forming the surface zone have been modified by recent erosional processes (involving both water and wind) and usually consist of thin layers of cross-bedded sand to depths of about 3.0 to 6.1 m (10 to 20 ft). The soil profile for a section through the library site is shown in Fig. 2.

The cohesive soils that dominate the profile are of stiff to very stiff consistency. The piling contractor had selected pile lengths that placed the base of the pile in the upper level of the till and the shafts predominantly in the clayey silt. Table 1 presents a summary of soil properties from tests conducted at this site and at nearby sites.

Configuration of Pile Heave

No detailed heave records were obtained until the potential problem had been recognized. At that time, 80 percent of the piles had been driven, including all of the piles in the central core. Spot checks that had been made by the contractor for three piles in the core showed heaving of 120, 229, and 813 mm (4.75, 9.0, and 32.0 in.). No trend was apparent from these results. The heave recorded on the top of the perimeter piles comprising the final 20 percent of the 292 piles showed a consistent trend, with the maximum total heave occurring at the center of the group in the range of 195 to 300 mm (8 to 12 in.). Heave was recorded for the piles only. Both the pile and the ground heaved with no apparent relative movement. The configuration of the recorded heave for two of the perimeter pile groups is shown in Fig. 3. Table 2 shows the incremental heave of the driven pile due to installation of others in the group.

The contractor had maintained excellent records of driving time and driving sequence for all the piles driven. This data permitted extrapolation of the detailed heave records obtained for the last 20 percent of the piles to an assessment of all pile groups.

Heave was recorded for all piles previously driven up to the driving of the

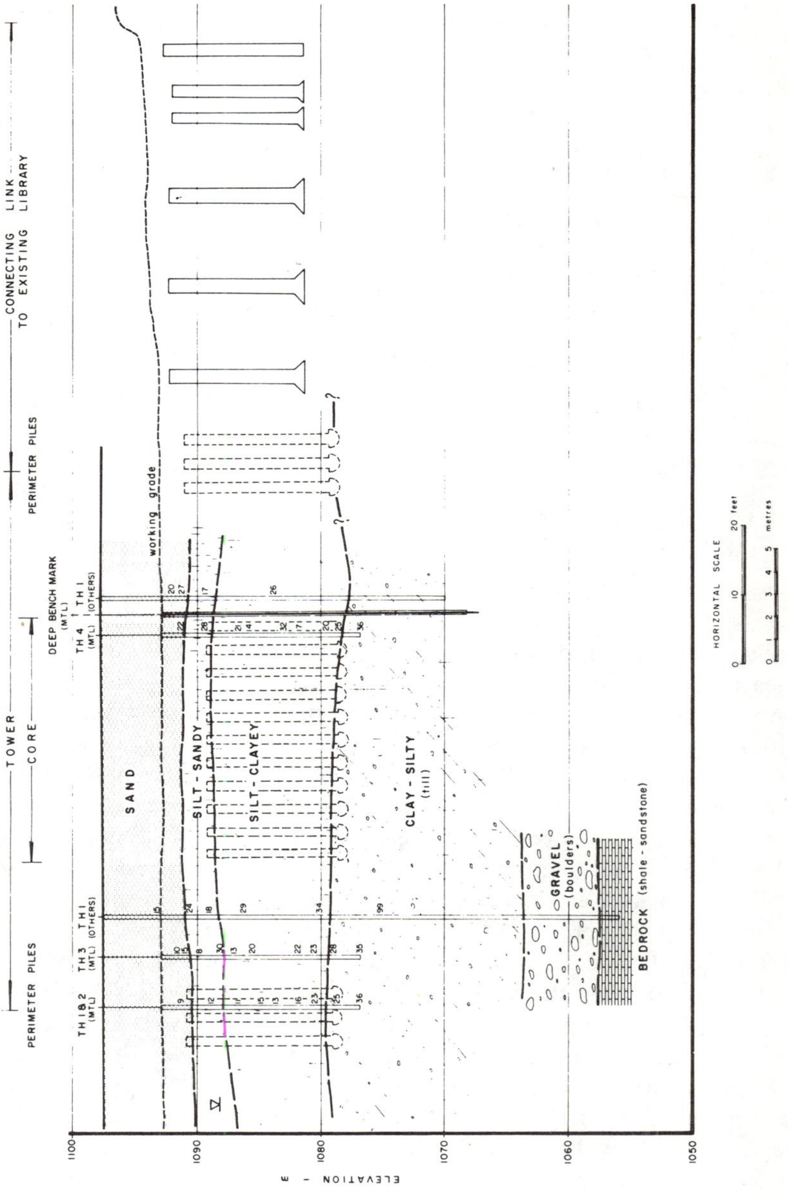

FIG. 2—*Soil profile through centerline of building.*

TABLE 1—*Summary of soil properties for library and nearby sites.*

SOIL TYPE	LABORATORY TESTS		U OF C LIBRARY	U OF C SCIENCE COMPLEX	McGILL APARTMENTS	QUEENS APARTMENTS
SILT sandy silts and silty sands (LACUSTRINE) ML-CL	GRAIN SIZE	Sand		20	52	70
		Silt		64	38	20
		Clay		15	10	10
	DENSITY (dry) kN/m^3		15.0			16.0
	MOISTURE CONTENT %		20-28	20-30	14-28	20-28
	ATTERBERG LIMITS	Plastic		18.5		
		Liquid		77.4		
	UNCONFINED COMP.STR. (kPa)		74.0	54		49
	EFF.STRENGTH PARAMETERS	ø		36°	39°	
		c'		0	0	
	STANDARD PENETRATION (N)		11-27	18-26	7-34	7-20
	CONSOLIDATION (Comp. Index)					
CLAY silty with interbedded sand lenses (REWORKED TILL & LACUSTRINE) CL	GRAIN SIZE	Sand	23	23	3	10
		Silt	59	59	84	71
		Clay	18	18	13	19
	DENSITY (dry) kN/m^3		17			17
	MOISTURE CONTENT %		18-26	15-25	23-26	23-26
	ATTERBERG LIMITS	Plastic	15	15.7-18.0	15.7	
		Liquid	21	21.9-30.6	21.0	
	UNCONFINED COMP.STR. (kPa)		154.2			
	EFF.STRENGTH PARAMETERS	ø			37°	
		c'			0	
	STANDARD PENETRATION (N)		18-34	23-42	15-25	19-31
	CONSOLIDATION (Comp. Index)		.106			
CLAY silty with occasional gravel sizes (TILL) CL-CI	GRAIN SIZE	Sand			16	7.5
		Silt			65	65
		Clay			19	28.5
	DENSITY (dry) kN/m^3		17.9	17.9		18.2
	MOISTURE CONTENT %		17-21	15-20	14-25	14-20
	ATTERBERG LIMITS	Plastic	14.5	15.2	17.0	19.1
		Liquid	35	28.5	30.1	30.1
	UNCONFINED COMP.STR. (kPa)		267.7			241.0
	EFF.STRENGTH PARAMETERS	ø		33.5		
		c'		0		
	STANDARD PENETRATION (N)		34-99	31-93	28-77	32-46
	CONSOLIDATION (Comp. Index)			0.078-0.117		

final pile within that group. Upward movement of previously driven piles was measured as the casing of the next pile was driven, when the plug was expelled, when the base was formed and as the shaft was formed during the withdrawal of the casing. Figure 4*a* illustrates the average heave of an adjacent pile at a distance of 2.5 diameters center to center from the driven pile for piles driven the same day and for piles driven at least 48 h after the pile for which heave was recorded. Figure 4*b* shows the same relationship for piles driven at a center to center spacing of 4.1 diameters. The total heave of piles spaced at 2.5 diameters is approximately the same in both cases, but about 75 percent of the heave occurs during the driving of the casing for the adjacent shaft when it is driven the same day, compared to about 40 percent after the observed pile had been in place for 48 h or more. The heave of the piles spaced at 4.1 diameters was twice as much for adjacent piles driven the same day as it was for piles driven at least 48 h after the observed pile was driven. These observations led to the conclusion that a large amount of distortion occurred in the shafts of the piles that heaved the same day they were driven and that heave occurring from the expulsion of an adjacent pile base would not necessarily reflect through to the top of the pile due to an "accordion" effect within the shaft. Piles spaced at 2.5 diameters that maintained the integrity of the shaft appeared to experience greater pushing up from the base, rather than lifting as the

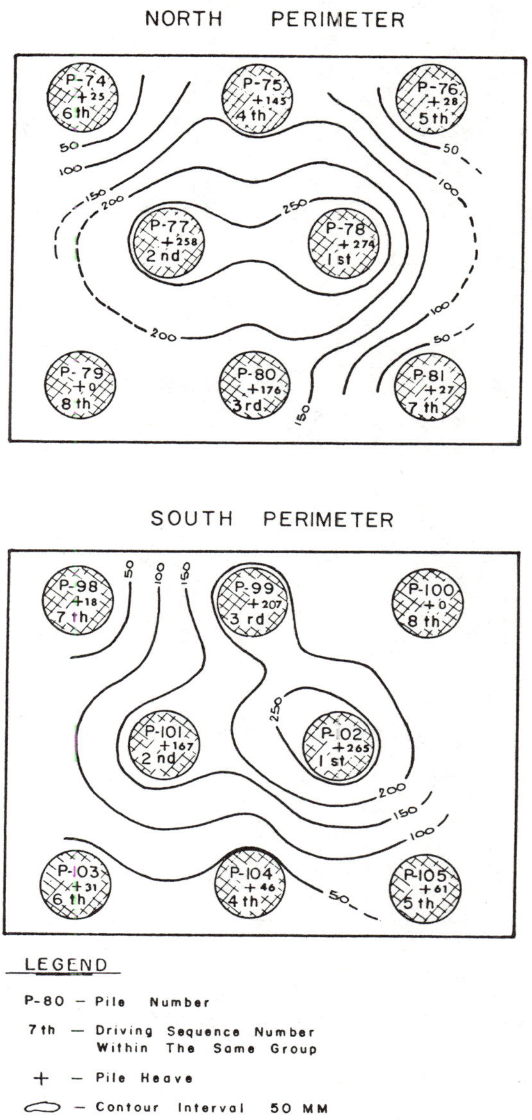

FIG. 3—*Heave configuration for two perimeter pile groups.*

TABLE 2—*Pile heave due to subsequent driving of piles in the same group.*

Order Driving	Pile That is Heaving	Pile Being Driven	Heave, mm
First	78	77	35
		80	37
		75	40
		76	47
		74	19
		81	64
		79	32
Total heave			274
Second	77	80	50
		75	35
		76	22
		74	46
		81	31
		79	74
Total heave			258
Third	80	75	19
		76	20
		74	13
		81	47
		79	77
Total heave			176
Fourth	75	76	70
		74	42
		81	21
		79	12
Total heave			145
Fifth	76	74	15
		81	8
		79	5
Total heave			25
Sixth	74	81	12
		79	13
Total heave			25
Seventh	81	79	27
Total heave			27

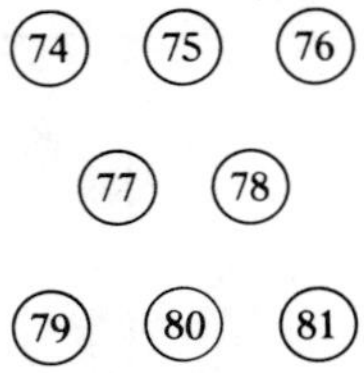

Pile Configuration

adjacent casing was driven. Piles at 4.1 diameter spacing demonstrated a much greater resistance to heaving after they had had time to develop some strength in the shaft.

Preliminary Evaluation of Piles

An analysis of the heave records led to the conclusion that many of the piles had experienced severe distortion and would not be able to support the design load. The records of time and sequence of driving maintained by the contractor, in conjunction with the heave records secured by the

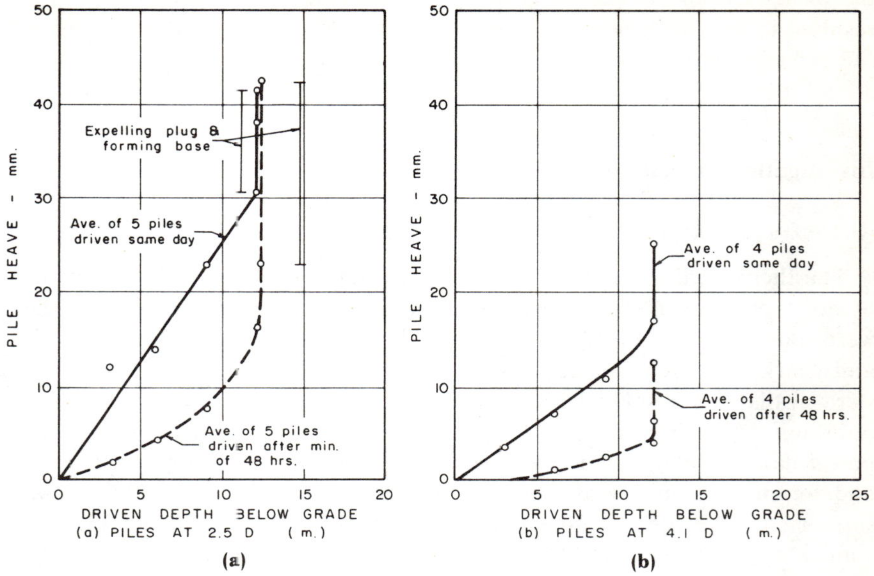

FIG. 4—*Average heave versus depth of adjacent driven pile* (a) *piles at 2.5 diameters and* (b) *piles at 4.1 diameters.*

consultant for the last 20 percent of the piles driven, allowed a broad classification of the piles into three groups. These groups were (1) probably damaged, (2) possibly damaged, and (3) probably sound. The first two groups comprised 75 percent of the piles driven.

The piles for which heave records had been obtained were readily classified, but the majority of the piles were much more difficult to assess. The driving sequence was not always the same for each group, and the time between driving varied considerably. For the first screening, all piles that had had an adjacent pile driven the same day were classified as probably damaged, and those piles that were the last to be driven in each group were classified as probably sound. The heave records were analyzed in conjunction with the driving sequence and time of driving. Although a consistent trend was evident for the heave records obtained, there were anomalies. Piles that heaved less than about 25 mm (1 in.) were classified as sound. This amount of heave usually occurred only when there was at least a day between driving of adjacent piles and where the center-to-center spacing was at least 4 diameters. Piles that consistently showed a trend of more than 75 mm (3 in.) heave were classified as probably damaged. The remainder were classified as possibly damaged.

The specifications for the piling contract spelled out that the owner could require the contractor to carry out one or more load tests. The cost of the load test would be borne by the contractor if the pile failed and by the owner if the test was successful. The high design capacity of the piles resulted in load tests being very expensive; therefore, the owners authorized the consultants to carry out a more detailed investigation before specifying a load test.

Investigation Procedure

Analysis of Capacity

The theoretical capacity of the piles was determined to indicate whether or not they were likely to have the required load-carrying capacity if they were sound. The analysis was based on the previous work of Meyerhof [*1*][2] and Clark and Meyerhof [*2*]. In addition, load tests for similar pile types were available for several projects in the immediate vicinity. Two nearby buildings on the university campus had been instrumented, and performance records had been maintained for each during construction and for more than a year after construction. One of these buildings was supported by drilled cast-in-place concrete piles and the other on piles similar to those of the library tower but constructed in much smaller groups because of lower column loads.

The theoretical analysis and assessment of available load tests and building performance records led to the conclusion that the piles would support the design loads provided that they had not been structurally damaged by the heaving that had occurred. The problem then reduced to one of assessing the structural integrity of the piles.

Sonic Tests

A sonic probe method was selected for assessing the structural integrity of the pile shafts. It had not previously been used for this purpose, but it was widely used in the oil industry to test well casing and to determine the structure of the rock in the immediate vicinity of the casing [*3–5*]. The sonic probe consisted of a sound source and two receivers at distances of 0.9 to 1.5 m (3 to 5 ft) from the source. A very sophisticated readout system provided an immediate picture of compression waves and shear waves on scopes with a filtering system designed to separate the two. Both travel time and waveforms were recorded for the 0.9-m and 1.5-m (3 and 5-ft) spans as well as the Δt. The probe was inserted in a 63-mm (2.5-in.) diameter cored hole drilled to the bottom of the pile by means of a diamond drill. The probe was raised in increments of 13 mm (0.5 in.). The waveform was photographed for each increment, and darkroom facilities in

[2] The italic numbers in brackets refer to the list of references appended to this paper.

the truck housing the equipment provided for immediate development of the recorded data and on the spot assessment.

The waveforms for each increment were plotted on a high-density chart that showed immediately the areas of low velocity, the breaks and voids in the pile, and the sound sections of pile. Figure 5 shows typical printouts of waveforms for two sections of the same pile, one for a solid section having a velocity of approximately 4 267 m/s (14 000 ft/s) and the other for a

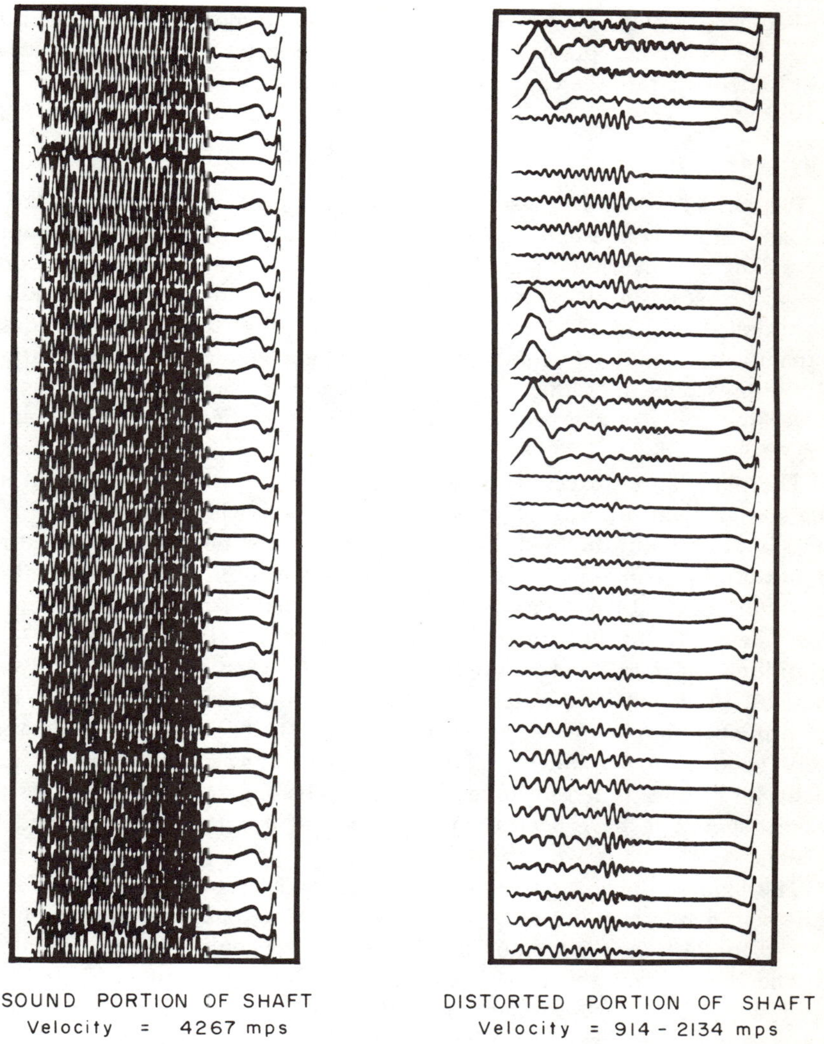

FIG. 5—*Waveforms.*

severely distorted section with velocities ranging from less than 914 to about 2 134 m/s (3 000 to 7 000 ft/s) with severe attenuation of the sonic pulse. Each section represents about 457 mm (18 in.) of pile shaft.

Six piles were tested with the sonic probe. The testing time took less than 2 h. All the piles tested were from the group that had been classified as probably damaged by heave, and all tests indicated severe damage to the piles. The contractor was reluctant to accept the conclusions, since the method had never been used before to test piles. It was learned much later that a somewhat similar technique to test for breaks or soil inclusions in drilled cast-in-place piles had been reported a few months before in France [*6*]. The owner accepted the results of the sonic tests and ordered a load test to demonstrate the capacity of a selected pile.

Pile Load Tests

Pile no. 94, which had been tested with the sonic probe, was selected for a load test. This pile had heaved 267 mm (10.5 in.). The load was applied by means of a hydraulic jack acting against a reaction member held down by drilled cast-in-place concrete piles. The testing procedure was in accordance with ASTM Method D 1143-69 and the maximum design capacity determined from the test was in accordance with the National Building Code of Canada (1965), which had been specified as the governing code for the project. The test load specified was 200 percent of the design load.

The pile showed excessive settlement, and the test was terminated at a load equivalent to 175 percent of design load when it was no longer possible to sustain the load. Following the test, the pile was excavated to a depth of approximately 6.1 m (20 ft) and was examined for structural discontinuities. Distinct separations could be observed at some points along the pile shaft, as illustrated by the photographs (Figs. 6 and 7). Two additional piles selected at random from those classified as probably and possibly damaged were tested, and a fourth pile from the group classified as probably sound was subjected to a load test. The sound pile showed a very good load settlement relationship, which was consistent with the behavior of piles tested for other projects in the general area. The piles that were believed to have been damaged failed at less than the specified test load.

The load settlement curves for the four load tests are shown in Fig. 8. The sound pile was the only pile for which the final test load was maintained for 24 h, as the load on the damaged piles could not be sustained.

The second and third load tests on damaged piles (piles 62 and 221) were selected by the contractor. After failure was recorded at 25 percent of the test load (which was 50 percent of the design load), the contractor accepted the fact that extensive damage had occurred as a result of the

FIG. 6—*Overall view of approximately the upper 6 m (20 ft) length of pile shaft.*

heaving and that the piles could not support the design load. The owner authorized the foundation consultant to work with the contractor to devise a method of providing a satisfactory foundation for the structure.

Several alternative solutions were investigated. These included a raft foundation, supplementing the existing piles with additional cast-in-place concrete piles or driven piles, and rehabilitation of the existing piles. The alternative of a raft foundation was discarded because it was suspected that the heaving had probably opened large fissures in the soil between the piles, and as a result, settlement could be excessive. Supplementing the existing piles with new driven piles was considered feasible, but the costs were not attractive. It was therefore decided to attempt to rehabilitate two of the damaged piles by pressure grouting. If this procedure proved to be successful and if it was more economical than other alternatives, all of the damaged piles would be rehabilitated by the same technique.

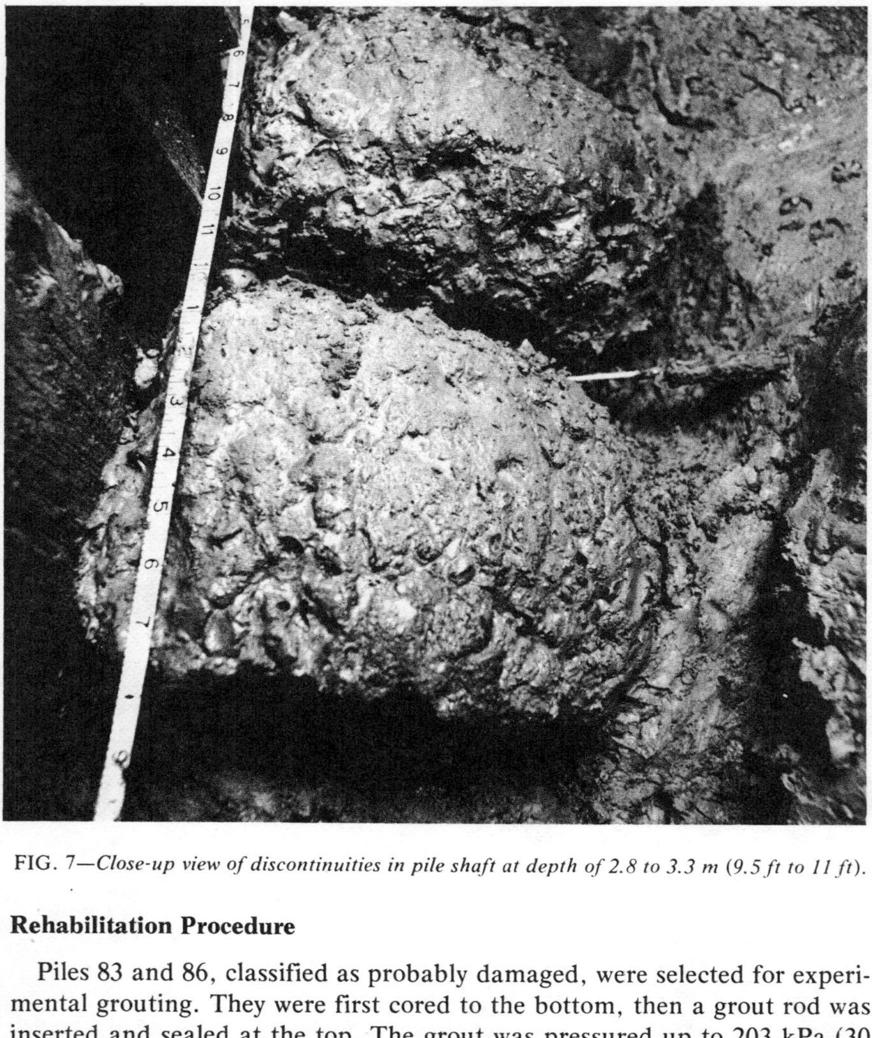

FIG. 7—*Close-up view of discontinuities in pile shaft at depth of 2.8 to 3.3 m (9.5 ft to 11 ft).*

Rehabilitation Procedure

Piles 83 and 86, classified as probably damaged, were selected for experimental grouting. They were first cored to the bottom, then a grout rod was inserted and sealed at the top. The grout was pressured up to 203 kPa (30 psi), and the pressure was maintained until grout penetration ceased. The piles were load tested 3 days after grouting. The load settlement relationship was substantially better than for the damaged piles but not as good as that of the sound piles. Figure 9 shows the load settlement curve for pile 86, which was almost identical to pile 83, and for comparison shows the load settlement curve for pile 94, which was the best of the damaged piles tested, and pile 89, which was the sound pile tested.

The experimental grouting indicated that it was likely that the piles could be rehabilitated, but the grouting procedure used was not considered to be adequate. The procedure adopted for the next piles consisted of

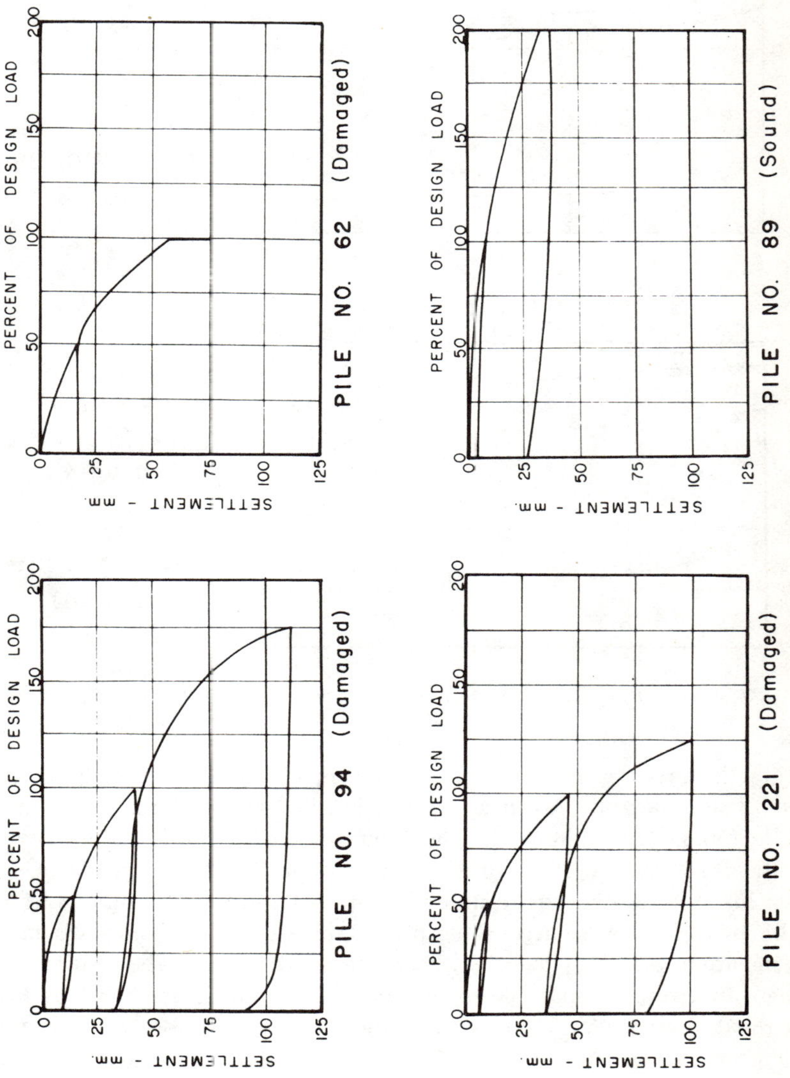

FIG. 8—*Results of load tests on damaged piles and sound pile.*

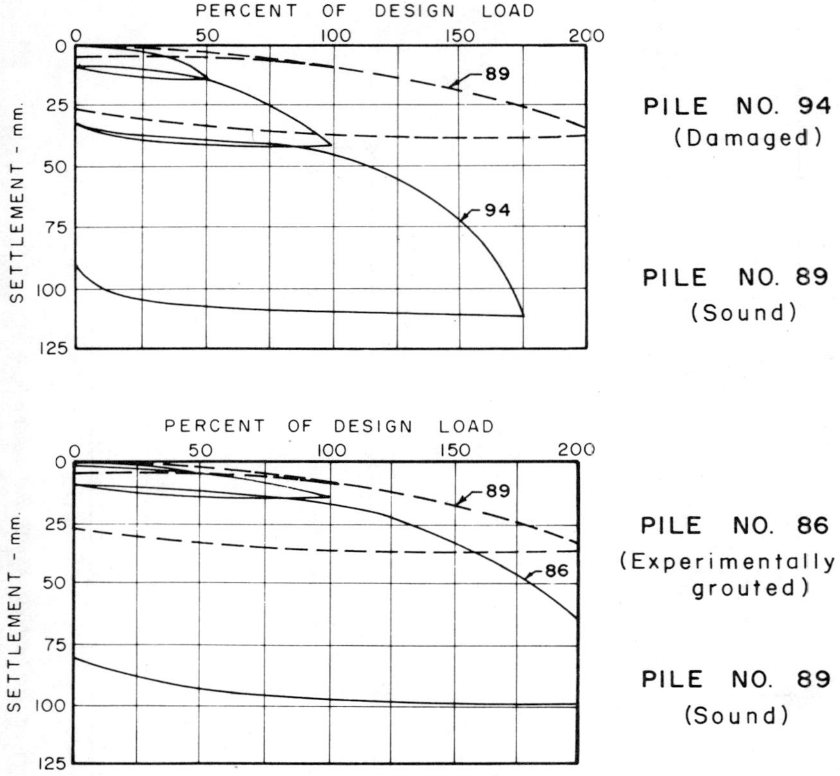

FIG. 9—*Load settlement curve for experimentally grouted damaged and sound pile.*

inserting the grout rod with a packer 1.5 m (5 ft) above the tip to the bottom of the pile and pressuring up to 270 kPa (40 psi) for 2 min then to 406 kPa (60 psi) until refusal. The grouting pressure was reduced with each lift of the rods to 270 and 203 kPa (40 and 30 psi), respectively, for the last lift. Lifts were approximately 1.5 m (5 ft) near the ground surface and 0.8 m (2.5 ft) elsewhere. This procedure was followed for two more piles, including pile 94, which had been previously tested. The results were excellent and were comparable to the sound pile tested, as shown by Fig. 10 for rehabilitated piles 94 and 56 and the sound pile 89 for comparison.

The grout specified consisted of one part water, two parts portland cement, and four percent of commercial bentonite based on the dry weight of cement. The holes for the remainder of the grouting were drilled by a percussion drill with an air and water flush. During the drilling of the holes, water was flushed out of previously drilled piles as much as 21.3 m (70 ft) away from the pile being drilled. This confirmed that there was

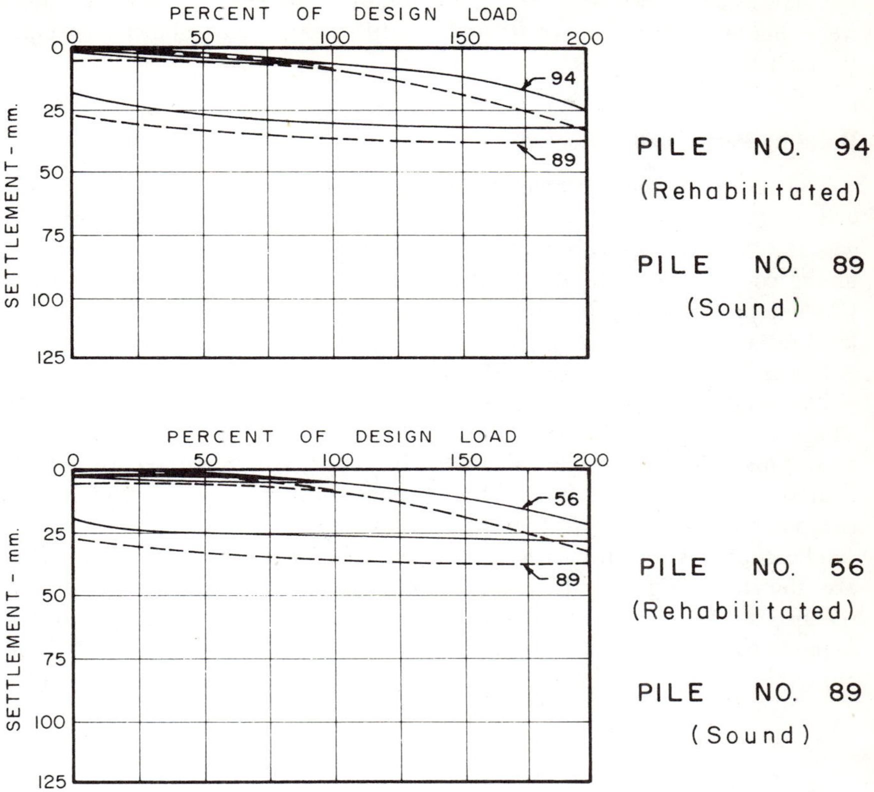

FIG. 10—*Load test results for rehabilitated piles.*

extensive fracturing of the soil between the piles and that fissures had been opened, particularly in the large core of the foundation, where pile density was highest. Most of the fracturing occurred between the depths of 9.1 and 10.7 m (30 and 35 ft) below the top of the pile (immediately above the till). A substantial amount of grout was used in filling these voids. The amount of grout ultimately used for the core of the foundation was equivalent to slightly more than a 0.3-m (1-ft) thick layer over the entire core area.

A total of 200 piles required rehabilitation. This group comprised all of the core piles except for the four corner piles, which were judged to be sound and 72 (45 percent) of the perimeter piles. Ten additional load tests were carried out on rehabilitated piles selected at random, and all proved to be acceptable according to the National Building Code of Canada (1965).

Although the grouting was considered to be acceptable and the foundation was judged to be rehabilitated, the obvious question as to how the pile foundation would perform still remained. The owner accepted a recommendation to fully instrument the building and settlement observations were made throughout construction (1971) and have been continued through to 1978.

Building Performance Records

The layout of the settlement plugs is shown on Fig. 11. A benchmark was set to a depth of 18.3 m (60 ft) below the bottom of the piles, where it was sealed in very hard basal till. The settlement plugs inserted at the base of the columns were similar to those developed by the National Research Council of Canada [7], but the settlement was recorded with a precise level rather than a water gage.

A prediction of the expected settlement of the 14-story structure was made using a finite element model and considering elastic settlements only. The value of Young's modulus used in the analysis was 276 MPa (40 000 psi). This value was back calculated from the settlement records of two nearby buildings (on similar foundations but smaller groups) on the campus that had been previously instrumented.

The contours of total settlement 1 year after completion of construction are shown in Fig. 12. Most settlement occurred in the core, and has reached a total of 52 mm (2.04 in.) 6 years after construction, compared to a predicted value of 55 mm (2.17 in.). Figure 13 shows the settlement along the centerline through the building. The settlement of the central

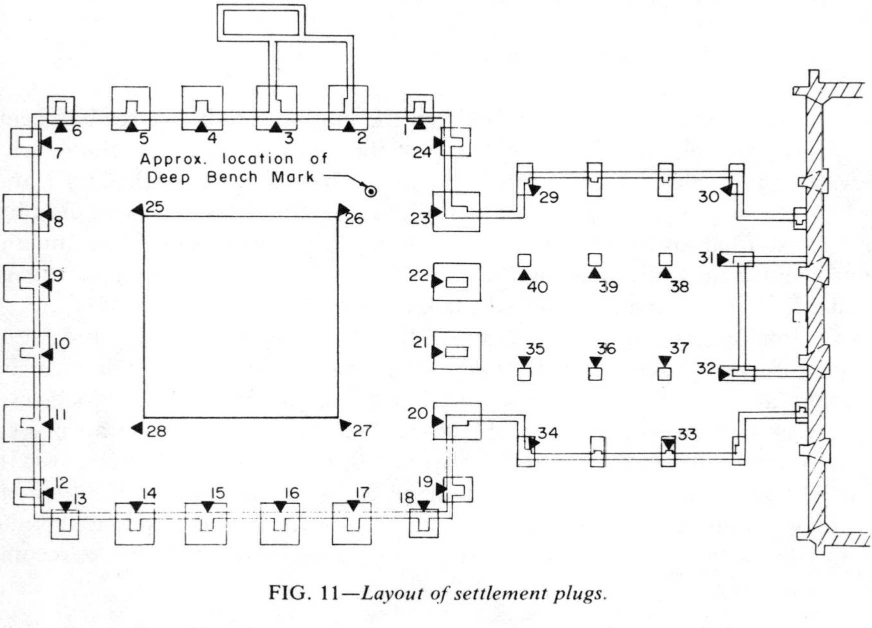

FIG. 11—*Layout of settlement plugs.*

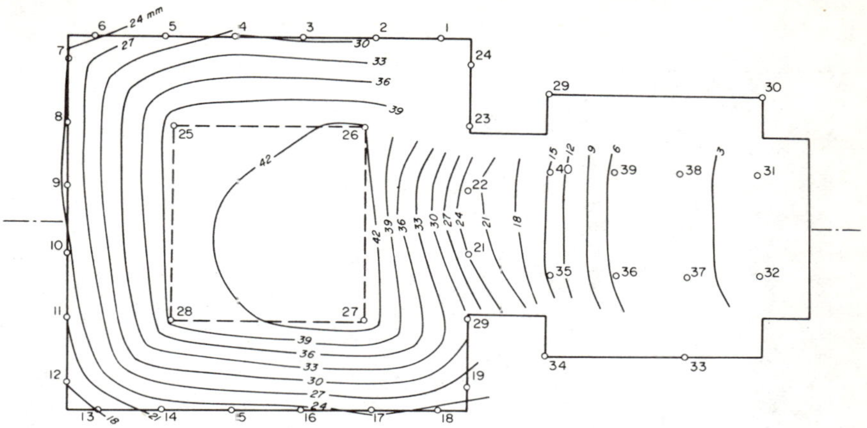

FIG. 12—*Contours of total settlement 1 year after completion.*

core plotted against the log of time during and after construction is shown in Fig. 14.

Discussion

The structure has performed successfully for almost 7 years since it was occupied. The settlements have been tolerable and very close to the predicted values. The performance is comparable to other buildings in the area that had been instrumented and where no foundation problems had been experienced.

The piling contractor was able to successfully rehabilitate the pile foundation, which experienced severe distress during construction. Very close control of grout quality and grouting procedures was required to obtain a satisfactory end result.

It is very likely that the entire problem could have been avoided by preboring the dynamically cast-in-place concrete piles or by initially selecting another pile type or raft foundation. The decision not to construct a raft foundation after the problem had been discovered proved in hindsight to be a wise one, as excessive settlement could have been experienced, as the large horizontal fissures in the soil would likely have closed as the foundation was loaded. The grouting was successful in filling the voids between piles as well as strengthening the pile shafts to a point where they could carry the design loads.

Conclusions

1. The sonic probe proved to be an economical and reliable method of testing the structural integrity of pile shafts.

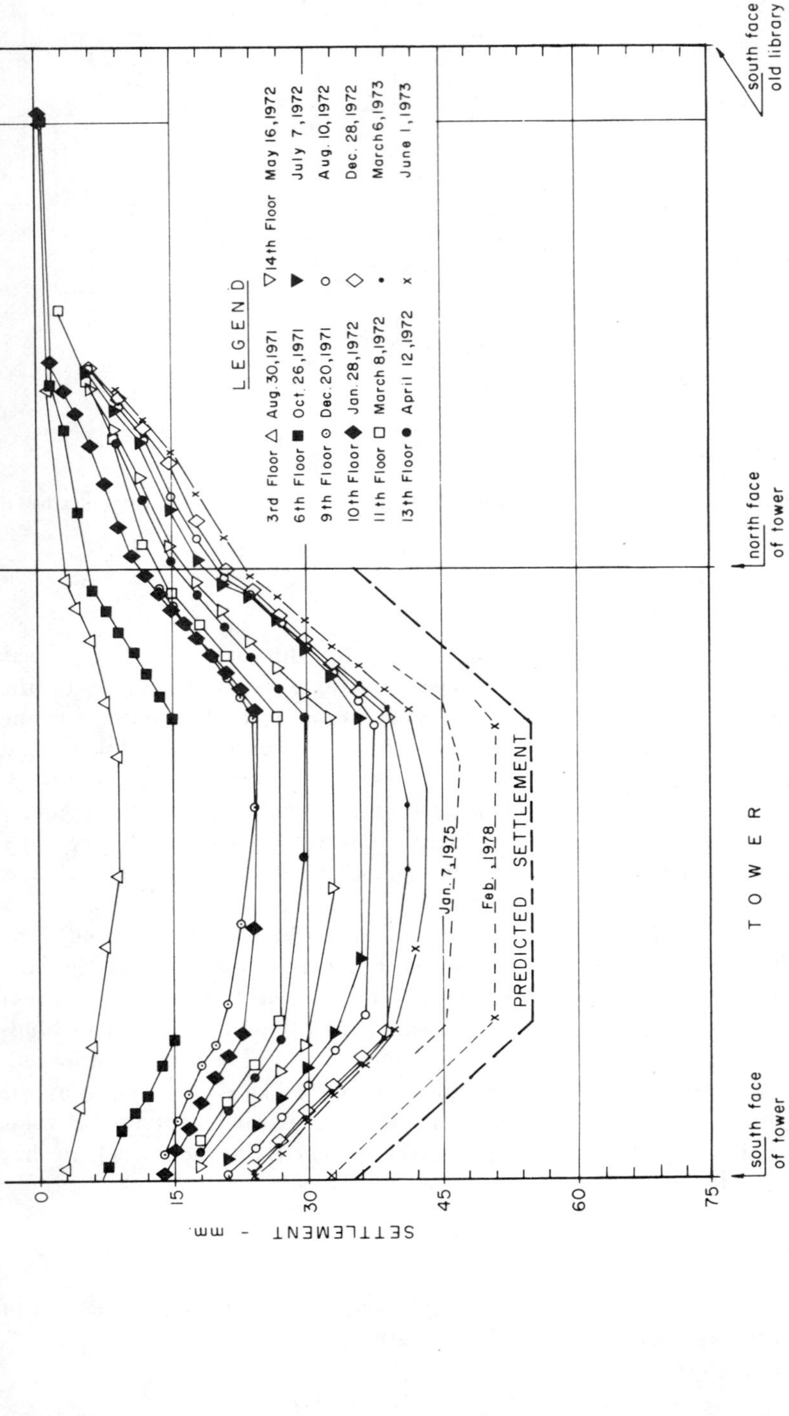

FIG. 13—*Settlement along centerline during and after construction and predicted settlement.*

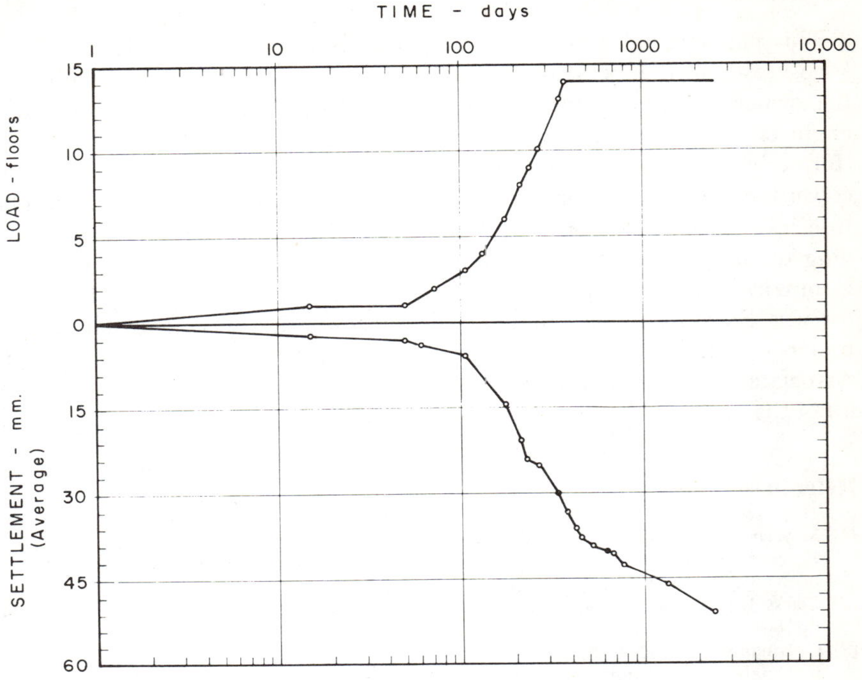

FIG. 14—*Load–settlement versus time.*

2. The pile heave records indicated that heave occurred as the casing for the adjacent piles were driven and as the plug was expelled and the base formed. No heave was observed as the shaft of the adjacent pile was constructed.

3. The greatest distortion of the pile shaft occurred when adjacent piles were driven the same day. When a pile was heaved the same day it was constructed, the shaft integrity was destroyed.

4. Heave due to the forming of an adjacent expanded compacted concrete base is not serious, since the heaved pile is pushed up through the soil. Heave due to driving of an adjacent casing is very serious, as it breaks up the shaft or separates the shaft from the base.

5. In order to prevent damage to the concrete shafts of previously driven piles, specifications should require preboring for the shafts and a staggered driving sequence to allow at least 24 h between the driving of adjacent piles.

6. Grouting procedures can be successful in rehabilitating piles damaged by heaving.

Acknowledgments

This paper is published with the permission of Dr. H. A. R. de Paiva, Vice-President (Services) of the University of Calgary, acting on behalf of the owners. Architect for the project was W. G. Milne, and consulting engineers were Carswell Engineering Limited. R. M. Hardy and Associates (formerly Materials Testing Laboratories) were engaged as foundation consultants after the problem had been encountered. The original foundation investigation for the project was carried out by others. The sonic logging of the piles was done by Schlumberger of Canada. The Foundation Company of Canada was the general contractor, and the piling subcontractor was Franki-Canada Limited. The rehabilitation work was carried out under the direction of H. A. Baragar, P. Eng., of R. M. Hardy and Associates. The writer gratefully acknowledges the helpful participation of all of the above parties in successfully solving the problem.

References

[*1*] Meyerhof, G. G., "The Design of Franki-Piles with Special Reference to Groups in Sands," Proceedings of Symposium on the Design of Pile Foundations, International Association of Bridge and Structural Engineering Congress, Stockholm, Sweden, 1960.

[*2*] Clark, J. I. and Meyerhof, G. G., *Canadian Geotechnical Journal*, Vol. 10, No. 1, 1973, pp. 86–102.

[*3*] Grosmangin, M., Kokesh, F., and Majani, P., "The Cement Bond Log—A Sonic Method for Analyzing the Quality of Cementation of Borehole Casings," Paper 1512-G, Society of Petroleum Engineers of American Institute of Mining, Metallurgical, and Petroleum Engineers, Dallas, Tex., 1960.

[*4*] Kokesh, F., Schwartz, R., Wall, W., and Morris, R. L., *Journal of Petroleum Technology*, Mar. 1965, pp. 282–268.

[*5*] Morris, R. L., Grine, D. R., and Arkfeld, T. E., The Use of Compressional and Shear Acoustic Amplitudes for the Location of Fractures, Paper SPE-723, Society of Petroleum Engineers of American Institute of Mining, Metallurgical and Petroleum Engineers, Dallas, Tex., 1963.

[*6*] Paquet, J., "Contrôle des Pieux par Carottage Sonique," *Annales de l'Institut Technique du Bâtiment et des Travaux Publics*, Oct. 1969. No. 262.

[*7*] Peckover, F. L., "A New Water Tube Level for Measuring the Settlement of Buildings," Technical Report No. 11, Division of Building Research, Ottawa, Ont., 1952.

R. W. Cooke[1]

Influence of Residual Installation Forces on the Stress Transfer and Settlement Under Working Loads of Jacked and Bored Piles in Cohesive Soils

REFERENCE: Cooke, R. W., "**Influence of Residual Installation Forces on the Stress Transfer and Settlement Under Working Loads of Jacked and Bored Piles in Cohesive Soils,**" *Behavior of Deep Foundations, ASTM STP 670,* Raymond Lundgren, Ed., American Society for Testing and Materials, 1979, pp. 231–249.

ABSTRACT: This paper discusses some of the causes of residual installation forces in piles and the effects of the forces on the subsequent behavior of piles under structural or test loading. In the cases of jacked and driven preformed piles, residual forces arise as a result of the different rates of mobilization of bearing forces at the base and frictional forces on the shaft. The magnitudes of these residual forces can be large, and for this reason they may have a dominant effect on the mechanism of load transfer to the supporting soil at working loads. Residual forces are likely to be much smaller in the case of bored, cast *in-situ* concrete piles.

Some loading tests of jacked tubular-steel piles and bored cast-in-situ, concrete piles are described. All the piles were instrumented so that the distribution of load transfer could be examined, and all were in London clay. Results are presented to show that, as a result of the existence of residual forces, the proportions of shaft and base resistance mobilized at working loads are different for the two types of pile. Some evidence is presented to show that the settlements at working loads may also be influenced by residual forces.

KEY WORDS: bored pile, driven pile, jacked pile, clay, friction, negative friction, load test, settlement, instrumentation, stress distribution

The installation of a pile inevitably causes changes in the state of stress in the soil in the vicinity of the pile. Boring a hole for a cast *in-situ* pile in clay, for instance, causes relief of stress so that moisture migrates toward the unloaded areas, and swelling frequently occurs. Driving piles into clay

[1] Principal scientific officer, Building Research Station, Garston, Watford, Herts WD2 7JR, England.

soils causes extensive displacement and remolding of the soil, and the resulting pore water pressures take time to revert to their original levels. Granular soils are normally compacted by pile driving, and stress levels are consequently increased. All these phenomena affect the stress-strain relationships and strength of the soil and influence the behavior of piles when they are subsequently loaded. While many designers are aware of these aspects of pile design and installation, it is probably not widely appreciated that changes in the soil stresses during and after installation of the pile will automatically be accompanied by the development of related forces in the pile itself. Equilibrium between the soil and pile forces, when no external load is applied to the pile, constitutes a residual force system relative to which the effects of structural or test loads on the pile must be measured. Many of these residual forces vary as a result of small changes in pile installation technique and are thus difficult to measure and to take account of. One set of residual forces, however, depends on the relative rate of mobilization of shaft, and base resistance on piles in clay and is amenable to repeated measurement. These forces appear to have considerable relevance to the mechanism of load transfer from jacked or driven piles to the supporting soil under working load conditions.

An opportunity to study the magnitude and distribution of residual forces and their effect on pile behavior was presented when the first of three instrumented tubular steel piles was being jacked into London clay at Brent Cross for a series of experiments on pile interaction. The residual force aspect of this work forms a major section of the present paper. In order that the significance of the effect of residual loads on the load transfer mechanism may be fully appreciated, the observations at Brent Cross have been contrasted with analogous observations made earlier on bored piles in similar soil. These observations formed part of the tests of large-diameter bored piles carried out by Whitaker and Cooke [*1*][2] in London clay at Wembley.

In a separate paper to this symposium, some recent measurements of load transfer from bored piles formed in London clay under bentonite slurry are briefly described. These measurements, which were included in a study of the effects of forming piles under bentonite on the friction resistance mobilized on the shafts, were sponsored by the Construction Industry Research and Information Association (CIRIA). They show the mechanism of load transfer in rather more detail than did the tests at Wembley. The effects of the method of forming the piles on the ultimate shaft friction have been reported by Fearenside and Cooke [*2*].

The present paper attempts to show how load transfer under working load depends on the residual forces set up during the installation of driven and bored piles, and suggests that the magnitude of the settlement under working load may also be related.

[2] The italic numbers in brackets refer to the list of references appended to this paper.

Present Knowledge

In the report of their tests of large bored piles in London clay, Whitaker and Cooke [*1*] pointed out that on unloading from a test load, each pile shaft remained in compression under the action of a residual load at the base and a balancing negative friction force on the shaft. This equilibrium condition in a bored pile after it has settled under load is analogous to that existing immediately after the installation of a driven pile. At about the same time, Hanna [*3*] using rod extensometers to record axial deformations of long piles detected similar residual loads subsequent to unloading from a test. Hunter and Davisson [*4*] noticed large apparent tensile forces at a pile base when compressive loading tests of piles in medium sand were followed by pulling tests. Since the occurrence of such tensile forces at the base was not possible, Hunter and Davisson concluded that unexpected residual compressive forces had not been taken into account. They then used the magnitude of the apparent tensile force to correct the load distribution in the shaft measured in the compressive loading tests. Hunter and Davisson showed that, although residual forces at the pile base could amount to 80 percent of the anticipated base load when piles were driven with a conventional hammer, residual base loads were negligible when piles were driven by a vibratory hammer. This is because vibratory hammers are extremely effective in minimizing shaft friction during pile installation, but residual base loads only occur when adequate negative friction forces can be mobilized to oppose them. As a result of these important observations, Hunter and Davisson concluded that, if residual loads were ignored in load transfer measurements, serious errors in the division of load between base bearing and shaft friciton could result. The total pile loads, they suggested, would not be affected.

The significance of the residual force system with respect to long piles and particularly to piles subjected to both compressive and tensile loading was discussed in detail by Hanna and Tan [*5*]. The same authors later compared the concepts they had presented with the results of model tests of long, extensively instrumented piles in deep beds of sand in which two different installation techniques were employed [*6*]. From the test results, it was concluded that the shape and magnitude of the load-settlement diagram for a pile were influenced by the residual stress state and thus that the behavior of a pile depended on its previous load history. Hanna and Tan further suggested that advances in the science of pile analysis would occur only when the complete stress systems in the vicinity of piles were studied and the extent of residual stressing along pile shafts had been established.

Development of Residual Forces

It was stated earlier that the residual forces that exert an important influence on the behavior of driven piles arise because of differences in the rates of mobilization of resistance on the shaft and at the base as a pile is displaced, either during installation or under compressive loading. The development of these residual forces is shown in idealized form in Fig. 1*a*. Elastic compression of the shaft under the installation force is ignored for simplicity. It is assumed that at the end of installation the necessary driving force, which is equal to the sum of the frictional and base resistances, falls to zero at point *A* on the displacement abscissa of the diagram. As the installation force falls to zero and the pile rebounds, the frictional force is presumed to fall at the same rate with respect to upward displacement of the shaft that it develops relative to downward displacement when the pile is loaded. The base resistance is similarly presumed to fall with rebound at the same rate that it increases during the application of load, and this, as has often been demonstrated experimentally, is at a lower rate than that associated with the devlopment of friction. For equilibrium to be established at point *A*, where the applied force is zero, a friction force R_n given by *AC* is necessary, equal in magnitude but opposite in sign to the residual base load Q_r given by the ordinate *AB*. Cooke and Price [7] showed that the magnitude of these forces could be as much as 50 percent of the base bearing capacity. The probable form of the soil displacement for the development of the residual forces as the pile rebounds is indicated in Fig. 2.

When the pile is test loaded to failure, the overall load-settlement curve follows the path *AL* in Fig. 1*a*, while the base resistance increases along *BJ* and the shaft friction along *CK*. At points *L, J,* and *K*, the loading paths rejoin the installation force envelope and its components and any additional loading results in an increase in the embedded length of the pile. For con-

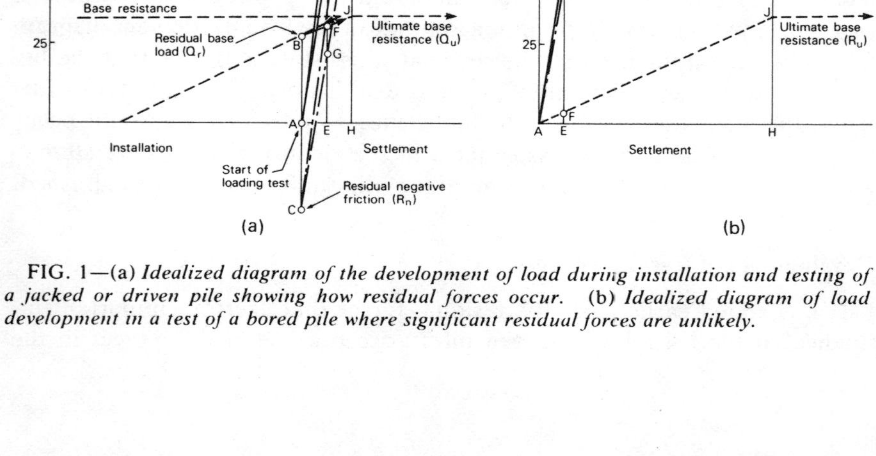

FIG. 1—(a) *Idealized diagram of the development of load during installation and testing of a jacked or driven pile showing how residual forces occur.* (b) *Idealized diagram of load development in a test of a bored pile where significant residual forces are unlikely.*

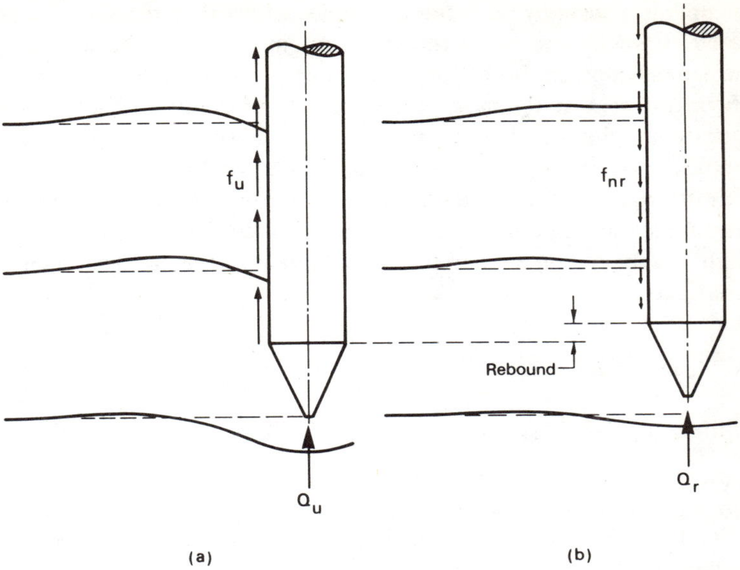

FIG. 2—*Probable form of the soil displacements near the base of a jacked or driven pile* (a) *during the application of the installing force and* (b) *immediately after its removal.*

venience, in the analyses described later in this paper, the working load P_w of the piles is taken to be one-half of the ultimate test load P_u. In Fig. 1*a*, such a working load will be given by point *D* on the overall load-settlement curve corresponding to a settlement *AE* measured from the point *A* to which rebound had occurred. It will be seen that at a settlement *AE*, the base load *EF* is a very large proportion of the ultimate base resistance Q_u, while the friction *EG* is less than one-third of the ultimate frictional resistance R_u. At working loads given by a larger load factor than 2, the friction would be less than *EG* and could even be negative if the settlement *AE* were so small that the point *G* was below the settlement axis *OH*. If Fig. 1*a* adequately represents pile behavior, it is clear that residual forces can be large in the case of jacked or conventional driven piles and that the base and frictional components of support at the working load are highly dependent upon them.

In Fig. 1b, idealized load-settlement relationships are presented for a bored pile having the same ultimate values of base resistance and friction and the same load-settlement characteristics as the driven pile discussed in Fig. 1*a*. At the working load *ED* in Fig. 1*b*, which is the same magnitude as *ED* in Fig. 1*a*, and the settlement *AE*, the frictional component *EG* is a larger proportion of R_u, and the base component *EF* is extremely small. Figure 1*b* probably adequately represents the behavior of bored piles without enlarged bases. It was discussed in detail by Whitaker and Cooke and clearly contrasts strongly with the probable behavior of driven piles shown in Fig. 1*a*. It should be noted that the settlement *AE* at the working load has the same value in both diagrams. These simple idealizations would only show different settlements if the slopes or origins of the component curves were to change. For example, concrete shrinkage or inadequate clearing of debris at the base of a pile bore hole would correspond to a new origin to the right of *A* for the base load curve *AFJ* in Fig. 1*b* and thus to a slightly larger settlement at the working load. Pile loading tests throwing some light on the residual force systems outlined above are described in the following sections.

Jacked Pile at Brent Cross

Test Equipment and Site Details

The pile was formed of steel tube 0.17 m in diameter and 5 m long with strain-gaged, pillar-type load cells dividing the pile shaft into four equal segments. Each load cell unit consisted of eight pillars, and completely independent electrical circuits connected the gages on alternate pillars. Thus, there were effectively two load cells at each level, but since the agreement between them was always good, only the mean has been quoted in the results.

Six months before the pile was installed, a well-supported trench, 5.5 m deep and 2 m wide, was excavated 2 m from the intended position of the pile. Trains of horizontal inclinometers described by Cooke and Price [*8*] were placed in 50-mm-diameter bore holes drilled from the trench face to the pile shaft surface. A section through the trench and the pile when jacked to the maximum penetration of 4.6 m is shown in Fig. 3. The figure also shows part of the cantilevered reference truss, from which soil heave measurements were made as the pile was installed and settlements were measured in vertical loading tests.

The pile was jacked into the ground from a driving frame supported on four bored piles, which also anchored the main beams providing the reaction for the loading tests. These piles were 300 mm in diameter, 4.5 m long, and they were formed at the corners of a rectangle so that their axes were approximately 20 test pile diameters from the test pile.

London clay at the site extended for considerable depth below the ground surface, and the undrained shear strength as determined from tests of 98-mm-diameter samples increased uniformly from 35 kN/m^2 near the surface to 78 kN/m^2 at the level of the pile base with a mean value $\bar{c}$ of 56 kN/m^2.

Observations During Installation

Because of the limited travel of the ram of the hydraulic jack used to

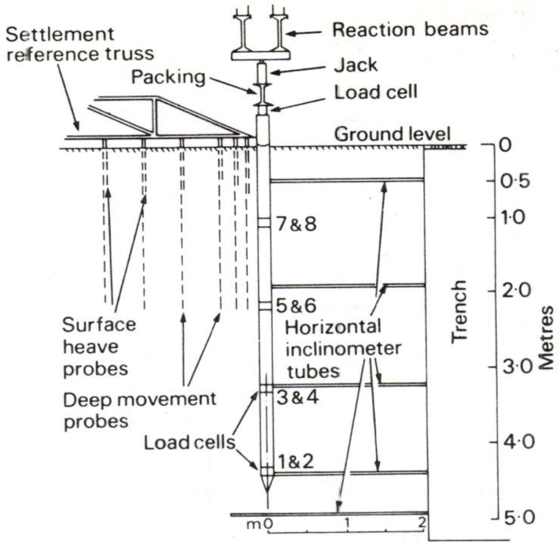

FIG. 3—*Vertical section through the experimental site at Brent Cross showing the pile and observation trench and the positions of the main instrumentation.*

install and test load the pile, it was necessary to unload the pile at penetration stages of about 0.1 m for the crosshead of the driving frame to be lowered. Thus, it was possible not only for the complete installation force-penetration diagram to be obtained, but also for residual loads to be observed at intervals of 0.1 m. Figure 4 presents this information and shows continuous readings of the pairs of load cells in the pile shaft from the points where they passed beneath the ground surface. At large penetrations, the residual loads at the base (cells 1 and 2) were approximately 75 percent of the maximum base load.

Each time the pile was unloaded for the jacking crosshead to be reset, the rebound was measured by a steel tape used for recording the overall penetration of the pile. Readings of the tape were only accurate to 0.5 mm, but Fig. 5 shows how the rebound was between 2 and 4 mm at penetrations less than about 1 m, where the friction that could be mobilized to oppose rebound was small. The rebound of the pile at larger penetrations appeared to stabilize at about 2 mm when sufficient negative friction was being developed on the shaft to prevent larger movements.

Typical observations of the displacements of the clay corresponding to this rebound of the pile are presented in Fig. 6. When the penetration of the pile base was six pile diameters (1 m), the soil displacements recorded by an inclinometer train at a depth of 0.45 m, during the application of the installing force were as shown by the full line in the upper curves of Fig. 6*a*. After the installing force was reduced to zero, the soil displacements were as shown by the dashed line in that figure. The soil recovery on unloading was clearly consistent with a pile rebound of about 2 mm. The

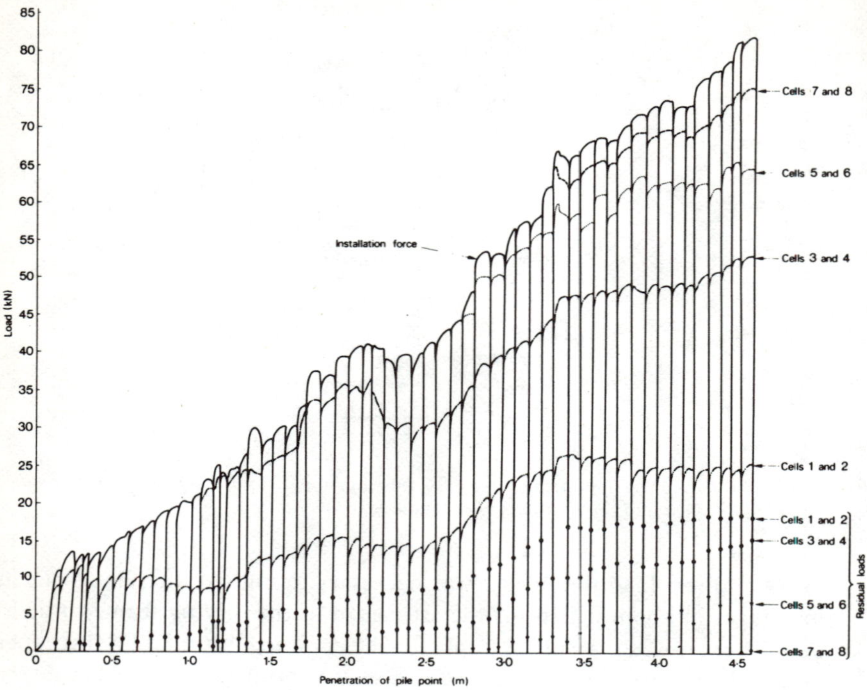

FIG. 4—*Installation force envelope and pile loading at each load cell position, showing the residual forces resulting from each unloading of the pile.*

lower curves of Fig. 6*a* show the soil displacements and their recovery on a plane some distance below the pile base at the same installation stage. Important similarities are seen to exist between the observations presented in Fig. 6*a* and the force-displacement system idealized in Fig. 2.

At a penetration of 15 pile diameters (2.5 m), when the pile base had passed the level of the second train of inclinometers, the soil displacements were as shown in Fig. 6*b*. At the lower level, probably because the pile base had not penetrated far beyond them, the inclinometers again showed a recovery from the fully deformed state consistent with a 2-mm rebound of the shaft. In contrast, the upper curves of Fig. 6*b* show that the soil near the ground surface was no longer being displaced to the same extent during installation of the pile. Nevertheless, when the installing force was reduced to zero, the soil recovery was 100 percent of the displacements occurring while the pile was being driven.

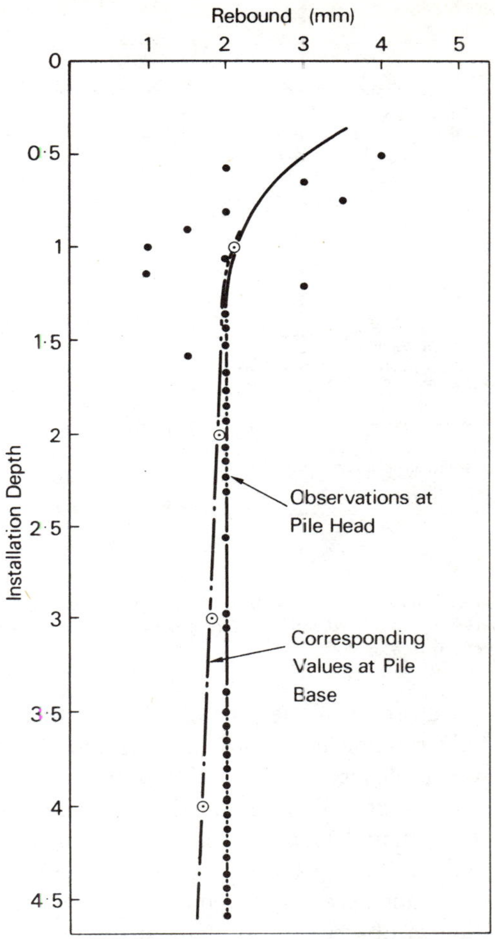

FIG. 5—*Magnitude of the rebound at every stage during installation of the pile at Brent Cross.*

Incremental Loading Tests

On a number of occasions during installation of the pile, a pause was made for incremental loading tests to be carried out. Before starting these loading tests, the residual loads recorded by those of the cells that had passed beneath the ground surface were noted. These residual loads have been plotted in Fig. 7 for six different values of the penetration–diameter ratio from 6 to 27.4. It will be seen that at every value of the penetration: diameter ratio the residual loads fall close to the curve through all the load cell

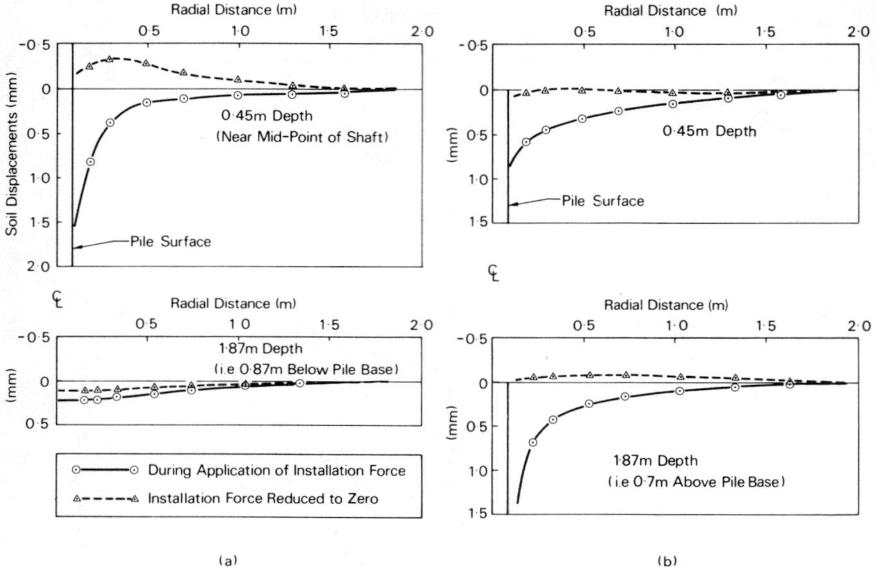

FIG. 6—*Soil displacements at two levels in a vertical plane through the pile shaft during application of the installing force and immediately after its removal* (a) *at a pile penetration of 1 m,* (b) *at a pile penetration of 2.5 m.*

readings for full penetration of the pile. This curve therefore represents a unique residual load relationship indicating the initial shaft loading condition for any length of pile less than the maximum.

Corresponding values of load and pile head displacement (settlement) measured in the incremental loading tests are shown in Fig. 8. Since the ultimate load in each test was given by a point on the installation load envelope (Fig. 4), it was not necessary to continue any of the tests beyond 50 percent of this value to observe the settlement at the working load corresponding to any length of pile. Values of settlement at the working load expressed as a percentage of the pile diameter are plotted against pile length in Fig. 9.

It is worthy of note that when the embedded length was small and thus the proportion of the working load carried by the base was large compared with that mobilized on the shaft, the settlement-diameter ratio was greater than 0.3 percent. As the embedded length was increased, the settlement at the working load fell. When the length-diameter ratio exceeded about 10, the pile assumed what might be called friction pile characteristics, and the settlement-diameter ratio stabilized at about 0.22 percent.

During installation of the pile and throughout the incremental loading tests, loads in the shaft had been continuously recorded. In Fig. 10, the distribution of residual load at the beginning of each incremental test and

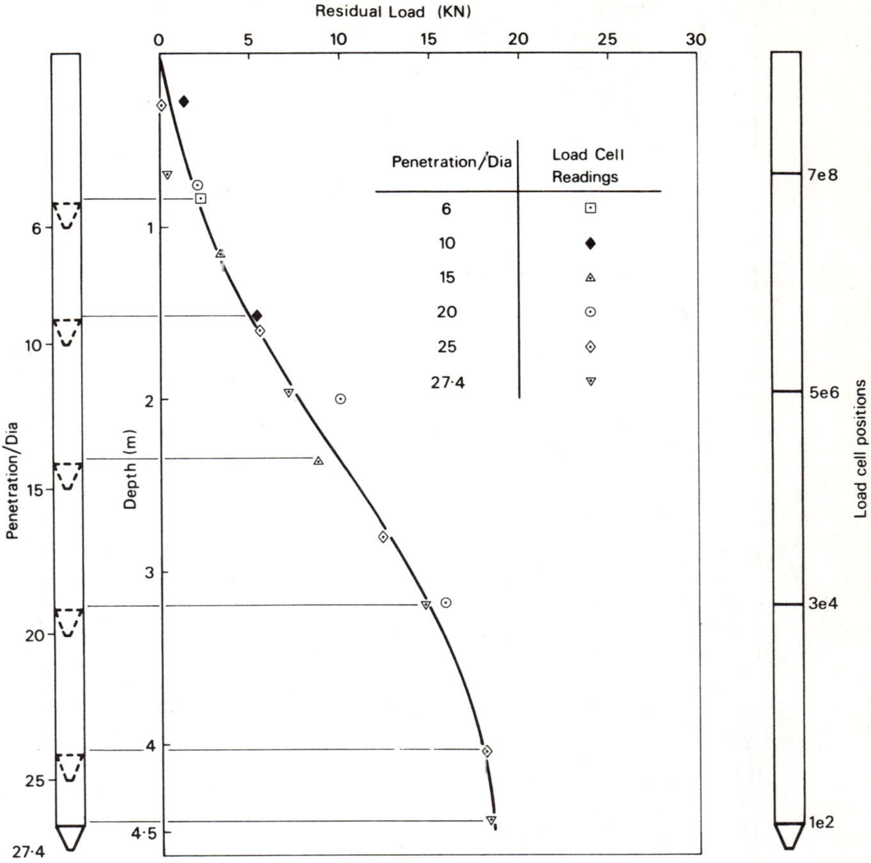

FIG. 7—*Residual load profile for all penetrations of the pile.*

the corresponding distributions during the application of each increment are shown for five different values of the length-diameter ratio. The corresponding load distributions observed during installation (that is, the ultimate load condition) at approximately the same penetrations are also shown. For the working loads, calculated assuming a load factor of 2 in each case, the distributions of shaft load have been obtained by interpolation. From the slopes of the curves at working loads, the distributions of stress transfer (that is, the shear stresses on the shaft surface) have been plotted. It will be seen that values of the stress transfer are negative at the top of the shaft in two of the five examples and that positive values are generally less than 20 kN/m^2. That the friction over any part of a pile shaft can be negative under loads as large as 50 percent of the ultimate bearing capacity is a most unexpected observation. However, the smoothly changing pattern

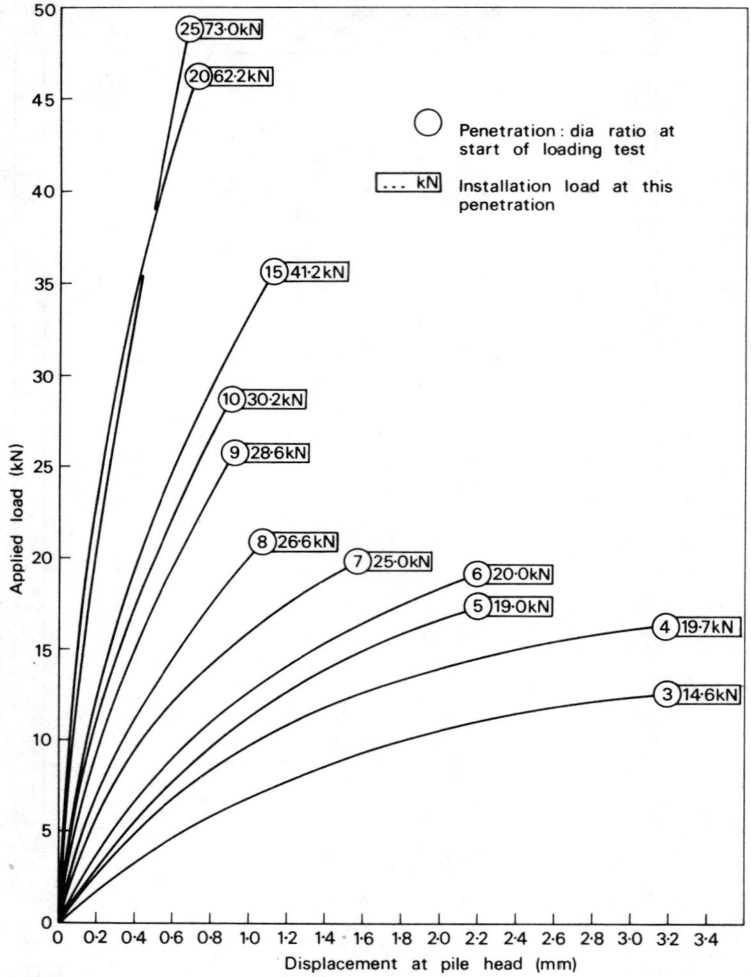

FIG. 8—*Results of incremental loading tests to working load made at a range of penetrations during pauses in the installation of the pile.*

of the load distribution curves with increasing applied load suggests that there is little basis for doubting the reliability of these results. Nevertheless, the point to be emphasized is that at the chosen working load, in every case, the mean value of the frictional stress transfer was low, while the proportion of the ultimate base resistance mobilized was high. In each diagram of Fig. 10, the values of the base load mobilized under working conditions is indicated.

If $\bar{f}$ is the mean stress transfer from the shaft and $\bar{c}$ is the mean shear

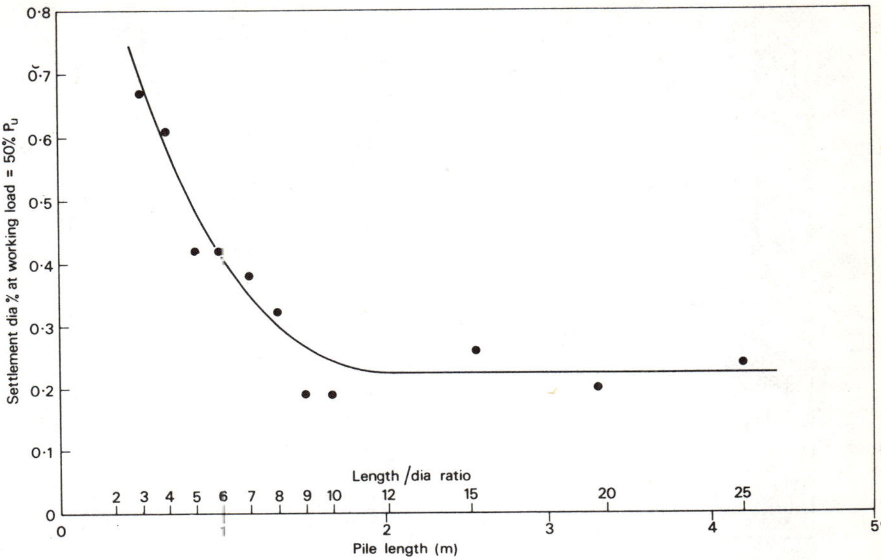

FIG. 9—*Settlement/diameter percent at the working load* (P/P_u = *0.5) for a range of penetrations of the pile.*

strength of the clay over the shaft length, the value of the mobilization factor $\bar{f}/\bar{c}$ at working load given by a load factor of 2 in the final test at full penetration was 0.11. At the base, if q is the bearing pressure developed at the working load and q_u is the ultimate bearing pressure, the factor q/q_u in the same test was 0.74. These figures will be compared with corresponding values observed in tests of bored piles later in this report.

Bored Cast-in-Situ Piles at Wembley

Brief Details

In this investigation, described in detail by Whitaker and Cooke [*1*], 13 piles having a range of shaft diameters and shaft lengths were loaded by increments to beyond the working load and by the constant rate of penetration (CRP) method to failure. Eight of the piles had bases enlarged by underreaming to twice the shaft diameter, and all of the piles were fitted with strain-gage load cells immediately above the bases so that the frictional and base-bearing components of resistance could be separated. The piles had shaft diameters of 0.63, 0.77, and 0.94 m, and the lengths varied from 9.3 to 15.2 m.

The shear strength of the clay at the site increased from about 70 kN/m^2 at the clay surface to 190 kN/m^2 at a depth of 15 m with a mean value $\bar{c}$ of 110 kN/m^2.

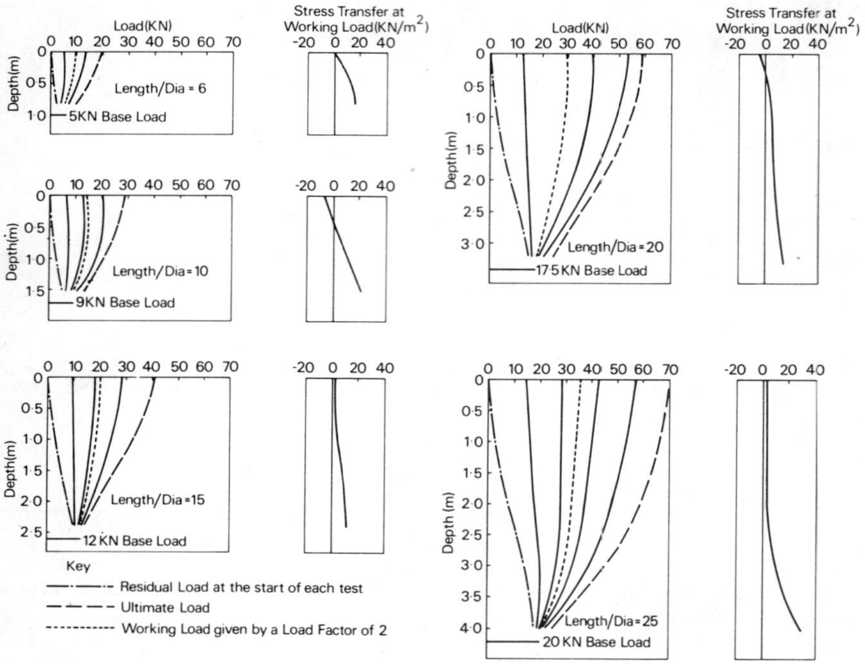

FIG. 10—*Distribution of pile shaft loading during the incremental tests at five penetrations. Attention is drawn to the distribution at the chosen working load for each penetration and to the corresponding stress transfer from the shaft surface.*

Observations During Installation

Reexamination of the test data has revealed that the vertical forces indicated by the load cells at the pile bases immediately after the concrete was poured were between one-third and one-half of the weight of wet concrete. In all but two of the 13 piles, these forces decreased during the period prior to the start of each incremental loading test. This is attributed to increased arching between the concrete and the clay around the shaft, which could have been initiated by shrinkage of the concrete away from the clay beneath the base. At the start of each loading test, the force at the pile base had fallen to a value close to the zero measured before the cell was lowered into the bore hole. Touma and Reese [*9*] pointed out that volumetric strains due to temperature changes and concrete shrinkage could affect the relative vertical positions of points on a pile shaft and in the surrounding soil, resulting in shear forces being developed.

Residual forces resulting from shrinkage of concrete in bored piles will be tensile in the shaft, in equilibrium with compressive forces in the surrounding soil. They will thus tend to be of opposite sign to those shown diagrammatically in Fig. 1*a* for a jacked or driven pile. The forces will reduce the magnitude of the base component *EF* shown in Fig. 1*b*, which idealizes the testing of a bored pile.

Incremental Loading Tests

From the results of the incremental tests, Whitaker and Cooke tabulated values of the applied load P, the base load Q_p, and by differences, the friction R_p, together with corresponding values of the settlement. The table also listed the maximum values P_u and Q_{pu} measured in the CRP test. By plotting P/P_u percent against settlement-diameter (Fig. 11) for each of the five straight-shafted piles, values of the settlement-diameter at loads corresponding to the working load, that is, $P/P_u = 50$ percent, have been obtained. These have been plotted against the corresponding length-diameter ratio for each pile in Fig. 12 and the mean is shown to be a little less than 0.30 percent.

It was possible to determine the base load Q_p corresponding to the working load from a plot of the variation of Q_p with P using the tabulated values given for each pile. These values of Q_p have been plotted at the appropriate depth for each pile in Fig. 13*a*. The applied loads causing these base loads are plotted in the figure at ground surface level. It will be seen that in every case the base load mobilized is only a small proportion of the applied (working) load.

Since no load cells or strain gages were installed in the shafts of these piles, it was not possible to determine the distribution of shaft loading. However, as tests of instrumented bored piles in London clay at Edmonton, to be described elsewhere, showed that the stress transfer did not depart significantly from proportionality to the clay strength at each level, it has been assumed that the behavior at Wembley was of this form. The curves shown dashed in Fig. 13*a* therefore indicate the probable distribution of load in each pile shaft.

Values of the mean stress transfer from each pile shaft, calculated from the difference between the applied load and the corresponding base load, are plotted in Fig. 13*b* at the depth of the midpoint of each pile. The mean values do not, of course, depend on the assumed load distributions shown in Fig. 13*a*, but likely forms of the distribution of stress transfer consistent with Fig. 13*a* are shown dashed in Fig. 13*b*.

Taking the mean value of the shear strength $\bar{c}$ as 110 kN/m², the overall mean of the mobilization factor $\bar{f}/\bar{c}$ at working load was 0.32. The corresponding value of q/q_u was 0.09.

Discussion and Conclusions

It is clear from the results of the tests of bored cast-in-place piles at

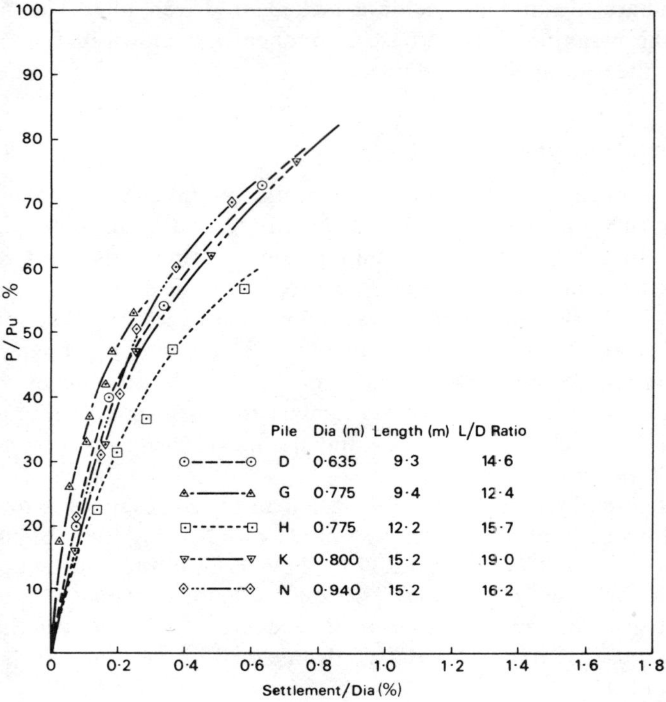

FIG. 11—*Dimensionless plot of the results of the incremental loading tests of straight bored piles at Wembley.*

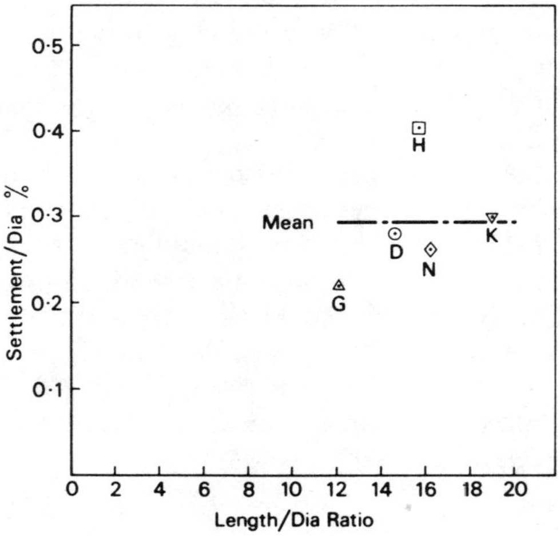

FIG. 12—*Settlement/diameter percent at the working load* (P/P_u = *0.5*) *in the Wembley bored pile tests.*

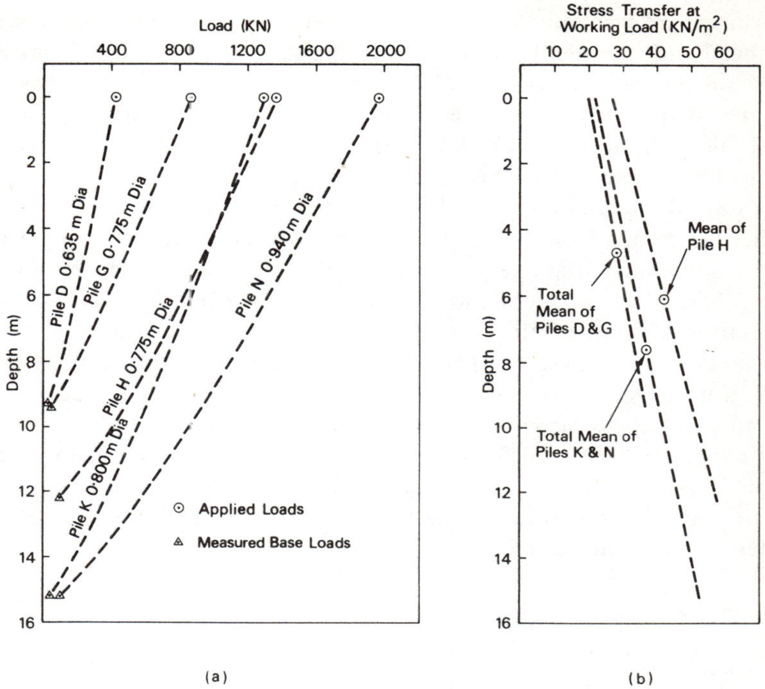

FIG. 13—(a) *Applied load and corresponding base load observed at one-half of the ultimate load in the tests of straight-shafted bored piles at Wembley showing the probable form of the load distribution down each shaft.* (b) *Mean values of the stress transfer from piles of similar lengths and the probable variation with depth corresponding to the load distribution shown in Fig. 13a.*

Wembley that, under working conditions, the proportion of the applied load reaching the pile base is extremely small. Residual forces in these piles must also be small. The magnitude of the residual forces would probably be less than the weight of the wet concrete of the pile shaft if they were of compressive sense, but residual forces in piles due to concrete shrinkage would be tensile. However, as the load cells in the Wembley piles were not sufficiently rigidly connected to the concrete bases to respond to tensile forces, it is not known whether these forces occurred.

In contrast to that of bored piles, residual installation forces have an important influence on the behavior of jacked, and probably of driven, piles in stiff clay. The tests at Brent Cross demonstrated that, at working loads, the friction near the top of a jacked pile may only just have changed from negative (downward) to positive (upward). As a result of the negative friction existing over the whole of the shaft prior to loading, the overall mean value of the positive friction at working load is small. In a test at large penetration (length-diameter ratio 25), the mean transfer stress $\bar{f}$ from the shaft at working load was only $0.1\bar{c}$ compared with $0.3\bar{c}$ for the bored piles. On the other hand, the mobilization of base support at working load was high because of the residual compressive forces at the base and q/q_u was 0.74 compared with 0.09 for the bored piles at Wembley. The settlement of the jacked pile at Brent Cross was 0.22 percent of the diameter compared with a mean of 0.30 percent for bored piles at Wembley.

The data for jacked and bored piles discussed above have been brought together for comparison in Fig. 14. For the two sets of tests, the spread of values of settlement-diameter percent at working load and the corresponding mean values of q/q_u are shown in Fig. 14*a*. In Fig. 14*b*, values of the mobilization of shaft friction $\bar{f}/\bar{f}_u$ at working loads are similarly related to settlement-diameter percent. The term $\bar{f}/\bar{f}_u$, where $\bar{f}_u$ is the ultimate value of $\bar{f}$, has been introduced here instead of the term $\bar{f}/\bar{c}$ used previously in order to facilitate comparison between Figs. 14*a* and 14*b*. Markedly different modes of load transfer between jacked and bored piles are clearly indicated. Whether differences in the settlement characteristics of the two types of pile exist that can be associated with the contrasted modes of load transfer remains unresolved.

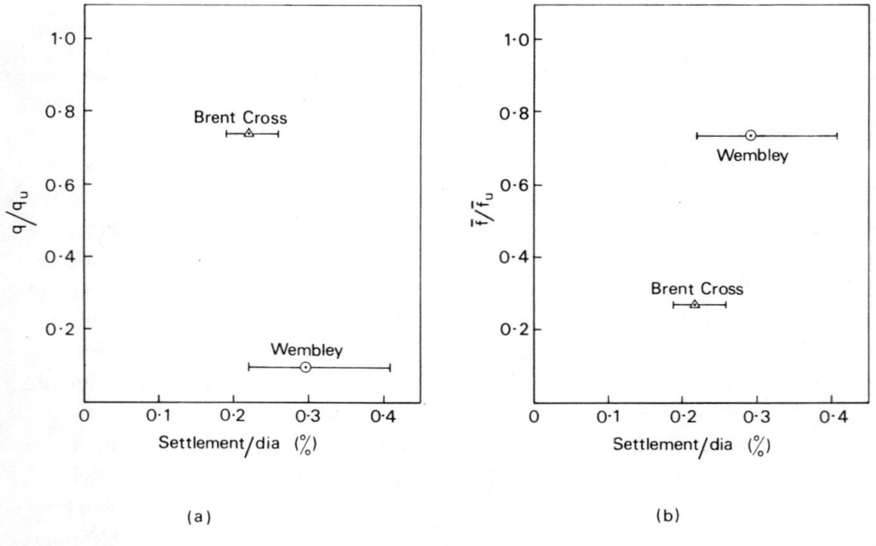

FIG. 14—*The mobilization of* (a) *base resistance* (q/q_u) *and* (b) *shaft frictional resistance* (f/f_u) *at working loads for jacked and bored piles related to the settlement/diameter ratios at the same working loads. The spread of settlement/diameter ratios in the tests at the two sites is indicated.*

Acknowledgments

The work described in this paper forms part of the research program of the Building Research Station and is published by permission of the Director of the Building Reasearch Establishment. The author particularly wishes to acknowledge the contribution of his colleagues G. Price, K. Tarr, and M. Gooch in the work at Brent Cross.

References

[1] Whitaker, T. and Cooke, R. W. *Symposium on Large Bored Piles.* Institution of Civil Engineers, London, 1966, pp. 7-49.

[2] Fearenside, G. R. and Cooke, R. W., "The Skin Friction of Bored Piles Formed in Clay Under Bentonite," CIRIA Report 77, Construction Industry Research and Information Association, London, 1978.

[3] Hanna, T. H., *Ontario Hydro Research Quarterly,* Vol. 18, No. 4, 1966, pp. 1-7.

[4] Hunter, A. H. and Davisson, M. T., *Performance of Deep Foundations, ASTM STP 444,* American Society for Testing and Materials, 1969, pp. 106-117.

[5] Hanna, T. H. and Tan, R. H. S., *Journal of Materials,* JMLSO, Vol. 6, No. 3, 1971, pp. 532-554.

[6] Hanna, T. H. and Tan, R. H. S., *Canadian Geotechnical Journal,* Vol. 10, No. 3, 1973, pp. 311-340.

[7] Cooke, R. W. and Price, G., Strains and displacements around friction piles. *Proceedings of the 8th International Conference on Soil Mechanics and Foundation Engineering,* Moscow, Vol. 2.1, 1973, pp. 53-60. (Also BRE Current Paper CP 28/73)

[8] Cooke, R. W. and Price, G., "Horizontal inclinometers for the measurement of vertical displacement in the soil around experimental foundations," *Proceedings,* Symposium on Field Instrumentation, 1973. (Also BRE Current Paper CP 26/73)

[9] Touma, F. T. and Reese, L. C., *Journal of the Geotechnical Engineering Division, Proceedings,* American Society of Civil Engineers, Vol. 100, No. GT7, 1974, pp. 749-761.

R. W. Cooke[1]

Load Transfer from Bored, Cast-*in-Situ* Piles in London Clay

REFERENCE: Cooke, R. W., **"Load Transfer from Bored, Cast-*in-Situ* Piles in London Clay,"** *Behavior of Deep Foundations, ASTM STP 670*, Raymond Lundgren, Ed., American Society for Testing and Materials, 1979, pp. 250–263.

ABSTRACT: This paper presents some measurements of the distribution of load transfer from the shafts of instrumented, bored, cast-*in-situ* concrete piles formed in London clay using a limited number of installation techniques. The measurements were included in a study of whether the use of bentonite slurries to stabilize open bore holes prior to concreting inhibits the development of friction on the shafts of piles in stiff clays. The instrumentation consisted of rod extensometers of various lengths and vibrating-wire gages clamped to the reinforcing cages. Devices attached to the bottom of each reinforcing cage prevented the development of any bearing forces at the pile bases. A method of correcting the extensometer observations for bending of the laterally unsupported length of shaft cased through fill is described, and the results are compared with a finite element analysis of the pile-soil system.

KEY WORDS: bored pile, clay, friction, load test, settlement, instrumentation, finite element method, stress distribution

The tests of which the measurements presented in this paper form a part were undertaken to investigate whether bentonite slurries, increasingly used to stabilize open boreholes prior to concreting, had a deleterious effect on the development of friction on the shafts of piles in stiff clays. In order to study the development of friction forces with the maximum of accuracy, support at the pile bases was eliminated by the installation there of soft toes formed of water-filled bags vented to the atmosphere. The pile shafts were instrumented by means of rod extensometers and by vibrating-wire strain gages so that the distribution of load and the efficacy of the soft toes could be determined.

When bentonite slurry is employed, the level in the bore hole is kept topped up as boring proceeds, and the concrete is placed at the bottom of

[1] Principal scientific officer, Building Research Station, Garston, Watford, Herts WD2 7JR, England.

the shaft by tremie pipe so that contamination of the concrete by bentonite is reduced to a minimum and the slurry floated off for reuse. Since it was thought that placing the concrete by tremie might of itself alter the frictional characteristics of the pile shaft, it was treated as a variable that should be studied independently of the effects of bentonite. For this reason, seven piles were tested:

1. Piles 7 and 8 were formed in dry conditions using a flight auger and the concrete poured from the top through a hopper.
2. Piles 9 and 11 were similarly augered in the dry but concreted by tremie from the base of the bore hole.
3. Piles 12 and 13 were excavated under bentonite using a drilling bucket and the concrete was placed by tremie.
4. Pile 10 was excavated by auger in the dry and the bore hole then filled with bentonite before concreting by tremie.

The investigation was carried out under the aegis of the Construction Industry Research and Information Association (CIRIA) and the effects of the four different methods of forming the piles on the ultimate frictional resistance have been reported by Fearenside and Cooke.[2] That paper dealt with incremental loading tests to low working loads and with constant rate of penetration (CRP) tests to failure. In the present paper, only the stress transfer characteristics and pile settlements under working conditions are considered.

Site and Soil Details

The site was at Edmonton in North London, where London clay is covered by 2.5 to 3 m of compact gravel and fill. The mean shear strength-depth profile for the London clay given by undrained triaxial tests of 100-mm diameter samples is shown in Fig. 1*a*. Samples were obtained from two site investigation bore holes and from the bores of three of the anchor piles. Because of the existence of the gravel and fill, the top 4.5 m length of each pile shaft was doubly cased as shown in Fig. 1*a* so that no friction was developed over this portion. Figure 1*a* also shows the position of the soft toe at the bottom of each shaft and the relationship between the clay shear strength profile and the active length of the pile. To prevent debris falling between the inner and outer casings causing the development of friction there, the annulus was filled with a weak bentonite suspension and covered before each pile was concreted.

[2]Fearenside, G. R. and Cooke, R. W., "The Skin Friction of Bored Piles Formed in Clay under Bentonite," CIRIA Report 77, Construction Industry Research and Information Association, London, England, 1978.

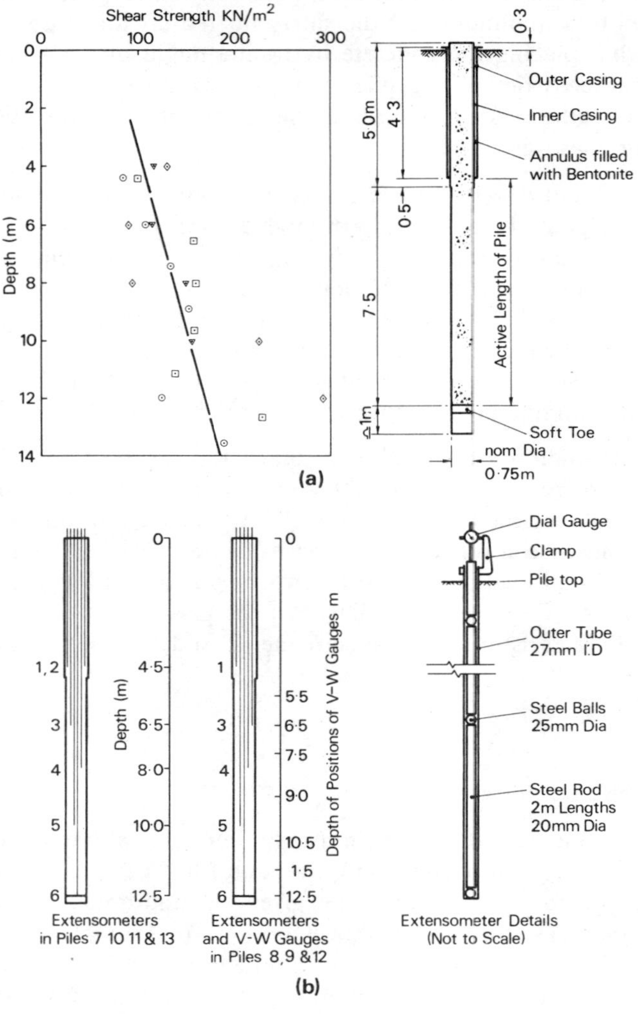

FIG. 1—(a) *Shear strength profile and the form of the piles at Edmonton.* (b) *Relative positions of the extensometers and vibrating-wire strain gages and details of the extensometer.*

Pile Instrumentation

The compression of each 0.75-m diameter pile shaft was measured by dial gage extensometers of the form shown in Fig. 1*b*, arranged in a ring around the cross section of the pile. The extensometers terminated at equal intervals between the levels of the bottom of the casing of each shaft and the upper surfaces of the soft toes. Each extensometer tube was filled with thin lubricating oil. Vibrating-wire strain gages clamped to short lengths of angle iron tied to the longitudinal reinforcement were installed in piles 8, 9, and 12 at the levels shown in Fig. 1*b*.

The soft toes to prevent the development of bearing forces at the pile bases consisted of cylindrical, water-filled, rubber bags between top and bottom steel plates. The bags could be expanded to the diameter of the bore hole by applying a head of water to a standpipe. During each test, the water was allowed to escape freely up the standpipe. Details of the soft toe have been given by Fearenside and Cooke (see footnote 2), who also described the loading arrangements.

Loading Tests

Settlements at Working Loads

In the incremental loading tests, each load was maintained for at least 1 h, and the maximum load of 350 kN was maintained overnight. The final settlement measured under each load increment is shown in Fig 2 for all the piles except pile 7. Insufficient load increments below 350 kN were applied to pile 7 for a satisfactory load-settlement curve to be obtained, and this pile has not been included. The results of the CRP tests to failure are shown in Figs. 3, and in Fig. 4 the data are replotted in the dimen-

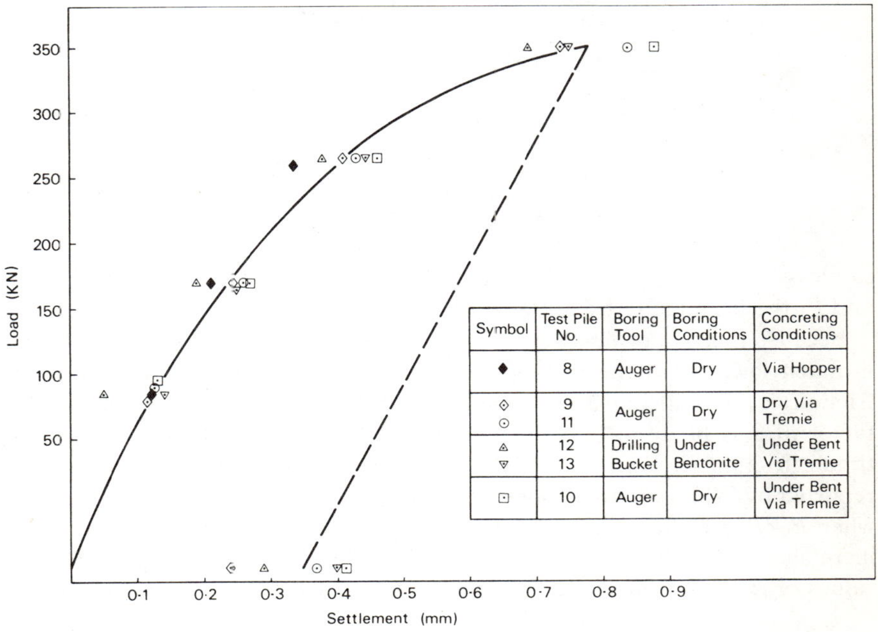

FIG. 2—*Mean load-settlement curve obtained in the incremental loading tests of bored piles at Edmonton.*

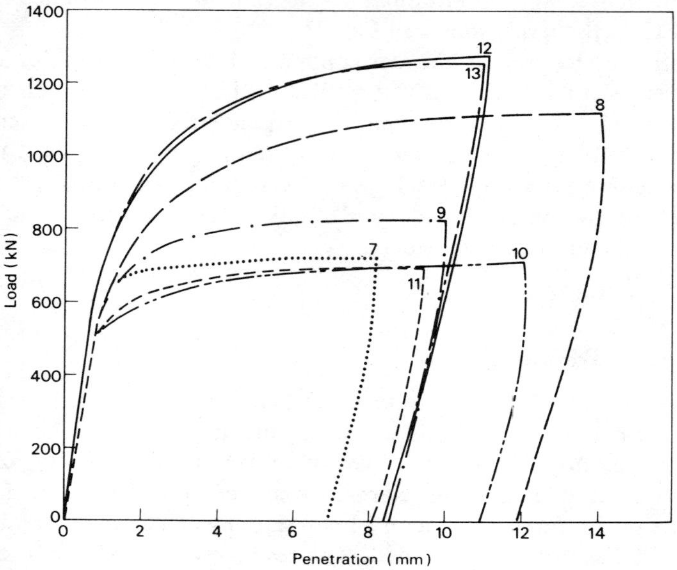

Curve	Test pile no	Boring tool	Boring conditions	Concreting conditions
··············	7	Auger	Dry	Dry, via hopper
— — —	8			
—·—·—	9	Auger	Dry	Dry, via Tremie
- - - - - - -	11			
————	12	Drilling Bucket	Under Bentonite	Under Bentonite via Tremie
-——-——-	13			
——··——··	10	Auger	Dry	Under Bentonite via Tremie

FIG. 3—*CRP test results at Edmonton.*

sionless form P/P_u percent versus settlement; where P is the applied load and P_u the ultimate test load for each pile.

Because the rate of development of resistance at the base of a bored pile is much lower than the rate of development of friction on the shaft, the friction R_w mobilizing at a working load P_w given by an overall safety factor of 2 is likely to be substantially greater than 50 percent of the ultimate friction R_u. In the tests of bored piles in London clay at Wembley reported by Whitaker and Cooke[3], the mean value of R_w/R_u was 75 percent when P_w/P_u was 50 percent. At Edmonton, no attempt was made to apply increments as high as 75 percent of R_u, and due to the unexpected high values of R_u shown in Fig. 3 for piles 8, 12, and 13, the incremental loads

[3]Whitaker, T. and Cooke, R. W., *Proceedings* of Symposium on Large Bored Piles, Institution of Civil Engineers, London, England, 1966, pp. 51–71.

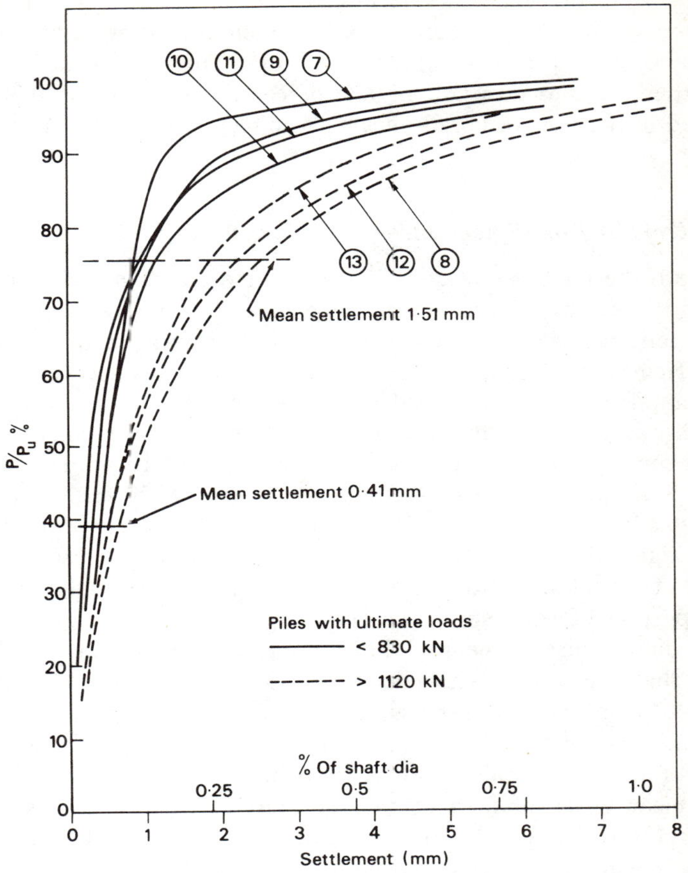

FIG. 4—*CRP test results in dimensionless form* P/P$_u$ *versus settlement and settlement/diameter percent.*

were not applied even to 50 percent of R_u in these cases. Thus, an observed mean value of the settlement at P_w cannot be presented for the tests at this site. An estimate of the settlement at P_w may, however, be arrived at by assuming that the ratio between settlements measured in the incremental loading test and the settlements (penetrations) recorded in the CRP test is constant at all values of P/P_u. The largest load in the incremental test, that is, 350 kN, was 39 percent of the mean value of R_u, and at that load, the mean settlement shown in Fig. 2 was 0.78 mm. Since the development of base resistance was prevented by the soft toes in these piles, $R_u = P_u$ and $R/R_u = P/P_u$. Thus, Fig. 4 shows that in the CRP tests the mean penetration at $R/R_u = 39$ percent was 0.41 mm and at R/R_u = 75 percent was 1.51 mm. A reasonable estimate of the mean settlement at R/R_u = 75 percent in incremental loading tests is therefore given by 0.78 × 1.51/0.41 mm = 2.87 mm. This settlement corresponds to a settlement/diameter ratio of 0.37 percent, compared with 0.30 percent recorded in the tests of bored piles at Wembley reported by Whitaker and Cooke.

Stress Transfer from Extensometer Observations

In both the incremental and CRP tests of the piles at Edmonton readings of the extensometers measuring compression between the pile head and various points in the shaft were generally proportional to the applied loads. Hence, for simplicity, shaft compression values corresponding to a single applied load were considered thereafter. The load chosen for this analysis was 1000 kN, because although this is larger than the ultimate load in some of the tests, the data can more readily be applied elsewhere. In Fig. 5, the value of the shaft compression recorded by each extensometer scaled to a load of 1000 kN is plotted for every pile. The curves in this figure will be referred to later. No clear pattern of compression can be detected for any of the piles, and this is due to bending of the laterally unsupported cased portion of each shaft. Eccentricities of applied loading, probably mainly due to misalignment of the jack and load cells, can cause bending of the shaft above the active length in contact with the clay that considerably modifies the extensometer behavior. Fortunately, since the extensometers were arranged in a ring about the axis of the pile, corrections for bending could easily be applied.

The mean values of shaft compression given by all the extensometers at each level in the seven piles under an applied load of 1000 kN are shown in Fig. 6. It will be seen that the points now follow a regular pattern, indicating that the eccentricities of loading were random about the extensometer positions. These mean values are compared with two curves of shaft compression (A and B) calculated from the mean values of the elastic modulus for the concrete of the piles. The observations are extremely close to curve A calculated assuming that the load transfer is uniform over the entire active length of the shaft and that no resistance is mobilized at the pile base. Curve A has therefore been added to Fig. 5, and corresponding curves for the theoretical compression at extreme extensometer positions under eccentricities of loading of 50 mm and 100 mm have also been included. Curve B in Fig. 6 showing the shaft compression for a purely end-bearing pile, clearly does not reflect the behavior of the piles taken collectively.

From the elastic modulus for the concrete of each pile, curves were computed similar to curve A and to the related eccentric compression curves that could be employed to correct each extensometer reading. In

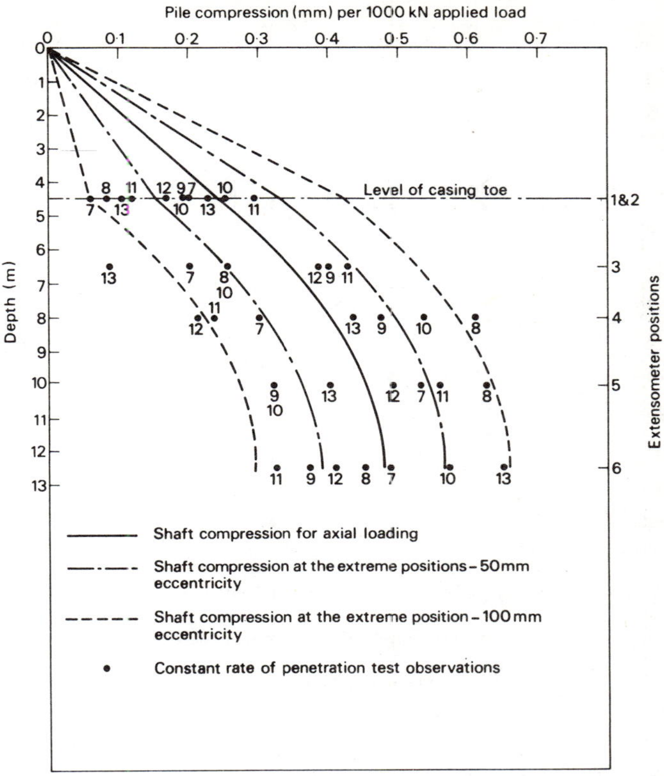

FIG. 5—*Compression of the pile shaft at the level of each extensometer base at an applied load of 1000 kN in the CRP tests of the Edmonton piles. The expected shaft compression due to axial loading and probable effects of eccentricity of loading are also shown.*

this way, it was possible to determine the magnitudes and directions of the eccentricity of loading on the piles and hence to derive values for the compression of the axis of each pile shaft at the level of each extensometer. A corrected curve for the compression of each pile shaft is shown in Fig. 7. It will be seen that the connected curves for piles 12 and 13 are curved downward near the pile base less markedly than the remainder. A trend toward the behavior represented by curve B in Fig. 6 is therefore apparent in the case of these two piles, suggesting small end-bearing components of resistance at or near the pile bases.

In Fig. 8*a* the curves of mean shaft compression for pairs of piles formed by similar techniques have been plotted. The results of a finite element study of a pile having a small gap at the base and cased from the surrounding clay over the top 4.5 m is also shown. For this study, the mean elastic modulus of the concrete of the seven piles tested was used for the

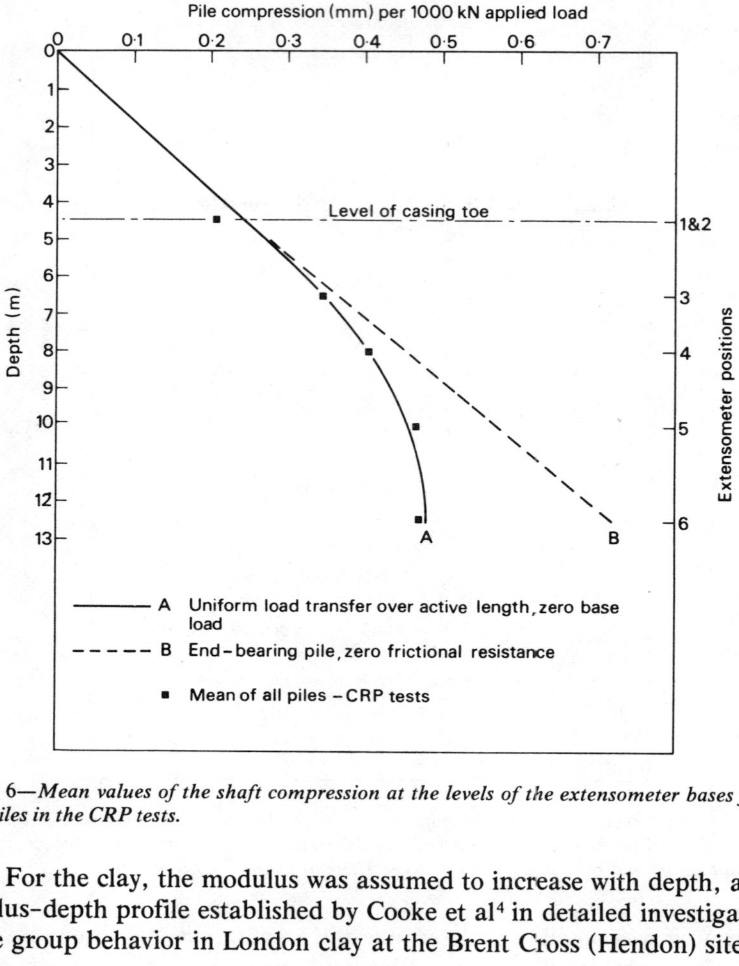

FIG. 6—*Mean values of the shaft compression at the levels of the extensometer bases for all seven piles in the CRP tests.*

shaft. For the clay, the modulus was assumed to increase with depth, and a modulus-depth profile established by Cooke et al[4] in detailed investigations of pile group behavior in London clay at the Brent Cross (Hendon) site was used.

The mean distributions of shaft loading necessary to cause the overall compressive strains shown in Fig. 8*a* are compared with the finite element analysis in Fig. 8*b*. The magnitudes of possible vertical bearing forces developed near the bases of piles 12 and 13 are evident in this figure. From the slope of each shaft load distribution curve, the distribution of stress transfer with depth can be obtained, and mean values for pairs of similar piles are given in Fig. 8*c*.

Because of the possibility of errors introduced at each stage of the process leading to the preparation of Fig. 8*c*, it would be prudent not to

[4]Cooke, R. W., Price, G., and Tarr, K., *Geotechnique*, Vol. 29, No. 2, 1979, pp. 113–147.

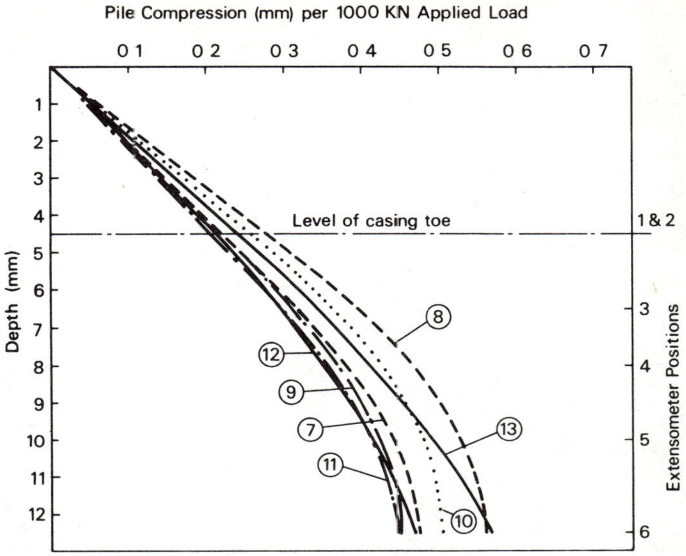

Symbol	Pile	Boring Tool	Boring Conditions	Concreting Conditions
– – – –	7 & 8	Auger	Dry	Via Hopper
—·—	9 & 11	Auger	Dry	Via Tremie
·······	10	Auger	Dry	Under Bent Via Tremie
———	12 & 13	Drilling Bucket	Under Bentonite	Under Bent Via Tremie

FIG. 7—*The shaft compression curves for each pile corrected for eccentricity of loading.*

infer too much from small differences between the curves of Fig. 8*c*. However, the stress transfer from the shafts of piles 12 and 13 bored through bentonite may be marginally less than that from the other five piles, and there is a possibility of some support from bearing forces near the bases. If the soft toes were fully effective in preventing load transfer at the bases, such support could only develop from narrow horizontal surfaces formed in the borehole sides by the boring tool. Fearenside and Cooke (see footnote 2) concluded that the high ultimate loads measured in the tests of these two piles might be attributable to differences in the shaft surface resulting from the use of a drilling bucket instead of an auger.

The curves of stress transfer from the shafts are generally in agreement with the results of the finite element analysis, and the overall mean value, corresponding to an applied load of 1000 kN, is 51 kN/m^2. This is equivalent to a mean value of 36 kN/m^2 at a working load for the piles given by

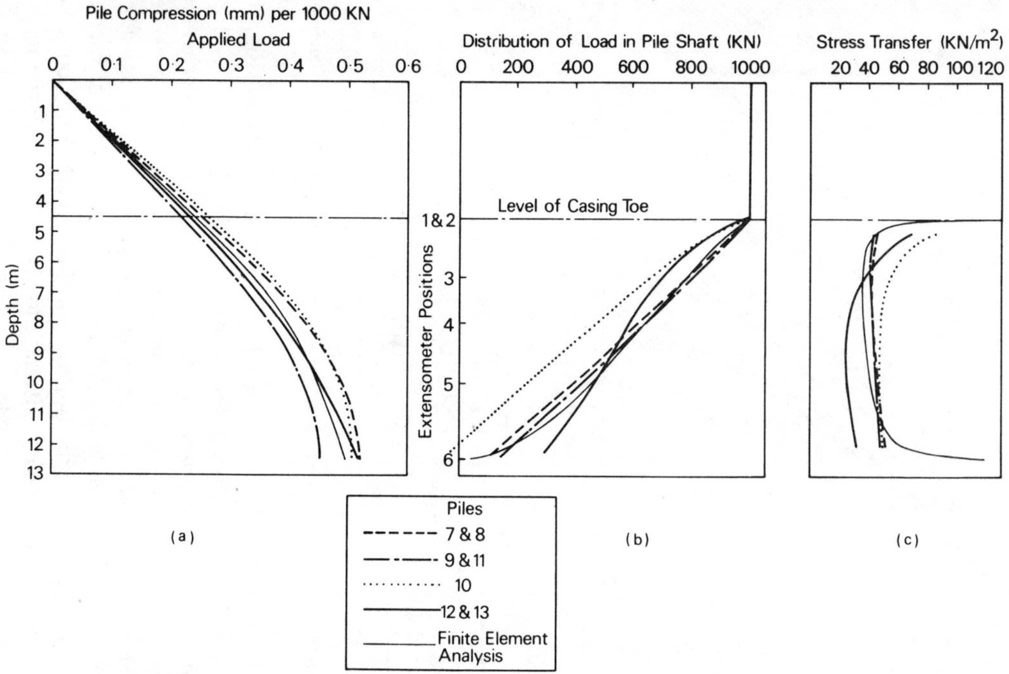

FIG. 8—(a) *Mean values of shaft compression for piles formed using similar techniques.* (b) *Corresponding distributions of shaft loading.* (c) *Distribution of stress transfer.* NOTE: *Values given by a finite element analysis of the pile-soil system are included in each of the figures for comparison.*

$P_w/P_u = 75$ percent. Taking the mean value of the shear strength of the clay $\bar{c}$ as 140 kN/m^2, the overall mean value of $\bar{f}/\bar{c}$ at working load is 0.26.

Strain Gage Observations

While the readings of the strain gage in piles 8, 9, and 12 were broadly proportional to increasing applied load, considerable scatter was experienced when the values were plotted with respect to the gage positions in the pile shafts. In Fig. 9*a*, strains scaled to an applied load of 1000 kN have been plotted at the gage levels and straight lines drawn through the points for each pile. It was assumed that the strains were zero at the level of each pile base because of the existence of the soft toe. In addition, the strain immediately below the casing was calculated from the measured elastic modulus for the concrete of each pile, the applied load, and the area of the shaft section. The constant strains shown for the cased length of shaft are less than the strains at the top of the active length, because, as shown in Fig. 1, the cross-sectional area of pile shaft within the casing was larger than that of the active length below it.

The mean values recorded by the strain gages in all three piles at each

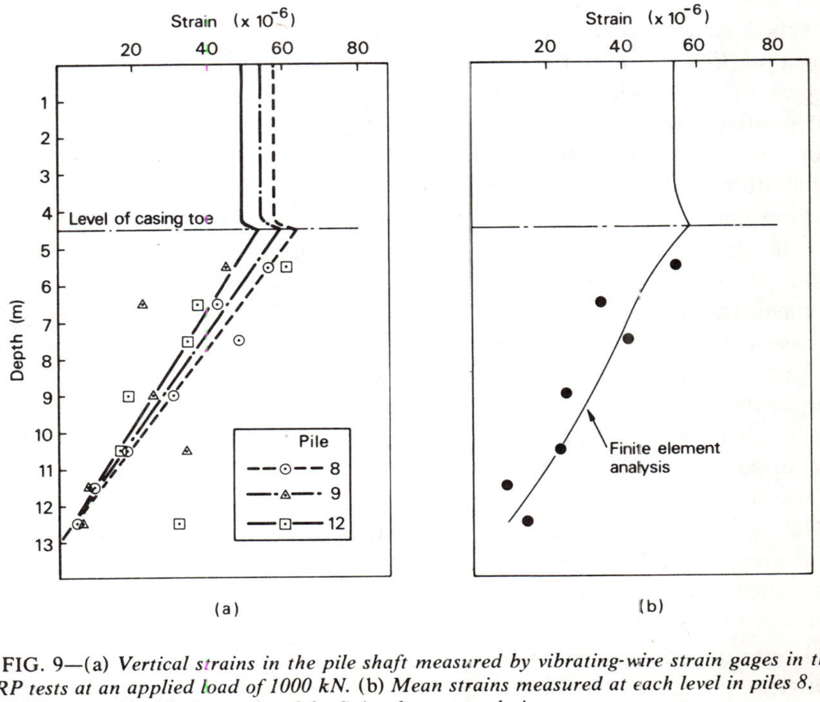

FIG. 9—(a) *Vertical strains in the pile shaft measured by vibrating-wire strain gages in the CRP tests at an applied load of 1000 kN.* (b) *Mean strains measured at each level in piles 8, 9, and 12 compared with the results of the finite element analysis.*

level are compared with the strains predicted by the finite element analysis, using the mean of the measured E values for the concrete of the shafts in Fig. 9*b*.

Discussion and Conclusions

It is clear that very considerable difficulties are inherent in using either extensometers or strain gages to determine the mechanism of load transfer from cast-in-situ concrete piles to the supporting soil. In addition to the experimental errors, each stage of the computation from strain or shaft compression to stress on the shaft surface is liable to introduce significant errors. Thus, it is unreasonable to attempt to draw distinct conclusions from the Edmonton test results on the effects of the four different methods employed for forming the piles. All that can be said on this aspect is that drilling buckets, normally used when piles are bored through bentonite suspensions, may score the borehole wall in a different way from flight augers and cause small changes in the load transfer characteristics.

With regard to stress transfer from bored piles generally, that is, irrespective of details of their formation, the extensometer and strain gage observations do permit useful conclusions. Since the piles at Edmonton were separated from the clay beneath them, residual installation forces must have been minimal, and thus the stress transfer should have approached closest to that predicted by theory. The stress transfer results shown in Fig. 8*c* and the mean shaft strains presented in Fig. 9*b* do suggest that the behavior of the piles considered collectively was close to that predicted by the finite element study for London clay.

In the tests at Wembley, where no attempt was made to determine the distribution of load in the pile shafts, the load reaching the base under working load was invariably small. There is thus every reason to presume that the stress transfer from the shafts of the piles at Wembley was similar to that measured at Edmonton. At Wembley, at the mean working load, the mean transfer stress on the shaft ($\bar{f}$) was $0.32\bar{c}$. Because of the absence of base resistance on the piles at Edmonton, deciding the appropriate working load presented difficulties. However, in view of the small base component of resistance usually measured in tests of bored piles, to presume it corresponded to $R/R_u = 75$ percent, the figure observed at Wembley, appears reasonable. At $R/R_u = 75$ percent, the mean transfer stress ($\bar{f}$) was $0.26\bar{c}$.

The data for these two groups of tests therefore suggest that, at working loads given by a load factor of 2 applied to the ultimate resistance, the mean shear stress on the shaft of a bored pile in stiff clay is likely to be of the order of $0.3\bar{c}$. The settlement at the working load measured in a short-term test can be expected to be between 0.3 and 0.4 percent of the shaft diameter.

Acknowledgments

The work described in this paper formed part of the research program of the Building Research Station and is published by permisision of the Director of the Building Research Establishment. The study of skin friction of bored piles formed under bentonite was carried out as a collaborative project between BRE and the Construction Industry Research and Information Association, and the piles were installed and tested by Cementation Piling and Foundations Ltd. The assistance given by the staff of CIRIA and Cementation Ltd is gratefully acknowledged.

E. F. Diekmann[1]

Timber Piles in Standards, Codes, and Practice

REFERENCE: Diekmann, E. F., **"Timber Piles in Standards, Codes, and Practice,"** *Behavior of Deep Foundations, ASTM STP 670,* Raymond Lundgren, Ed., American Society for Testing and Materials, 1979, pp. 264-281.

ABSTRACT: This paper presents a limited review of the standard and code provisions relating to timber piles from the turn of the century to the present. The provisions relative to axial stress in ASTM Establishing Design Stresses for Round Timber Piles (D 2899-74) are discussed and compared to some of the background experimental data. The conclusion is drawn that ASTM Method D 2899-74 is generally satisfactory regarding axial stresses but has specific weaknesses requiring changes. While no changes are forseen in the immediate future in the use of timber piles for 335 to 445 KN (40 to 50 ton) loads, it is suggested that timber piles may be utilized for higher capacity requirements. The view is expressed that limitations on pile material stresses are not proper in a building code, that such limitations are more appropriately placed in guidelines for foundation engineers selecting piles for specific applications.

KEY WORDS: wooden piles, standards, building codes, stresses, mechanical properties, dimensions, evaluation

Better knowledge of soil site conditions and soil mechanics and the widespread availability of a variety of driving equipment has made it possible to drive smaller piles deeper than ever before. The stress capabilities of the pile material is increasingly a limiting factor in establishing pile capacity. A recently published method for establishing design stresses in round timber piles [ASTM Establishing Design Stresses for Round Timber Piles (D 2899-74)], has been gaining acceptance throughout the country. Use of the new procedures in this standard has been held to produce design stresses for timber piles higher than those previously codified. The new procedures have been criticized as deficient, misleading, and dangerous and controversy is continuing.

This paper presents a limited review of the past and current standards and codes governing timber pile selection and design. Some provisions

[1] President, GFDS Engineers, San Francisco, Calif. 94111.

of ASTM Method D 2899-74 are examined in relationship to supporting data. Some suggestions are then made as to continued and future use of timber piles. The viewpoint is that of a West Coast structural engineer involved in code approval procedures, the study of wood technology, and the selection of piling types for specific projects.

Standards Development

At the turn of the century, no standards existed for timber foundation piles. Typical of the practice of the time are these words from a 1905 engineering text [*1*][2]

> Wooden piles for building foundations should be spaced not more than 36-inches nor less than 20-inches from center to center and they should be of a size that the least dimension at the small end is five-inches and the greatest dimension at the large, or butt end, is 12-inches. Where piles are over 20-feet long the butt end should be at least 20-inches in diameter.

By 1915, there was a standard for piles and for 55 years thereafter wood piles were specified under provisions of ASTM Specification for Round Timber Piles (D 25), first published in 1915. Timber piles were classified under these specifications into three general divisions according to the use intended as follows:

Class A: Piles suitable for use in heavy railway bridges and trestles.
Class B: Piles suitable for use in docks, wharves, highway work, and general construction.
Class C: Piles suitable for use in foundations that will always be completely submerged and for coffer dams, falsework, and sundry temporary work.

In 1970, ASTM Method D 25 underwent significant revisions. Classes A, B, and C were eliminated, and separate tables of dimensions were provided for each new class of pile—friction and end-bearing. In 1973, the tables in ASTM Method D 25 were revised so that the pile circumferences were tabulated rather than the diameters. This change facilitates measurements of the pile.

Coincident with the development of ASTM Method D 25-70, ASTM Committee D07 on Wood and Wood Based Products published Tentative Method for Establishing Design Stresses for Round Timber Piles (D 2899-70T). ASTM Method D 2899 became a full standard in 1974 as ASTM Method D 2899-74. This standard sets forth formulas based on the clear-wood strength values published in ASTM Establishing Clear-Wood Strength Values (D 2555) to determine the working stress for piles in (a) compression parallel to the grain, (b) bending, (c) horizontal shear, and (d) compression perpendicular to the grain. ASTM Method D 2899-74 has been the generator of controversy, aspects of which will be discussed later.

[2]The italic numbers in brackets refer to the list of references appended to this paper.

Code Evaluation

The 1927 Uniform Building Code [*2*] (UBC) included the following statements:

> The diameter of wood piles at the point shall not be less than six-inches and at the butt shall be not less than ten (10)-inches for piles twenty-five (25)-feet in length and shall be not less than twelve (12)-inches at the butt for piles more than twenty-five (25) feet in length. . . . The allowable load on a wood pile shall in no case exceed twenty-five (25)-tons. . . .

In 1937, the UBC still contained the dimensional limits of the 1927 UBC, but in addition contained the following stress limitation:

> No wooden pile shall be loaded in excess of 500 pounds per square-inch of the right section of the pile at mid-length.

In 1949, the UBC contained the following:

> The allowable stress in compression parallel to the grain of round wooden piles shall not exceed 60% of the basic stress for clear material as recommended in UBC Standard No. 25-2 and in no event shall the stress exceed 1000 pounds per square-inch.

In 1970, the UBC published a table of allowable stresses that are still in use (Table 1). Note that allowable stresses for Douglas fir and southern pine in compression parallel to the grain are 1200 pounds per square-inch. The 1977 edition of the National Design Specification for Wood Construction (NDS) [*3*] includes Table 2 with the comment:

> Design values given in the Table were determined in accordance with ASTM Designation D 2899-74, "Standard Method for Establishing Design Stresses for Round Timber Piles."

Note that the table is for "treated round timber piles." Timber pile allowable stresses in our building codes have obviously increased over the 40-year period in this review.

TABLE 1—*Allowable unit stresses (psi for normal duration of load) for round timber poles and piles.*

Species	Extreme Fiber in Bending	Compression Parallel to Grain (L/D = 11 or less)	Compression Perpendicular to Grain	Horizontal Shear	Average Modulus of Elasticity (in millions)
Southern pine	2150	1200	260	130	1 600 000
Douglas fir (coast)	2150	1200	260	110	1 600 000
Western larch	2150	1200	260	110	1 600 000
Red oak	2000	1100	400	150	1 500 000
Ponderosa pine	1200	830	200	100	1 000 000
Lodgepole pine	1200	800	180	80	1 000 000
Red (Norway) pine	1550	850	180	100	1 200 000

[a] Extreme fiber in bending values include 18 percent increase allowed for round shape.

TABLE 2—*Design values (psi for normal duration of load and wet conditions of use) for treated round timber piles.*

Species	Compression Parallel to Grain F_C	Extreme Fiber in Bending F_b	Horizontal Shear F_v	Compression Perpendicular to Grain $F_{C\perp}$	Modulus of Elasticity E
Pacific coast Douglas fir[a]	1250	2450	115	230	1 500 000
Southern pine[b]	1200	2400	110	250	1 500 000
Red oak[c]	1100	2450	135	350	1 250 000
Red pine[d]	900	1900	85	155	1 280 000

[a] Pacific coast Douglas fir values apply to this species as defined in ASTM Specification for Pressure Treatment of Timber Products (D 1760). For fastener design, use Douglas fir-larch design values.

[b] Southern pine values apply to longleaf, slash, loblolly, and shortleaf pines.

[c] Red oak values apply to northern and southern red oak.

[d] Red pine values apply to red pine grown in the United States. For fastener design, use northern pine design values.

The provisions of ASTM Method D 2899-74 have been incorporated by reference in the 1975 BOCA Basic Building Code [*4*] and in the 1976 National Building Code [*5*]. It seems probable that the specific NDS values or ASTM Method D 2899-74 will be adopted by other code-writing agencies. The basic values arrived at using ASTM Method D 2899-74 are not significantly different from current code accepted values. Comparison of the values in Tables 1 and 2 shows that some of the concerns regarding ASTM Method D 2899-74 leading to much higher design stresses than those in current use are unfounded.

Material Standards

ASTM Method D 2899-74 was prepared by ASTM Committee D07 as a material standard. As such, it was intended to provide methods for determining the capabilities of the material in a wood pile, either pressure-treated or untreated. The method does not address either the driveability or capacity of installed timber piles, as these may be determined by limiting factors outside basic material capabilities. Material standards routinely address themselves solely to material capabilities and not to the possible modifications of those capabilities as dictated by a given usage. For example, the establishment by ASTM Specification for Structural Steel (A 36) of a steel yield point of 248 MPa (36 000 psi) has never been argued as establishing that a 1-in. square bar of steel 6 m long will carry a 160-kN (36 000-lb) compressive force before failure. Similarly, establishment of a 8.62-MPa (1250-psi) allowable stress for wood pile material should not be interpreted as establishing that a 20-cm diameter wood pile always has an allowable load of 279 kN (31.4 tons). Material capability defines an upper bound of usage. Knowledgeable professionals must determine whether that upper bound can be achieved in a given situation.

ASTM Method D 2899-74 contains the following statement:

> 13.1 No formal factor of safety has been included in the formulas for working stresses in compression parallel to the grain and in bending.

If no "formal" factor of safety is included, then some look need be taken at the informal factor of safety that is implied as present.

Material Quality

ASTM Method D 2899-74 presents the following formula for calculating compression parallel to the grain

$$C_{\parallel} = (S - 1.645\text{SD})/1.88$$

where

$C_{\parallel}$ = working stress in compression parallel to the grain for green untreated timber piles,

S = average small clear crushing strength from ASTM Method D 2555, and

SD = standard deviation of small clear crushing strength from ASTM Method D 2555

The working stress at the tip for untreated piles would be:

Coast Douglas fir

$C_{\parallel} = 3784 - 1.645(734)/1.88$

$C_{\parallel} = 9.45$ MPa (1370 psi)

Southern pine (longleaf, slash, loblolly, and shortleaf)

$C_{\parallel} = 3430 - 1.645(617)/1.88$

$C_{\parallel} = 8.86$ MPa (1285 psi)

Red oak (northern and southern)

$C_{\parallel} = 3030 - 1.645(545)/1.88$

$C_{\parallel} = 7.83$ MPa (1135 psi)

The calculations for both southern pine and red oak have been made using the lowest characteristics of the group. Averages, weighted or otherwise, that may conceal poor material do not seem appropriate for piling where each piece may be structurally critical.

ASTM Method D 2899-74 prescribes adjustments to the working stresses to account for conditioning effects. After processing, the working stresses for treated material would be:

Coast Douglas fir (Boulton)	0.9 (9.45) = 8.51 MPa (1233 psi)
Red oak (Boulton)	0.9 (7.83) = 6.66 MPa (1022 psi)
Southern pine (Steam)	0.85 (8.86) = 7.53 MPa (1092 psi)

ASTM Method D 2899 also permits for Douglas fir and southern pine the increase of the working stress based on the location of the section under analysis measured from the tip. The increase is 0.2L (percent), where L is the distance (in feet) from the tip. The increase is based on data from a study of 50-ft piles. There is an unanswered question as to whether the increase can be applied to piles over 50 ft in length. Forty-five feet from the tip, the working stress is treated material would be:

Coast Douglas fir	$C_{\parallel} = 8.51\,[1 + 0.002(45)] = 9.27$ MPa (1344 psi)
Southern pine	$C_{\parallel} = 7.53\,[1 + 0.002(45)] = 8.21$ MPa (1190 psi)

ASTM Method D 2899 drew upon a study [*6*] by the U.S. Forest Products Laboratory (USFPL) into the strength of timber piles of Douglas fir, southern pine, and red oak. In the course of this study, actual full-size sections of piles 3 ft in length were taken from both the pile tip and butt and loaded to failure in axial compression. The tests showed the tip to be the weakest segment of the pile. This is a reasonable finding, since a greater portion of the total material volume will be in knots near the tip rather than at the butt, and the tip contains less dense material. Crushing failures occurred at the knots in the laboratory tests. The crushing strength at the tip or butt as shown by the USFPL tests is shown in Table 3 with a "factor of safety" tabulated, representing the short-term crushing strength divided by the working stress as determined above. Examination of the table data results in several general conclusions:

Douglas Fir

1. Two piles, specimen 1 and 7, would appear to have failed a load test carried to twice design if the entire load was transferred in the vicinity of the tip. Because the test specimens exceeded the specification minimum diameter, in actuality, the factor of safety on design would have been 2.76 and 2.12, respectively. The stresses developed using ASTM 2899-74 appear satisfactory; however, it appears that the more severe restrictions on knot size found in ASTM Method D 25 prior to 1970 should be used in specifying tip-bearing Douglas fir piles to guard against the possibility of an absolute minimum diameter and the worst knot combination occurring in the same pile.

2. The stresses developed from ASTM D 2899-74 provide a factor of safety in excess of 2.0 for all test specimens at the butt.

TABLE 3—*Test crushing strength versus design strength.*

Specimen No.	Tip Crushing Strength, psi	Tip Factor of Safety	Butt Crushing Strength, psi	Butt Factor of Safety
		Douglas Fir		
1	2270	1.84	3310	2.46
2	2530	2.05	2900	2.16
3	2850	2.31	2860	2.13
4	3460	2.81	3970	2.95
5	3280	2.66	3690	2.75
6	3420	2.77	4260	3.17
7	2090	1.70	3510	2.61
8	2810	2.28	3780	2.81
9	3320	2.69	4150	3.09
10	3160	2.56	3770	2.81
11	3490	2.83	3640	2.71
12	3300	2.68	4420	3.29
13	2580	2.09	3300	2.46
14	3310	2.68	3090	2.46
15	2630	2.13	3220	2.40
Average	2960	2.40	3590	2.67
		Southern Pile-Steam Conditioned		
1	1940	1.78	3040	2.55
2	1660	1.52	2830	2.38
3	1380	1.26	2720	2.29
4	1910	1.75	2860	2.40
5	2030	1.86	3100	2.61
6	1970	1.80	2980	2.50
7	2200	2.01	3230	2.71
8	2000	1.83	2820	2.37
9	1730	1.58	2650	2.23
10	1640	1.50	2360	1.98
11	1730	1.58	3000	2.52
12	2090	1.91	3160	2.66
13	2010	1.84	2840	2.39
14	1660	1.52	3210	2.70
15	1400	1.28	3390	2.85
Average	1820	1.68	2950	2.48

Southern Pine

1. The factors of safety derived from ASTM Method D 2899-74 stresses are inadequate at the tip based on the steam-conditioned test specimens. The tests also show current code stress values for southern pine to be too high for steam-conditioned material. Steam-conditioned southern pine tip-bearing piles should be designed for two-thirds ASTM Method D 2899-74 stresses.

2. ASTM Method D 2899-74 provides a factor of safety in excess of 2.0 for all test specimens at the butt, except for the one specimen at 1.98. The

TABLE 3—(*Continued*).

Specimen No.	Tip Crushing Strength, psi	Tip Factor of Safety	Butt Crushing Strength, psi	Butt Factor of Safety
		Southern Pine—Kiln Dried		
B	3180	2.91	3910	3.41[a]
E	4130	3.78	4400	3.84[a]
H	4540	4.16	4640	4.05[a]
K	3230	2.96	3870	3.37[a]
N	3840	3.52	3820	33.3
Average	3784	3.47	4128	3.60[a]
		Red Oak		
OT 1	3700	3.62	3630	3.55
2	3710	3.63	3400	3.33
3	3150	3.08	3620	3.54
4	3240	3.17	3700	3.62
5	3220	2.95	3560	3.48
6	3800	3.72	3680	3.60
7	3200	3.13	3570	3.49
8	3760	3.68	3870	3.79
9	3290	3.22	3400	3.33
10	3370	3.30	3600	3.52
11	3380	3.31	3810	3.73
12	3570	3.49	3470	3.40
13	3650	3.57	4030	3.94
14	3720	3.64	3520	3.44
15	3210	3.14	3400	3.33
Average	3460	3.39	3620	3.54

Conversion factor—1000 psi = 6.9 MPa.
[a] 25 ft pile length 1 092 [1 + 0.002 (25)] = 1147 psi (7.91 MPa) working stress.

method provides a factor of safety in excess of 2.9 for all specimens kiln dried prior to treatment.

Red Oak

1. ASTM Method D 2899-74 provides a factor of safety in excess of 3.0 at both the tip and the butt.

Duration of Loading

One of the concerns voiced about timber piles from foundation engineers involves the effects of long-term load on wood. Strength as determined from short-term timber tests is currently converted to 50-year strength by multiplying by a factor of 9/16 (0.565). A short-term factor of safety of 1.78 (16/9) seems to be required merely to overcome this apparent degrading of wood strength with time under load. An initial design factor of safety of 2.0 on a timber pile loaded continuously at design load would thus shrink to about 1.3 (2/1.78) after 50 years if wood was indeed subject to such a strength degradation. Such a long-term factor of safety would be unacceptable.

Recently developed evidence suggests strongly that such strength degradation will not in fact occur in timber piling. The duration of load data was developed by USFPL using small, clear specimens loaded under constant stress in bending in a range of 55 to 83 MPa (8000 to 12 000 psi), far beyond the stress range contemplated for timber piles. Bending tests by Madsen [*7*] using 2 by 6 material led to the conclusion that no time effects were discernible below actual stresses of 17.93 MPa (2600 psi). Proposed pile design stresses are in the range of 8.62 to 10.34 MPa (1250 to 1500 psi). Corollary evidence exists in the form of an evaluation of piles pulled and tested after 85 years under load [*8*] that showed specific gravity 12 percent lower than published averages. Strength in compression parallel to the grain has been found to be directly proportional to specific gravity. This latter data is only suggestive, as the piles were not tested prior to being driven and direct comparisons cannot be made.

Pile Practice

Timber piles have moved from the range of 133 to 222 kN (15 to 25 ton) capacity in the thinking of most West Coast foundation engineers into the 355 to 455 kN (40 to 50 ton) range at the present time. Random projects exist utilizing timber piles for design loads of 534 to 667 kN (60 to 75 tons) and pile tests have been carried up to 2090 kN (235 tons) on timber piles. These limited experiences indicate that with proper soil conditions and properly selected equipment, timber piles can be utilized above 335 to 445 kN (40 to 50 tons).

Recent timber pile tests in Utah [*9*] and Nevada [*10*] arrived at design capacities of 578 kN (65 tons) and 623 kN (70 tons) (622 720 N), respectively. On the Nevada sites utilizing friction piles, a production pile with a 20-cm minimum tip and a 38-cm minimum butt was selected and driving is now in progress. On the sites utilizing tip-bearing piles, a minimum 32-cm outside diameter tip has been specified. Final selection has not yet been made by the contractor. The tests were particularly interesting as piles of different materials were driven and comparisons of results were possible between the different types of piles. The wooden piles performed well with some test loads exceeding 1779 kN (200 tons). The lowest test load on a timber pile was 1192 kN (134 tons). On each of the Nevada sites, timber outperformed at least one type of pile associated with high capacity, including monotube, step taper, pipe, prestressed concrete, and H-piles and achieved these results with significantly less overall driven length.

Building Codes and Pile Stress

Much effort is expended by partisan interests in arguments before code bodies over the question of whether allowable pile stresses in building codes should or should not be increased. Such arguments are raging currently over whether pile steel stresses of 50 percent of yield stress should be allowed. Part of the problem arises from the lack of the basic agreement as to what pile stresses they are that are being set and indeed even why the stresses are being set.

Two different problems are involved. The first problem involves questions of the practicality of driving piles of certain types and sizes on specific sites with available equipment to achieve a desired capacity. It is this question that interests soils and foundation engineers. The answer to this complex question, which directly involves the soil-pile interaction, has in the past and is to a large extent currently answered on the basis of empirical observations. The foundation (soil) engineer knows from experience that a certain size and type of pile driven with a certain hammer to a set number of blows per inch in soil of the nature of that on the site will have a certain capacity. He also knows that a larger hammer will probably damage the pile. The stress in the pile has not entered into this empirical balance, and most soil engineers cannot answer as to the stress in the pile occasioned by either driving or resulting from the tonnage capacity placed on the pile. Numerous empirical observations of obtained capacity have been converted into stress in the pile material, and this "allowable stress" is currently embodied in building codes. It is not an allowable stress, however, as the term is usually applied to materials, since it has arisen from measurements of the soil-pile interaction, not the capabilities of the pile material itself.

The advent of wave equation analysis has made it possible to look at this problem in an analytical way without relying solely on empirical observations. These analyses [*11*] have resulted in recommendations for static design stresses for steel, concrete, and wood by backing off from predicted capacity. These recommendations and analyses have been used in opposition to increased stress values in building codes.

This "allowable stress" approach to pile selection is wrong, whether the stress is based on the older empirical approach or the newer wave equation analyses. It is wrong because the single number is wrong. A single number does not cover the myriad combinations of soils, piles, and installation equipment and techniques. The single number will be too low for especially favorable circumstances and too high for adverse conditions. Codification of an allowable stress for a given pile material as a basis for pile selection offers no advantage to the foundation engineer, who is well aware of and technically qualified to consider the numerous factors involved in proper pile selection. On the other hand, the allowable stress is at least potentially restrictive to such an engineer in the utilization of favorable conditions.

An allowable pile stress is potentially dangerous in the hands of the unskilled, as a building code is not a design manual and offers no commentary as to those conditions and/or circumstances under which the pile "allowable stress" is or is not appropriate in selecting a particular pile.

The second question involves the capabilities of the piling after it has been installed, that is, its performance in the completed structure. This question, however, is already dealt with in the appropriate materials chapters of our building codes. Imposition of an allowable stress intended to control driving problems or based on driving considerations leads to design confusion when installed pile capability is assessed. Typical of the confusion is the imposition of a maximum stress of $0.33\,f'_c$ on concrete piles when concrete sections are now assessed by ultimate strength, which uses strain compatability, not stress, as the governing criterion. Similar confusions arise over the limitation of steel pile design stresses to $0.33F_y$ when steel material stresses are elsewhere set at $0.60F_y$. The designer interested in the ability of a particular pile cross section to resist some combination of axial load and bending is left to his own devices to reconcile these confusions. The result may be designs much more conservative than are actually demanded by either soil or material requirements.

The Structural Engineers Association of Northern California (SEAONC) has discussed proposing the elimination of all allowable stresses from the foundation chapter of the UBC so that all questions of material capabilities would have to be referred to the chapters dealing with the specific materials. Timber piling is already handled in this way in the UBC. The proposal was shelved as untimely in the midst of the pile proposals already being considered for the UBC, but the viewpoint involved is very much alive.

Foundation engineers should solve pile driveability problems through a combination of technical education and experience. If codification of driveability criteria is considered necessary, then such code provisions should be clearly identified as applying to piles during driving. Such driveability criteria should properly address ultimate dynamic material stress and not static design stress levels. The foundation (soil-pile) factor of safety should be established by the foundation engineer. The pile (material) factor of safety should be established by the structural designer. Neither factor of safety should be developed or exist without mutual concurrence and understanding of both engineers.

Conclusions

1. Developments in standards as reflected in building codes have evolved and allow higher stresses in timber piles.

2. ASTM Method D 2899-74 adequately reflects axial stress tests on pile materials except: (a) Conservatism is warranted in limiting knots in Douglas fir piles; the knot limits in ASTM Method D 25-58 are preferable to those

in ASTM Method D 25-73. (b) Steam-conditioned, treated southern pine piles should be designed using two-thirds ASTM Method D 2899-74 steam-conditioned stress values or 0.56 of green values; kiln dried, treated southern pine may use the 0.90 recommeded for Boulton processing.

3. The time a wood pile is placed under load is probably not a concern at current design stress levels.

4. Timber piles warrant consideration for loads above currently considered load ranges. Increased experience with the higher loads coming into current practice will accomplish this. No restructuring of design stresses is required to justify higher loads.

5. Allowable pile stresses as a means of controlling pile driveability or setting pile capacity should not be in a building code. The pile section of a building code should address stresses on pile materials only if different factors of safety from those normally associated with design in the materials are desired. Pile driveability should be handled separately, preferably outside the building code.

References

[1] "International Library of Technology," International Text Book, Scranton, Pennsylvania, 1905, p. 57.

[2] "Uniform Building Code," International Conference of Building Officials, Los Angeles, 1927.

[3] "National Design Specification for Wood Construction," 1977 ed., National Forest Products Association, Washington, D.C., 1977, p. 27.

[4] "The BOCA Basic Building Code/1975," Building Officials and Code Administrators International, Inc., Chicago, 1975.

[5] "National Building Code 1976," American Insurance Association, New York, 1976.

[6] Wilkinson, T. L., "Strength Evaluation of Round Timber Piles," FPL 101, U.S. Forest Products Laboratory, Madison, Wisconsin, Dec. 1968.

[7] Madsen, B., *Forest Products Journal*, Vol. 23, No. 2, Feb. 1973, pp. 21-28.

[8] Bendtsen, B. A., "Bending Strength and Stiffness of Bridge Piles After 85 Years in Milwaukee River," FPL Research Note 0229, U.S. Forest Products Laboratory, Madison, Wis., 1974.

[9] "Pile Load Test Results, Wastewater Treatment Facility, Provo, Utah," Rollins, Brown and Gunnell, Inc., Jan. 1976.

[10] Keating, D. J., "Lovelock-Winnemucca Test Pile Study Interim Report," Nevada Highway Department, July 1974.

[11] Davisson, M. T., "Pile Load Capacity," Seminar, Design Construction and Performance of Deep Foundation, Berkeley, Calif., 1975, p. 14.

DISCUSSION

W. A. Norum[1] *(written discussion)*—Mr. Diekmann has researched the evolution of timber pile design and construction by reference to regulations

[1] District manager, National Forest Products Association, Mountain View, Calif.

beginning with the early editions of the Uniform Building Code and other model codes that are in continuing use throughout the United States, as well as by reference to a 1905 engineering text. He also traces the history of ASTM standards for timber piles back to the year 1915—all of which helps to document the long period of time that round-timber foundation piles have been successfully used. The author has also done an excellent job of revealing the degree of conservatism practiced using timber foundation piles. I can agree with most of his itemized conclusions. In my discussion, I intend to explore Mr. Diekmann's estimate of the "real" factor of safety for timber piles, explain my differences with his stated conclusions, and discuss the merits of his idea on how building codes should be written to allow progress while maintaining a satisfactory level of conservation in the design of pile foundations.

The author makes reference to "the timber pile controversy," which, I believe, needs some clarification before my discussion on the subject of timber pile strength properties. His comment relates to ASTM Establishing Design Stresses for Round Timber Piles (D 2899-74).[1] As the name of the standard indicates, it sets forth the method for deriving safe design stresses for timber piles. It has become the focal point of attention because it is now referenced in some of the model codes. The allowable design stresses derived by this method has resulted in higher values than previously used in some parts of the country, which makes the timber pile appear more competitive. It was pointed out by Diekmann, however, that the increase is not significantly different from the updated model codes, and he concluded that the concerns regarding ASTM Method D 2899-74 seem unfounded. He also concluded that "timber piles warrant consideration for loads above currently considered load ranges" and he further concluded the such designs with higher loads can be safely done without increasing the recommended design stresses. In other words, the design stresses are not controlling the design load of timber piles. Experiences with timber piles suggest that the ability to penetrate soil and develop the design capacity of the soil is more limiting than the allowable or recommended timber pile design stress. This fact should not, however, be a cause for reducing efforts toward improving current methods for deriving design stresses for timber piles, which may lead to even higher recommended values.

The industry is continually seeking ways on how to more accurately predict the strength properties of wood for use by design engineers. The method used for deriving stresses for all structural timber makes use of proven probability techniques combined with a large data base. The results of all statistical data obtained from laboratory tests, samples of wood taken from growing trees, and actual experience are printed in ASTM Establishing Clear-Wood Strength Values (D 2555). This method differs greatly from methods used to derive stresses for man-made materials such as steel and concrete for obvious reasons. An understanding by the design engineer on how timber design stresses are derived can be important to his proper and wise application of recommended or allowable design values. I agree with Diekmann: "The pile [material] factor of safety should be established by the structural designer." However, he can only exercise good judgment if he understands how timber stresses are derived. This information for timber piles is contained in ASTM Methods D 2899, D 2555, and D 245 (Establishing Structural Grades and Related Allowable Properties for Visually Graded Lumber).

In a quote taken from ASTM Method D 2899-74, Diekmann points out that no formal factor of safety has been included in the formulas for working stresses in compression parallel to grain and in bending. He then proceeds to establish a factor of safety by calculating the ratio of design stresses derived by the procedures of ASTM D 2899 for compression parallel to grain using Douglas fir (coast), southern pine, and red oak and relating these numbers to the crushing values of full-size round timber butt and tip test specimens for these same species reported by the Forest Products Laboratory, U.S. Forest Service, in Research Paper FPL 101, December 1968. Results of this analysis indicates an average factor of safety of 2.40, 1.67, 3.47, and 3.39 for preservatively treated Douglas fir (coast), southern pine (steam conditioned), southern pine (kiln dried), and red oak, respectively. It should be recognized that the author used the values for shortleaf southern pine, which has the lowest specific gravity of the southern pine group. It should also be recognized that the risk level as represented by these factors of safety are now approaching zero. This is a level of conservatism that may be justified by engineers for specific projects, but I do not believe such a restriction should be made mandatory for all projects. This should be kept a judgmatical decision and not made so restrictive by codes. Possibly the author did not intend to propose such a restriction, but it seemed implied by his comments. The calculated factors of safety are said to approach zero risk because the probability is extremely low that southern pine piles having the lowest specific gravity will be combined with the smallest acceptable tip size having the maximum allowable knots and installed in a manner that the maximum design stress is developed and maintained constant at the pile tip. It is for this same reason that I do not agree with the author's conclusion that the more severe restrictions on knot size specified in ASTM Method D 25 prior to 1970 should be specified for tip-bearing Douglas fir piles in the current ASTM Method D 25.

The formula for deriving the working stress in compression parallel to grain for green untreated piles was used by the author to estimate the "real" factor of safety. The formula needs some explanation because, I believe, it also represents a very conservative step in the process of developing reasonable compression stresses for use in design. The terms of the formula $C^{\parallel} = S - 1.645SD/1.88$ are defined in the author's paper. The coefficient 1.645 in the right side of the formula establishes the 5 percent exclusion limit for clear wood parallel to grain crushing strength. The divisor of 1.88 accounts for a reduction in strength to represent the pile tip values, a reduction in the standard deviation because round timber is less variable than sawn timber, and an adjustment for duration of load from short to normal duration. This factor was arrived at by relating the test information contained in U.S. Forest Service Research Paper FPL 101, December 1968, with data contained in ASTM Method D 2555. The following coefficients were calculated:

$$\frac{\text{Crushing strength of small clears from pile tips}}{\text{Crushing strength of small clears from D 2555}} = 0.90$$

$$\frac{\text{Standard deviation of small clears from pile tips}}{\text{Standard deviation of small clears from D 2555}} = 0.85$$

$$\frac{\text{Crushing strength of full-size tip sections}}{\text{Crushing strength of small clears from pile tips}} = 0.90$$

To convert ASTM Method D 2555 strength values to small clear values taken from pile tips and to convert D 2555 standard deviations to small clear standard deviations taken from pile tips, the above coefficients are multiplied by the listed values for small clears ($S \times 0.90$ and $\text{SD} \times 0.85$). The formula that expresses these conversions as well as the 5 percent exclusion limit reads $S \times 0.90 - 1.645 \times \text{SD} \times 0.85$. Now, to convert this to design stress values for full-size pile tip sections, the formula is multiplied by 0.90 to account for grade effect. To account for duration of load, it is converted from short-time test conditions to normal loading conditions by dividing the equation by 1.52. It now reads

$$C_{\parallel} = \frac{(S \times 0.90 - 1.645 \times \text{SD} \times 0.85)\, 0.90}{1.52}$$

In order to simplify the formula, the factor of 0.85 was changed to 0.90, which makes it even more conservative.

$$C_{\parallel} = \frac{S - 1.645\text{SD}}{1.88}$$

As mentioned, it includes a factor for duration of load, so that now it relates to normal loading conditions, which means 10 years of accumulated time at maximum stress conditions. A reduction to 95 percent of the derived design values relates it to 50 years, and a reduction by 90 percent converts it to a permanent status, which means maximum stresses are occurring continuously.

On the subject of building codes, the author concludes that design stresses for pile materials are best kept separate from any restrictions on soils and that pile driveability should be a subject for engineering manuals of recommended practice. I couldn't agree more! Foundation engineers should solve pile driveability problems outside the code arena. There are a variety of ways of overcoming difficult driving problems that cannot and should not be specified in a building code. Codes should allow a maximum of flexibility for the engineer to solve these problems so that the costs of all projects can be kept to a reasonable level. Engineers should not try to shift the responsibility for designing and specifying a safe construction project to building codes, which should be written in performance terms wherever possible so the engineer can do what he has been trained to do.

Before closing, I would like to update Mr. Diekmann's report on the recent pile project located at Winnemucca, Nev. The contractor has selected treated round-timber piles for all the 400 or more friction piles, which are designed for a capacity of 70 tons. They are being driven to a specified tip elevation through a highly compacted engineered fill and into the supporting material by use of predrilled pilot holes.

During the last 15 years, we have witnessed a significant change in the willingness of the design engineer to increase pile capacities. This, in my opinion, is directly attributable to his increased knowledge of soils used as a structural material. What improvements that have occurred in timber piles is limited to the selection of trees and processing by the manufacturer, including debarking, trimming, and treating with preservatives. A practice of applying metal bands around Douglas fir piles has helped overcome some driving limitations of this species. More importantly, though, foundation engineers have learned considerably more about soils, allowing them to act with confidence. The blind practice of using timber piles only for a capacity of 25 tons is disappearing.

T. L. Wilkinson[2] (*written discussion*)—I have reviewed the paper by Ed Diekmann to be presented at the ASTM Symposium on Behavior of Deep Foundations and have a few comments to make. I am most familiar with items connected with ASTM Establishing Design Stresses for Round Timber Piles (D 2899) and thus my comments for the most part are restricted to this area.

The small clear strength values for southern pine that Diekmann uses are those listed in ASTM Establishing Clear Wood Strength Values (D 2555-76). ASTM Method D 2555-76a lists an average crushing strength of 3511 psi and a standard deviation of 612 psi for the weaker of the four southern pine species listed. Use of these values will decrease slightly the factors of safety for southern pine listed in Table 3.

[2]Engineer, Engineering Design Criteria, Forest Products Laboratory, U.S. Department of Agriculture, Madison, Wis. 53705.

The statement is made that ASTM Method D 2899 drew heavily upon a study at the U.S. Forest Products Laboratory (USFPL). While this study was used, it was not as heavily used as other studies, such as "Factors Affecting the Variation in Compressive Strength of Southern Pine Piling" by W. S. Thompson (published in *AWPA Proceedings*, Vol. 65, 1969).

I think it should be pointed out that only one of the 15 pile tip specimens of southern pine from the USFPL study met the requirements for knots in the old ASTM Specification for Round Timber Piles (D 25). Thirteen of them meet the requirements of the present D 25. The Douglas fir specimens had four that failed the old knot requirements, while all of them pass the new requirements. I don't know how Diekmann can make the conclusion concerning specification of knots without some discussion of the knots present in the specimens.

I do not believe that the 0.85 reduction for steam conditioning should have been used to arrive at the safety factors for kiln dried southern pine in Table 3. Diekmann does finally state this in his conclusions.

E. F. Diekmann (*author's closure*)—Mr. Norum has performed a valuable service in presenting the various factors reflected in the derivation of the compression parallel to the grain formula found in ASTM D 2899. This derivation has been available previously only to selected individuals. The derivation was felt to be conservative as stated by Norum. The factors of safety in Table 3 are an indication as to whether the belief in the conservatism of the derivation was justified. The formula does not yield sufficiently conservative results for steam-conditioned southern pine tips. The factor for steam conditioning has been changed from 0.85 to 0.75 at the subcommittee level of ASTM Committee D07 and the change in the standard should be forthcoming. The tabulated average factor of safety for steam conditioned southern pine would be increased from 1.67 to 1.89 by this change in the standard.

ASTM D 2555 permits combining species utilizing volume weighting and stress values so computed may be used in the formulas in ASTM D 2899. Table 2 indicates that volume weighting and rounding yields a design stress of 1200 psi for a southern pine combination in lieu of 1092 psi based on shortleaf pine properties alone. The effect of volume weighting would be to reduce the steam-conditioned southern pine average factor of safety back to 1.72 from 1.89.

The recommendations in my report probably do approach a zero risk design situation. While no failure can be lightly regarded, a failure of a pile may be particularly serious since the entire structure may rest on it and the difficulties of repair and the attendant costs are so great. Conservatism seems warranted. ASTM D 2899 in a sense takes a less conservative approach to setting design stresses, but provides in Section 13.1 for an additional factor of safety of 1.25 in compression parallel to grain to be applied if additional safety is desired. Application of this added factor would increase the average factor of safety for steam-conditioned southern pine to 2.15. While this is an acceptable average value, the poorest two pieces of pile tested would have a factor of safety of only

$$1.26 \times \frac{0.85}{0.75} \times \frac{1092}{1200} \times 1.25 = 1.62$$

after applying the adjustments discussed herein. This is still too low on the basis of our current knowledge. My recommendation for permitting only two thirds of ASTM D 2899 values would adjust the factor of safety on the lowest material to 1.89 [1.26 × (1/0.66) = 1.89]. This recommendation was made without acknowledging the change in progress in the steam-conditioning factor in the standard to 0.75. The recommendation could be changed to three-quarters ASTM D 2899 values with the change in the standard. The poorest pile would then have a factor of safety of

$$1.26 \times \frac{0.85}{0.75} \times \frac{1}{0.75} = 1.90$$

With the added 1.25 factor suggested by ASTM D 2899, the factor of safety would go to 2.38 which is amply conservative for all cases as this factor would be for the poorest piece. I am unaware of any studies that relate the probabilities of lowest specific gravity, smallest acceptable tip size and maximum allowable knots occurring together in a single pile. The degree of conservatism suggested may well prove unwarranted should such a study be made.

The criticism by both Wilkinson and Norum regarding the recommendation in the paper governing knot limitations is partially justified. The data in Ref. *6* does not explain the recommendation. A separate study of knot/strength relationships was responsible for the recommendation so that it appears somewhat out of context in the paper.

T. D. Dismuke[1]

Behavior of Steel Piles During Installation and Service

REFERENCE: Dismuke, T. D., **"Behavior of Steel Piles During Installation and Service,"** *Behavior of Deep Foundations, ASTM STP 670,* Raymond Lundgren, Ed., American Society for Testing and Materials, 1979, pp. 282–299.

ABSTRACT: Field conditions to which steel piles are exposed during installation and service, along with strain measurements made during driving, are presented. The mechanical properties of the steel in the pile butt after driving were determined to see if impact of the ram affected the steel. A review of existing data on the strength of mild steels subjected to high strain rates showed that, in the range of strain rates imposed by impact drivers, the steel yield strength varies from 1.4 to 1.7 times the specification yield strength. The relationship between the specified steel strength and that of a full-size specimen was reviewed. It is concluded that (*a*) the driving stress may vary from 1.4 times the yield strength to the ultimate strength of the steel, (*b*) creep phenomena do not affect mild steels at ground temperatures, and (*c*) the strength of rolled specimens may be determined using specification strength.

KEY WORDS: steel piles, driving stress, mill acceptance tests, creep, corrosion

The pile foundation system functions properly only when both components—the soil and the pile—can support the imposed loads. The pile must be installed without a reduction in its load-carrying capability to a depth that ensures support and then withstands the effects of long-term loading and exposure to the surrounding elements.

Depending on the intensity and nature of loading and exposure conditions, the common failure modes of rolled steel shapes could include overstressing, deterioration, creep, fracture, and fatigue. These various modes are examined with regard to steel when used as piles. The conditions of use, field behavior, and practice, together with a review of some related existing laboratory data, are included. The relationship between mill acceptance tests and actual material behavior is considered.

The information in this paper is limited to mild steels, from which most

[1]Consulting engineer, Technical Services, Engineering Department, Bethlehem Steel Corporation, Bethlehem, Pa. 18016.

bearing piles are produced. These steels are designated as ASTM A36, A252, and A572 high-strength low alloy (HSLA).

Field Conditions

Steel piles are subjected to one-point handling, flame cutting, and welding operations prior to and during driving. The pile undergoes repeated loadings during driving, resulting in high localized stress levels for short time periods. After installation, the pile is exposed to a mild level and range of temperature fluctuation [*1*],[2] various loadings, and the deteriorationg effect of corrosion. The ground temperatures encountered do not significantly change the mechanical properties of mild steel relative to properties determined at room temperature.

Installation Forces and Pile Stresses

Driving by impact hammer subjects piles to forces that are larger and of a different nature than those normally encountered *in situ*. The material's mechanical properties at high loading and unloading rates and section geometry are important factors that affect pile behavior. Section geometry is important with regard to local buckling, as gross or overall buckling rarely occurs, because of the short loading period.

Figure 1 shows the recorded force-time curves at two locations on an H-pile near termination of driving. The strain rates under increasing loads are 1.14 s^{-1} at point A and 0.70 s^{-1} at point B. Strain rates due to impact by several diesel and other steam-air hammers with stiff (>10 MN/mm, 57 000 k/in.) cap blocks give values up to 7 s^{-1}. Strain rates imparted by drop hammers with large falls of approximately 5 m (16 ft) have not been determined, but probably approach a strain rate of 10 s^{-1}. The data given in Fig. 1 were taken from a test described elsewhere [*2*].

Typical distribution of peak stresses along an instrumented H-pile is given in Fig. 2 [*3*]. The stresses are given for two blows—one at the beginning of driving and the other at blow 439 near the termination of driving. The tensile stress for blow 1 is approximately one-half of the compressive stress and practically disappears for blow 439. Contact stress at the butt appears to be larger than that induced by the propagating wave. The high butt stress faded approximately 0.6 m (24 in.) from the butt. The top strain gage was located 200 mm (7.9 in.) from the butt where, for blow 439, the stress was 351.2 N/mm^2 (50.8 ksi); whereas, at the strain gage 530 mm (20.9 in.) from the top, stress was 255.8 N/mm^2 (37.1 ksi). The strain gages were mounted on one side of the web on the centerline of the H-piles for the data given in Fig. 1 and 2.

[2]The italic numbers in brackets refer to the list of references appended to this paper.

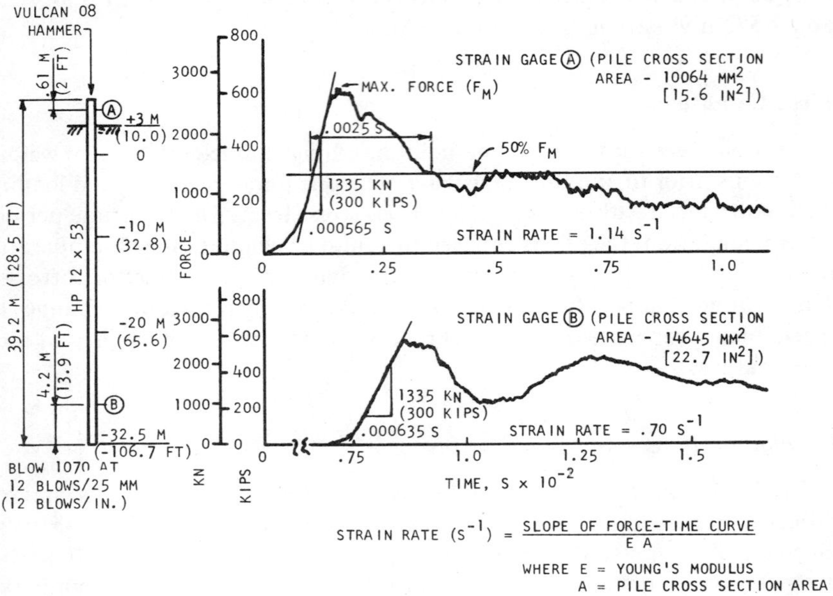

FIG. 1—*Maximum strain rates at two locations on an embedded H-pile.*

Peak wave-propagated compressive stress levels, which occur under each hammer blow, decay rapidly. Of the identifiable compressive stress fluctuations per blow at a point, only several may be 50 percent or more of the maximum stress level. As shown in Fig. 1 at strain gage A, the maximum stress existed for less than 0.0005 s, and at or above 50 percent of the maximum stress the time of loading was approximately 0.0025 s.

Installation Damage

Damage sustained by steel piles during installation is usually confined to the tip or the butt. At the tip, an H-pile flange can bend either inward or outward and occasionally results in distortion of the entire section or separation of the flange from the web. Figure 3 is a photo of an HP 10 × 42 pile with a cast steel tip that was driven into a skull (a dense mixture of iron and slag) by a Vulcan 08 hammer. The tip is intact, but the pile tip was rotated, as the surface of the skull was at an angle to the horizontal, causing the initiation of an accordian-type failure. Pipe piles can develop the accordian-type failure through the same mechanism as H-piles—eccentric loading at the tip, which causes the development of bending stresses. However, concentric overloading of thin-wall pipe piles can produce a similar configuration. Flange tearing or flange and web bending occurring at the tip can result from concentrated

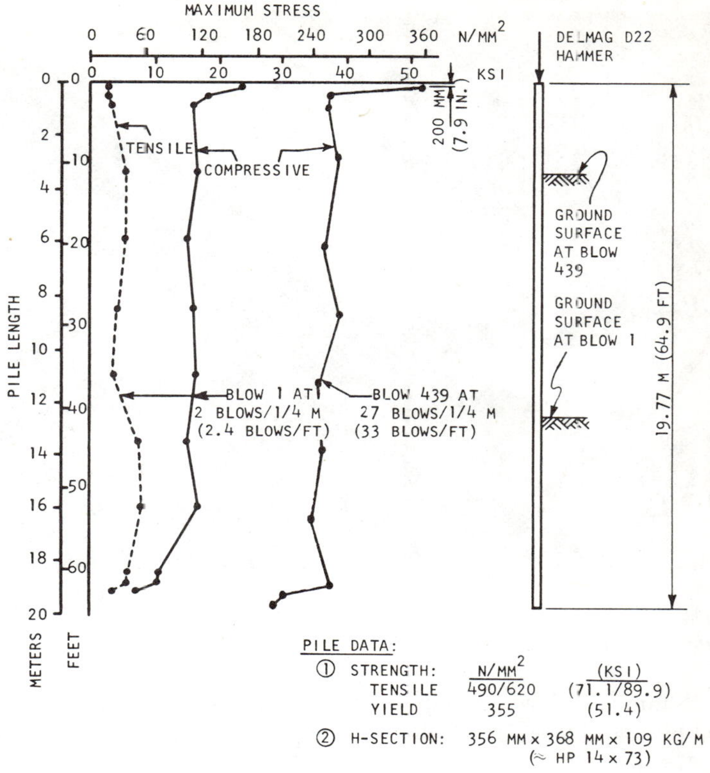

FIG. 2—*Maximum stress versus pile length for two hammer blows.*

and/or eccentric loading. This type of failure can usually be prevented by flange corner clipping, using a heavier section or attachment of cast steel tips [*4*]. In most cases examined by the author, eccentric loading at the tip was believed to have been the primary cause of damage.

At the butt, immediately below the helmet, the flanges and web may start to roll over under high-impact ram velocity, and wavy flanges and webs appear. Pile damage at the butt is usually due to misalignment between the pile and driving elements. Figure 4 shows the butt of an H-pile where local instability has occurred and further driving is useless. This configuration is generally associated with pile hammer misalignment and high (>4.57 m/s, 15 ft/s) ram velocity.

The potential for occurrence of driving damage at splices has been studied. One investigation conducted by the author involved the use of partial butt welds on H-piles. Figure 5 is a photograph of the cross section of a partial butt weld after driving. The pile was impacted 1100 times and driven to 12

FIG. 3—*Tip of HP 10 × 42 steel pile driven into an inclined skull which initiated an accordian-type configuration.*

blows/25 mm (12 blows/ in.) by a Vulcan 50C hammer, using a steel plate and Babbitt cap block. The weld and the pile did not show any visually apparent distress.

A study of the effects of driving HP 10 × 42 piles of A36 steel to strong and weak rock was reported by Goble et al [*4*]. At the test location where hard rock existed, all the pile tips without cast steel points were deformed as a result of overdriving. As determined by load testing, where flange warping or folding occurred, the structural capacity of the piles was not reduced. One pile, which had buckled into an accordian-type configuration and shortened at least 0.9 m (3 ft) during driving, still carried a load of 1575 kN (354 k), which is equivalent to a stress of 196.3 N/mm^2 (28.5 ksi) on the cross section. H-pile flange warping without accompanying reduction in pile strength was observed in the tests presented elsewhere [*2*].

In Situ *Conditions*

Piles—particularly long piles—are rarely driven perfectly straight. The H-piles referred to in Ref *2* had gradual sweep profiles that resulted in tip offsets of less than 2 percent. Others have reported profiles of H- and pipe piles with much larger offsets [*5*]. The structural capacity of steel piles is usually not affected as much by the curvature as by the ability of the soil to sustain the loading of a pile with large offsets.

FIG. 4—*Buckled flanges and web at the pile butt.*

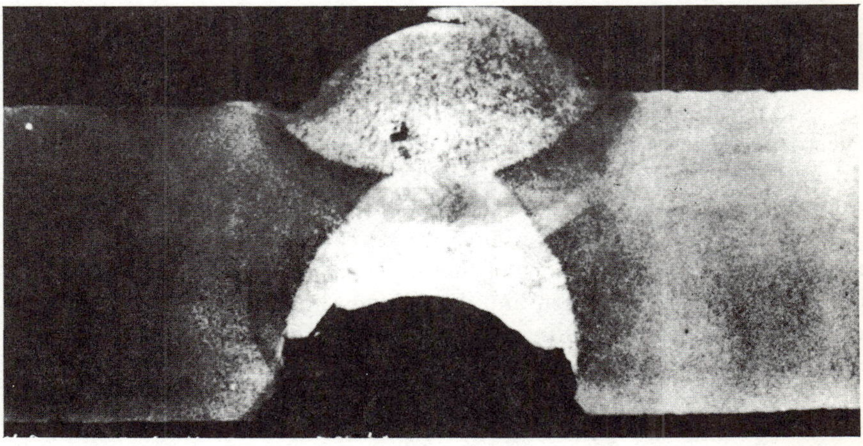

FIG. 5—*Cross section of an intact undersized weld after 1100 blows by a vulcan 50C Hammer.*

Pile tests are usually conducted using the maintained load or constant rate of penetration methods. Occasionally, a quick load test is made to simulate more rapid loadings. Dead loads usually comprise the largest part of the load, and therefore the strain rate is static and creep properties of the pile material must be known. The live load and rate of loading can vary from almost static values to rates exceeding 4448 N/s (1 k/s). Rates usesd in applying loads for maintained load tests fluctuate, but are in the neighborhood of 8896 N/s (2 k/s). For H-piles this results in strain rates around 10^{-7} s^{-1}.

Soils are extremely variable with regard to composition, strength, etc., but are relatively uniform with regard to corrosivity of steel piles. Exposure conditions, such as the location of the water table and the composition and age of penetrated fills, are important when evaluating the general condition as it relates to pile durability.

Strength of Mild Steel Piles

Considerable laboratory and field data are available concerning the behavior of mild steels subjected to varying strain rates, stresses, temperatures, and environmental exposure. Most of the strain rate data are for steels other than constructional and shapes other than pipe or H. Also, the specimen geometry varied from test to test. Although these variations exist, the strain rate data provide a reasonable basis for establishing the effect of strain on constructional steel. The mechanical properties and susceptibility of steel to various failure modes are presented in this section.

Mechanical Properties

The mill acceptance tests, which conform to ASTM Methods and Definitions for Mechanical Testing of Steel Products (A 370), can result in a range of strain rates [*6*]. A strain rate of approximately 0.0011 s^{-1} which conforms to ASTM Method A 370 was used for most of the tests reported [*6, 7*]. Mill acceptance tests, of necessity, are made at or near the maximum strain rates allowed. For H-piles, loading rates of 1335 kN/s (300 k/s) are necessary to approach the strain rates at which the mill tests are conducted.

Structural H-shapes have residual stresses caused by differences in cooling rates over the various portions of the section after rolling, as shown in Fig. 6*a*. The flange tips cool first, and the web and interior portion of the flange cool at a slower rate, resulting in the tips developing a compression residual stress while most of the interior is pulled into tension. To define the effect of residual stress and strain rate in the plastic region, Beedle and Tall [*7*] tested stub columns and compared the resulting stress–strain curve to that of a coupon, as shown in Fig. 6*b*. The reasons for the difference between the curves are shown in Figure 7. Mill acceptance tests may produce curves A or B (roundhouse) using a web coupon. The yield stress for curve B is determined by the 0.2 percent offset method, or at a strain of 0.5 percent (ex-

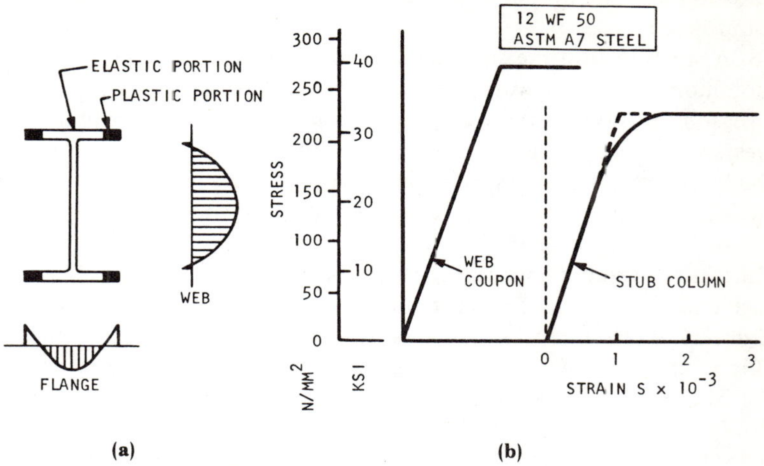

FIG. 6—*Influence of residual stress on stub column stress–strain curve.* (a) *Cross section residual stress pattern.* (b) *Stub column stress–strain curve for as-delivered material.*

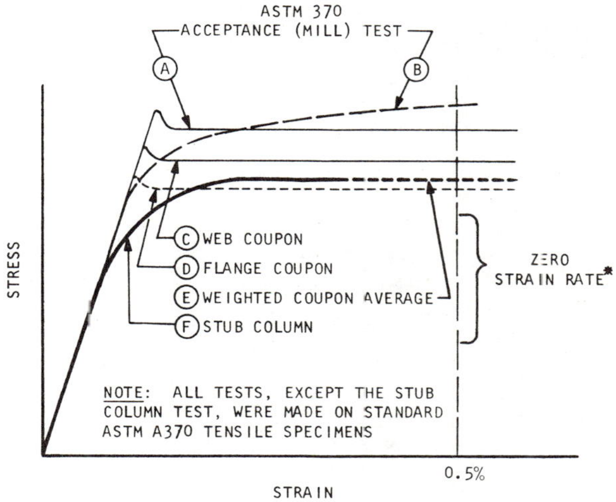

FIG. 7—*Influence of several variables on yield stress level. Asterisk refers to following: lower yield stress is the yield stress at zero strain rate.*

tension under load method). The yield level for curve C is determined by reducing the strain rate to zero in the plastic range. Curve D is the curve for the coupon taken from the flange, with a zero strain rate in the plastic region. Taking into account the various areas of the wide flange, with their respective residual stress, the weighted coupon average yield (curve E) agrees well with the stub column yield strength (curve F). The effect of the residual stress is to lower the proportional limit, but the zero strain rate yield strength is not affected.

HSLA steels exhibit the same general magnitude of residual stress as does the common structural grade A36 for a given section geometry [*8*]. The probable reason for this is that the cooling rates for the different portions of rolled shapes are the same, regardless of steel grade.

The foregoing applies to specimens tested to failure in one cycle of loading. Available substantial evidence [*9*] shows that repeated stressing may alter and fade the original residual stress.

Some of the reports concerning the behavior of steel specimens subjected to a wide range of strain rates are given elsewhere [*10–14*]. Most of the data were developed from tension specimens. One set of data by Manjoine (Fig. 8 [*10*]) is a representation of the influence of strain rate on the tensile properties of mild steel specimens. The strain rates that are of interest for driven steel piles are noted on the figure. The ratio of yield stress determined at a strain rate of 10^{-3} s^{-1} to that at 10^0 and 10^1 s^{-1} is 1.4 and 1.7 (ultimate strength), respectively. A compilation of existing data on the influence of strain rate in steels at room temperature was made by Garner in 1970 [*4*]. Steels with ultimate strengths of up to 552 N/mm² (80 ksi) exhibited greater sensitivity to strain rate than steels of higher strength. All the constructional steels commonly used in piling are in the former category. Figure 9 is a plot of yield strength ratio versus strain rate. The yield strength ratio, as determined from the best fit low-strength curve, ranges from 1.4 to 1.7 for the pile driving strain rates of 10^0 s^{-1} to 10^1 s^{-1}.

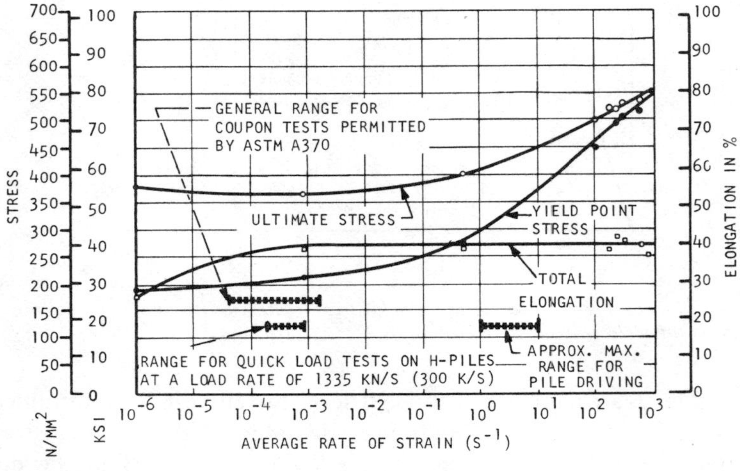

FIG. 8—*Influence of rate of strain on tensile properties of mild steel at room temperature and strain rates for various coupon and pile loading.*

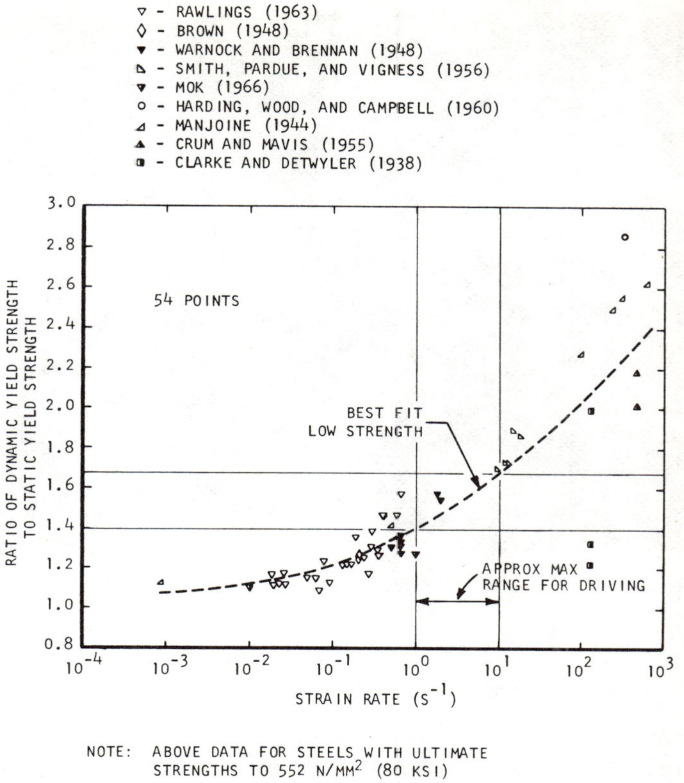

FIG. 9—*Yield strength ratio versus strain rate.*

Steel Properties After Driving

Figure 10 is a photograph of the butt of an HP 14 × 73 pile of ASTM A36 steel. It was impacted by a Vulcan 08 hammer with a close-fitting ±25.4 mm (1 in.) flat-faced helmet to 30 blows/25 mm (30 blows/in.). Tensile coupons were taken from the H-pile to obtain the mechanical properties exhibited by the steel after impact. Figures 11 and 12 give the results of these tests. The top transverse tensile specimens were cut as close to the top surface as possible. Lower transverse specimen strength was expected because of the direction of impact loading. The large difference in elongation values between the top and bottom transverse specimens of the pile may be due to the effect of cold working. The data shown in Figure 12 indicate that the effect of cold working could be detected by hardness testing as far as 63 mm (2.5 in.) below the impacted surface. The microstructure of the steel, as shown in Fig. 13, shows a grain structure disturbance of approximately 0.32 mm (0.012 in.).

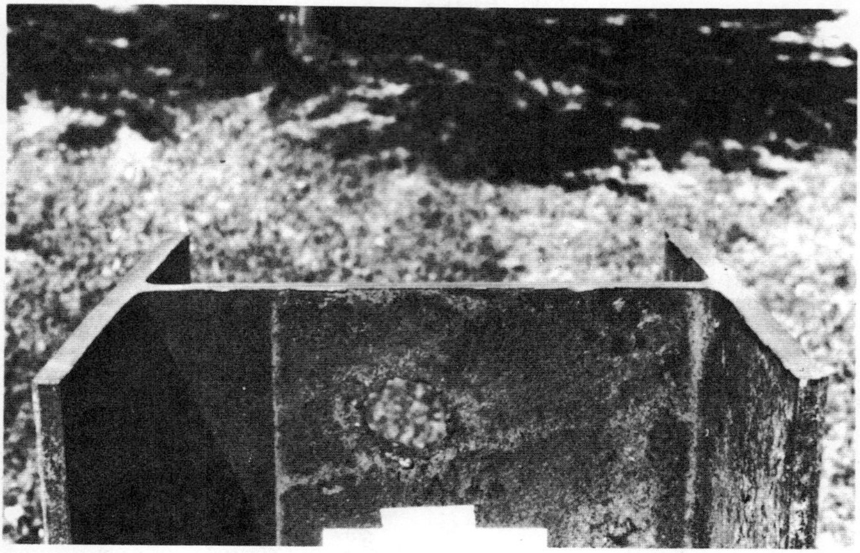

FIG. 10—*Impacted butt of HP 14 × 73 steel pile of A36 steel. 39.6 m (130 ft.) long; impacted 1899 times with a Vulcan 08 hammer; terminal driving 30 blows/25 mm (30 blows/in.).*

Load, Moment and Interaction

As stated before, piles are rarely driven completely straight. If substantial curvature has occurred during installation, the strength of the pile may be investigated by means of interaction relationships. Figure 14 is the interaction curve with nondimensionalized ordinate (axial load) and abscissa (plastic moment) for the ultimate strength of a W 8 × 31 section [*15*]. Since gross or overall buckling is not a failure mode for embedded piling, the curve for the L/r_x (L - unbraced pile length r_x - radius of gyration on the x axis) ratio of 0 is shown. The overall dimensions of the W 8 × 31 section are similar to those of an HP 8 × 36 pile section. The principal difference between the sections is that the HP section has a heavier web than the wide flange section. The curve given is reasonably accurate for determining the interaction relationships of not only the HP 8 × 36 pile section but of all the HP sections.

Based on sweeping pile curvature that appears to be the general condition, the theoretical bending stress is low ($\approx$7 to 14 N/mm^2, 1 to 2 ksi), and the axial pile capacity is virtually unaffected.

Creep Deformation, Brittle Fracture, and Fatigue

Little information is available on creep deformation on steels at temperatures below 200°C. The lack of data suggests that there have been

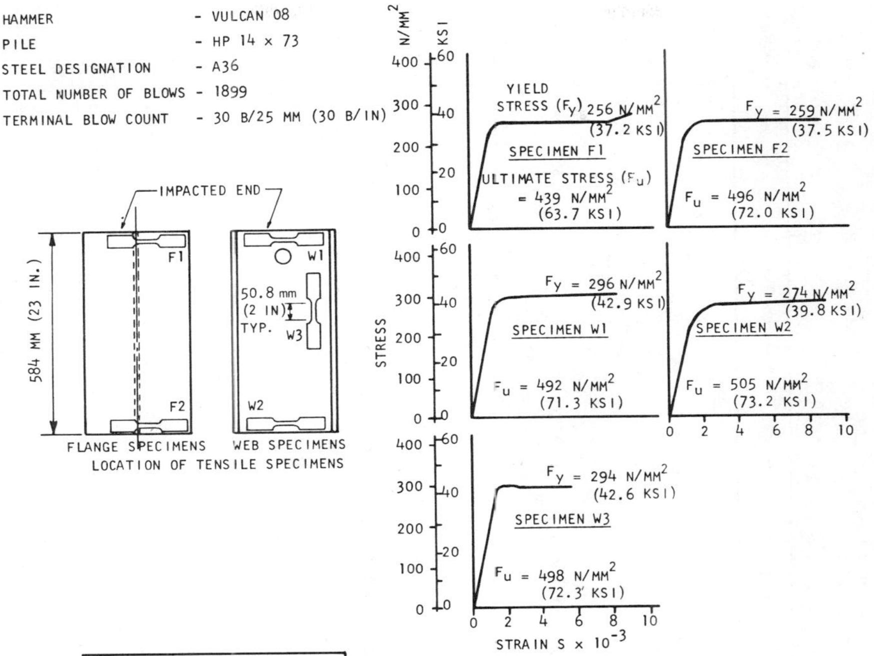

% ELONGATION				
F1	F2	W1	W2	W3
35.0	52.0	33.0	50.0	38.0

FIG. 11—*Stress–strain and elongation data for tensile specimens taken from an impacted pile butt.*

few, if any, problems concerning creep of steel used in various applications where temperatures are less than 200°C. Piles are subjected to a narrow temperature range, with upper limit at approximately the ambient temperature. Campus [*16*] reports some results of creep tests on steel bars made at a temperature of 21 ± 1°C (70 ± 1°F). The critical creep stress—the smallest sustained stress producing fracture—for the steels tested was slightly below the tensile (ultimate) strength.

There is no reported evidence that failure modes, such as brittle fracture and fatigue, occur during or after driving. The observation that few breaks have occurred at welded splices during driving indicates that even these discontinuities are not susceptible to failure.

Corrosion

Extensive field investigations of the condition of *in situ* steel piling have been made by the National Bureau of Standards [*17*]. The reported corrosion

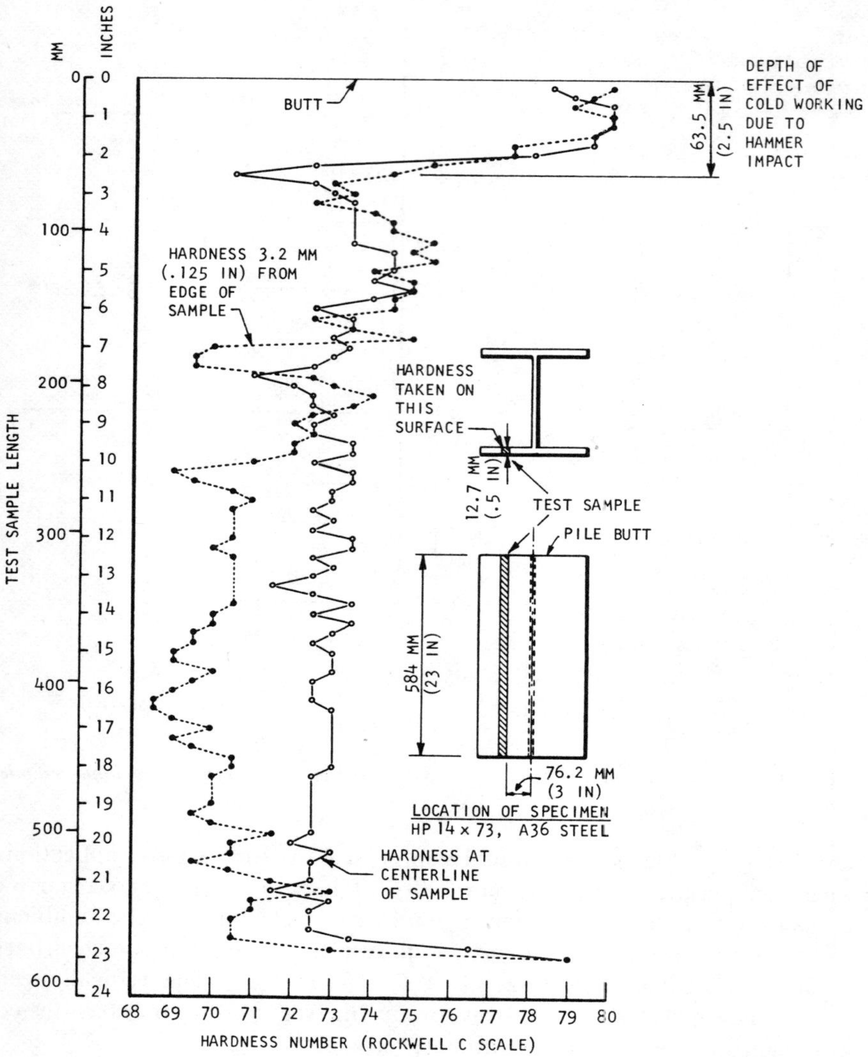

FIG. 12—*Effect of pile hammer impact on H-pile butt as indicated by hardness.*

rates are negligible for piles driven into undisturbed earth. The data also show that there is little corrosion when piles are exposed to both the fill and undisturbed earth encountered at the investigated sites. The investigated piling were made of mild constructional steels and included A7. The service conditions that prevented corrosion on A7 steel are lack of free oxygen and a favorable anode-to-cathode ratio. The presence of stresses does not usually affect the general corrosion behavior of metals to any significant extent [*18*].

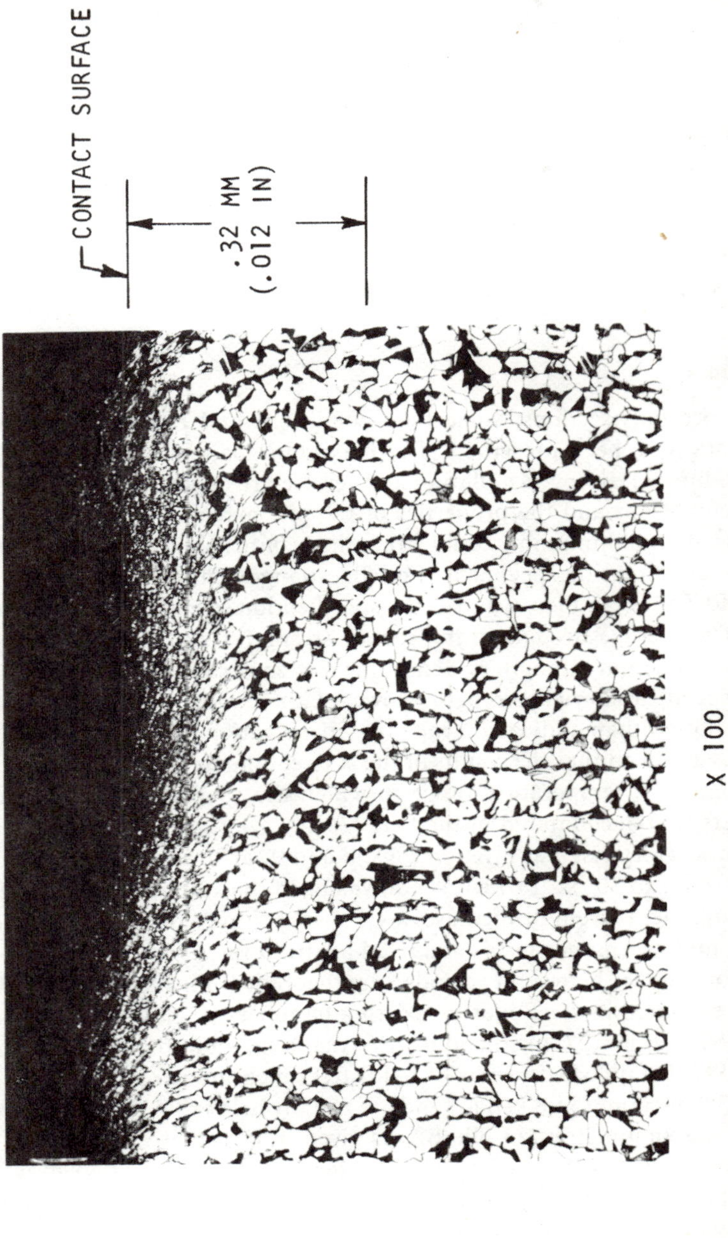

FIG. 13—*Microstructure showing depth of grain disturbance due to hammer impact at pile butt* (× *100*).

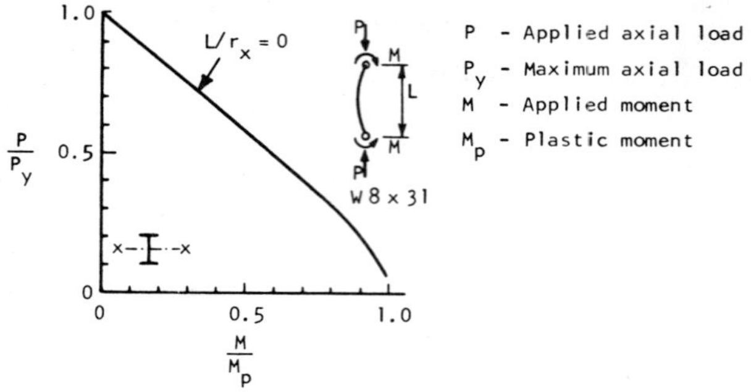

FIG. 14—*Ultimate-strength interaction curve for equal end moments.*

Discussion

In the previous sections, some field conditions and results of pertinent testing data have been presented. The maximum load levels to which piles should be limited during installation and service are dependent on the effect of these conditions and the properties of the pile material.

The specified minimum material strength is confirmed by mill acceptance tests on coupon tension specimens. For wide-flange sections, as indicated by Beedle and Tall [*7*] in Figs. 6 and 7, the specification strength is generally exceeded. The average strength of full section strength tests (when the full section or stub column strength is determined at zero strain rate) was higher than the specification yield strength. However, there were some tests that had strengths lower than indicated from the specification yield strength. The author has assumed that the results of the stub column tests are applicable to HP sections. The temperature range in which the specification strength is valid covers that encountered in the field [*10*].

From laboratory tests, as given in Figs. 8 and 9, at strain rates of approximately 10^0 s^{-1} to 10^1 s^{-1}, which is equivalent to that range encountered in pile driving, yield strength increases by 1.4 to 1.7 over the strength determined by mill acceptance tests conducted at strain rates near 10^{-3} s^{-1}.

Based on impact between two rods, the theoretical maximum stress that can be propagated in steel piles by existing hammers (excluding drop hammers) is 189 N/mm^2 (27.4 ksi) and is twice that amount at the tip if the tip is fully fixed [*19*]. Published results of field stress measurements report values less than 378 N/mm^2 (54.8 ksi) [*2*] and are usually below 207 N/mm^2 (30 ksi) [*19*]. The maximum stress in pile B2, as given in Fig. 1, is 276 N/mm^2 (40 ksi). Contact stress at the butt may be greater that the propagation stress as shown in Fig. 2 [*3*]. The existence of high butt stress levels above that due to the propagating wave is no indicator of pile damage. Figures 10 through 13 show that A36 steel H-piles can be subjected to high impact values without distortion and retain the specified yield properties. The steel had apparently been stressed higher than its dynamic yield strength in the top several inches. Darragh and Bell report that an indicated driving stress of 345 N/mm^2 (50 ksi) usually corresponds to pile driving damage for pipe piles made of 241 N/mm^2 (35 ksi) yield steel [*20*]. This corresponds to a ratio of driving stress to specification yield stress of 1.4. Field data qualitatively corroborate the laboratory data in that strain rate effects substantially increase the yield strength of mild steels. Damage at the pile tip is primarily due to eccentric and concentrated loading and is not as sensitive to impact variations as the butt.

There are several possible results when pile tip damage occurs. One is that if the pile is damaged prior to reaching a bearing stratum, it cannot be driven further, the soil cannot support the pile load, and the structural capacity may be reduced. The other possibility is that if the pile reaches the bearing stratum and is then damaged, the structural capacity of the pile may be reduced. Figure 3 is an example of a pile that could not be driven further and had to be pulled or abandoned. Reference *4* is an example where structural capacity of some of the piles was reduced when the H-piles were driven to bearing and then deliberately overdriven. In the latter case, the deformation of the H-pile, which consisted of flange warping and folding, did not constitute damage, as the structural capacity was not reduced.

Butt damage usually occurs as a result of hammer-pile misalignment, uneven cross section engagement due to worn driving helmets, or both. Figure 4 is an example of the effects of driving with a misaligned heavy ram that strikes the butt at a high velocity.

Structural capacity of a steel pile may be affected due to pile configuration. An evaluation of the structural capacity may be made by using interaction relationships similar to that given on Fig. 14. Long pile sweeps usually result in small bending moments, but when dog legs occur, an extensive evaluation of the pile and soil capacity may be necessary.

As indicated elsewhere [*16*], steel piles are not subject to creep under service loading. No strength reduction is necesary. However, deformation due to long-term loading may occur because the supporting soil may be subject to creep.

The deterioration potential of steel piles *in situ* is almost nonexistent in undisturbed soils [*17*]. When used in new fills, well-aerated soils, or where the water table is close to the surface, an evaluation should be made as to the advisability of providing a pile coating. Stress should not be affected by a consideration of deterioration.

The relation of the driving stress indicated by the wave equation to actual measured values has not been consistent. It has underestimated the stress level in several cases where the stress has been measured [*2*] and exceeded it in other cases. If an allowable driving stress is to be a useful parameter in preventing pile damage, the accuracy of the theoretically determined wave-propagated driving stress needs to be improved. Also, a verified correlation has to be established between driving stress and various types of pile damage.

Based on measured stress values and laboratory tests, the allowable wave-propagated driving stress could reasonably be established at values varying from 1.4 times the specification yield stress to the ultimate strength. The lower range would apply to low pile-to-ram weight ratios.

Conclusions

The information in this paper contains a review of some of the existing data on properties of steel and some results of field and laboratory tests. The conclusions reached from the review of the data are:

1. Specification yield strength may be used to determine the minimum full section strength.
2. The minimum tentative allowable driving stresses may be 1.4 times the specified yield strength.
3. Creep does not occur in steel piles below the yield stress in the temperature ranges encountered underground.
4. Flange warping and folding does not, in all cases, reduce H-pile capacity.
5. Steel piles are not subject to brittle fracture and fatigue failure modes.
6. Improved alignment of driving components and the pile would result in less pile driving damage.

References

[1] Penrod, E. B., Walton, W. W., and Terrell, D. V., *Journal of the Soil Mechanics and Foundations Division*, American Society of Civil Engineers, Vol. 84, No. SMI, 1537, Feb. 1958, pp. 1537-1-21.
[2] High Capacity Steel H-Piles, American Iron and Steel Institute, Washington, D. C., 1976.
[3] Farid, M. U., "An Investigation of the Dynamic and Static Stresses in Universal Bearing H-Piles," Report No. T/PROD/858/77/C, British Steel Corporation, Teeside Laboratories, 24 Mar. 1977.
[4] Goble, G. G., Litkins, G., and Teferra, W., "Tests on H Piles Driven to Rock with Five Different Hammers," Piletalk Seminar, Associated Pile & Fitting Corporation, Miami Beach, 1978, pp. 85–98.
[5] Hanna, T. H., *Canadian Geotechnical Journal,* Vol. 5, 1968, pp. 150–172.
[6] Rao, N. R. N., Lohrmann, M., and Tall, L., *Journal of Materials*, Vol. 1, No. 1, March 1966, pp. 241–262.
[7] Beedle, L. S. and Tall, L., *Transactions*, American Society of Civil Engineers, Vol. 127, Part II, 1962.
[8] Tebedge, N., and Tall, L., *Construction Métallique*, No. 2, 1974, pp. 37–48.
[9] Horger, O. J. in *Metals Engineering Design*, 1st ed., O. J. Horger, Ed., American Society of Mechanical Engineers Handbook, McGraw-Hill, New York, 1953, Chapter 4, pp. 42–62.
[10] Manjoine, M. J., *Transactions*, American Society of Mechanical Engineers, Vol. 66, 1944, pp. A211–A218.
[11] Smith, R. C., Pardue, T. E., and Vigness, I., *Proceedings*, Society of Experimental Stress Analysis, Vol. 13, No. 2, 1956, pp. 183–197.
[12] Davidson, T., and Meier, J. H., *Proceedings*, Society of Experimental Stress Analysis, Vol. 4, No. 1, 1946, pp. 88–111.

[*13*] "Protective Construction," Defense Civil Preparedness Agency, Technical Report 20, Vol. 4, May 1977.

[*14*] Garner, R. R., *Journal of Materials*, Vol. 5, No. 3, Sept. 1970, pp. 618-632.

[*15*] Galambos, T. V., in *Structural Steel Design*, 2nd ed., L. Tall, Ed., Ronald, New York, 1974, Chapter II, pp. 369-405.

[*16*] Campus, F., "Creep and Relaxation of Steel at Room Temperature," International Association for Bridge and Structural Engineering, Conference, Portugal, 1957, pp. 312-315.

[*17*] Romanoff, M., and Schwardtfeger, W. J., *NBS Papers on Underground Corrosion of Steel Piling, 1962-1971*, NBS Monograph 127, National Bureau of Standards, Washington, D.C., 1972.

[*18*] Proctor, R. P. M in *Metal/Enviroments Interactive*, L. L. Shreer, Ed., Vol. 1, Newnes-Butterworth, London, 1976, p. 1.47.

[*19*] Rausche, F., and Goble, G. G., *Journal of the Construction Division*, American Society of Civil Engineers, Vol. 98, No. C02, 9188, Sept. 1972, pp. 201-218.

[*20*] Darragh, R. D., and Bell, R. A. in *Load Tests on Long Bearing Piles, ASTM STP 444*, American Society for Testing and Materials, March 1969, pp. 41-67.

T. D. Dismuke[1]

Influence of Codes and Standards on the Use of Steel Piles

REFERENCE: Dismuke, T. D., **"Influence of Codes and Standards on the Use of Steel Piles,"** *Behavior of Deep Foundations, ASTM STP 670,* Raymond Lundgren, Ed., American Society for Testing and Materials, 1979, pp. 300–305.

ABSTRACT: A brief review and evaluation of the provisions of existing codes and standards regarding steel piles is presented. The reasoning behind the current proposal for an allowable bearing load of 50 percent of the steel yield strength is presented.

The use of existing codes and standards to provide adequate guidance for preventing foundation failures is questioned. Codes based on load factor concepts and improved field inspection are thought to be necessary.

KEY WORDS: codes, standards, allowable stress, load testing, durability, load factor concept

The basic purpose of the pile regulations in a building code is the establishment of those legal requirements that are essential for public safety. Building code regulations should provide for the use of all kinds of foundation piles—steel, concrete, and wood—with assurance of adequate and comparable safety, exercising due care that unwarranted penalties on building costs be avoided. The soil and conditions at the site will generally determine the kind of pile that can be used to best advantage. The code should provide for use of each within the full scope of its qualifications.

Most provisions in pile regulations, such as pile spacing, minimum number of piles, or installation procedures, are common for all pile types. Those provisions that bear directly on the individual pile type include the allowable stress and load, dimensions, splices, and durability. The effect of these provisions on the use of steel piles and the available evidence for the current allowable stress proposals are discussed in the following sections.

[1]Consulting engineer, Technical Services, Engineering Department, Bethlehem Steel Corporation, Bethlehem, Pennsylvania 18016.

Codes and Standards

Existing codes and standards are a combination of descriptive and prescriptive requirements, with the former predominating. Usually the code is written with the implicit acceptance of the existence of the most unfavorable soil, rock, and construction conditions as the basis of consideration. Also, the competency of engineering is assumed to be minimal. As a result, the allowable load or unit stress of the pile material is reduced irrespective of the support conditions, installation procedure, and quality of engineering and inspection. This approach may be unduly costly, as pile foundations are often overdesigned.

In general, pile regulations are not complete in the descriptive sense, as they do not cover several important facets of foundation design and have wide variations of allowable values of material stress. No code establishes requirements or limitations that cover all of the essential considerations for structure support, and some are very restrictive in the descriptive provisions.

To consider all provisions of existing pile regulations is beyond the intent of this paper, but a brief comment on the load testing provision that is contained in one form or another in all codes follows. The existing regulations place reliance on pile test loading to verify or establish the allowable load. The pile test is not as effective in this regard as the reliance indicates because of many factors. In some jurisdictions, the support conditions are such that the pile load test is considered as a means of determining the upper bound limits against a plunging-type failure. In others, the test is construed as a code requirement without physical meaning. In rare cases, the settlement data are used to predict structure settlement. The pile load test is, of necessity, an almost time-independent test made on a small element of the supporting medium. The pile load test can be a valuable tool for control and design purposes, but its limitations should be recognized.

Steel piles were first mentioned in the 1930s in the National Building Code and were included in most of the other codes by the 1940s. Figure 1 shows the stress that a number of codes allowed for steel piles beginning in 1943. The introduction of high-strength low-alloy (HSLA) steels, beginning in the 1960s, introduced another factor to be considered. This is shown in Fig. 1 by the vertical lines, which reflect the constructional grade of steel used. Each increase in allowable stress or load in any code was justified by considerable field experience.

While the majority of onshore structures are much smaller than offshore structures, and therefore the foundation loads are less, there is application for piles that will support loads from 200 to 300 tons to above 1000 tons. The driving and handling equipment is available for such piles. While most current codes do provide for the introduction of higher loads,

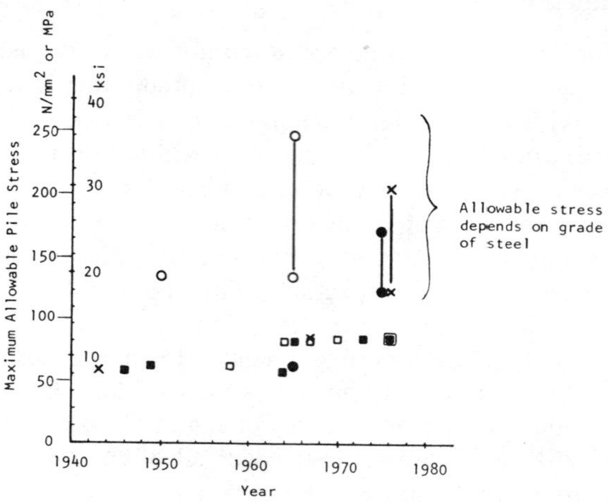

×	National Building Code	Maximum stress = $0.5 f_y$.
■	Standard Building Code	Maximum f_y = 36 ksi; higher loads allowed with suitable documentation
□	Uniform Building Code	Maximum f_y = 36 ksi; may be $0.5 f_y$ if substantiated data submitted and approved.
○	Basic Building Code	Maximum load cannot exceed 200 tons
●	New Orleans Building Code	Maximum f_y = 50 ksi

FIG. 1—*Allowable steel H-pile stresses as given in various codes.*

it is difficult to increase stress levels in many jurisdictions. The reluctance of code-writing groups to increase allowable stress is primarily because the probability of failure becomes greater as loads increase and redundancy decreases. In the past, steel piles were generally capable of carrying considerably higher loads than were allowed by codes and used in practice.

The continually increasing capability of driving equipment has enabled driving of existing pile sections so that the material properties, as well as section properties, have to be considered as possible limitations on pile capacity.

In the near future, the supporting capacity of the soil or the structural integrity of the *in-situ* pile can be exceeded if adequate engineering practice and installation procedures are not utilized. To provide for safety and the economic use of materials, the concept of load factor design should be considered. The development of codes that reflect that concept, along with a major revision of the field inspection provisions, may be the next required step in ensuring foundation safety.

Allowable Steel Pile Stress

Existing codes provide a certain degree of freedom for innovation by means of variance procedures. These provisions, where and when implemented, allow for use, with documentation, of more highly loaded piles. Experience with the higher-loaded piles is thus gained within the jurisdiction of the code. This experience, together with that obtained by testing and use elsewhere, provides the cumulative evidence necessary for proposing higher allowable stress or loads. Also, the quality of the evidence has to be such that the designer is satisfied that the allowable stress or load is suitable.

At the present time, experience with the steel pile stress level of $0.5f_y$ is considered to be sufficient for inclusion as an allowable stress. It is expected that this stress will be treated by the designer as a maximum allowable to be varied where necessary to be consistent with job site conditions and safe practice. The experience is primarily composed of driving and load testing, durability, and material behavior studies. As a result of the above, a number of major structures have been supported by high-capacity steel pile foundations [*1,2*].[2]

Driving and Load Testing

Load tests have been conducted to the yield point of the steel on numerous steel piles in many different geographical areas [*3*]. Various driving systems and geological conditions have been encountered, and a number of contractors have installed the high-capacity piles. Welded splices and standard H- and pipe pile sizes were used in the tests. The specification yield strength of the steel ranged from 228 N/mm² (33 ksi) to 345 N/mm² (50 ksi). A selected summary of the tests is given in Table 1. The summary is intended to indicate the range of test conditions and loads.

Durability

Acceptable data on the durability of piling is obtained by observing the condition of *in-situ* piling. For this reason, considerable time may elapse before an evaluation of the durability can be made or where existing piling can be located and exposed. The American Iron and Steel Institute has supported durability studies of steel piling at the National Bureau of Standards for a number of years. The Bureau has published several monographs concerning these studies [*4,5*]. The reported rates of corrosion on steel piles are negligible in undisturbed soils. Where piles are driven through new fills or well-aerated soils or where the water table is close to the surface, the necessity for protection measures should be reviewed.

[2] The italic numbers in brackets refer to the list of references appended to this paper.

TABLE 1—*Selected summary of high capacity steel pile tests.*

No.	Pile	Length, m (ft)	Steel Designation	Hammer Energy, j (ft-lbs)	Soil Type[a]	Maximum Test Load, kN (tons)
1	HP10 × 42	25.8 (84.75)	A7	40 979 (30 225)	very dense hardpan	2135 (240)
2	PP12¾ × 0.375 (closed end)	33.5 (110.0)	A252 grade 3	55 045 (40 600)	dense sand and gravel	4003 (450)
3	HP14 × 73	42.0 (137.9)	A7	20 337 (15 000)	clayey gravel	3025 (340)
4	HP14 × 89	54.6 (179.0)	A572 grade 50	44 064 (32 500)	limestone boulders and hardpan	5604 (630)
5	HP14 × 89	73.1 (240.0)	A36	86 636 (63 900)	dense sand, silt and clay lenses	4092 (460)
6	PP36 × 0.75	49.4 (162.0)	A53 grade B	178 966 (132 000)	dense sand	13344 (1400)

[a] Type of soil in which the pile terminates.

Long-term Loading

As structures are usually used for many years, the long-term loading behavior of steel piles is of interest. As reported elsewhere [*6*], creep in steel within the elastic range and below a temperature of 200°C is nonexistent or very low.

Summary

The existing pile regulations in codes were developed when steel pile loads were lower than at the present time. The actual capacity of the driven steel pile was generally higher in relation to the allowable load than current capacity. Limits of the capacity of the soil and piles to support large loads have been approached or reached on a number of occasions. Factors of safety are diminishing; therefore, another approach, such as that recently developed for structural design toward providing safe foundations, is desirable.

References

[*1*] Blessey, W. E., *High Capacity Long Steel Piles*, American Iron and Steel Institute, Washington, D.C., 1969.

[*2*] American Iron and Steel Institute, *High Capacity Steel H-Piles*, Washington, D.C., 1976.

[*3*] American Iron and Steel Institute, *Steel Pile Load Test Data*, Washington, D.C., 1975.

[*4*] Romanoff, M., and Schwardtfeger, W. J., *NBS Papers on Underground Corrosion of Steel Piling, 1962-1971*, NBS Monograph 127, National Bureau of Standards, Washington, D.C., 1972.

[*5*] Schwardtfeger, W. J., and Romanoff, M., *Corrosion Rates on Underground Steel Test Piles at Turcot Yard, Montreal*, NBS Monograph 128, National Bureau of Standards, Washington, D.C., 1972.

[*6*] Dismuke, T. D., "Behavior of Steel Piles During Installation and Service," this volume.

W. L. Gamble[1]

Capacity of Reinforced and Prestressed Concrete Pile Sections

REFERENCE: Gamble, W. L., **"Capacity of Reinforced and Prestressed Concrete Pile Sections,"** *Behavior of Deep Foundations, ASTM STP 670,* Raymond Lundgren, Ed., American Society for Testing and Materials, 1979, pp. 306–322.

ABSTRACT: Information needed to compute the nominal strengths of various pile cross sections that are subjected to combined compressive thrust and bending moments is presented, with examples of moment-thrust interaction diagrams for several typical kinds of pile sections.

The components of the factor of safety are examined, with particular emphasis on the reliable strength of the pile section. The use of a "strength reduction factor," a multiplier that is less than 1.0, to establish a reliable or minimum expected failure load that is somewhat lower than the nominal failure load to be expected from an ideal member is discussed.

This strength reduction factor may vary with the number of piles in an individual footing and should be higher for piles that can be inspected than for those that cannot be inspected after completion. Precast piles and piles that are cast in permanent metal cylinders are examples of members that can be inspected to different extents, while piles cast-in-place in unlined holes or that have the liner pulled out of the hole after the concrete is cast are examples of members that cannot be easily inspected after construction.

KEY WORDS: concrete piles, reinforced concrete, prestressed concrete, precast concrete, structural analysis, structural design, eccentric loadings, failure, safety factor

Nomenclature

A_g Gross area of member

A_s Total area of steel in cross section (for P_0 calculation)

E_c Young's modulus of concrete

E_s Young's modulus of steel

f_c' Compressive strength of standard 150 by 300 mm (6 by 12 in.) concrete test cylinder

f_{ce} Effective precompression in concrete

[1] Professor of Civil Engineering, University of Illinois at Urbana-Chapaign, Urbana, Ill. 61801.

f_s Steel stress
f_{se} Effective prestress in steel
f_y Yield stress of reinforcing steel bars
P_0 Nominal capacity of concentrically loaded pile
ϵ_{ce} f_{ce}/E_c
ϵ_{se} f_{se}/E_s
ϕ Strength reduction factor
P_n Nominal capacity of eccentrically loaded pile
LL Live load force
DL Dead load force

This paper is concerned with the strength of reinforced and prestressed concrete pile sections and is not concerned with the larger problem of the strength of the pile-soil system. However, it is necessary to understand fully the strength and behavior of each of the components in a system in order to evaluate the entire system, and this paper is directed toward the strength of the pile alone.

It will be helpful to remember from the beginning that a pile remains a reinforced or prestressed concrete member even after it has been driven into the ground and that as such its strength and behavior are predictable using general methods of structural analysis. This basic fact appears to have been forgotten in many cases, especially where there is a significant moment in addition to the thrust and when specifications have been written only in terms of allowable stresses and then considering only thrust.

The emphasis in this paper is placed on strengths of members. It has been recognized at least since the 1936 American Concrete Institute (ACI) Code that service load stresses in reinforced concrete compression members were relatively meaningless because of the redistribution of internal forces between concrete and steel with time as a result of creep and shrinkage of the concrete. In 1936, the profession was not ready for a strength design code, and that code disguised a strength design provision.

The material that follows is based on the assumptions outlined in the 1977 ACI Building Code[2] requirements for the evaluation of the strengths of short column sections, in spite of a statement that the 1977 code does not apply to the parts of piles or drilled piers that are embedded in the ground.

The "Recommendations for Design, Manufacture, and Installation of Concrete Piles"[3] prepared by ACI Committee 543 refers to "fundamental differences between foundation piles and columns." However, it is not believed that there are any fundamental differences between piles and

[2] "Building Code Requirements for Reinforced Concrete," ACI Standard 318-77, American Concrete Institute, Detroit, 1977.

[3] "Recommendations for Design, Manufacture, and Installation of Concrete Piles," ACI Committee 543, *ACI Manual of Concrete Practice*, 1977, Part 3, pp. 543-1 to 543-40.

columns, and the remainder of this paper is based on the assumption that they are very similar, if not identical.

Moment–Thrust Interaction

The thrust capacity of a pile is strongly dependent on the moment that accompanies that thrust. The moment may be a result of the design loading conditions on the pile or pile footing, or it may be the result of an accidental eccentricity of the thrust.

A moment-thrust interaction diagram for a particular pile is shown in Fig. 1. This is an envelope of all combinations of moment and thrust that would cause failure. Several features stand out. The thrust capacity is reduced significantly by the presence of relatively small bending moments. If the thrust is small, increasing the thrust increases the moment capacity, but if the axial force is tensile, the moment capacity is drastically reduced as the force increases. This last observation is particularly important in the design of tension piles.

The points for this curve were computed for a series of assumed strain distributions similar to that shown in Fig. 2. The failure criterion is that the section fails when the concrete compressive strain reaches 0.003 and the neutral axis position has been varied by steps. The largest moment occurs for the strain distribution giving the limiting strain of 0.003 at the compression face of the member and the yield strain for the opposite layer of steel. This point is often referred to as the balance point. The reinforcement is assumed to have an elastic-perfectly plastic stress-strain curve. The concrete has a distinctly nonlinear stress distribution but can be approximated by the equivalent rectangular stress distribution shown in the figure [2].

The upper end of the solid curve corresponds to a neutral axis depth equal to the total depth of the cross section, and points along the broken portion of the curve cannot be calculated using the equivalent rectangular stress distribution.

The maximum load P_0 that can be resisted by a reinforced (not prestressed) section can be computed as

$$P_0 = 0.85 f_c'(A_g - A_s) + A_s f_y \tag{1}$$

This is simply the sum of the strengths of the concrete and steel areas present. The 0.85 is an experimentally determined factor related to the differences that normally exist between the concrete in the structural member and in the small test cylinders used to determine f_c'. This is documented in references [*1,2*][4] and applies to loadings applied in a few hours or less.

[4] The italic numbers in brackets refer to the list of references appended to this paper.

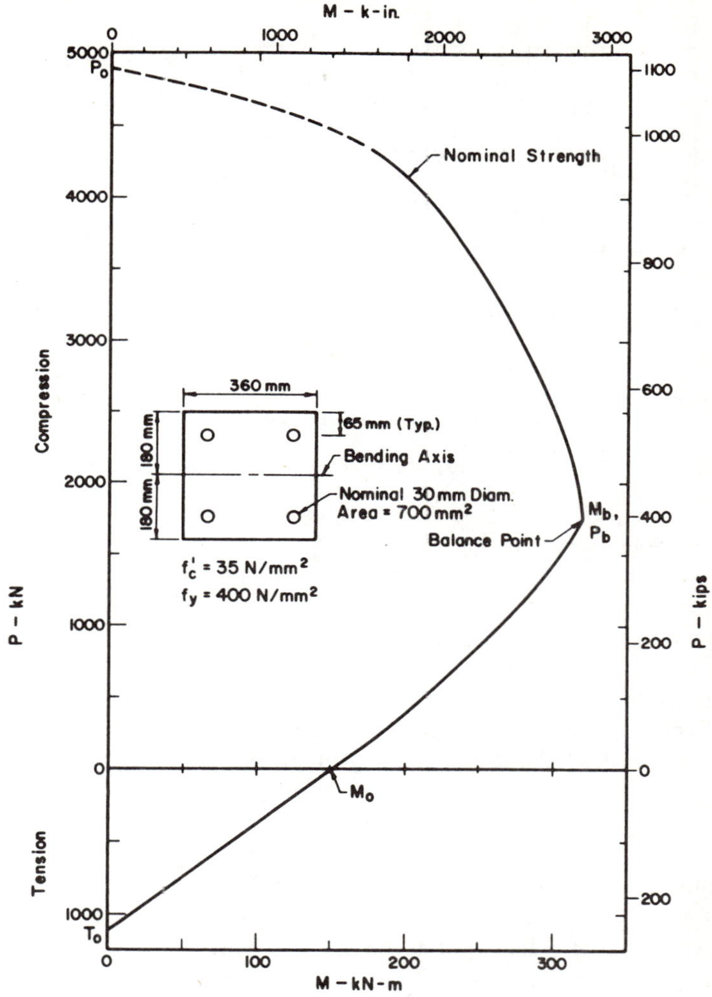

FIG. 1—*Moment-thrust interaction diagram for square pile.*

The concept of the construction of the moment-thrust interaction curve has been confirmed by the results of extensive tests [*3*] and is presented in most textbooks and many handbooks on structural concrete.

The curve shown in Fig. 1 represents the ideal strength, which would be achieved in a perfect test specimen in which the strengths of the concrete and steel were exactly as desired, all dimensions were correct, the reinforcement is in its correct locations, and the load is applied with the assumed relationship between moment and thrust.

The evaluation of a factor of safety must include the possibilities of both

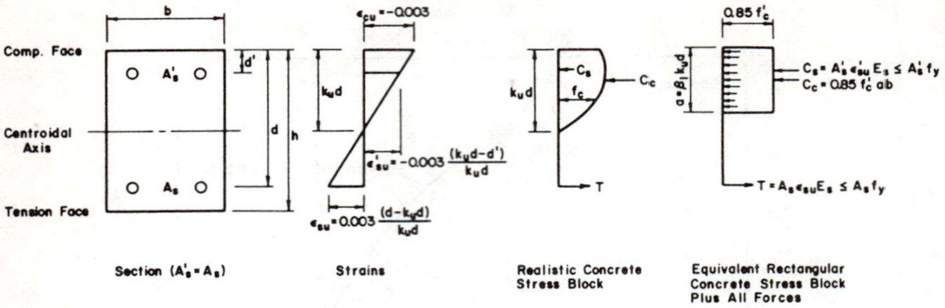

$$\text{Thrust} = C_s + C_c - T$$

$$\Sigma M_{centroid} = C_s\left(\frac{h}{2} - d'\right) + C_c\left(\frac{h}{2} - \frac{a}{2}\right) + T\left(d - \frac{h}{2}\right)$$

$$\beta_1 = 0.85 \text{ if } f_c' \le 4000 \text{ lb/in.}^2 \text{ (28 N/mm}^2\text{)}$$

$$\beta = \left[0.85 - 0.05\left(\frac{f_c' - 4000}{1000}\right)\right] \ge 0.65 \text{ if } f_c' > 4000 \text{ lb/in}^2$$

$$\approx \left[0.85 - 0.007\,(f_c' - 28)\right] \ge 0.65 \text{ if } f_c' > 28 \text{ N/mm}^2$$

FIG. 2—*Summary of method of calculation of moment–thrust interaction diagram.*

having overloads and having members with capacities smaller than expected. The ACI code uses a "strength reduction factor" to reduce the nominal capacity of a member. This factor, ϕ, is currently taken as 0.7 for a tied column, although it increases to 0.9 as the moment becomes large and the thrust small as the flexural case is approached. The 0.7 factor is considerably lower than the 0.9 used for flexure for at least two reasons. One reflects the importance of the member to the structure—a single column can hardly fail without causing great additional damage. A second reason reflects the sensitivity of the strength member to the compressive strength of the concrete. Equation 1 can be used to show the great dependence of P_0 on f_c'.

A number of factors need to be considered in establishing suitable ϕ factors for reinforced and prestressed concrete piles. It may be desirable to use a lower value of ϕ for a pile to be used individually or in a small group than for a pile to be used in a large group. The failure of a single high-capacity pile in a four-pile or smaller group is as serious as the failure of a column, while the failure of one in a group of 25 would ordinarily be much less serious.

It appears necessary to use different ϕ factors for different types of piles. The key difference lies in the ability to inspect the completed pile. A factory precast pile, whether reinforced or prestressed, can be inspected for defects after it is manufactured but before driving. Inspection after driving is possible only by indirect means. A pile that is cast-in-place in a metal shell that is left in place represents another case in which inspection is possible. After the shell is driven, it can be examined to determine that it is straight and open all the way to the bottom. The concrete cannot be inspected after casting, but there are not serious problems in filling a long vertical tube with concrete unless there is an excessive amount of reinforcement. Batter piles may present filling problems, and metal shells may leak soil and water, which present inspection problems.

A pile that is cast-in-place in an unlined hole or a pile that is cast and the liner then pulled out represents a case in which there is much greater uncertainty about the concrete quality and continuity. Such a pile should be designed using a considerably lower ϕ factor than for the previous two types. There is adequate evidence with various specific types of piles cast in unlined holes to support the validity of this viewpoint, although relatively few cases become public knowledge because of legal problems.

The AASHTO bridge specifications[5] provide a precedent for using different ϕ factors for precast and cast in place construction.

The above paragraphs are statements of a philosophy of design. Specific numbers are not going to be suggested as ϕ factors for the various cases, except for the use of $\phi = 0.7$ in the following example.

Interpretation of Moment–Thrust Interaction Data

The moment-thrust interaction curve shown in Fig. 1 is repeated in Fig. 3, with some additional information added. First, the nominal strength has been limited to $0.8P_0$, as required by the 1977 ACI code. (The limit is 0.85 P_0 for a column with a spiral meeting ACI requirements.) The previous ACI codes used a concept of a minimum eccentricity of the load that resulted in a comparable reduction from P_0. This was based on the assumption that all columns are likely to have at least some moments, even when normal structural analysis methods might indicate otherwise. The same argument can be made for piles. A pile is normally rigidly connected to its pile cap, and loading of the cap must produce at least some bending in most of the piles. Even if the loading were applied with a zero eccentricity, the pile itself may not be perfectly concentric and the center of resistance may not coincide with the geometric centroid of the section. The $0.8P_0$ value has been adopted rather than using the minimum eccentricity concept primarily to simplify the calculations.

The inner curve on Fig. 3 is what might be termed the reliable strength, and is the nominal strength reduced by multiplying by the ϕ factor, taken as 0.7 in this case (except that it increases to 0.9 as zero thrust is approached).

A straight sloping line from the origin is also shown. This represents a loading path for the specific case of a load at an eccentricity of 65 mm (2.56 in.), that is, a load located 65 mm from the centroid of the pile sec-

[5]"Standard Specifications for Highway Bridges," American Association of State Highway Officials, 12th ed., 1977, Sec. 1.6.5.

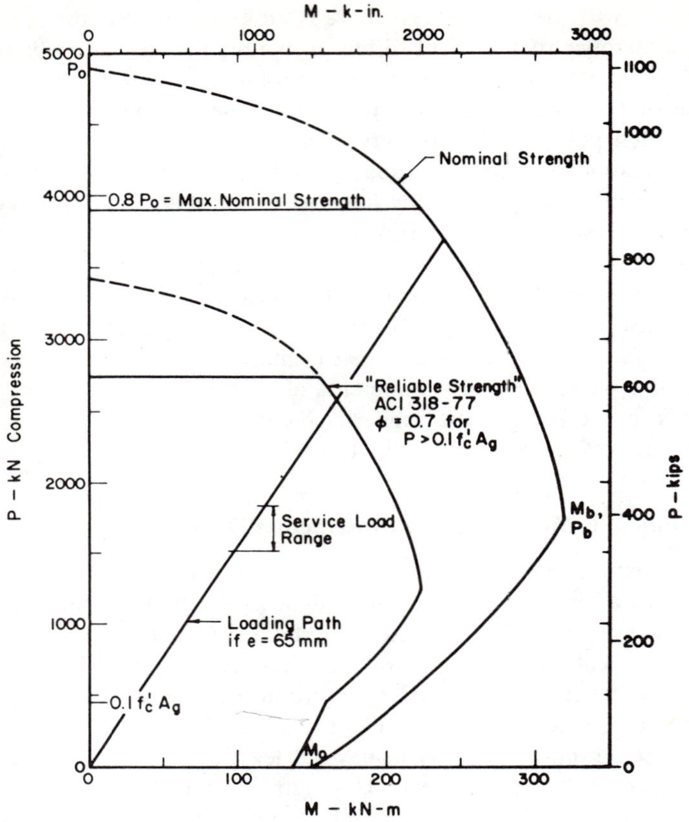

FIG. 3—*Moment-thrust interaction diagram including strength reduction factor.*

tion and lying on a major axis. This can also be visualized as the loading path for a situation in which each 1000 kN (225 kips) load is accompanied by 65 kN·m (576 kip-in.) moment. This line intersects the nominal strength curve at a thrust of about 3700 kN (832 kips), and the reduced strength curve at about 2575 kN (579 kips).

Following the ACI code philosophy, the 2575 kN force becomes the force, ϕP_n, with which the factored applied loads are compared. The design is acceptable if

$$1.4 \text{ DL thrust} + 1.7 \text{ LL thrust} \leq \phi P_n \qquad (2)$$

Consequently, a service load for this pile would be somewhere between 1840 and 1515 kN (414 and 340 kips), depending on the relative values of the service dead and live loads.

The forces quoted above are probably larger than can be resisted by the pile–soil system. Recommendations in Ref *4* indicate that a service load stress of 11 N/mm^2 (1.6 kip/in.^{2}) on the gross area is a reasonable maximum that can be obtained from pile-driving considerations on the basis of no soil freeze. This leads to a maximum force of 1430 kN (321 kips), which is somewhat less than would be obtained from the structural calculations alone. Using this logic and set of overload and strength reduction factors, the overall factor of safety against the nominal capacity of the pile would be between 2.0 and 2.4.

Since the various components of the factor of safety vary with different building codes and specifications, and sometimes with conditions of construction, and the nominal capacities of sections do not, it is convenient to collect all of the safety factor components on one side of the equation. Equation 2 may be rewritten

$$\frac{1.4 \text{ DL thrust} + 1.7 \text{ LL thrust}}{\phi} \leq P_n \qquad (3)$$

This allows all comparisons to be made with the nominal strength curves.

Handbooks and textbooks on reinforced concrete contain moment–thrust interaction information in various forms that may be useful to the pile designer. The only caution that should be added is that the designer must fully understand whether the graphical or tabulated values include the ϕ factors or not.

Moment–Thrust Diagrams for Various Cross Sections

The moment–thrust interaction curve discussed above was for a particular square cross section. Similar information can be developed for other cross sections, and the curves will have the same general forms but contain differences in details.

As one example, if the pile shown in Fig. 1 had either smaller cover over the reinforcement or reinforcement with a lower yield stress, the portion of the curve for thrust greater than the balance point thrust would be closer to a straight line connecting the balance point and the point at P_0. The portion of the curve at thrusts lower than the balance point would have more convex outward curvature. The usual textbook illustrations show cases more nearly like those just described.

Figure 4 contains three interaction curves for circular pile sections reinforced with six bars of three different sizes. The nominal strength has been plotted, so the ϕ factor is specifically not included in this example or any of the following ones. The curves for the circular pile do not have as pronounced a discontinuity at the balance point as is shown in Fig. 1 as long as there are six or more bars, and the moment capacity is not sensitive to the direction of bending relative to the group of reinforcing bars.

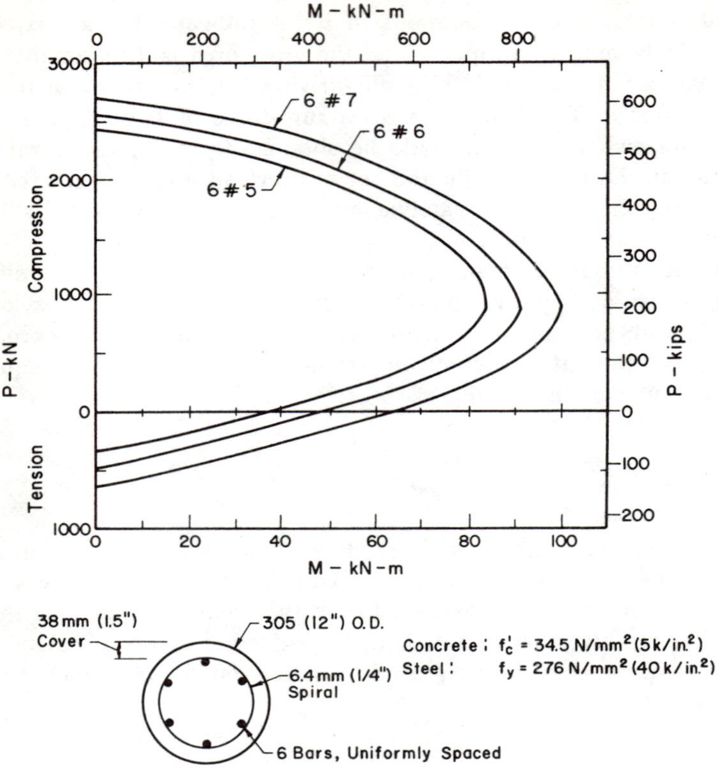

FIG. 4—*Moment-thrust interaction diagrams for typical round piles.*

The computation of points on the interaction curve is more complex for a circular section than for a rectangular one. There are more layers of reinforcement, and the stress in each layer must be evaluated. The force in the concrete, and the location of the centroid of that force, is a more difficult problem, which can be solved in any practical design situation only by the use of the equivalent rectangular stress block. This is illustrated in Fig. 5. The area and centroid of the shaded compression area, which is subjected to a uniformly distributed stress of $0.85f_c'$, can be determined with the aid of a graph published as Fig. D.7 of Ref *5*.

Figure 6 shows interaction curves for an unusual small circular section reinforced with a single reinforcing bar. The bar was first assumed to be located at the center of the section, and the real question that lead to the development of the figure was "But what if the bar is not at the center?" Because of the method of construction of the pile, a maximum eccentricity of the bar from the center of the pile could be established. Then, with the bar in this position, moment-thrust interaction curves were developed for

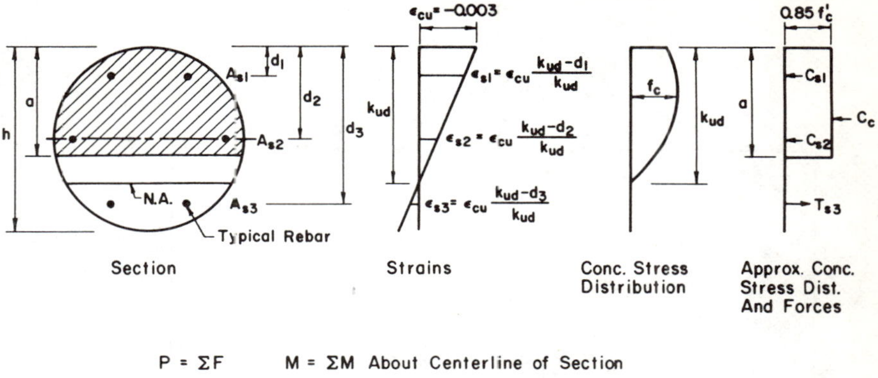

FIG. 5—*Strain and stress distributions considered for a round cross section.*

bending both with the bar as close to the tension face as possible and again with the bar as close to the compression face as possible.

Once the curves had been plotted, the effects of having the bar off center could be evaluated. In this case, the desired thrust was known, and the moment capacity that accompanied that thrust was the most critical quantity. At the design thrust of 196 kN (44 kips), it quickly became apparent that the position of the bar had little effect on the moment capacity. If the thrust had been lower, at about 100 kN (22 kips), displacing the bar could have caused a serious reduction in the moment capacity, depending on the direction of the bending.

Prestressed Concrete Piles

Figure 7 shows moment-thrust interaction diagrams for 356-mm (14-in.) square piles that contained six, seven, or eight 12.7-mm diameter (½ in.) seven-wire prestressing strands spaced uniformly around a 229-mm (9-in.) diameter circle. The strand had a minimum breaking stress of 1860 N/mm^2 (270 kip/in.^{2}). The concrete compressive strength was 41.4 N/mm^2 (6 kip/in.^{2}). Part of a linear interaction diagram representing initiation of flexural cracking is also shown.

Prestressed piles are used because the precompression in the concrete enables one to control tensions set up by reflected tension stress waves that may occur during driving, because they have relatively high and reliable cracking moment capacities and hence can be handled easily in long sections, because relatively large moment capacities can be obtained, and for economic reasons.

The primary force requirement for this particular job was 890 kN (200 kip) compression and 67.8 kN·m (50 ft·kip) moment, and an overall

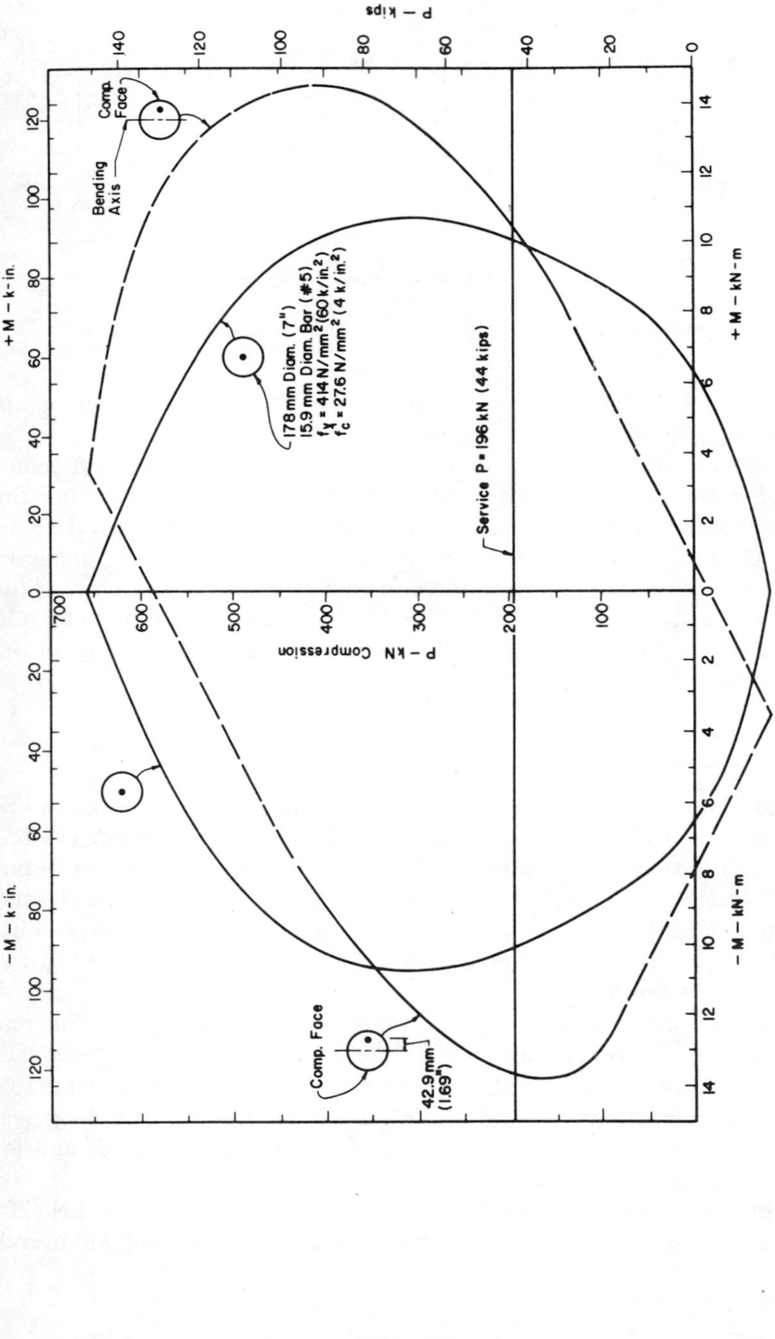

FIG. 6—*Moment-thrust interaction diagrams for round section with single bar.*

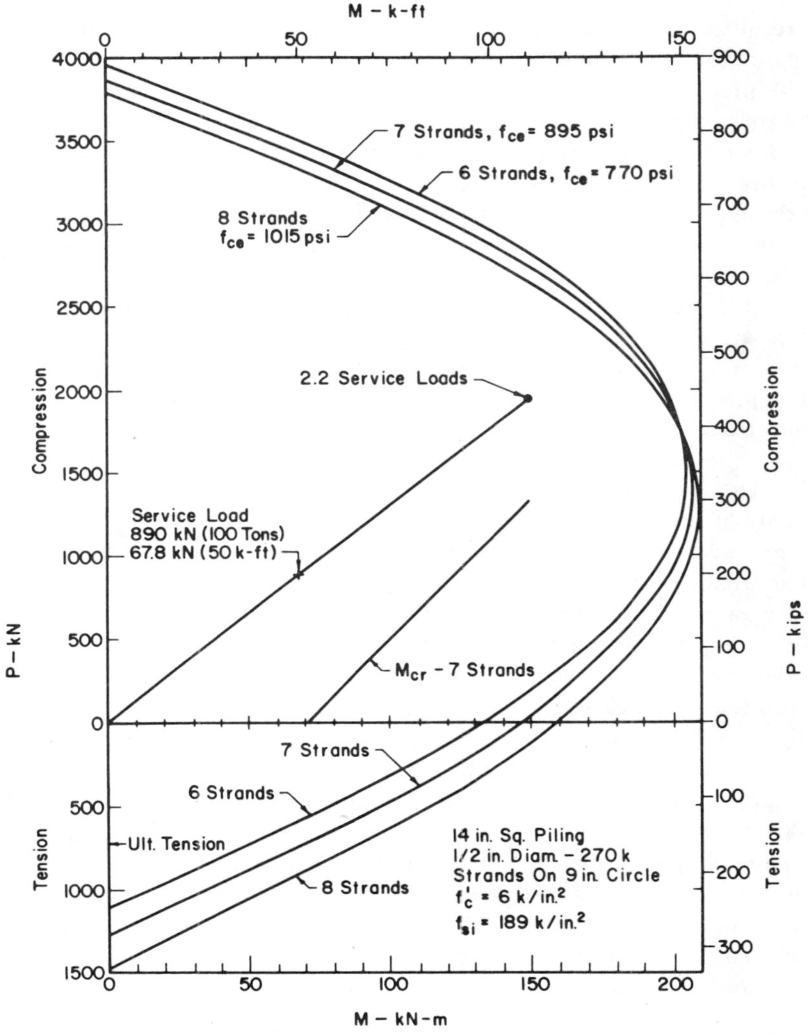

FIG. 7—*Typical moment-thrust interaction diagrams for pretensioned concrete piles.*

factor of safety of 2.2 had been specified. In addition, a minimum ultimate tension capacity of 712 kN (160 kip) was specified.

It is clear that the strength requirements could be satisfied by either six, seven, or eight strands, and the choice then involves the driving requirements. At the specified eccentricity of 76 mm (3 in.), the pile with six strands is theoretically the strongest by a small margin, but the eight-strand pile is strongest for pure bending or any loading including tension.

The pile with seven strands was eventually chosen for this installation, and the results of a load test on an installed pile are described in another paper prepared for this symposium [*6*].

The precompression in the concrete decays continuously with time as a result of creep and shrinkage of the concrete and relaxation of steel stress in the strands. Various methods of estimating the final precompression have been given[6,7] [*7*]. The precompression stresses noted in the drawing are those existing at the time of driving, 1 to 2 months after casting of the pile, and the final values will be slightly lower.

The effects of any credible change in the precompression on the strength of the pile are very small. An extreme case is illustrated in Fig. 8, where curves are shown for the six-strand pile with a precompression of 5.31 N/mm^2 (770 $lb/in.^2$) and with zero precompression. For eccentricities ranging from zero up to about 110 mm (4.3 in.), the pile with zero prestress is the stronger one. The basic reason for this is that all steel is in tension, even at the time of compression failure, if the normal levels of prestress exist. This may be thought of as "using up" some of the compression capacity of the section just in maintaining internal equilibrium. If there is no prestress, the steel is able to act in compression when the member fails.

The analysis of the strength of a prestressed section is more involved than of a normal reinforced concrete section because of the presence of the precompression stresses and strains and because the stress-strain curve for the reinforcement cannot be idealized as an elastic-plastic case, but instead the actual curve must be taken into account.

Figure 9 illustrates the strain and stress conditions that exist in the section as failure under some combination of moment and thrust is reached. The sum of the prestrains in the steel and concrete will be of the order of 0.006 tension, while the maximum compressive strain in the concrete at one edge of the section at failure will be about 0.003. The strain in each layer of steel at ultimate is the sum of the initial steel and concrete strains plus or minus the appropriate change in strain, but the steel strains will always remain tensile. The stress in each layer is determined from the stress-strain curve for the material after the strain has been found. The stress-strain curve shown in Fig. 10 was used in the preparation of Figs. 7 and 8.

Once all of the forces have been found, they are summed to give the thrust. Moments of the forces are summed about the middepth axis to find the accompanying moment, and the moment due to the concrete force will dominate. Care must be exercised with the signs of the various force and moment components.

[6] "Standard Specifications for Highway Bridges," American Association of State Highway and Transportation Officials, 12th ed., 1977, Sec. 1.6.7 (B).

[7] "Recommendations for Estimating Prestress Losses," PCI Committee on Prestress Losses, *Journal of the Prestressed Concrete Institute*, Vol. 20, No. 4, July-Aug. 1975 (See also *Discussion*, Vol. 21, No. 2, March-April 1976).

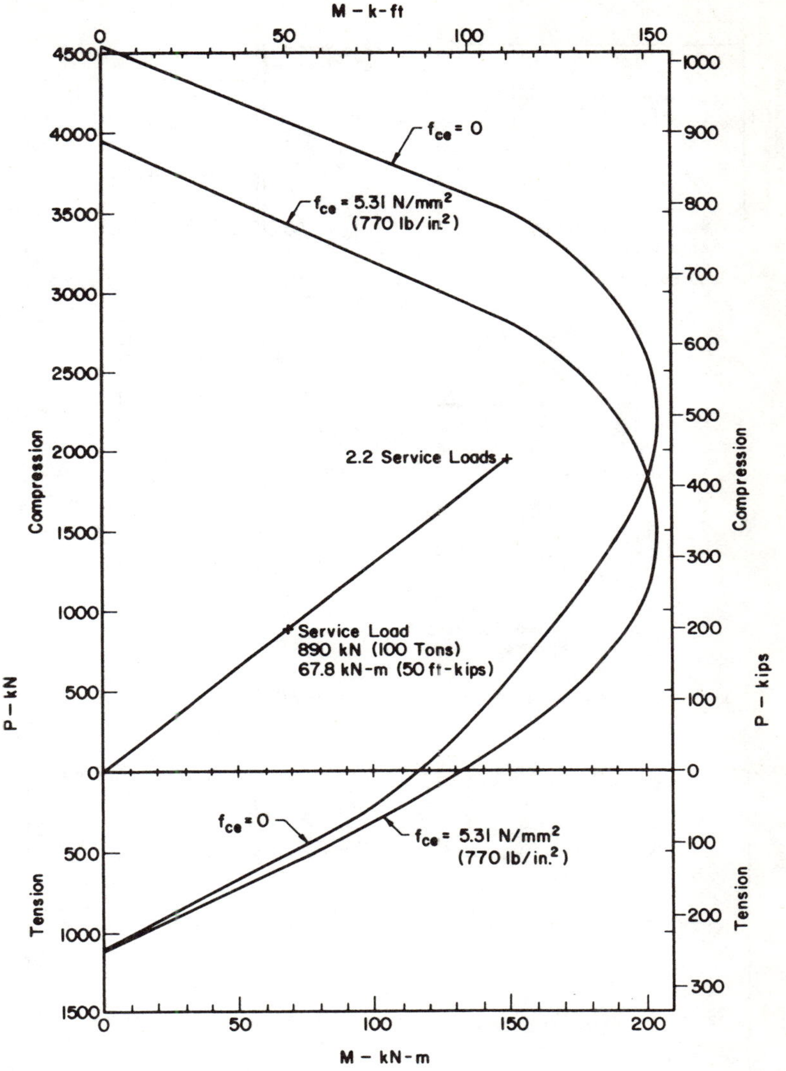

FIG. 8—*Moment-thrust diagrams for piles with two different precompressions.*

For the special case of concentric compression of a prestressed pile, the force at failure may be computed approximately as

$$P_0 = (A_g - A_s)\,0.85f_c' - A_s f_s \tag{4}$$

where $f_s = E_s(\epsilon_{se} + \epsilon_{ce} - 0.002)$ = tensile stress at failure of section. The

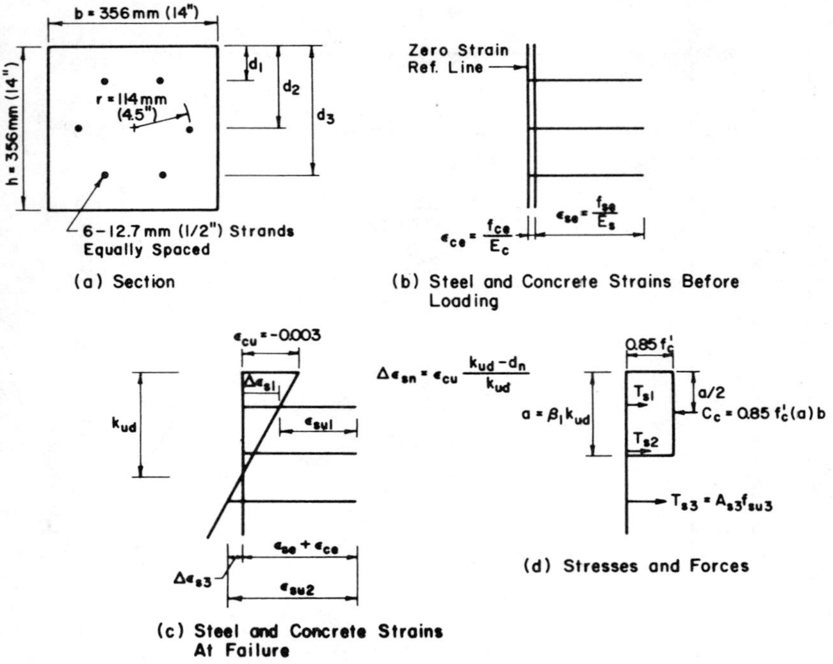

FIG. 9—*Strain and stress distributions at failure of a pretensioned concrete pile.*

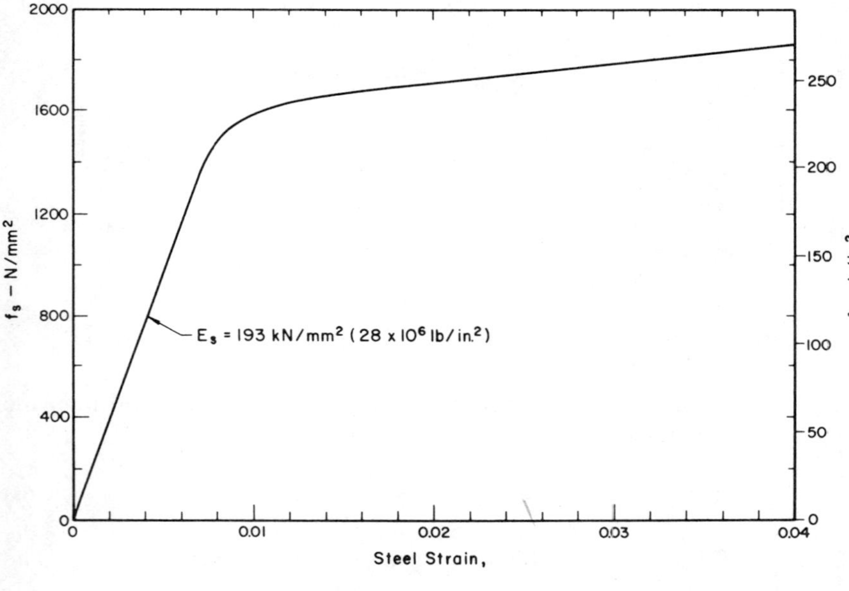

FIG. 10—*Stress-strain curve for prestressing steel.*

0.002 is approximately the compressive strain at failure in a section subjected to axial compression.

Effects of Driving Stresses

The effects of driving stresses can be divided into two categories, which have been studied separately. First is the effect of the stress or strain rate and the second is the effect of a relatively low number of repeated stress cycles to the driving stress level. The pile itself is the principal test specimen in which both effects have been applied at the same time, and that knowledge is largely empirical.

It is well known that high stress or strain rates lead to increased compressive strength in concrete. Stress rates of 21 kN/mm²/s (3 by 10^6 lb/in.²/s) have been implied from time-force measurements made during pile driving [*8*]. On the basis of information in Refs *9* and *10*, this stress rate can be expected to lead to a 40 percent increase in compressive strength if the normal, slow-loading strength is 38 N/mm² (5.5 k/in.²). It is important to note that the peak stress existed in the pile test measurements for only about 0.001 s, and time plays an important part in the failure of concrete.

The combined effects of stress rates and fatigue have been studied [*11*]. Stress rates comparable to those encountered in pile driving were not reached, but extrapolations to very high rates were made. It appears reasonable to expect that 73 percent of the single loading strength can be sustained for about 2000 cycles, 80 percent for about 1000 cycles, and 90 percent for no more than 475 cycles.

In simplistic terms, one may conclude that the beneficial effects of having a high stress rate are offset by the low-cycle fatigue effects. One may also conclude that this particular problem needs extensive additional study before it is completely understood.

Very slow loading rates or sustained high loads tend to cause concrete to fail at lower compressive stresses than in short-term tests, and this may be significant in some pile load tests. The greatest reduction from the short-term loading strength appears to be about 15 percent [*12*], and this affects the concrete alone. This reduction is in addition to the reduction from the test cylinder strength shown in Eq 1, and results in a minimum failure load of

$$P = 0.72 f_c'(A_g - A_s) + A_s f_y \tag{5}$$

where f_c' is the cylinder strength at the time of loading. This is a lower threshold limit, and a pile would normally be expected to carry this load for at least several hours and possibly several days before failure. This has not been extensively studied with large test specimens, and the picture may be complicated by the gain in strength of the concrete that occurs if the concrete was young when loaded.

There is another facet to the long-term strength question. Reference *1* includes reports of tests in which columns were subjected to their service loads for a year and then loaded to failure in a short-term tests. The sustained service loads did not alter the failure loads.

Conclusions

Methods for assessing the strength of reinforced and prestressed concrete piles have been presented. Moment-thrust interaction curves of strength have been used in the same way that they are used in evaluating reinforced concrete columns.

The components in the overall factor of safety have been discussed, with recommendations made for distinguishing between piles that have been precast or are cast in permanent metal forms on one hand and piles cast in uncased holes or in holes from which the casing has been withdrawn on the other hand. It was also suggested that it may be appropriate to use a higher factor of safety for a pile used as one of a small group than for the same pile used in a large group.

The nature of the moment-thrust interactions diagrams for different types of pile was illustrated by suitable examples of square and round reinforced and prestressed concrete pile sections.

References

[1] Richart, F. E. and Brown, R. L., "An Investigation of Reinforced Concrete Columns," Bulletin 267, Engineering Experiment Station, University of Illinois, Urbana, 1934.

[2] Richart, F. E., Draffin, J. O., Olson, T. A., and Heitman, R. H., "Effect of Eccentric Loading, Protective Shell, Slenderness Ratios, and Other Variables in Reinforced Concrete Columns," Bulletin 368, Engineering Experiment Station, University of Illinois, Urbana, 1948.

[3] Hognestad, E., "A Study of Combined Bending and Axial Load in Reinforced Concrete Members," Bulletin 399, Engineering Experiment Station, University of Illinois, Urbana, 1951.

[4] Davisson, M. T., *Proceedings,* Design, Construction and Performance of Deep Foundations, American Society of Civil Engineers-University of California, Berkeley, Aug. 1975.

[5] Ferguson, P. M., *Reinforced Concrete Fundamentals,* 3rd ed., Wiley, New York, 1973.

[6] Lacey, H. S., "Load Testing of Instrumented 225-Ft-Long Prestressed Concrete Piles," this volume.

[7] Hernandez, H. D. and Gamble, W. L., "Time-Dependent Prestress Losses in Pretensioned Concrete Construction," Civil Engineering Studies, Structural Research Series No. 417, University of Illinois at Urbana-Champaign, May 1975.

[8] Davisson, M. T. and McDonald V. J. in *Performance of Deep Foundations, ASTM STP 444,* American Society for Testing and Materials, 1969, pp. 295–337.

[9] Watstein, D., *Proceedings,* American Concrete Institute, Vol. 49, April 1953, pp. 729–744.

[10] McHenry, D. and Shideler, J. J. in *Speed and Testing, ASTM STP 185,* American Society for Testing and Materials, 1956, pp. 72–82.

[11] Awad, M. E., and Hilsdorf, H. K., "Strength and Deformation Characteristics of Plain Concrete Subjected to High Repeated and Sustained Loads," Civil Engineering Studies, Structural Research Series No. 372, University of Illinois at Urbana-Champaign, Feb. 1971.

[12] Rusch, H., *Proceedings,* American Concrete Institute, Vol. 57, July 1960, pp. 1–28.

B. C. Gerwick, Jr.[1] *and H. A. Brauner*[2]

Design of High-Performance Prestressed Concrete Piles for Dynamic Loading

---

REFERENCE: Gerwick, B. C., Jr. and Brauner, H. A., **"Design of High-Performance Prestressed Concrete Piles for Dynamic Loading,"** *Behavior of Deep Foundations, ASTM STP 670,* Raymond Lundgren, Ed., American Society for Testing and Materials, 1979, pp. 323–334.

ABSTRACT: Although prestressed concrete piling have given exemplary performance as foundation piling and there exists ample guidance as to their design, manufacture, and installation for such purposes, recent extensions of their application to more demanding situations requires a renewed look at the criteria for longitudinal prestress (precompression) and spiral confinement.

Problem areas include the occasional occurrence of transverse cracking under heavy driving into erratic soils, longitudinal cracking during and after installation, and the need for ductile response to imposed deformations due to earthquakes.

A more rational analysis is attempted to explain and quantify these phenomena and the stresses and strains developed. Recommendations are then made giving values of prestress and steel requirements that will enable these severe criteria to be met. Some of these recommendations are supported by tests that are as yet unreported elsewhere.

It is believed that these recommendations will enhance the capabilities of prestressed concrete piling and thus make practicable their extension to more demanding applications.

KEY WORDS: prestressed concrete piling, seismic behavior of piling, dynamic behavior of concrete, minimum reinforcement

Prestressed concrete piles have been widely utilized for the foundations of buildings, bridges, and harbor structures (Fig. 1). They have proven to be practicable and economical and have given generally excellent performance during installation and service.

The American Concrete Institute (ACI) has published a report by Committee 543 on "Recommended Practice for Concrete Piles" [*1*],[3] and the

[1] Professor of Civil Engineering, University of California, Berkeley, Calif. 94720.
[2] Plant manager, Prestressed Division, Santa Fe-Pomeroy, Inc., Petaluma, Calif. 94952.
[3] The italic numbers in brackets refer to the list of references appended to this paper.

FIG. 1—*Prestressed concrete hollow-core piling.*

Prestressed Concrete Institute (PCI) has published its "Recommended Practice for the Design, Construction, and Installation of Prestressed Concrete Piling" [*2*]. Particular attention is directed to the moment-axial load interaction curves contained in reference [*2*]. PCI has participated with American State Highway and Transportation Officials (ASHTO) and American Railway Engineering Association (AREA) in developing standards for piling.

The above documents are generally adequate and up to date for most foundation piles. However, in recent years, the success of these prestressed concrete piling under "normal" conditions has led to their employment under much more demanding conditions of installation and use. In particular, they are being driven through more erratic strata, with heavier hammers and deeper required penetrations. In service, they are being asked to accept high lateral loads and deformations, including those imposed during earthquake.

Many of these new and additional requirements can be taken care of adequately by carefully controlled installation procedures and by standard design procedures. However, in a disturbing number of recent cases, cracking has developed during and after installation. Calculations for maximum imposed curvatures during landslip and earthquake have indicated the potential for brittle failure and sudden loss of carrying capacity, and indeed, some such failures have occurred during local landslides and during the San Fernando earthquake.

Finally, there is the potential employment of prestressed concrete piles in offshore terminals and deepwater bridges and fender piling, for which current designs are inadequate. Hence, there is a need to examine the special design-related aspects that can ensure safe, ductile, and long-term satisfactory performance.

Regarding the dynamic loads that occur during installation, these can be definitely reduced by the use of special installation techniques, such as thicker softwood pile head cushion, selection of proper hammer, predrilling, and jetting. However, these will not be treated further here since they are not directly design-related. Attention is directed to Refs *2–4* for further discussions on installation practice.

Similarly, tip design has an effect on installation stresses. Pointed and wedge-shaped tips may introduce bending stresses during driving, especially for batter piles. Tip design is discussed in Refs *2* and *4*.

Thus, this paper will direct its attention to two major areas—driving stresses during installation and longitudinal splitting—with a view to developing improved design recommendations, particularly for piling that must carry dynamic loads and severe imposed deformations.

Driving Stresses

For a number of years, it has been recognized that hammer blows are transmitted as sharp compressive waves moving down through the pile at the speed of sound in concrete. Upon reaching the tip, this wave is reflected within the pile either as a compression wave (if the tip is on hard material) or as a tensile wave (if the tip is in soft material). This reflected wave then travels to the top, where it once again reflects. Thus, it is somewhat analogous to water hammer.

This wave phenomenon is damped by the internal damping in the material. Thus, reinforced nonprestressed piling will have more damping than a prestressed concrete pile due to microcracking. This wave is also damped by radiation damping in the soil.

When the tensile wave in a precast pile exceeds the tensile capacity of the concrete, including precompression due to prestress and self-weight, then the pile cracks [*3*]. Further hammer blows extend the crack and may lead to yield of the steel. While cracking increases the damping, nevertheless, when the crack becomes of sufficient width, then reflections tend to take place from the crack itself, and brittle fracture of the steel may ensue due to low cycle fatigue.

What about piles that only crack but do not rupture? Usually such cracks occur in the early stages of driving and are visible. Due to the dead weight of the pile, they tend to occur in the upper one-third of the pile. Some may occur below ground. What is their effect on long-term performance?

In the chemically stable oxygen-free environment, corrosion is generally not a problem. In fact, there is considerable evidence that small cracks transverse to the principal steel do not lead to corrosion even in a seawater environment [5]. Further, the prestress and the subsequently applied axial dead load create a state of compression across this zone. In the presence of water or even a very humid environment, such cracks will recement (autogenous healing). Where piles will later be subjected to many cycles of lateral load, cracking will reduce the fatigue endurance if the steel is stretched beyond yield. However, where the crack has reclosed under prestress, then adequate fatigue capacity is maintained [6, 7].

The important thing is that the steel in the cracked zone not be stressed beyond yield. This can be effectively obtained if the proportion of steel area (A_s) to concrete area (A_c) times the yield strength of the steel is equal or greater than the initial precompression in the concrete plus the direct tensile strength of the concrete

$$(A_s/A_c)f_y \geq f_{pc} + f_{ct} \tag{1}$$

where

f_y = yield strength of steel (0.2 percent offset for prestressing steel),
f_{pc} = precompression in concrete, and
f_{ct} = direct tensile strength of concrete.

However, the initial precompression is in itself a function of the steel percentage and the initial stress in the steel f_s:

$$f_{pc} = \frac{A_s}{A_c} f_s$$

$$\frac{A_s}{A_c} = \frac{f_{pc}}{f_y} \frac{f_{ct}}{f_y} \tag{2}$$

$$\frac{A_s}{A_c} = \frac{(A_s/A_c)f_s}{f_y} \frac{f_{ct}}{f_y}$$

$$\frac{A_s}{A_c}\left(1 - \frac{f_s}{f_y}\right) = \frac{f_{ct}}{f_y} \tag{3}$$

For the usual case, with prestressing steel having an ultimate strength of 1900 MPa (270 000 psi), an equivalent yield stress of 1520 MPa (216 000 psi), a 0.2 percent offset of 80 percent of this ultimate strength, an effective stress after losses of 1100 MPa (156 000 psi), and concrete with an instantaneous direct splitting strength of 3.6 MPa (500 psi)

$$\frac{A_s}{A_c}\left(1 - \frac{1100}{1520}\right) = \frac{3.6}{1520}$$

$$0.28 \frac{A_s}{A_c} = \frac{3.6}{1520}$$

$$A_s/A_c = 0.008 = \text{about } 0.8 \text{ percent}$$

This is equivalent to an effective precompression in the concrete of 0.008 × 1100, or about 8.5 MPa (1200 psi). This is 50 percent higher than the degree of prestress normally used in foundation piles and found entirely adequate on many hundreds of thousands of piles worldwide. The explanation appears to lie in the fact that in numerous cases measured by strain and stress gages, the reflected tensile wave seldom exceeded 8.5 MPa (1200 psi), and thus the value of 5.6 MPa (800 psi) plus the instantaneous direct splitting strength was adequate. However, it has been found absolutely essential to use proper head cushioning on the pile and to properly size hammer ram and impact velocity.

Further, in the case where an occasional crack or cracks occurred, as long as the ultimate strength (f) of the strands was not exceeded, the damping increased for subsequent blows, and most or all of them did not reach the critical values again.

Thus, for such piles, we can substitute f for f_y in Eq 3 and for the usual case,

$$A_s/A_c = 0.44 \text{ percent} \qquad \text{and} \qquad f_{pc} = \text{about 5 MPa (700 psi)}$$

However, for piles that will be subjected to extremely hard driving with heavy hammers under severe conditions, such as alternating layers of limestone and unconsolidated sands typical of subtropical regions, and that will subsequently be loaded primarily by cyclic lateral loads, such as wave loads, then a higher prestress value approaching 8.5 MPa (1200 psi) appears to be indicated [*8*]. Similarly, batter piles may be subjected to bending plus driving stresses during installation, so they, too, should be prestressed to the higher value.

In any event, the use of suitable head cushion is essential, since otherwise high peak stresses may occur well above the cracking resistance of even the more highly prestressed piles. Further discussion of head cushion is found in Refs *3* and *4*.

A number of technical articles and even some current specifications warn against "hard driving" or "overdriving." A number of recent cases have been reported as "failure under hard driving." The circumstances strongly suggest that these failures were probably low-cycle fatigue due to continued hammering on piles that had been previously cracked in the early, soft-driving phases [*7*] [*8*]. For such piles, the use of higher prestress values and better head cushions would appear appropriate. Hard driving per se should not cause failures, since the hammer-induced compressive stresses will not normally exceed 4000 psi, especially if proper cushioning is used.

Longitudinal Cracking

Longitudinal cracking may occur due to a number of causes, all of which tend to have superimposed and aggravated effects. These circumferential tensile stresses may have been partially built in during manufacture, for example, shrinkage stresses that may easily be at 2 MPa (280 psi) initially, diminishing with time due to tensile creep. Prestressing adds tensile stress due to Poisson's effect. Pile driving also causes a cyclic bulging due to Poisson's effect that is particularly serious at the head and tip [*6*]. For both solid and hollow-core piles, the geometrical constraints of Poisson's effect tend to produce circumferential tension on the exterior face.

High-strength concrete piles may be particularly prone to such splitting, due to our normal practice of loading them more highly or using heavier prestress and subjecting them to harder driving. The tensile strength of the concrete increases only as the square root of the compressive strength.

There are other installation phenomena that produce longitudinal splitting. These include water hammer effects, hydrostatic bursting stresses due to build-up water or semifluid mud inside as a result of the hydraulic ram effect in very soft soils or as a result of jetting, and the wedging effect of the plug of soil in the tip.

Water hammer is a serious concern only in special cases, one of those being when the ram impacts directly on the water column of a hollow pile. In that case, the instantaneous compressive wave will act as a hydraulic ram and cause the pile to burst.

In the more usual case of a hollow-core pile open at both ends but with the water level in the pile well below the top, water hammer occurs due to the acceleration of the pile faster than the water column, followed by the water column's deceleration upon reaching the plug that has formed in the tip. Various studies show that internal pressures of 0.7 to 1.0 MPa (100 to 150 psi) can theoretically build up inside. In actual practice, these are generally much lower due to the gradual deceleration provided by the cushioning effect of the soil in the plug.

Water hammer can be prevented by using a closed tip and keeping the pile interior dry, but this may increase driving resistance and prevent drilling ahead or socketing if that is required.

Jetting inside to break up the plug can raise the internal water level dangerously high. This should be prevented by the provision of holes formed in the wall at suitable intervals.

The hydraulic build-up inside during driving is due to the fact that the pile must displace an equal volume. If the material in the inside core of the pile is remolded and semifluid, the progressively displaced material will be forced into the core by each blow, raising the water level inside. This is particularly serious in very soft muds, such as those in the Gulf of Mexico. Such a build-up can be prevented by the provision of vent holes.

In firmer soils, such as sands, the plug may be wedged into the tip, and each succeeding blow tends to exert a splitting force on the pile tip.

Once the pile is in place, the potential difficulties are not completely over. If a concrete plug is to be poured inside a hollow core pile, there are the potential splitting problems due to the hydrostatic head of the fresh concrete and, also due to the heat of hydration of the core concrete.

In service, there are a number of phenomena tending to produce splitting and longitudinal cracks (for example, freezing). These problems are adequately dealt with in existing literature.

Such tensile splitting stresses result in cracks that are a function of cover and bar size, similar to the case of bending stresses. The Comité European du Beton and Fédération Internationale de la Precontrainte (CEB/FIP) formula for crack depth is $c + 7\phi$ where c is the cover and ϕ is the diameter of the spiral. Thus, we should attempt to provide enough steel area to replace that lost by tensile cracking of the concrete without exceeding yield in the steel. A suitable formula might be

$$(c + 7\phi)\,f_{ct}\,sp = f_y A_s \tag{4}$$

where

f_y = yield strength of spiral,
f_{ct} = concrete tensile strength, and
sp = spacing of spirals.

For the typical case where c = 50 mm, ϕ = 8 mm, and f_y = 400 MPa, then A_s = 50 mm^2, and the required spacing would be approximately 50 mm. In typical United States practice, this would lead to use of 9.4 mm (3/8-in.) spiral at 75 mm (3-in.) spacing, or 12.5 mm (½-in.) spiral at 150 mm (6-in.) spacing.

If the pile were a 600 mm (24-in.) pile with 350 mm (14-in.) hollow core or a cylinder pile with 125 mm (5-in.) thick walls, then the spiral percentage would be 0.7 percent.

These required values are significantly greater than those commonly used in current practice and which have proven satisfactory for many thousands of foundation piles. These new values approach those of column requirements, and are indicative of the spiral needed to develop the full structural capacity of the pile. Thus they are definitely relevant to some of the more sophisticated and complex structural uses of prestressed concrete piles.

The general validity of this was shown by static tests performed in Japan, for which 6-mm (¼-in.) spiral at 50-mm (2-in.) centers was found necessary to develop the full strength of a 300-mm (12-in.) diameter concrete pile with f_c' of 70 MPa (10 000 psi), whereas a smaller spiral at greater pitch led to premature failure by longitudinal splitting. [*6*]

Seismic Deformations

There is a second form of dynamic loading, that due to seismic displacement and distortion of the soil mass under earthquake. As described elsewhere [*9*], the piles are generally forced to deform to match the soil strata-displacements. This imposes curvatures on the pile that may well exceed normal values, especially if the boundaries between firm and soft soils are sharp.

It was proposed that additional spiral confinement be employed to enable the pile to accept inelastic deformations [*8*]. Since that date, tests have been run that dramatically confirm the ability of the piles with extra heavy confinement to accept curvatures up to 8×10^{-4} while still sustaining their design axial load. This is approximately three times the elastic curvature capability, and thus becomes a method of incorporating ductility within prestressed concrete pile design.

Such confinement does not prevent spalling outside the core. However, in the case of such an extreme earthquake as that contemplated, such spalling cannot be considered as a failure condition. Indeed, as a result of the passive resistance of the soil, the likelihood is that the cover will merely crack, without displacement.

In comparative tests of standard State of California piling and those with heavy wire spiral, the latter sustained four times the deflection of the first. Even then the final failure was ductile as compared with a brittle failure with the first case (Figs. 2 through 5).

Similar tests performed in Japan by H. Makita of Tokyu Concrete Pile Co. [*6*] showed increased deflection capacities as follows:

Spiral	Maximum Deflection at End of Test
3.2 mm at 100 mm pitch	31 mm
6 mm at 50 mm pitch	47 mm
9 mm at 50 mm pitch	62 mm

This approach to confinement also appears to be desirable where piles are designed to act as free-standing cantilevers or rigid frames, subjected to predominantly lateral loads, with high shear occurring at the same location as high moment, and where significant concentrated lateral loads or impact may occur, as in the case of fender piles for marine structures.

Summary and Recommendations

1. For normal prestressed concrete piling driven as foundation piling, subjected to primarily axial compressive loading, the existing prestress values of 5 to 6 MPa (700 to 800 psi) are entirely satisfactory, provided proper installation procedures are followed. Higher prestress may be required for marine piles subject to severe installation and operational forces.

FIG. 2—*Test piles*: (top) *heavy spiral*, (bottom) *standard spiral*.

FIG. 3—*Brittle fracture of pile with standard spiral.*

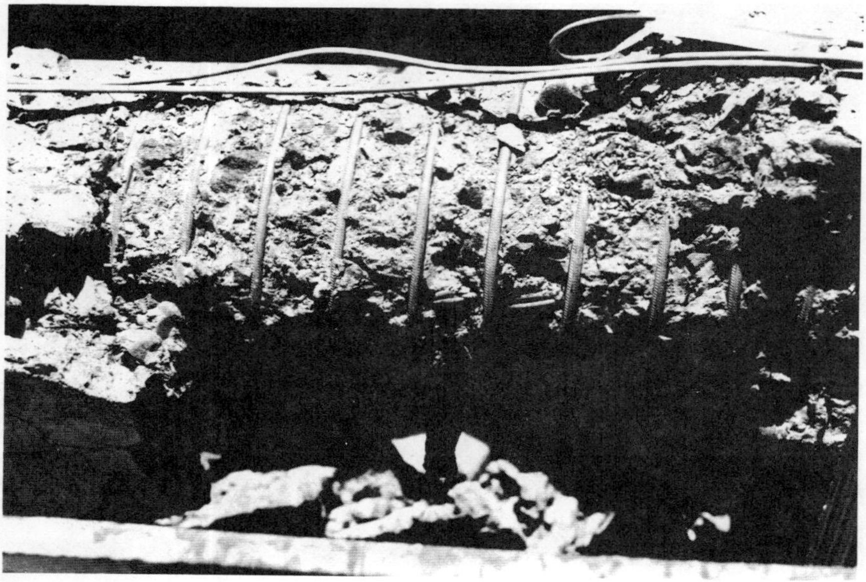

FIG. 4—*Ductile performance of pile with heavy spiral.*

2. The present standards for spiral reinforcement throughout the body of the pile are minimal values. Where no. 5 (5-mm diameter) spiral is now being employed on 150-mm (6-in.) centers throughout the body of the pile, a reduction to 75 mm (3 in.) would appear more rational and actually would add little cost. This is for standard piles (not hollow core) that are fully embedded.
3. For free-standing piles (piles standing above ground) and for hollow-core piles, the spiral should be increased to that required by Eq 4, which will generally raise the requirement to about 0.7 percent.
4. This same requirement of 0.7 percent or Eq 4, whichever is greater, should be applied for any piles installed in seismic zones throughout that portion of the pile potentially subjected to severe imposed curvatures.
5. For the top and bottom of hollow-core piles and in rigid-frame piles where maxima shears and moments occur simultaneously, a value of 1 percent steel spiral area should be provided over a length of 1 to 2 diameters.

The implementation of these provisions will convert our present standard prestressed concrete piling to "ductile" piles, capable of withstanding severe distortions, impact, and bending. By following these recommendations and by applying proper driving procedures, the annoying recurrence of cases of cracking and splitting during and after installation would be greatly minimized. Rare as these cases are, (estimated less than 1 percent of the

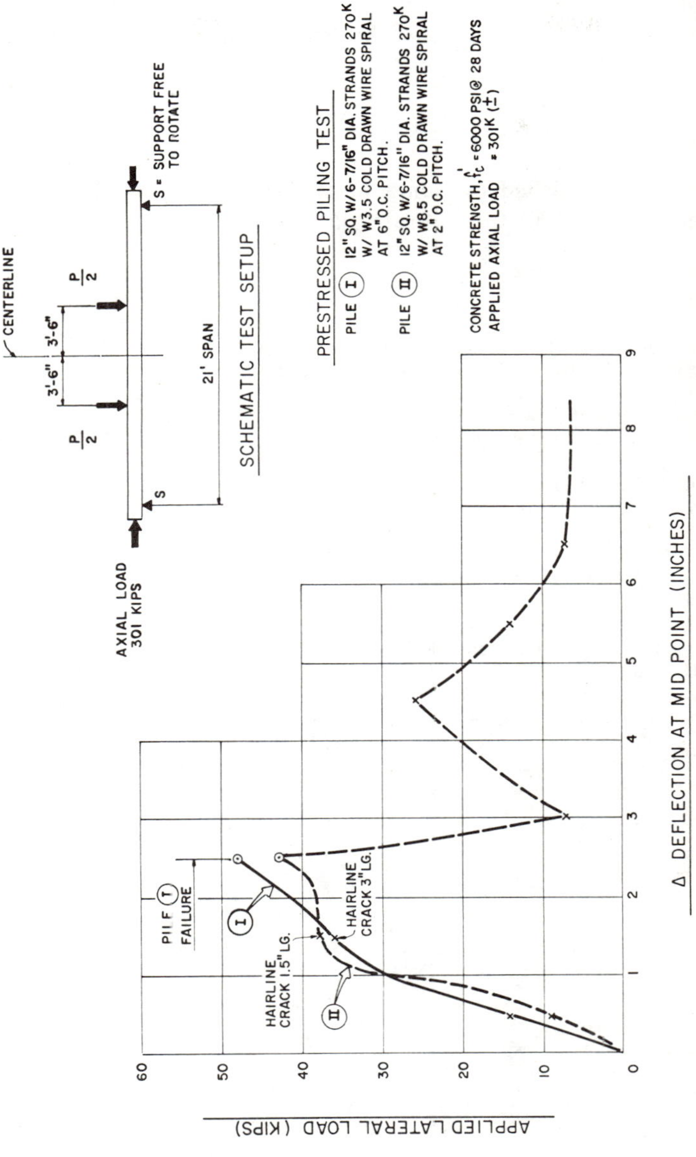

FIG. 5—*Comparative performance of prestressed concrete piles with varying amounts of spiral.*

projects), they are nevertheless a serious matter to the individual project on which they occur.

The resultant "ductile" prestressed concrete piles should find an even wider application, capable of withstanding forces and curvatures that in practice today can only be matched by heavy-walled steel pipe piles filled with concrete.

Acknowledgments

The test data furnished by Santa Fe-Pomeroy, Inc. and H. Makita of Tokyu Concrete Pile Co., Inc. is gratefully acknowledged.

References

[*1*] ACI Committee 543, "Recommendations for the Design, Manufacture, and Installation of Concrete Piles," *Journal of the American Concrete Institute,* Vol. 70, No. 8, Aug. 1973, p. 509 with revisions Vol. 71, No. 10, Oct. 1974, p. 477.

[*2*] PCI, "Recommended Practice for Design, Manufacture, and Installation of Prestressed Concrete Piling," *Journal of the Prestressed Concrete Institute.* Vol. 22, No. 2, Mar.-Apr. 1977, p. 20.

[*3*] Anderson, A. R. and Moustafa, S. E., *Civil Engineering*, Aug. 1971.

[*4*] Gerwick, B. C., Jr., *Construction of Prestressed Concrete Structures*, Wiley Interscience, New York, 1971.

[*5*] Beeby, A. W., "Cracking and Corrosion," Technical Report No. 1, *Concrete in the Oceans,* ISBN O 7210 11276, Cement and Concrete Association, Wexham Springs, Slough SL3 6PL, U.K. 1978.

[*6*] Makita, H., "Comparative Tests of High Strength Concrete Piling with Varying Degrees of Confinement," Tokyu Concrete Pile Co., Tokyo, 1977 (private communication).

[*7*] Taylor, H. P. and Sharp, J. V., *The Structural Engineer,* Vol. 56A, No. 3, March 1978.

[*8*] Hawkins, N. *Journal of the Prestressed Concrete Institute,* Vol. 22, No. 6 and Vol. 23, No. 1, Nov.-Dec. 1977, pp. 80, 40.

[*9*] Gerwick, B., Seed, H. B., Bertero, V., Brauner, H., Lin, T., and Fotinos, G., "Aseismic Design of Prestressed Concrete Piling," Paper presented at the Seventh Fédération Internationale de la Precontrainte Congress, New York, 1974.

D. M. Holloway,[1] *G. W. Clough,*[2] *and A. S. Vesic*[3]

A Rational Procedure for Evaluating the Behavior of Impact-Driven Piles

REFERENCE: Holloway, D. M., Clough, G. W., and Vesic, A. S., **"A Rational Procedure for Evaluating the Behavior of Impact-Driven Piles,"** *Behavior of Deep Foundations, ASTM STP 670,* Raymond Lundgren, Ed., American Society for Testing and Materials, 1979, pp. 335–357.

ABSTRACT: A numerical solution has been developed using a finite difference formulation to simulate the one-dimensional impact wave propagation behavior and subsequent static load test performance. Based on this solution, the importance of residual driving stresses in properly interpreting pile behavior under axial loads is examined. This paper (1) briefly describes the analytical procedure developed by the authors to incorporate residual stresses explicitly, (2) applies the procedure to a documented pile load test to demonstrate the concepts developed, (3) examines the consequences of ignoring residual driving stresses in interpreting/predicting pile behavior under axial loads, and (4) demonstrates the influence of various parameters on the predicted pile performance.

KEY WORDS: pile driving, pile load test, wave equation, residual stresses, prediction, interpretation

Nomenclature

E_{ram} Rated hammer energy, ft·lb

f_0 Unit shaft resistance, lb/ft^2

J_s, J_p Shaft, point damping parameters after Smith, s/ft

k_{tip} Point stiffness, lb/in

K_s^c, K_s^t Lateral earth pressure coefficient at failure in compression, tension

Q Pile load, lb

Q_s, Q_p Shaft, point resistance, lb

q_0 Unit point bearing resistance, lb/ft^2

[1] Geotechnical engineer, Woodward-Clyde Consultants, San Francisco, Calif. 94111.
[2] Professor of Civil Engineering, Stanford University, Stanford, Calif.
[3] Dean, School of Engineering, Duke University, Durham, N.C. 27706.

$QUAKE_p$ Point quake, in.
W_{ram} Ram weight, lb
δ Pile–soil interface friction angle, degrees
$\gamma, \bar{\gamma}$ Total, buoyant unit weight, lb/ft^3
κ Interface shear stiffness parameter
$\bar{\kappa}$ Interface shear stiffness parameter obtained from laboratory tests
ϕ Angle of internal friction for soil, degrees
$\tau_{max}{}^c, \tau_{max}{}^t$ Maximum unit shear stress at failure in compression tension, lb/ft^2

This paper focuses on the importance of modeling the installation process in a rational manner to better predict pile behavior under axial loading. The findings suggest a more careful interpretation of pile load test results may be needed to avoid undue conservatism in pile foundation design, especially for end-bearing piles. Misinterpretation may also result in important installation difficulties that should otherwise be anticipated.

The effects of residual pile driving stresses are generally ignored in design, since it is reasonable to assume that they have no significant influence on ultimate pile capacity. Residual stresses, however, can result in unanticipated deformations under tension loads. More importantly, the incorrect interpretation of pile load test results may have serious design implications when residual driving stresses are neglected, since errors in estimating the distribution of shaft versus point resistance could be magnified if used to determine capacities of alternative pile configurations.

Experimental Observations

Measurements of axial load distribution with depth for field and model piles in cohesionless soils indicate that unit shaft capacity f_0, appears to reach a limiting value at some critical relative depth [*1–3*].[4] The measurements show this tendency occurs at 10 to 20 pile diameters of depth, depending upon relative density. This behavior has been explained as an arching phenomenon along the pile shaft. Finite element analyses predict such behavior only within a few diameters of the pile point [*4–6*]. Hanna and Tan [*7*] performed laboratory model tests using relatively long piles and observed the critical depth approached 40 pile diameters.

The causes of the critical depth phenomena and constant f_0 well above the pile point remain somewhat of a mystery. A critical concern that accompanies these observations is the measurement of a limiting point bearing capacity that appears to remain constant at greater depths. The assumption of uniform shaft friction and constant point bearing capacity

[4] The italic numbers in brackets refer to the list of references appended to this paper.

values below a critical relative depth gives considerably more conservative results than assuming "triangular" distributions with unit shaft and tip capacities increasing with depth.

One possible cause of the critical depth phenomena noted in some cases is that the test pile was not loaded to failure, although some bearing capacity criterion was satisfied. In such cases, the shaft and point resistances at greater depths would not be fully mobilized. This is a common error in the interpretation and performance of pile tests. Nevertheless, in some cases, definite "plunging" characteristics are described by the load-displacement curve, while the instrumentation measurements describe the same limit reaching tendencies in skin friction distribution and point bearing values. Many prediction inconsistencies have been attributed to "residual driving stresses" in lieu of any other explanation. In the case of estimating unit point bearing and shaft friction values correctly, the existence of residual driving stresses plays a major role.

Residual Driving Stress Effects

Pile tests performed at Lock and Dam No. 4, Arkansas River Project, Corps of Engineers [*8*], were studied in detail by Hunter and Davisson [*9*]. Their paper represents the most prominent study of residual driving stresses for piles in cohesionless soils. Instrumented steel pipe piles, H-piles, and precast concrete piles were load tested in compression and thereafter in tension. Since the instrumentation was zeroed at the start of the compression load test, the unload of the residual (compression) load was measured as tension. Measured tension loads as high as 45 000 kg (100 kips) at the pile point represent residual compression loads after driving. A graphical procedure was used to estimate "corrected" load distribution curves based on the assumption that no residual loads remained at the end of the tension test.

It is important to note that unless strain gages are "zeroed" prior to installation or after extraction of the load test pile, actual residual load distributions cannot be determined. The correct values of point capacity and total shaft capacity can be obtained from subsequent tension load test data, provided continuous load-strain measurements are made near the pile point, since residual compression stress at the pile point would be observed as "tension load" at ultimate tension load. The shapes of unit skin friction distribution and pile load distribution curves at failure in both compression and tension remain indeterminate, however, in the absence of true zero gage readings.

A hypothetical example showing the effects of residual stresses on load transfer behavior in compression and tension is given in Fig. 1. A triangular distribution of skin friction τ at failure (curves 1 and 2) versus depth has been assumed for this example, with corresponding parabolic

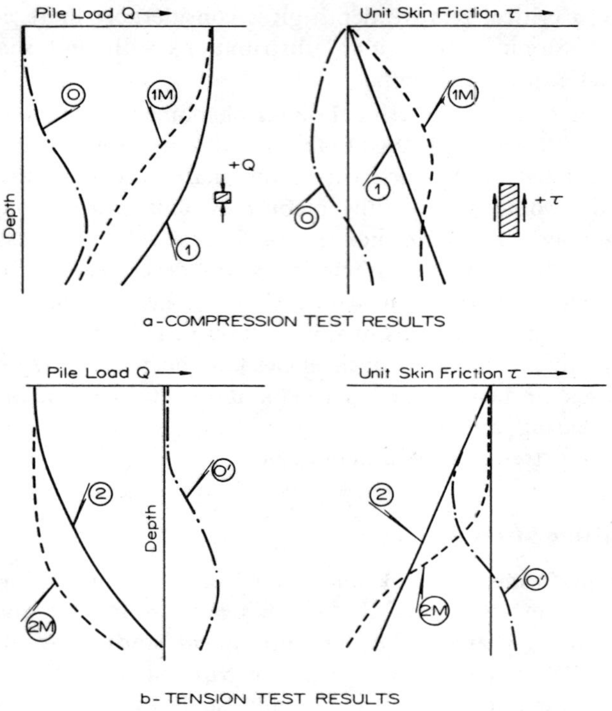

FIG. 1—*Residual load distribution effects. 0 and 0′ are residual Q, τ curves in compression-tension, 1 and 2 are "actual" Q, τ curves at failure in compression-tension, and 1M and 2M are "mobilized" or, "measured" Q, τ curves in compression-tension.*

load, Q versus depth curves at failure, since τ is directly proportional to the slope of the axial load versus depth curve. In other words, the slope of the load versus depth curve (divided by shaft perimeter) on the left-hand figure in each pair is plotted as the unit shaft friction τ in the companion figure (on the right) in Fig. 1.

Curves 0 and 0′ describe the residual load–shear stress distribution in the pile–soil system at the start of the compression and tension load tests, respectively. The shapes of these curves were assumed based on preliminary static analyses of cyclic load test behavior.

Compression load test behavior near failure is described in Fig. 1*a*. The differences between curves 1 and 0 are plotted as curves 1M. Strictly speaking, curves 1M describe forces and shear stresses "mobilized" in the pile–soil system at the current load level. In most instrumented pile tests reported in the literature, the measurements assume implicitly that residual loads due to driving are negligible; that is, that curves 0 coincide with the respective origins of the load distribution and skin friction distribution curves. Therefore, curves 1M describe the shapes of the (hypothetical) measured load and computed shear stress curves, uncorrected for the residual load distribution. Note in particular the unit shaft resistance curve 1M shows a linear increase of shear stress only at shallow depth, after which it drops to a near constant value. Likewise, the measured (uncorrected) value of point bearing capacity underestimates the true value by the magnitude of the residual point load.

For the tension test, curves 2 in Fig. 1*b* represent the true forces and shear stresses acting on the system at failure. Curves 0′ represent the residual load distribution for which the instruments were zeroed (for the tension test). Once again, curves 2M describe "mobilized" or measured data. Note the measured tension force at the pile point corresponds to the residual compression load at the tip at the time the gage was zeroed. If the strain gages were zeroed at the start of the tension test subsequent to a compression test, the shapes of curves 0 and 0′ would probably be different.

This example demonstrates that ignoring significant residual loads in analyses of pile test performance could lead to serious difficulties in applying the results in design. Skin friction curve 1M (Fig. 1*a*), describes a greater area beneath the curve and, thus, a higher shaft capacity and lower point capacity than are actually present. Extrapolating these results to design piles of different cross section or greater embedment could lead to serious errors in pile capacity estimates. As an example, underestimation of point bearing capacity at greater depths could cause installation difficulties and/or overly conservative design penetrations, both of which would be unnecessarily more costly.

Clearly, this example greatly simplifies the mechanism of the physical behavior. Nevertheless, the general observations would remain the same: by neglecting residual driving stresses in analyzing pile load test results one usually (1) overestimates pile shaft capacity, (2) underestimates point capacity, and (3) determines incorrectly the actual resistance distribution at failure.

In light of these observations, an analytical procedure that can include residual load effects in the formulation is certainly needed. Load test results for which significant residual loads are neglected are the rule rather than the exception in the geotechnical practice. A procedure has been developed that will approximate pile-soil interaction behavior throughout installation and load testing. The procedure is briefly described in the following paragraphs. An example case history is presented thereafter to demonstrate the use of the analytical procedure, and parametric variations from the example case are used to examine each parameter's influence on predicted behavior.

Approach to the Problem

In order to simulate impact pile driving and axial load testing in a complete procedure, a one-dimensional, discrete element formulation was developed in the computer program DUKFOR. Pile driving is modeled after Smith [*10*] using one-dimensional wave equation solution techniques, which obtain an approximate static equilibrium solution for a "complete" hammer blow. In this manner, the application of a series of single blows is modeled in order to establish the residual stress state at the end of driving. Upon completion of the impact driving sequence, an arbitrary axial loading path (monotonic, cyclic, load reversal, or a combination of these) may be applied, using the pile driving solution to define initial equilibrium conditions. Thus, the installation and load test behavior are simulated sequentially, although changes in soil resistance conditions (setup or relaxation effects) have been neglected in the formulation.

The DUKFOR algorithms apply piecewise linear approximations to the soil rheology using either bilinear (elastic–plastic) or nonlinear (hyperbolic) approximations of the static resistance and conventional Smith-type [*10*] viscous resistance components for pile driving analyses. Details of the elemental representations and solution algorithms employed in DUKFOR are described elsewhere [*6, 11*].

Case History Evaluation

General Description

A comprehensive series of pile tests was performed at lock and dam no. 4 (LD4), Arkansas River Navigation Project for the U.S. Army Engineer District, Little Rock, by Fruco and Associates [*8, 12*] The general subsurface conditions at the test site consist of three major soil strata—an overlying 4-m (13-ft) thick surface blanket of silts and fine sands; an intermediate stratum of relatively dense, medium to fine sand approximately 28m (93 ft) thick; and a basal stratum of Tertiary clay. The *in-situ* dry density of the deep sand strata ranged between 1440 and 1750 kg/m^3 (90 and 109 lb/ft^3), showing no consistent trend with depth. Twenty feet of soil was excavated prior to the pile tests. Piezometers indicated that the water table remained 0.6 to 0.9 m (2 to 3 ft) below excavation grade during pile-driving and load-testing operations.

Laboratory tests on the soils were performed by the U.S. Army Engineer Division Laboratory, Southwestern, Dallas, Texas [*13*]. Direct shear and interface shear tests were performed on the (SP-SM) sand samples prepared in the laboratory at selected densities. The angle of internal friction for the soil ϕ ranged between 31 and 35 deg, with an average of 32 deg. The interface friction angle δ for sand-on-steel measured between 23 and 30 deg, with the higher values associated with the higher relative densities. For the purpose of the analyses performed herein, the average value of ϕ was assumed ($\phi = 32$ deg); the value of δ was assumed as 30 deg, chosen because of the likely densification of the sand near the pile shaft. Hyperbolic deformation parameters were derived from the available interface shear test results and employed directly in DUKFOR analyses. The details of the hyperbolic formulation and the parameters thus obtained are given elsewhere [*6*].

In order to model shaft load transfer in DUKFOR one must approximate the pile-soil system response at the interface. Deformations in the adjacent soil should soften the measured pile load transfer relative to laboratory interface shear test behavior. The effects of pile driving on pile shaft behavior (soil densification and repeated load-unload cycles) should stiffen the measured response. It is assumed for the DUKFOR analyses presented herein that the laboratory interface shear test behavior represents the field load transfer mechanism adequately, which was verified in the study [*6*].

The test piles installed at LD4 included steel H-piles (HP 14 × 73), steel pipe piles (0.30-, 0.41-, and 0.51-m outside diameter) (12-, 16-, and 20-in. outside diameter), prestressed concrete piles (0.41- and 0.51-m square) (16- and 20-in. square), timber piles and steel sheet piles. For this paper test pile 2 (TP2), a 0.41-m (16-in.) outside diameter pipe pile, has been selected to demonstrate the application of DUKFOR and compare predicted with measured behavior. Analyses of the four remaining pipe piles gave comparable results [*6*]. The properties and installation data for LD4TP2 are given in Table 1. The details of the Vulcan 140C hammer assembly used to install this pile are listed in Table 2.

Pile Test Results

LD4TP2 was loaded cyclically to near "plunging" failure first in compression and thereafter in tension. The maximum applied load was assumed as the respective pile capacity. The results of the compression test are summarized in Table 3, and the tension test results are summarized in Table 4. Measurements of pile butt deflection versus load and load dis-

TABLE 1—*Properties and installation data, LD4TP2.*

Type and details	16-in. OD pipe, 0.312-in. wall, with 4, 4 ⊔ 7.25 protective channels—ASTM A-252, grade 2 steel pipes
Net cross section	23.86 in.^{2}
Effective perimeter	5.32 ft
Gross tip area	1.57 ft^2 (closed-ended)
Length	55 ft
Embedment	53 ft
Hammer type	Vulcan 140C
Blow count/final foot	38 blows/ft
Instrumentation	Strain rods and strain gages

Conversion factors—
1 in. = 2.54 cm, and
1 ft = 0.3048 m.

TABLE 2—*Impact hammer assembly data, LD4.*

Hammer—Vulcan 140C double-acting steam hammer	
Ram weight, W_1	14 000 lb
Rated energy	36 000 ft·lb
Efficiency (assumed)	0.78
Cap block—micarta: 10 × 1.0-in. thick disks, 17.5-in. O.D. with 4.5-in. bored center hole	
Cross-section	224.6 in.^{2}
Thickness	10 in.
Secant modulus (stress level dependent)	4.5×10^5 lb/in.^{2}
Coefficient of restitution (assumed)	0.80
Pile cap (no cushion block)	
Weight, W_2	1710 lb

Conversion factors—
1 in. = 2.54 cm,
1 ft = 0.3048 m, and
1 lb = 0.454 kg.

TABLE 3—*Compression test results, LD4TP2.*

Maximum compression load	502 kips
Mobilized compression capacity	
Point	150 kips
Shaft	352 kips
Adjusted[a] compression capacity	
Point	240 kips
Shaft	262 kips
Lateral earth pressure coefficient at failure[b], K_s[c]	
Mobilized	1.22
Adjusted	0.906

Conversion factors—
1 lb = 0.454 kg
1 kip = 454 kg and
1 ft = 0.3048 m.

[a] Adjusted for measured residual point load after Hunter and Davisson [*8*].

[b] Triangular distribution assumed proportional to effective overburden stress; interface friction angle, $\delta = 30$ deg assumed; WT at 2 ft depth, $\gamma = 125$ lb/ft^3, $\bar{\gamma} = 63$ lb/ft^3.

TABLE 4—*Tension test results, LD4TP2.*

Maximum tension capacity	232 kips
Lateral earth pressure coefficient at failure	0.802

Conversion factor—1 kip = 454 kg.

tribution versus depth were obtained throughout the test program. Hunter and Davisson [*9*] reported that LD4TP2 had 40 800 kg (90 kips) residual compression load at the pile point, interpreting the strain gage data consistent with the example described in Fig. 1. However, the actual distribution of residual pile loads above the pile point remains unknown, as the "true" zero load for the gages was never obtained. Hunter and Davisson approximated the "true" zero for the gages by assuming no residual pile stresses after unloading the pile from failure in tension when some magnitude of residual tensile stresses may have been present.

Analysis Results

In order to perform DUKFOR analysis, it is necessary to prescribe the unit shaft resistance distribution and point resistance values that correspond to the ultimate pile capacity observed. For the purpose of these analyses, it is assumed that the unit shaft friction at failure along the pile is directly proportional to the effective overburden pressure at the corresponding depth, a common, though greatly simplified assumption. Thus, one may directly compute the value of the coefficient of lateral earth pressure at failure K_s^c, which corresponds to the ultimate shaft capacity in compression for a particular pile.

In Table 3, two sets of capacity values are given; the first set gives "mobilized" capacity values that correspond to those obtained from strain gage measurements for which the zeroes were assumed at the start of the load test. These are the loads "mobilized" during the load test. The second set of values is "adjusted" to incorporate the 40 800 kg (90-kip) residual point load cited by Hunter and Davisson [*9*]. Note that the K_s^c value computed from "mobilized" loads is approximately 35 percent larger than that obtained using the "adjusted" shaft capacity, similar to the tendency noted in the hypothetical example discussed in Fig. 1.

Pile Driving—DUKFOR analyses were performed to predict the blow count and residual load distribution at the end of driving 16.2-m (53-ft penetration) assuming $K_s^c = 0.906$ and point resistance) $Q_p = 109\ 000$ kg (240 kips), values determined from "adjusted" resistance distribution. The initial tangent point quake (displacement at the intersection of the initial tangent and ultimate point load assymptote) was assumed to be 2.1 mm (0.083 in.), which approximately fits a hyperbola through the measured point load versus point displacement data. Unload-reload behavior was modeled as linear elastic with the same slope as the initial tangent value.

The first correlation study attempted to define the value(s) of point and shaft damping parameters (J_p and J_s, respectively) that would predict the measured blow count (125 blows/n) (38 blows/ft) correctly. These wave equation analysis results were found to coverge within three simulated blows so that both blow count and residual stress distribution remained unchanged at the end of each succeeding blow. In order to insure convergence, each data point shown in Fig. 2 represents the results of the fifth hammer blow.

It readily may be seen that a number of (J_p, J_s) combinations would satisfy the blow count correlations, indicating that J_p and J_s are by no means unique soil properties. Indeed, they are simply correlation factors accommodating a large variety of unknowns within the problems. The dashed lines in Fig 2 describe the comparable correlations using the bilinear soil resistance model after Smith [10] in the complete blow analysis. These results demonstrate that the predicted blow count is relatively insensitive to refinements in load transfer model assumptions. Predicted residual point load was relatively independent of soil damping value. Using the results in Fig. 2, values of J_p = 0.33 s/m (0.10 s/ft) and J_s = 0.16 s/m (0.05 s/ft) were assigned for 50 000-kg LD4TP2. This set of assumptions produced a predicted residual point load of (110-kips), reasonably close to that observed (40 800 kg) (90-kips).

Load Test Analyses—Using the correlated damping parameters, the dynamic analysis was repeated for LD4TP2 in order to establish the initial residual stress state at the end of driving. The maximum applied load to LD4TP2 was 502 kips, applied in cyclic compression to failure. Cyclic load testing was modeled using unload-reload cycles at 45 400 kg (100-kip) intervals (as used in the pile test) up to 215 500-kg (475 kips). Load-displacement predictions using DUKFOR with hyperbolic and bilinear deformation models are shown in Fig. 3 along with the field measurements.

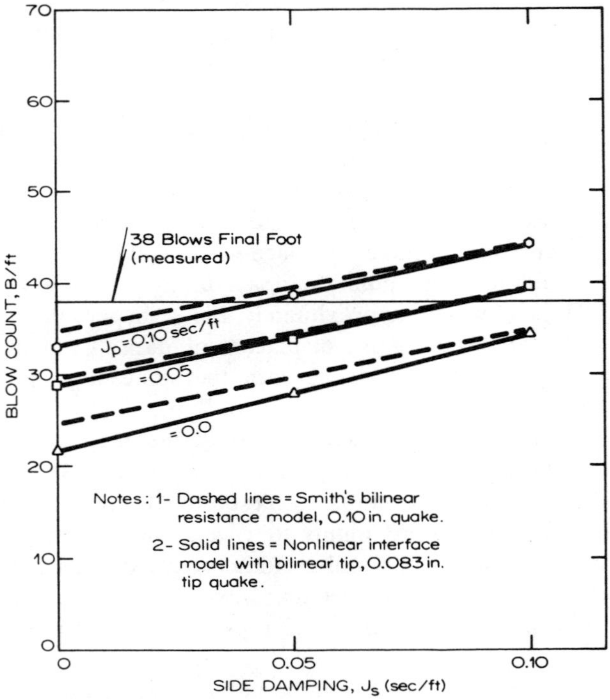

FIG. 2—*Damping coefficient correlation with measured blow count, LD4TP2 (1 ft = 0.3048 m).*

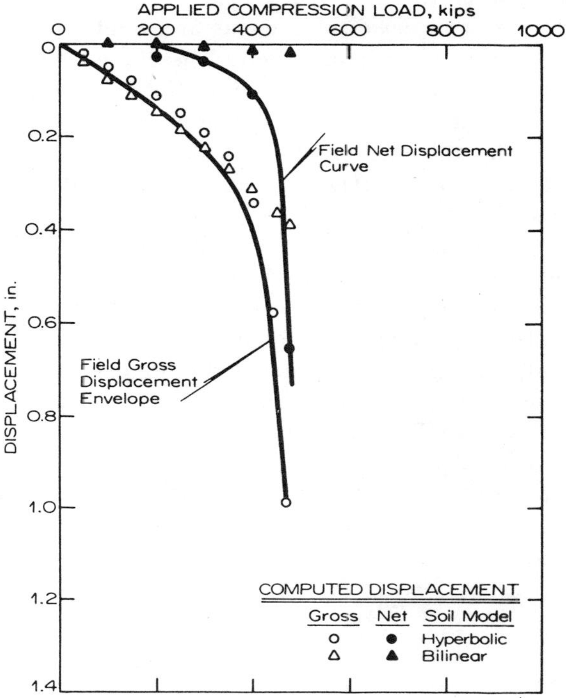

FIG. 3—*Cyclic compression test load-displacement behavior; bilinear versus nonlinear models, LD4TP2 (1 kip = 454 kg; 1 in. = 2.54 cm).*

The load-displacement predictions using either soil model indicate close agreement with field-measured gross displacements in the initial, linear portion of the curve. At higher load levels, the bilinear model is less accurate. However, the nonlinear model predicts response in close agreement with that obtained throughout the load range. In terms of net displacements (permanent settlement upon unload) the nonlinear deformation model yields acceptable agreement with field observations. The bilinear representation gives only a poor estimate of net settlements.

A second comparison with observed data may be made between the "mobilized" and "adjusted" bearing capacity parameters. In order to demonstrate the effects, a static DUKFOR analysis was made assuming an initially "stress free" pile using parameters determined based on observed "mobilized" loads [$K_s^c = 1.22$, $K_s^t = 0.802$, $Q_p = 68\,000$-kg (150 kips)]. One would expect that the stress free initial conditions and mobilized resistance parameters would give essentially the same load-displacement behavior as the driven pile case using parameters adjusted for residual point load, since the mobilized resistances to penetration at failure are distributed similarly under either condition. This expectation is realized in Fig. 4. Both gross and net displacements are predicted as almost identical values for each set of assumptions. These predictions corroborate the statements made previously regarding the effects residual loads have on load test interpretation (Fig. 1). Without adjustment for residual loads, the coefficient of lateral earth pressure at (compression) failure K_s^c is overestimated by 35 percent. The tip capacity Q_p is underestimated by 38 percent. Reliable design estimates of K_s^c and Q_p from LD4TP2 are only obtained by correctly accounting for residual driving load at the point.

Predicted and measured pile load distribution data during the compression load test are given in Fig. 5. Recall that the gages were zeroed at the start of the load test, assuming no residual driving stresses present. Curve 0 predicts the residual stresses immediately following driving. Curves 1 and 2 are DUKFOR predictions of the actual or "adjusted" load distributions corresponding to "virgin" compression loads of 90 700 and 181 400-kg (200 and 400 kips), respectively. Curves A and B are drawn through the strain gage-strain rod data. In order to predict Curves A and B using

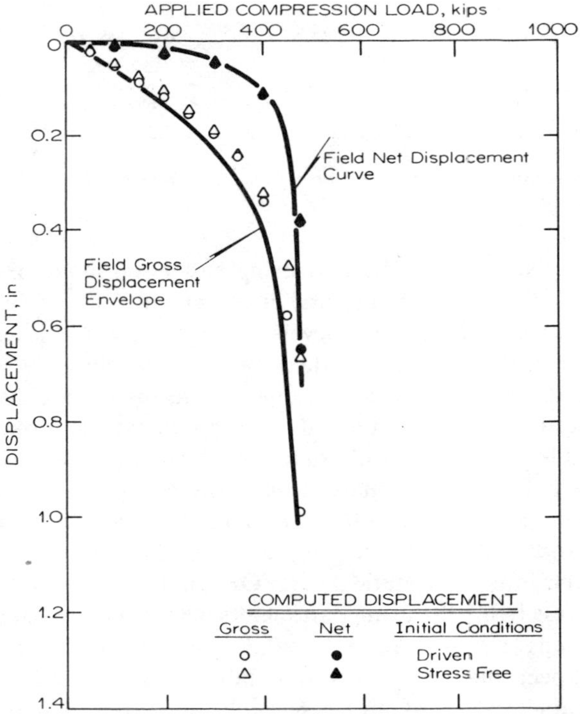

FIG. 4—*Cyclic compression test load-displacement behavior; driven versus stress free pile, LD4TP2 (1 kip = 454 kg; 1 in. = 2.54 cm).*

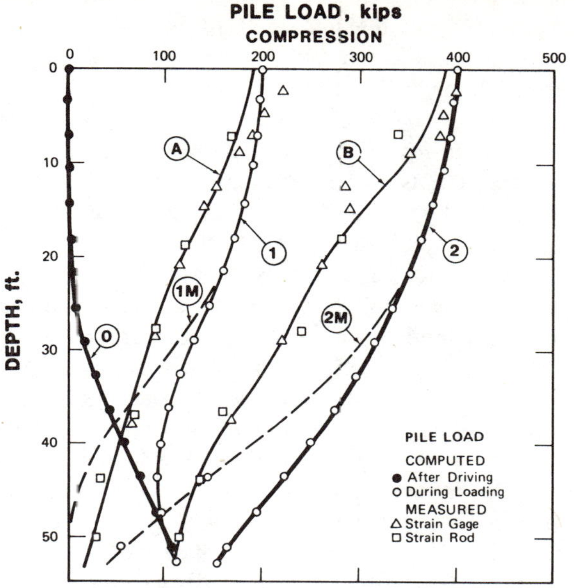

FIG. 5—*Cyclic compression test pile load distribution, LD4TP2. Numbers in circles denote loading sequence; M denotes "mobilized" curve. (1 kip = 454 kg; 1 ft = 0.3048 m).*

DUKFOR, one must subtract the amplitudes of load in curve 0 from curves 1 and 2, respectively, obtaining new "mobilized" curves 1M and 2M.

It appears that curves A and 1M (at the lower load level) are in better agreement than curves B and 2M. Indeed, curve B describes a steeper slope (larger unit skin friction values) at shallower depths than was assumed in the DUKFOR analyses. As the true zeros of the gages are unknown, the tendency for steeper slopes at shallow depth and flatter slopes at greater depth cannot be explained readily as residual stress effects, variations in K_s^c with depth, or otherwise. It is most likely due to some combination of these factors. Nevertheless, these data demonstrate an apparent "strain-hardening" of the upper soil strata (slope-unit shaft resistance increasing with increasing axial compression load between curves A and B), which the one-dimensional idealization in DUKFOR does not accommodate. Axisymmetric finite element analyses by the authors [*6*] demonstrated this "strain-hardening" tendency, but to a considerably lesser degree than observed.

Cyclic tension test analyses were performed using three separate sets of assumptions: (1) a sequence of driving/cyclic compression–cyclic tension field behavior (D/CC–CT); (2) assuming an initially stress free pile with mobilized K_s^c and Q_p values that is loaded first in cyclic compression, then in cyclic tension (SF/CC–CT); and finally, (3) an initially stress free pile that is loaded only in cyclic tension (SF/CT).

The results of the cyclic tension test analyses are plotted in terms of load–displacement behavior in Fig. 6. The load–displacement predictions for the first two cases, D/CC–CT and SF/CC–CT, are essentially similar, since the initial conditions at the start of the cyclic tension test are quite similar. These predictions agree reasonably well with observed behavior up to about −45 400-kg (−100 kips), whereafter the field behavior shows increasingly greater displacements than DUKFOR predicts. The tension test response of the DUKFOR model is apparently too stiff. There appears to be a softening tendency of the system at higher-tension loads–displacements, opposite to the tendency noted as "strain-hardening" in compression.

The third case (SF/CT) provides a direct estimate of the effect of residual compression stresses on tension test performance. Residual compression forces cause much of the skin friction in the upper section of the pile to be acting downward at the start. When tension load is applied, the upper portion of the pile cannot "mobilize" as much resistance, causing the mobilized shaft resistance to develop deeper, (Fig. 1*b*). For cases D/CC–CT and SF/CC–CT, the elastic deformations of the pile are greater near the top, causing noticeably greater measured displacements than the SF/CT (having no residual compression stresses) assumptions predict.

Further evidence to support this argument is offered by comparison of the SF/CT test predictions with the final field reload cycle. Prior to the

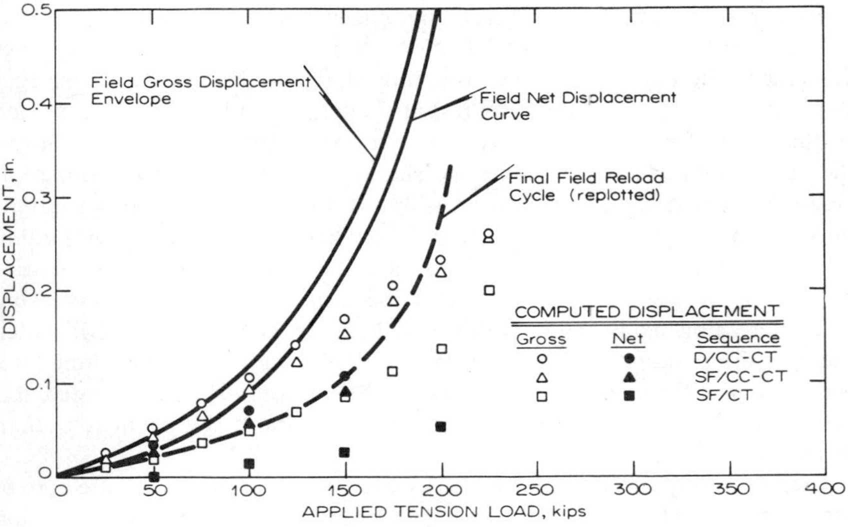

FIG. 6—*Cyclic tension test load-displacement behavior. LD4TP2 (1 kip = 454 kg; 1 in. = 2.54 cm).*

last load cycle, the pile has been loaded almost to failure in tension and then unloaded. As a result, the residual compression forces have been "unloaded" so that the initial conditions for the final tension load cycle would probably involve primarily residual tension forces at depth along the pile. The shaft friction sustaining residual tension forces would act upward (positive) on the pile near the pile butt. Subsequent application of tension load would yield substantially less displacement than the gross displacement curve. The SF/CT test predictions, therefore, best estimate the tension reload cycle behavior.

The comparison of measured versus predicted tension load distribution data is included in Fig. 7. Curve 4 describes the −68 000-kg (−150 kip) load results for the loading cycle determined in case D/CC-CT. As before, the field data should agree with the DUKFOR prediction curve minus the residual distribution curve corresponding to the gage zero stage of the test. Two occasions for zeroing the gages are possible: continuous reading from the start of the compression test (residual load curve 0) and zeroing values at the start of the tension test (residual load curve 3). Using curve 0, curve 4M is obtained. If curve 3 is used, one finds curve 4$\overline{\text{M}}$ instead. In either case, the data describe a "mobilized" load distribution curve. For

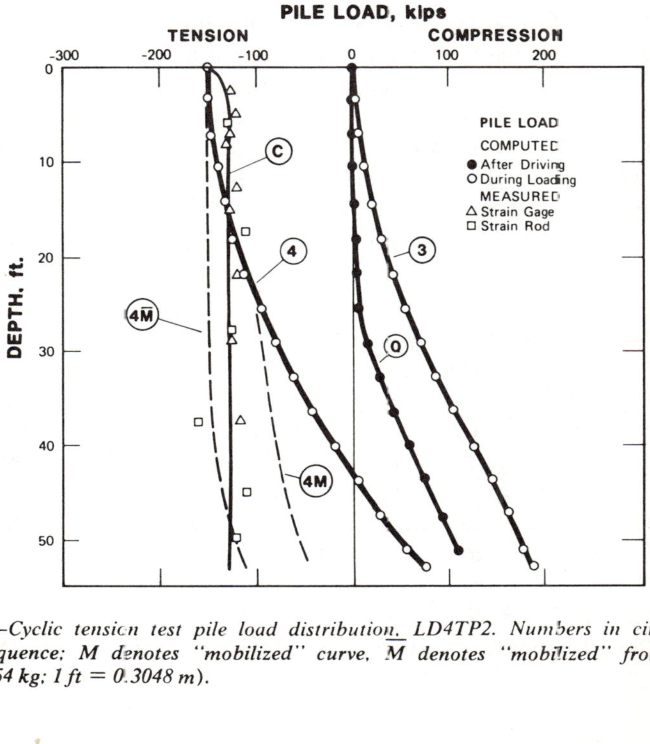

FIG. 7—*Cyclic tension test pile load distribution, LD4TP2. Numbers in circles denote loading sequence; M denotes "mobilized" curve,* $\overline{M}$ *denotes "mobilized" from curve 3. (1 kip = 454 kg; 1 ft = 0.3048 m).*

LD4TP2, it appears that the gages were rezeroed at the start of the tension test, since curve 4$\overline{M}$ provides the closer agreement with curve C.

Parametric Analyses

General Comments

In conjunction with the case history analyzed herein, it is useful to examine the sensitivity of analytical results to reasonable variations in input parameter values. For this purpose LD4TP2 is used as a "control" case, and throughout this section, the labeled parameter is varied individually with respect to the LD4TP2 inputs described in the previous section. The parametric variations described hereafter are by no means comprehensive in scope; however, understanding the requirements for "accurate" input data is vital in applying any engineering solution. The remainder of this section focuses on the influence of specific parametric variations on pile-driving solution results.

Interface Shear Stiffness

To determine the effects of the interface shear (IS), deformation model on pile driving performance the hyperbolic stiffness coefficient κ was varied. An increase in IS stiffness is analagous to use of a smaller elastic "quake" value in Smith's model [*10*]. Figure 8 compares the calculated LD4TP2 results (corresponding to $\kappa = \bar{\kappa}$, the interface shear test value) with three other cases $\kappa = 2\bar{\kappa}$, $\bar{\kappa}/2$, and $\bar{\kappa}/10$. The pile driving results (after the fifth blow) indicate that a stiffer IS model gives somewhat larger residual loads. Reduced stiffness gives a lower level of residual loads. This is logical, since the stiffer shaft resistance behavior locks in higher stresses near the tip upon rebound. For a much reduced stiffness ($\bar{\kappa}/10$), the would-be residual tip load is greatly dissipated.

The actual value of κ for LD4TP2 may be anywhere in this range. Curves 2, 3, and 4 describe the most probable range of values expected. The residual load distributions and particularly blow counts are quite similar. This is reassuring, since a precise value of κ is unknown. The deformations in the adjacent soil are, no doubt, interrelated with the strain-hardening and strain-softening phenomena noted previously. The influence of compaction and large shear deformations in the soil and at the pile–soil interface make accurate determinations of the load transfer behavior impossible *a priori*. These results suggest that the analyses are not very sensitive to the interface shear model. In terms of soil profile these results indicate that "soft" deposits should develop lower residual driving loads.

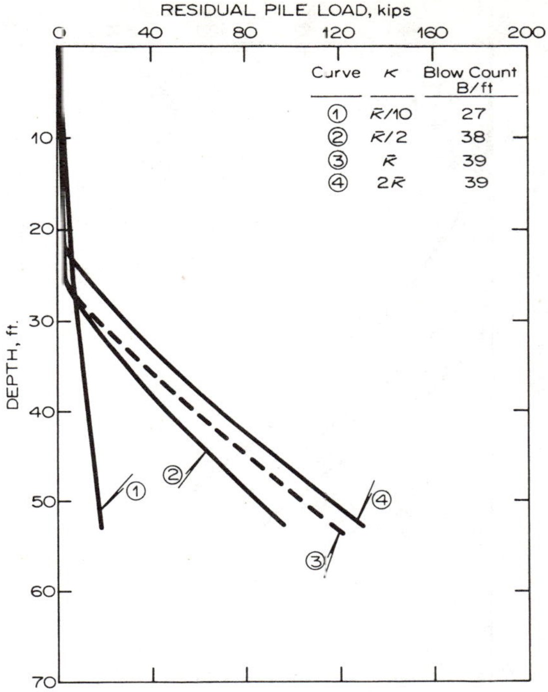

FIG. 8—*Parametric analyses: Interface shear stiffness effects (1 kip = 454 kg; 1 ft = 0.3048 m).*

Point Resistance Stiffness

The effects of point resistance stiffness on LD4TP2 performance are demonstrated in Fig. 9. Lower values of point quake $QUAKE_p$ result in a stiffer point response (stiffness $k_{tip} = Q_p/QUAKE_p$). Interestingly, the effects on blow count are minor once more; however, the stiffer point-bearing response gives smaller residual load. In reaching equilibrium after the blow, the tip "spring" unloads. Along the shaft interface, this point unload results in (negative) friction loading. The stiffer (unloading) point spring permits more load to be released from the point upon rebound, thus reducing the residual pile stresses.

Figures 8 and 9 demonstrate that the measured residual point load for LD4TP2 could be better "predicted" by a stiffer point response (decreasing $QUAKE_p$) or softer shaft response (decreasing κ). With no field measurements from which to develop load transfer relationships (correctly including measured residual load distributions), the actual causes of the discrepancies are not known. Nevertheless, the compression load-displacement predictions

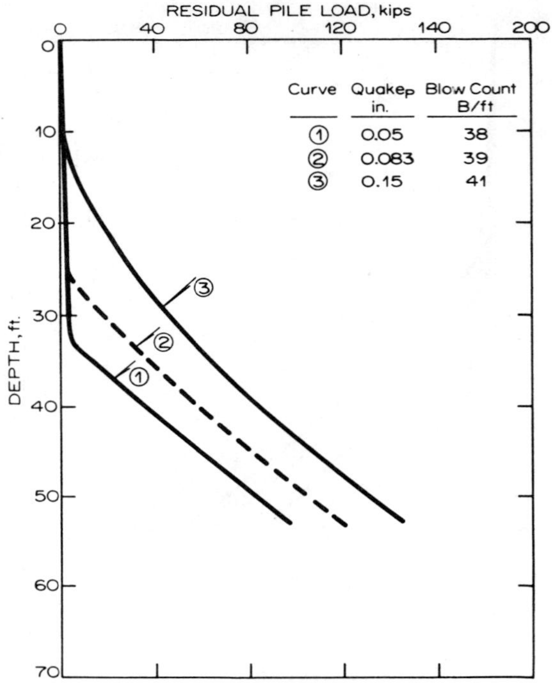

FIG. 9—*Parametric analyses: Tip resistance stiffness effects (1 kip = 454 kg; 1 in. = 2.54 cm; 1 ft = 0.3048 m).*

indicate that such load transfer accuracy is not needed for practical purposes, at least for the pile test analyzed herein.

Soil Damping

Blow count increases almost linearly with an increase in J_p or J_s, as shown in Fig. 2. The data plotted in Fig. 10 correspond to results shown in Fig. 2. The effects of damping on predicted residual driving load distributions are rather small. A general trend suggests that greater damping results in somewhat lower residual pile stresses overall, but differences are not significant. The minor effects on residual stresses indicate that the predicted load-displacement behavior should be relatively independent of soil damping used in the dynamic analyses. Damping greatly affects the predicted blow count; obviously, bearing capacity predictions from field-measured blow counts are sensitive to damping assumptions.

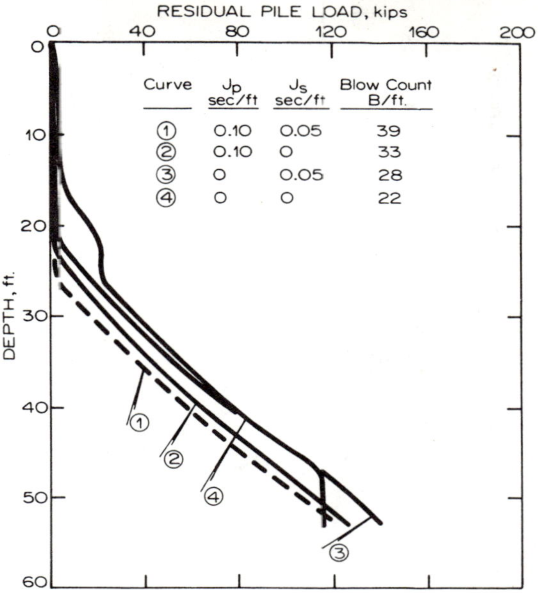

FIG. 10—*Parametric analyses: Damping resistance effects (1 kip = 454 kg; 1 ft = 0.3048 m).*

Pile Stiffness

Pile stiffness affects were analyzed using the LD4 driving system, and the results are presented in Fig. 11. Two alternate cross sections were analyzed, 64.5 and 451.6 cm² (10 and 70 in.²), in addition to the example pile 154.8 cm² (24 in.²). The 451.6 cm² (70 in.²) steel section has approximately the same impedance as 45.7 cm square (18-in. square) prestressed concrete pile.

The data given in Fig. 11 provide an interesting comparison of blow counts. For the low-impedance pile (curve 1), the blow count is nearly three times that of LD4TP2. Moreover, the peak driving stress exceeded 3586 kg/cm² (51 ksi). This indicates that the ram would probably destroy such a light section. LD4TP2 analyses predict peak driving stresses of about 1760 kg/cm² (25 ksi). Finally, for the heaviest cross section, only a small increase in penetration per blow over the LD4TP2 section (approximately 15 percent) is predicted. This indicates that a point of diminishing return exists in terms of driving efficiency with respect to steel costs, which lies somewhere within the range given in Fig. 11.

The residual driving load distributions for the three pile impedance cases are presented in Fig. 11. The stiffest pile section has the lowest residual point load, and the lightest section develops the greatest residual

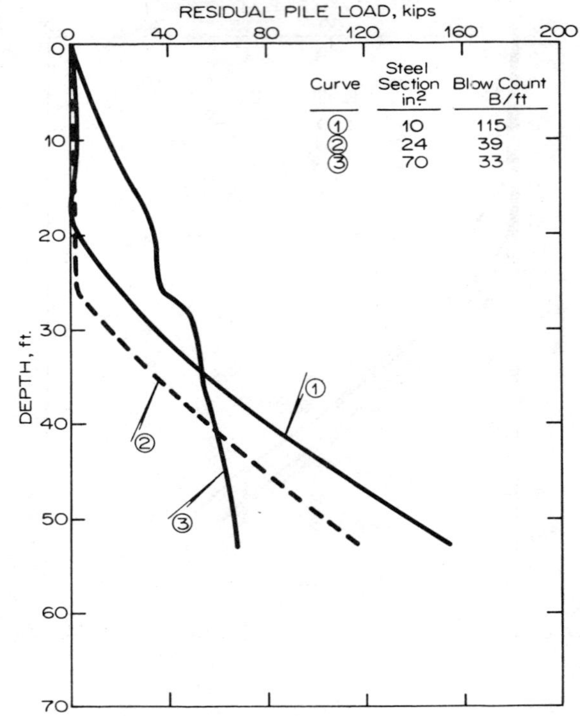

FIG. 11—*Parametric analyses: Pile stiffness effects (1 kip = 454 kg; 1 in. = 2.54 cm; 1 ft = 0.3048 m).*

point load. Thus, both blow count and residual load distribution may be significantly affected by changes in pile section stiffness.

Driving System

Table 5 presents the results of analyses of LD4TP2 driving behavior, assuming four different pile drivers were used. Case 1 represents the actual driving system (Vulcan 140C) used for the test pile. Case 2 applies the same rated energy (4 977 m kg) (36 000 ft·lb) as the Vulcan 140C, but a lighter ram was assumed, requiring higher impact velocity. Case 3 uses the same ram weight (6 350 kg) (14 000 lb) as the Vulcan 140C at a lower rated energy (2 765 m kg) (20 000 ft·lb), and thus a reduced impact velocity. Case 4 "scales up" the Vulcan 140C by increasing the impact energy (6 913 m kg) (50 000 ft·lb) and ram weight (8 822 kg) (19 450) and thereby maintaining the same impact velocity.

The differences predicted in pile driving response are very interesting. First, the residual load data for all the cases plotted almost exactly on each

TABLE 5—*Driving System Affects on pile penetration rate.*

Case	E_{ram}, ft·lb	W_{ram}, lb	Blow Count, blows/ft
1[a]	36 000	14 000	39
2	36 000	6 000	15
3	20 000	14 000	121
4	50 000	19 450	39

Conversion factors—
1 lb = 0.454 kg, and
1 ft = 0.3048 m.
[a] Vulcan 140C, used at LD4.

other, indicating that the residual load distribution is essentially independent of impact pile driver and is primarily a function of the pile-soil system properties only.

Second, case 2 results in an increase in impact velocity and peak stresses of about 50 percent [to 2 600 kg/cm² (37 ksi)] with a decreased stress pulse duration when compared with case 1. Case 2 results indicate a significant increase in driving efficiency that can only be attributed to the difference in peak stress amplitude with respect to case 1. In order to minimize the potential for pile damage a softer cap block cushion would probably be necessary, which would reduce the driving efficiency (penetration per blow) somewhat. The results of case 4 (larger rated energy with the same peak stress) show a negligible improvement in driving performance relative to case 1. These data demonstrate that an optimum driving system for LD4TP2 would deliver as large a magnitude stress pulse as the pile can safely sustain without damage.

Conclusions

The behavior of impact-driven piles in response to applied loads may be greatly influenced by the installation process, specifically in terms of the residual pile-soil system stresses induced by driving. An analysis procedure has been developed to incorporate these installation effects explicitly. The computer program DUKFOR was used to evaluate the results of an example case history and selected parametric variations of this case from which several conclusions can be drawn.

For impact-driven displacement piles, a significant residual point load often develops, depending primarily on the pile-soil system (independent of the pile driver). Interpretations of load transfer data from subsequent pile load tests frequently ignore these residual pile stresses, resulting in an underestimation of actual point bearing capacity and overestimation of the total shaft bearing capacity. This is of little importance in "proof" testing of previously designed piles, since little understanding of the load transfer mechanism is needed in such cases. On the other hand, basing alternative pile-configuration designs on pile test results generally requires proper interpretation of the actual load transfer mechanisms. Neglecting residual pile point load may lead to expensive design measures when a minimum point capacity requirement must be satisfied, for example. In load test cases where residual driving stresses may impact the design solutions, proper interpretation of the load transfer behavior is essential.

The DUKFOR program provides a useful tool in predicting the pile-soil system conditions following impact-driving and the subsequent pile response to axial load based on reasonable foundation engineering assumptions generally applied in pile design. In this manner, alternate designs can be compared directly in terms of driveability and response to applied load. The example case presented in this paper demonstrates the application of DUKFOR in interpreting pile test behavior, and the results demonstrate satisfactory agreement with field-measured behavior.

Acknowledgments

The authors gratefully acknowledge the U.S. Army Engineer Waterways Experiment Station (WES) and Duke University in supporting the research reported herein. Permission of WES to publish this paper is appreciated.

References

[*1*] Vesić, A. S., *Proceedings,* Conference on Design and Installation of Pile Foundations and Cellular Structures, Lehigh University, Bethlehem, Pennsylvania, April, 1970, pp. 47-73.

[*2*] Kerisel, J., *Proceedings,* North American Conference on Deep Foundations, Mexico City, 1964, pp. 5-44.

[*3*] Leonards, G. A., *Pile Foundations,* Highway Research Record No. 333, 1970, pp. 55-59.

[*4*] Ellison, R. D., *An Analytical Study of the Mechanics of Single Pile Foundations,* Ph.D. thesis Carnegie-Mellon University, Pittsburgh, Pa. 1969.

[*5*] Desai, C. S., and Holloway, D. M., "Load-Deformation Analysis of Deep Pile Foundations," *Proceedings,* Symposium on Applications of the Finite Element Method to Problems in Geotechnical Engineering, U.S. Army Engineer Waterways Experiment Station, 1-4 May 1972, Vicksburg, Miss.

[*6*] Holloway, D. M., Clough, G. W., and Vesić, A. S., "The Mechanics of Pile-Soil Interaction in Cohesionless Soils," (FINAL) Contract Report S-75-5, U.S. Army Engineer Waterways Experiment Station, Vicksburg, Miss., Dec. 1975 (also Duke Soil Mechanics Series No. 39).

[*7*] Hanna, T. H. and Tan, R. H. S., *Canadian Geotechnical Journal,* Vol. 10, No. 3, August 1973, pp. 311-340.

[*8*] Fruco and Associates, "Pile Driving and Loading Tests: Lock and Dam No.4, Arkansas River and Tributaries, Arkansas and Oklahoma," Report for the U.S. Army Engineer District, Little Rock, Corps of Engineers, Sept. 1964.

[*9*] Hunter, A. H. and Davisson, M. T. in *Performance of Deep Foundations, ASTM STP 444,* American Society for Testing and Materials, 1969, pp. 106-117.

[*10*] Smith, A. E. L., *Journal of Soil Mechanics and Foundations Division,* Vol. 86, No. SM4, Aug. 1960, pp. 35-61.

[*11*] Holloway, D. M., Clough, G. W., and Vesić, A. S., "The Effects of Residual Driving

Stresses on Pile Performance under Axial Loads," Offshore Technology Conference 3306, May 1978.

[12] Mansur, C. I. and Hunter, A. H., *Journal of Soil Mechanics and Foundations Division*, Vol. 96, No. SM5, Sept. 1970, pp. 1545–1582.

[13] U.S. Army Engineer Division Laboratory, Southwestern, "Results of Tests in Foundation Materials, Lock and Dam No. 4, Arkansas River Navigation Project, Little Rock District," SWDGL Reports No. 7920 and 7932, Dallas, Tex., 1962.

H. S. Lacy[1]

Load Testing of Instrumented 225-Foot-Long Prestressed Concrete Piles*

REFERENCE: Lacy, H. S., "**Load Testing of Instrumented 225-Foot-Long Prestressed Concrete Piles,**" *Behavior of Deep Foundations, ASTM STP 670,* Raymond Lundgren, Ed., American Society for Testing and Materials, 1979, pp. 358–380.

ABSTRACT: Piles driven through 62 to 77 m (200 to 250 ft) of low-strength clay penetrated up to 23 m (75 ft) into compact fine sand with little increase in driving resistance. At other locations, the piles met refusal after only about 1 m (3.3 ft) of penetration into the sand. Borings indicated a consistently compact bearing stratum. This test program was developed to establish whether piles driven to very low driving resistances at moderate penetration into the lower compact sands would support the 91-Mg (100-ton) design load. The overlying clays provided no permanent pile support, as in some areas they are underconsolidated as a result of site filling.

The test program included casting eight electric strain gages and a 7.6-cm (3-in.) diameter hole in 36-cm (14-in.) square prestressed concrete piles. The piles were driven in three sections to low driving resistances with a continuous record of driving stresses. Anchors were installed in the center hole, with rods extending to the surface. The top of the piles were loaded in increments to 3½ times the design load to overcome the loss of load into the thick clay layer. A procedure of performing the load test a short time after the pile was driven and cyclic loading after reaching maximum test load was used to increase the load reaching the lower compact sand stratum. Full-scale laboratory load tests were preformed on a section of pile to establish the variation in elastic modulus to permit accurate estimates of pile load at various depths.

The pile load tests demonstrated that the deep compact sand layer was capable of supporting the design loads, even though penetration was terminated at low driving resistances. Substantial portions of the load during the short-term tests were resisted by adhesion in the thick clay layer. The amount of clay adhesion was directly related to the time after driving of the piles.

KEY WORDS: concrete piles, soil properties, modulus of elasticity, pore pressure, adhesion, instruments, load tests, standard penetration tests

Expansion of the Metropolitan Sewage Treatment Plant at Syracuse, N.Y., required support of the structures on piling 64 to 82 m (210 to 270

*Original experimental data were measured in customary inch-pound units.

[1]Associate, Mueser, Rutledge, Johnston and De Simone, New York, N.Y. 10017.

ft) long. Three pile load tests were used to demonstrate pile capacities in the lower bearing strata for the approximately 4500 contract piles.

The pile load test program was undertaken to determine that the piles could provide adequate support in compact soils encountered at a depth generally greater than 61 m (200 ft) without depending on support of the overlying lower-strength clays. The capacity of the bearing strata could not be predicted with confidence due to low final driving resistances at limited penetration into these soils and uncertainties of the accuracy of static methods of analyses of piles at these depths. Model tests by Vesić [*1*][2] suggest that a critical depth is reached where the confining pressures against the pile perimeter remains constant with increasing depth. It was also necessary to establish the minimum penetration that would reliably support the 91-Mg (100-ton) design load in order to make use of prestressed concrete piles feasible.

It was not practical to perform the load test within a cased hole in order to remove the load support of clays above the bearing strata because of the extreme length of the piles. Various methods were used to measure and minimize effects of clay adhesion in transmitting the test load to the bearing strata. In evaluating the test results, information was developed that could be of interest to others. This included the importance of use of multiple telltales rather than a single telltale in piles of this length and measurement of the variation in mobilized pile adhesion at various applied loads and the increase in adhesion with time after the pile was driven.

The reduction in undrained shear strength of clay due to remolding close to piles following driving has been measured by others [*3,4*]. Recovery of undrained shear strength around piles has also been measured [*4,5*]. High local positive pore pressures develop following driving of piles into normally consolidated clays [*4,6–9*]. Instrumentation of piling to determine the end-bearing and skin friction resistance of piles has been often reported [*2,10, 11*].

Site Description

Site History

Onondaga Lake lies in a preglacial river channel that was deepened by ice gouging of southerly advancing glaciers to below sea level. As the glaciers receded, outwash sands and glacial lake varved silts and clays were deposited followed by late pleistocene clays that are slightly organic. At the turn of the century, the site was a marsh at the south end of Onondaga Lake. Chemical waste and granular fill were placed so long ago that the deep clay deposits have largely consolidated under this 3.7-m (12-ft) rise in

[2]The italic numbers in brackets refer to the list of references appended to this paper.

the grade above the adjacent lake. An additional 0.6 to 1.5 m (2 to 5 ft) of fill was placed over part of the site in the late 1950s, causing 0.3 to over 0.9 m (1 to 3 ft) of settlement of the ground surface. In these areas, up to 15 cm (6 in.) of additional settlement is anticipated over the next 20 years. Analyses indicated that long piles to underlying compact materials were the only reliable means of supporting the plant structures. Piles driven previously to support plant structures have penetrated to widely varying tip elevations.

Soil Profile

Figure 1 illustrates the typical soil profile. Soil borings were made in a 10 cm (4-in.) inside diameter hole using drilling mud. It was necessary to mix salt with the drilling mud and to monitor carefully the mud specific gravity to assure that the weight of the column of drilling fluid exceeded artesian pressures in the lower sands in order to prevent these sands from moving up the hole. Standard penetration test (SPT) resistances were mea-

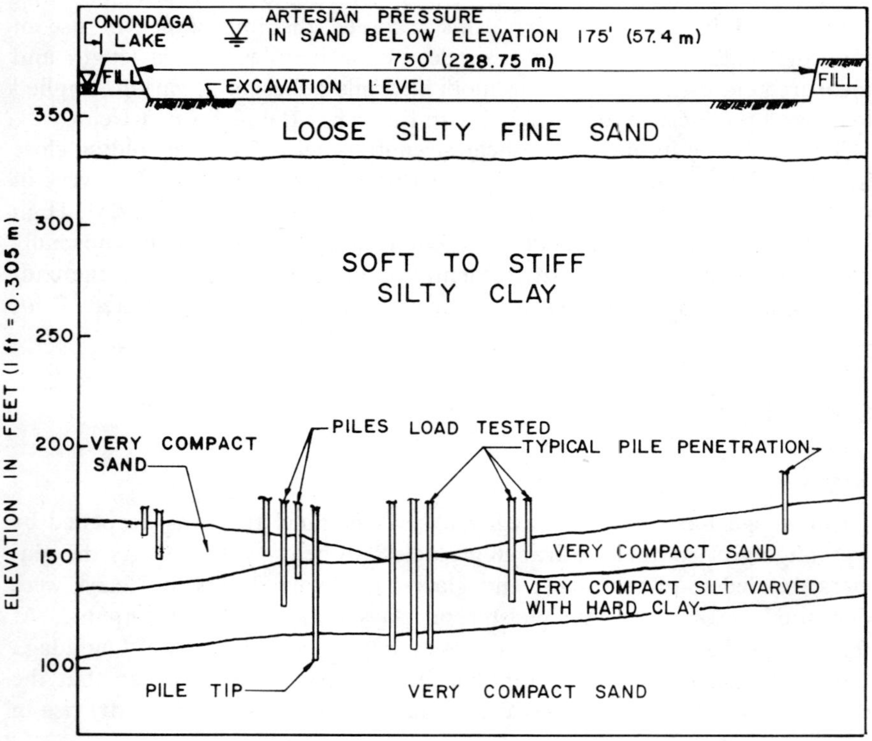

FIG. 1—*Typical profile through site.*

sured using the ASTM Penetration Test and Split-Barrel Sampling of Soils (D 1586). General excavation for the treatment plant structures extended 1.5 to 4.5 m (5 to 15 ft) below ground surface. Varying thicknesses of highly compressible chemical waste and other fill materials extended below the excavation level at some locations. The deep deposits of soft to stiff silty clay was slightly organic at higher elevations and contained occasional random layers of loose, fine sand. The bearing strata consist of an upper layer of very compact (SPT of 50 blows per foot or more) silty fine to medium sand in varying thicknesses underlain by a very compact silt varved with hard clay. The lowest pile-bearing layer, consisting of a very compact silty fine to medium sand, extends at least 18 m (60 ft) deeper. Both of the sand strata contain zones of medium sand with traces of coarse sand and occasional traces of fine gravel. At several locations, the fine sand layers grade into a compact nonplastic silt with some fine sand. Hard clay varves represent an average of 25 percent of the very compact varved silt stratum. Average properties of the soils are shown on Table 1. While ground water levels in the shallow granular soils were 0.9 to 1.5 m (3 to 5 ft) below orig-

TABLE 1—*Summary of Soil Properties.*

Elevation, ft	Soil	Natural Water Content, % Range	Natural Water Content, % Average	Liquid Limit,[a] %	Plastic Limit,[a] %	Undrained Shear Strength, psf
170-330	Soft to stiff silty clay, slightly organic at higher elevations, containing layers of clayey silt, sandy silt, and occasional layers of loose fine sand	16-70	44	50	26	330-800 at depths up to 50 ft
105-150	Very compact silt	10-20	19	26	17	4100-9300
	Varved with hard clay	21-28	23.5	31	18	

Elevation, ft	Soil	SPT, blows/ft Range	SPT, blows/ft Average	Gradation-Typical Values D10 MM	D50 MM	D60 MM	D85 MM
130-170	Very compact silty fine to medium sand	27-200	78	0.07	0.46	0.48	0.70
				0.18	0.70	1.03	2.80
				...	0.082	0.11	0.17
				...	0.28	0.30	0.37
60-135	Very compact silty fine to medium sand	45-175	80	0.0065	0.16	0.18	0.22[b]
				0.026	0.27	0.31	0.50
				...	0.34	0.40	0.63
				...	0.12	0.14	0.21

[a] Typical values.
[b] Most common value.

inal ground surface, artesian pressures in the sandy bearing strata were measured at 2.4 to 3.0 m (8 to 10 ft) above ground surface or 5.5 to 7.0 m (18 to 23 ft) above excavation level.

Planning of Pile Foundations

Several methods of analysis of static bearing capacity for pipe piles indicated high end-bearing resistance in the sand strata as a result of the large vertical effective stresses in these deep layers. The primary method used is shown in figure 13–2 of the NAVFAC Manual [*12*]. Estimated end bearing resistance was such that the design capacity could be obtained with only shallow penetration into these materials. Even with greater penetration, side friction in the sands contributed only 10 to 15 percent of the total support. Piles that extended through the upper compact sand and were founded in the very compact silt varved with hard clay had a lower estimated bearing capacity due to the limiting shear strength of the clay varves.

Contract documents for this project included alternative pile types for a 32 by 0.95 cm (12¾ by ⅜ in.) steel pipe pile, 12 BP 74 and 36 cm (14 in.) square prestressed concrete piles. All piles were required to be spliced in a manner that would develop the full bending moment of the pile section and in certain piles to resist design uplift loads of 35 Mg (40 tons). For bid comparison purposes, an estimated length of H-piles was assumed that averaged 12 m (40 ft) longer than other alternatives. The contractor submitted a significantly lower price for the use of prestressed concrete piles. Pile load testing was required in the specifications for this pile type. The pile used had seven or eight strands of 1.27 cm (½ in.) diameter reinforcement to prestress pile sections to 62 and 66 kg/cm^2 (877 and 943 psi).

Drive Test Pile Program

Piles were driven using a Link Belt model 660 diesel hammer with a maximum rated energy of 61 020 N·m (45 000 ft·lb). Hammer energy was reduced to about 36 612 N·m (27 000 ft·lb) when low driving resistance was experienced in the low-strength clays to prevent damage from reflected tension waves. Piles were driven in sections in lengths varying from 15 to 26 m (50 to 85 ft.) Twelve drive test piles were driven at widely spaced locations throughout the site prior to selection of locations for load test piles. At some locations, piles reached refusal (10 blows per in.) after only about 1 m (3.3 ft) several feet of penetration into the upper very compact sand stratum, while in other areas, piles extended as much as 23 m (75 ft) into the bearing strata without encountering the specified driving resistance of 6 blows per inch.

The pile-driving criteria was established using analyses of static capacity, pile-driving formulas, and wave equation analyses. Piles were required to be driven a minimum of 3 m (10 ft) into the bearing strata to final minimum driving resistance of 6 blows per inch or to a maximum penetration into the bearing strata to be confirmed by the load tests. Soil borings had been made at each of the locations where test piles were driven. Consistantly high SPT values, generally 50 to 100 blows per foot had been observed in each of these soil borings when the bearing strata was encountered following SPT values of 1 to 10 in the overlying soft to stiff silty clays. Average SPT values were nearly as high in the very compact silt stratum, but much more variable when compared with the very compact sand strata.

As illustrated in Fig. 2, at some locations, piles experienced no measurable increase in driving resistance upon entering the bearing strata, while SPT values exhibited a marked increase. The interruption of pile driving for 1 to 2 hs after penetrating the bearing strata resulted in very high redrive resistances. The apparent reason for the low initial driving resistances within the bearing stratum is illustrated on the right-hand panel of Fig. 2. A piezometer was installed in Boring No. B-1 with its intake between eleva-

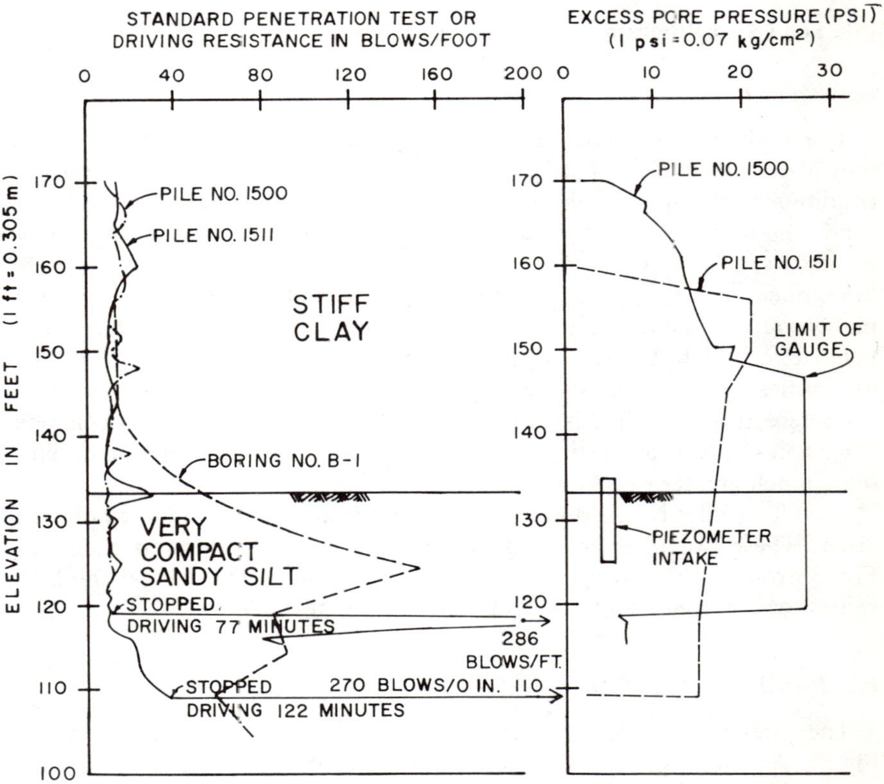

FIG. 2—*Excess pore pressure during pile driving.*

tion 125 and 135 using heavy drilling fluid and a pneumatic packer and by grouting the annulus around the riser pipe to the ground surface. The piezometer was located about 8 ft horizontally from Piles Nos. 1500 and 1511. The variation in pore pressure was measured during pile driving. The pore pressures shown are only those values exceeding the static artesian pressures present within this stratum. Pore pressures during driving of Pile No. 1500 exceeded the limit of the gage before the pile tip reached the bearing stratum. While this value is only approximately 45 percent of the effective overburden stress, it suggests that pore pressures immediately around the pile tip exceeded existing overburden stress. This caused soils near the pile tip to behave more like a liquid when loaded rapidly during pile driving. These high pore pressures largely dissipated over a period of approximately 1 h. Efforts to develop a consistant relationship between grain size of the sand at various locations within the bearing stratum or the order of driving piles with the degree of resistance as a pile penetrated the bearing stratum were generally unsuccessful.

Pile Load Test Program

Driving of Piles

The load test piles were located where the upper compact sand layer was only about 10 ft thick. This stratification appeared to be the most critical condition at the site. While these piles could have penetrated to the lower sand, there are locations where slightly greater thicknesses of the upper compact sand caused the piles to meet near refusal within this layer. There are numerous areas where borderline thicknesses of the upper sand yielded highly variable penetration resistances within the upper sand. Strict adherence to a high driving resistance criterion would have obviously created difficulties in estimating the lengths of these concrete piles. It would have been necessary to maintain a separate stockpile of variable-length concrete extensions as well as casting extra splice sections on pile tops or accepting a much greater waste in pile cutoff.

Load Test Piles Nos. 1 and 2 were driven within about 9 m (30 ft) of each other. These two piles are used primarily to illustrate this test program. Fig. 3 presents the driving resistance for these test piles as well as the SPT values for a boring about 4.6 m (15 ft) from the test piles.

Pile Instrumentation System

The piles were driven in three sections using the splice illustrated in Fig. 4. A 7.6 cm (3-in.) PVC pipe was cast centrally in the load test pile to permit installation of telltales. A full wheatstone bridge configuration of epoxy-molded electrical resistance strain gages was placed prior to casting

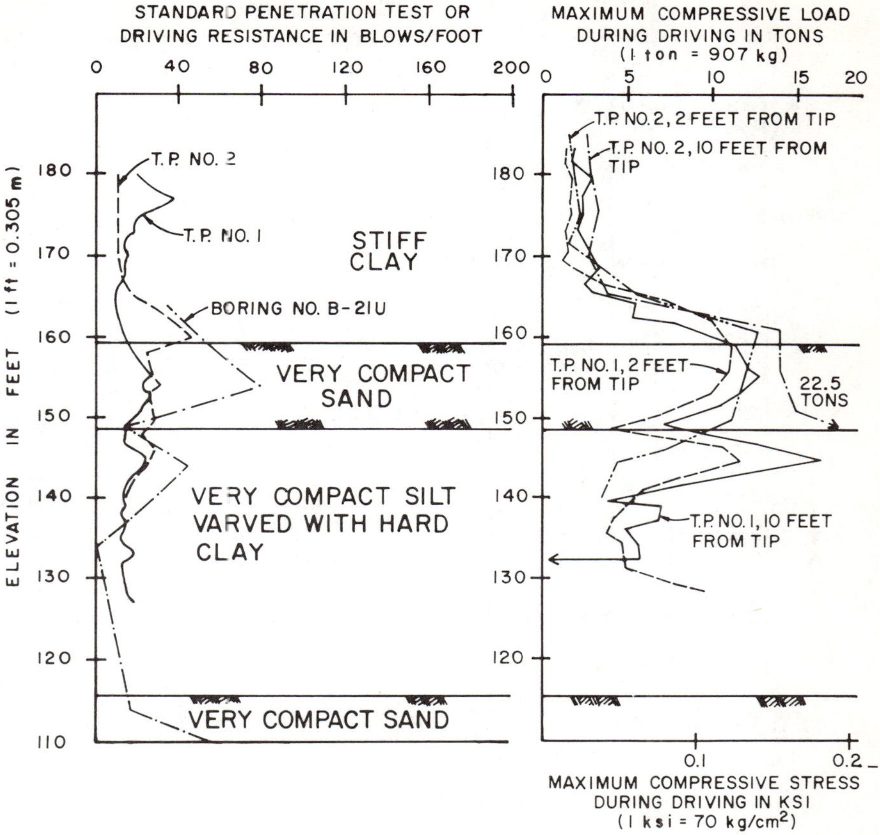

FIG. 3—*Load test pile driving resistances and stresses.*

each pile 0.6 and 3 m (2 and 10 ft) from the pile tip. Each group consisted of two active gages and two dummy gages isolated from concrete strain by encapsulation in a steel tube. The active gages were positioned and wired to cancel bending strains. The cable extended inside of the pile to a point about 0.9 m (3 ft) below the first splice. The point at which the cable protruded from the pile side was shielded by a steel angle. As the pile was driven, the cable was fastened to the pile with a metal clip at intervals using a Ramset gun. With some practice, the cable could be fastened without interrupting pile driving. Pile strain was monitored with a Gulton model no. TR 444 four-channel strip chart recorder during driving. Compressive load measured during driving two of the piles for the last 12.3 m (40 ft) is shown on the right-hand side of Fig. 3 and indicates a marked increase as the pile enters the bearing stratum. The readings also show the residual compressive load in Test Pile No. 2 at the end of driving near the top of the bearing

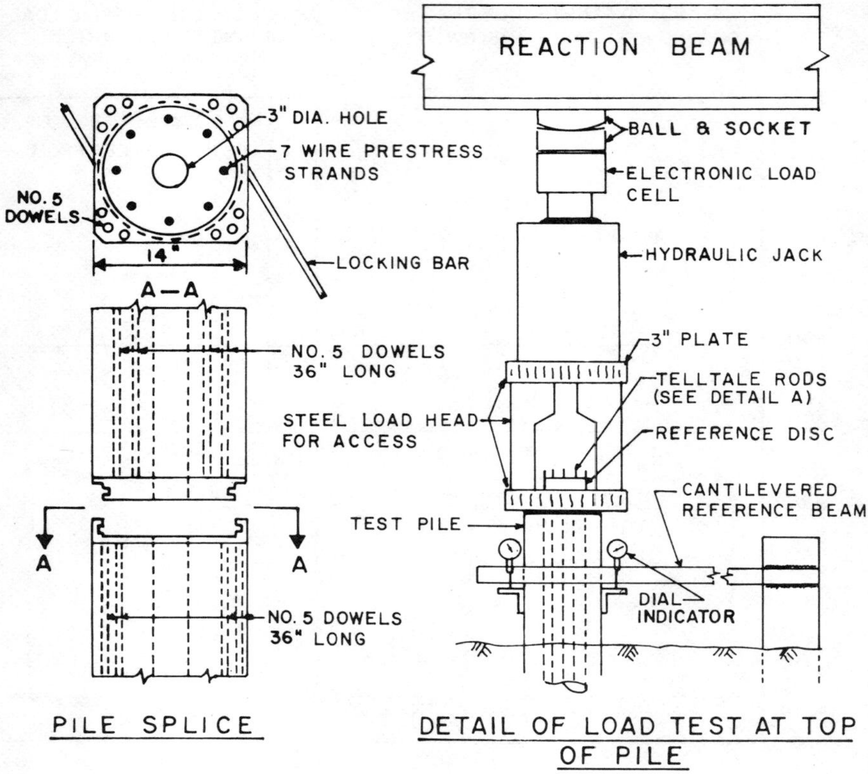

FIG. 4—*Pile load test details.*

stratum was approximately 20 Mg (22 tons). However, the indicated residual compressive load dropped to 2.7 Mg (3 tons) 2.5 m (8 ft) lower or 0.6 m (2 ft) above the pile tip, where the pile tip had penetrated into the varved silt stratum. In general, residual stresses following pile driving were low in the range of 2.1 to 15.4 kg/cm^2 (0.03 to 0.22 ksi).

Pile top movement was monitored using dial indicators and a taut wire reference and was checked by optical survey. The applied load was measured using a strain-gaged load cell and checked with the hydraulic jack readings. A ball-and-socket bearing plate assembly was used to insure loading parallel to the pile axis.

Deviation of Load Test Piles from Vertical

After each test pile was driven, measurements of the exposed pile top showed that none of the test piles exceeded the allowable 3 percent out-of-plumbness. A biaxial torpedo slope inclinometer was then lowered down the central pipe to determine the sweep in each pile. Fig. 5 illustrates the results. After the experience with Pile No. 1, the remaining piles were plumbed much more carefully before driving the initial pile sections. The amount of pile sweep decreased with the latter piles. Maximum pipe tip lateral displacement was 5.7 m (18.6 ft) for Pile No. 1.

Installation of Telltales

Telltale anchors were installed within the central pipe at various levels. The anchor and telltale configuration are shown in Fig. 6. After the telltales were lowered to the proper level, the surrounding space was cement grouted. Fig. 4 illustrates details of the top of the load test pile including

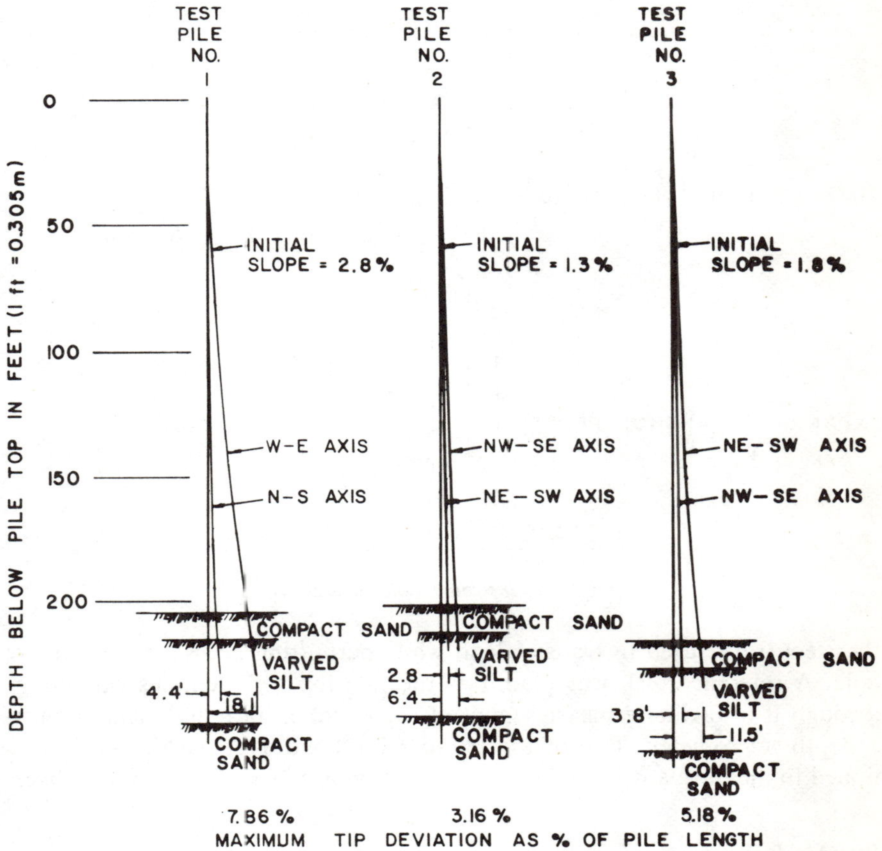

FIG. 5—*Deviation of load test piles from vertical.*

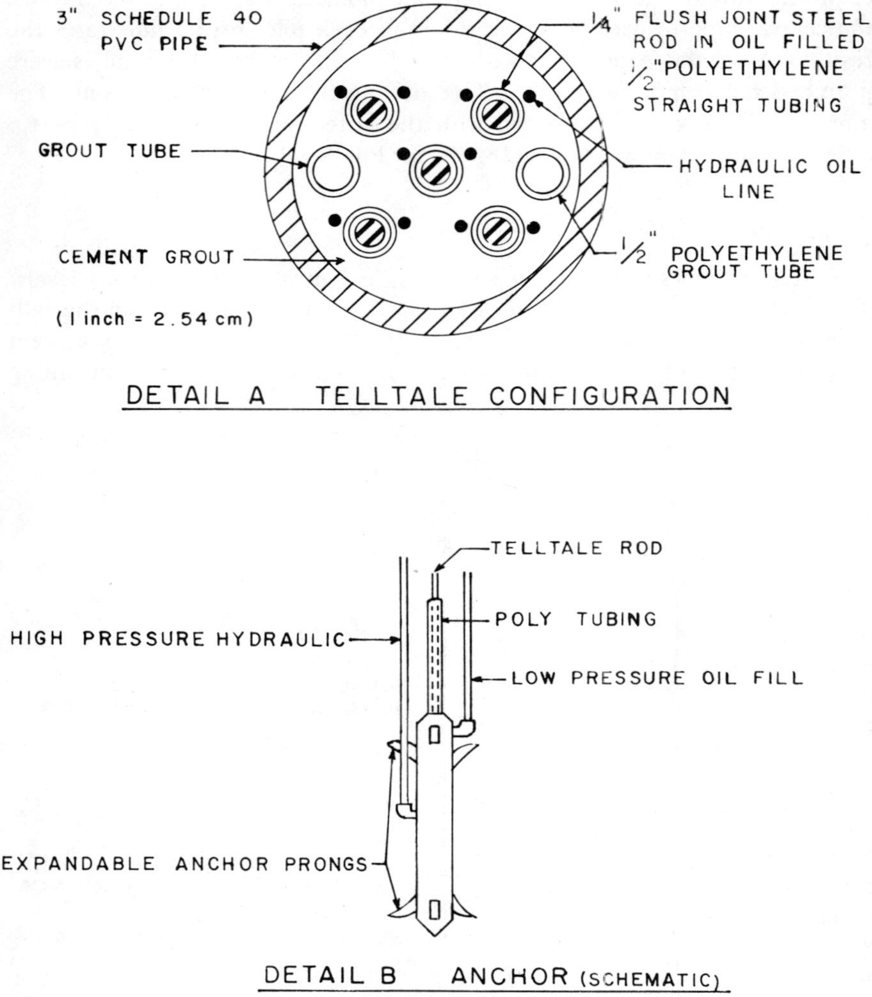

FIG. 6—*Instrumentation details.*

the steel frame used to transfer load while permitting access to the center hole. A reference disk was placed at the pile top with telltales extending through it to facilitate measurement of the variation in telltale length with a depth micrometer. Two of the telltales for Load Test No. 2 were positioned to measure any irregularity in deflection across the lowest pile splice.

Pile Modulus

Three-metre (10-ft) long sections of pile were cast with imbedded electrical strain gages as part of the same pour for both Load Test Piles Nos. 1 and 2. Concrete cylinders were also obtained from these pours. The pile sections were load tested at the University of Illinois. Variation in elastic modulus with concrete stress is shown in Fig. 7 for a 1.3-m (50-in.) mechanical strain gage and for the electrical strain gages. These tests were conducted approximately 1 and 2 months after the sections were cast or as close as possible to the time of the pile load tests. The tests on concrete cylinders 1, 2, and 3 months after casting also show higher modulus values with increasing time. The two moduli tests on full-size sections using the

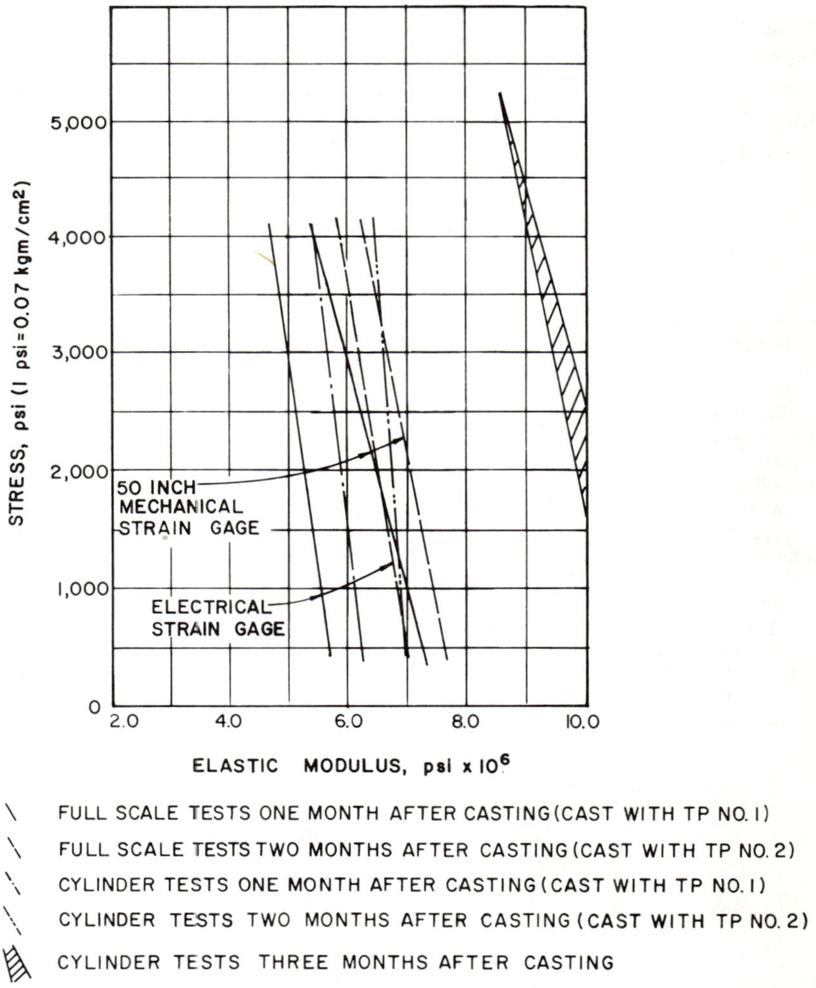

FIG. 7—*Elastic modulus of pile.*

1.3-m (50-in.) mechanical strain gage produced values that were in close agreement with the average of four 25.4-cm (10-in.) mechanical strain gages on each section. The lowest moduli for the 1.3-m (50-in.) gage were used to estimate pile load from the telltale movements.

Pile Load Test Results

The pile load test procedure was in general conformance with the ASTM Standard Test for Load-Settlement Relationship for Individual Piles Under Static Load (D 1143-69) [presently Testing Piles Under Axial Compressive Load (D 1143-74)]. Pile load was applied by jacking against a frame loaded with sheet piling. Results of pile load tests are summarized in Table 2. The magnitude of load reaching the bearing strata increased directly with the length of time after the pile was driven until the maximum load was attained. Figure 8 illustrates the telltale reading for Test Pile No. 2 as pile shortening between the pile top and the telltale location. A comparison of

TABLE 2—*Summary of pile load test data.*

	Test No. 1	Test No. 2	Test No. 3
Pile length, ft	237	221	234
Penetration in bearing stratum, ft	33	20	18
Final blows/ft	18	15	18
Horizontal deviation of tip, ft	18.6	7.1	12.4
Maximum test load	315	350	280
Top settlement, in.	1.55	1.59	0.84
After cycling load 30 times		2.12	
Tip settlement, in.			
Initial	0.60	0.35	0
After 30 cycles		0.63	
Estimated maximum load at top of bearing strata, tons			
Initial	184	0	0
After holding maximum load for 3 days		80	
After cycling load 30 times		165	
Time after pile was driven that maximum applied load was reached, h	49	79	186
Reason for termination	pile broke	350 tons cyclic load	Pile top cracked

Load-Deflection Summary, tons	Pile Top Deflection (in.)		
	Test No. 1	Test No. 2	Test No. 3
70	0.17	0.14	0.11
140	0.49	0.37	0.27
245	1.06	0.87	0.66
280	1.21	1.03	0.84
315	1.55	1.22	. . .
350	. . .	1.59	. . . .

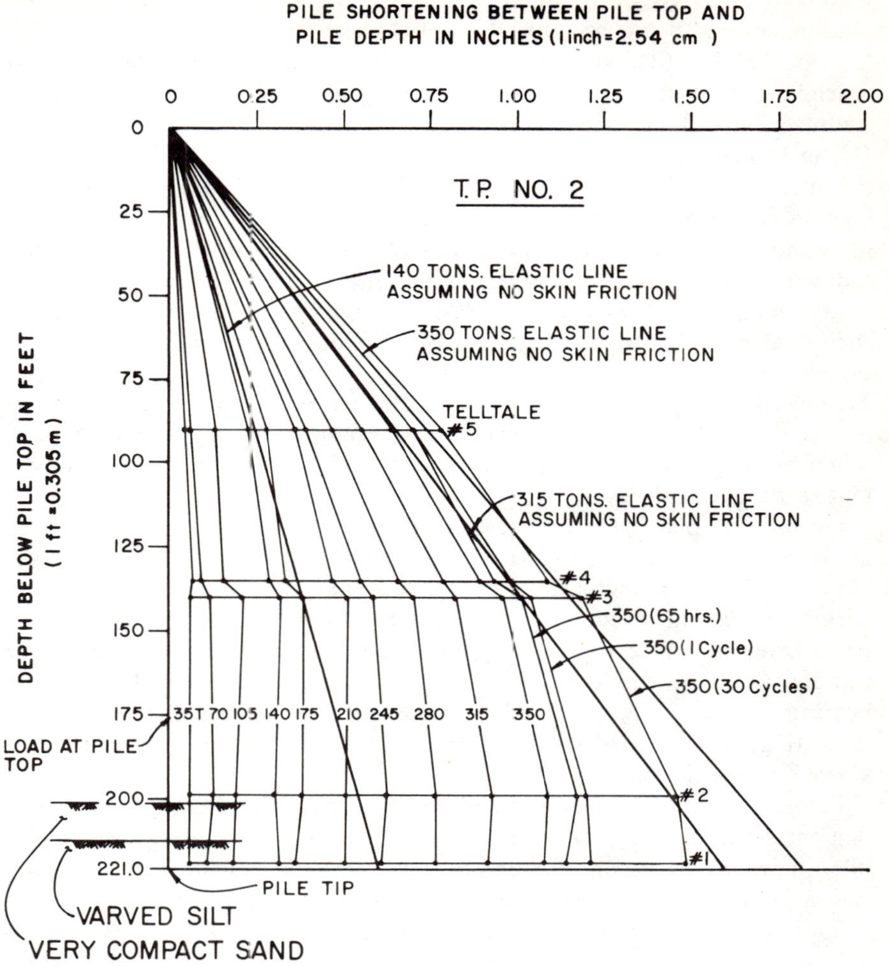

FIG. 8—*Pile shortening at various pile loading—test pile no. 2.*

the elastic line assuming no skin friction with the telltale readings at the several selected load levels illustrates the load lost into the compressible soils overlying the bearing strata. The elastic modulus used in computing the elastic line was obtained from Fig. 7 using the full-scale tests for the appropriate stress level. The slope of the line between telltales is an indication of the rate at which load is being lost into the soils above the bearing strata. The uniform progression of the telltale observations is an indication of the reliability of the results. The pronounced break between Telltale Nos. 3 and 4 indicates a closure of a small gap across the splice plate at a load of about 95 Mg (105 tons). At the 127-Mg (140-ton) load, almost no load penetrated below a depth of 43 m (140 ft). It was not until a pile top load of 222 Mg (245 tons) was attained that the telltales indicated load reached greater depths. At higher loads, the slope of the line between Telltales 2 and 3 increased.

The lack of apparent load between Telltales Nos. 1 and 2 is partially the result of load transfer into the very compact sand layer through side friction in the pile while the pile tip in the compact silt did not build up resistance as readily. In addition, residual tensile stresses from pile driving indicated a marked load loss between Telltales 2 and 1. As the pile top load was applied, these residual stresses would have to be overcome before the telltales would indicate any increase in pile loads at this level. As additional load was transferred to the pile tip, some load increase between Telltales 1 and 2 was recorded. The indicated tension at some load levels between telltales 1 and 2 are only partially accounted for by residual stresses. Slight binding of the telltales must be responsible for much of the inconsistent readings at this level.

During the pile load test, several methods were used to increase the amount of load reaching to the bearing strata. After applying the 318 Mg (350 ton) load, an increase in pile shortening was observed over a 65-h period, indicating a reduction by creep in adhesion in the overlying soils, transfering more load to a lower level. The pile load was then released and rapidly cycled 30 times, producing a substantial increase in load at the bearing strata. The telltale readings extend slightly beyond the elastic line, indicating that there was also some closure during loading in the splice above Telltale No. 5.

Figure 9 compares telltales readings for Test Piles Nos. 1 and 2 and demonstrates that, by conducting the pile tests sooner after driving the pile, there was less reconsolidation of soils above the bearing strata, permitting a larger load to reach bearing level. The 127 and 286 Mg (140 and 315 ton) applied load are typical of the other load increments.

Figure 10 shows the estimated load distribution throughout each pile length at maximum load. The plotted points represent estimates based on assuming uniform pile adhesion between telltales.

Analysis Method

Estimated pile load was calculated using three methods as identified in Fig. 10.

Method 1—The reduction in load from the top of the pile to the first telltale was calculated using the relationship

$$\Delta_x = \frac{(Q_t + Q_x)L_x}{2A_pE_p}$$

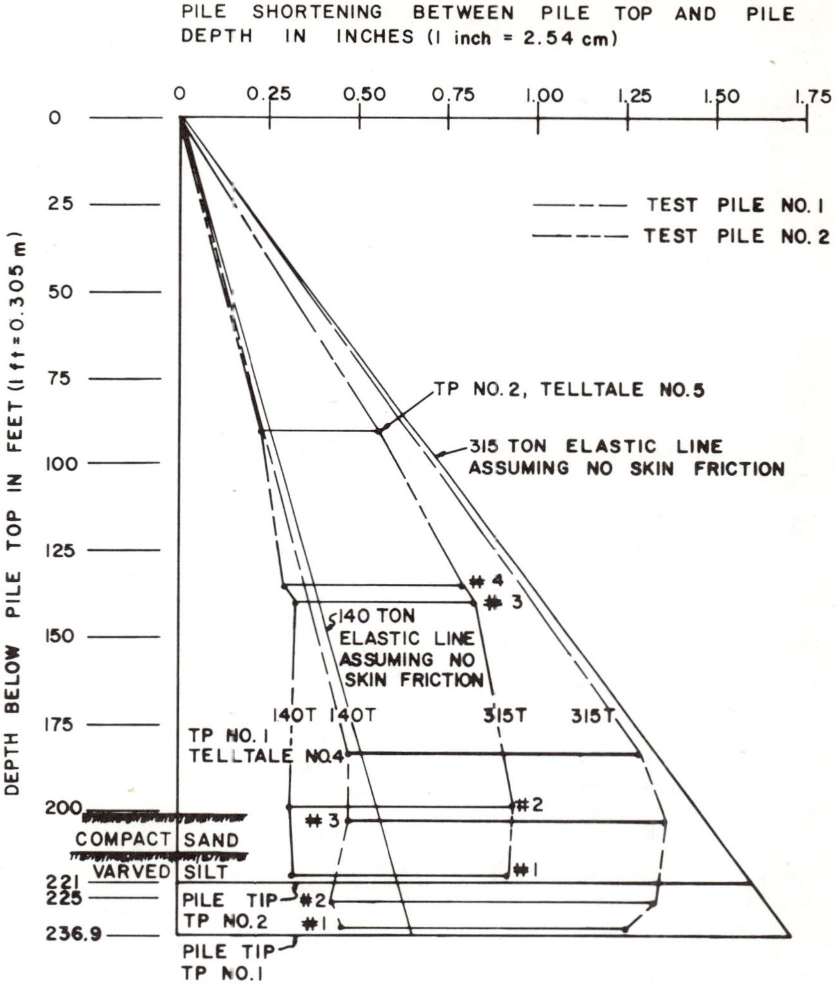

FIG. 9—*Comparison of pile shortening—test piles 1 and 2.*

where

Δ_x = elastic shortening of pile with uniform adhesion,
L_x = length between pile top and point of measurement,
A_p = pile cross-sectional area,
E_p = modulus of elasticity of pile,
Q_t = load at the top of pile, and
Q_x = load at the point of measurement.

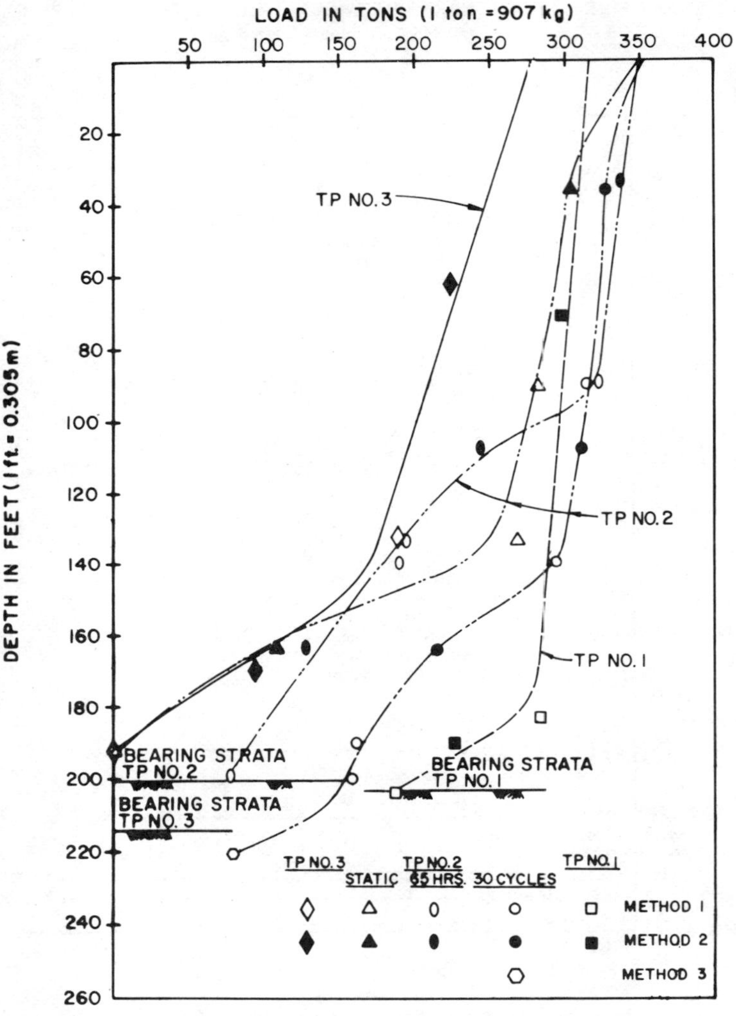

FIG. 10—*Load distribution in test piles at maximum load.*

The above formula can be expressed as

$$Q_x = \frac{\Delta_x \, 2A_p E_p}{L_x} - Q_t$$

Reduction in loading was then calculated progressively for each succeeding lower telltale. Using this formula, Q_x is calculated at each level and becomes Q_t for the next calculation.

Method 2—Loads at the midpoint between two adjacent telltales were calculated by using the difference between the two telltale movements and treating the pile as a load cell at that level.

Method 3—Loads near the pile tip were determined using electrical strain gages employing moduli from strain gages embedded in the full-scale test sections. These values proved useful as check against the lowest telltale measurements that appeared to be unreliable, particularly in Test Pile No. 1.

Multiple Telltales

A comparison of the use of more than one telltale versus a single telltale extending to the pile tip illustrates the variation in estimates of pile load reaching the top of the bearing strata using method 1. Telltale readings for Test Pile No. 2 were used after the 317-Mg (350-ton) pile top load was cycled 30 times (see Table 3). This variation in estimated load at the bearing level is the result of different assumed distribution of adhesion in the overlying thick deposit of low-strength clay.

Variation in Pile Adhesion

The mobilized pile adhesion during application of the maximum test load can be estimated at various depths from the slopes of the lines on Fig. 10. Figure 11 shows this variation in mobilized adhesion at 21 and 49 m (70 and 160 ft) depths for the three load tests plotted against time after the piles were driven. At the shallower depth, adhesion at maximum pile test load increased generally with increasing time. The values in parentheses are adhesion expressed as a percentage of estimated shear strength at this depth. Undrained shear strength of this soil is about 34 percent of effective overburden stress. While mobilized pile adhesion did not increase significantly at 49-m (160-ft) depth in Test Pile No. 1, much larger values were measured initially in Test Pile No. 2 at this depth. Holding the maximum load for 65 h caused a large drop in mobilized adhesion at this depth as more of the pile load was resisted near the 30.5-m (100-ft) depth and within the bearing strata. Cycling of the load resulted in a slight decrease in mobilized adhesion at 21-m (70-ft) depth, a pronounced decrease at 100-ft depth, with almost no change at 49-m (160-ft) depth.

TABLE 3—*Telltale readings for Test Pile No. 2 (load cycled 30 times).*

Number of Telltales	Estimated Load at Bearing Level
1 at pile tip	266 Mg (293 tons)
1 at pile tip and 1 at midpoint	181 Mg (200 tons)
5 actually used	150 Mg (165 tons)

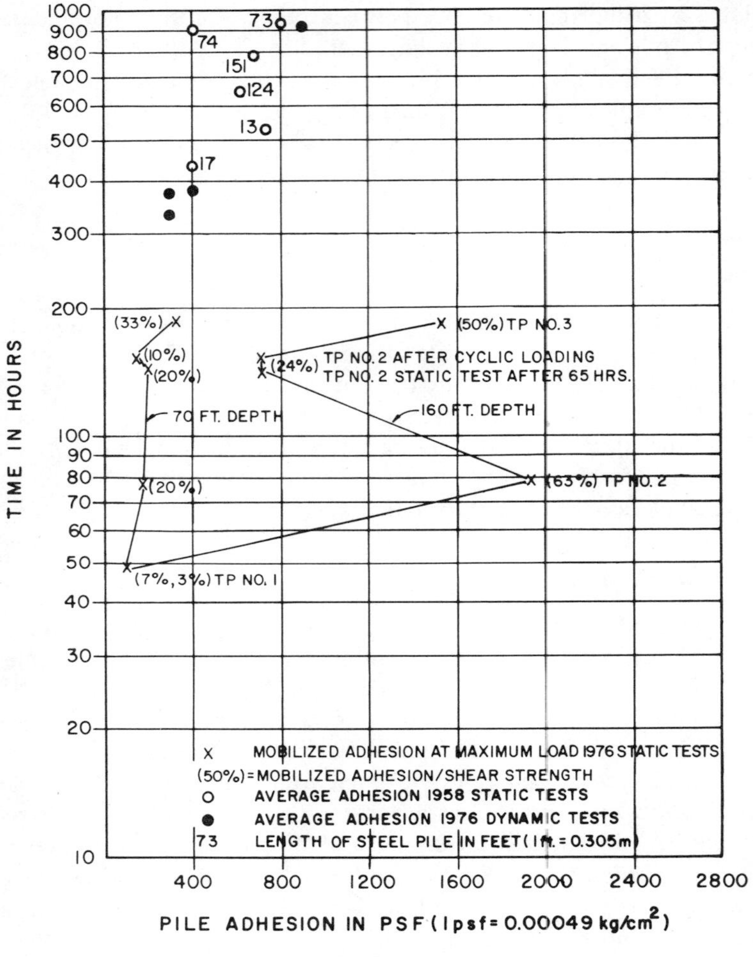

FIG. 11—*Average pile adhesion versus time.*

Figure 11 also shows estimates of pile adhesion during redriving resulting from measurement of dynamic response of piles from instrumentation attached near the pile top. In 1958, several steel shell and pipe piles were driven to varying penetrations into the clay layer without reaching the bearing strata. Average pile adhesion determined from these tests are, as expected, below the trend indicated for the concrete piles. It can be concluded from this plot that mobilized pile adhesion increased from 0.07 times the shear strength two days after Test Pile No. 1 was driven to 0.33 times the average strength of the clay five days after Test Pile No. 3 was driven at 21-m (70-ft) depth. Mobilized adhesion at lower levels varied widely, possibly due to the presence of sand lenses.

Piles that experienced an interruption in driving while their tips were still above the bearing strata drove at increasingly higher resistances for larger interruption periods. This trend is illustrated in Fig. 12. Most of these piles were penetrating the upper soils at a resistance of five to 10 blows per foot prior to interruption of driving. Even a delay of 15 min to an hour while the pile was spliced resulted in a marked increased in driving resistance for the first foot of driving, with progressively reduced driving resistances for continued driving thereafter, eventually approaching the original low values. The wide variation in driving resistance is a result of

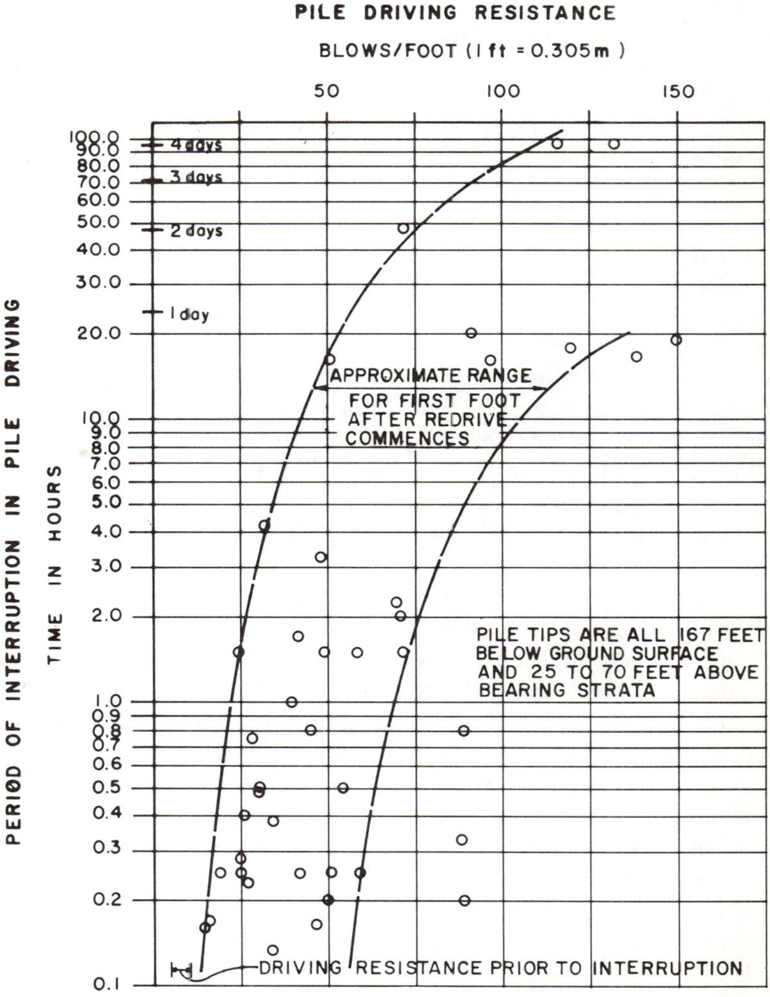

FIG. 12—*Redrive resistance versus time with pile tip in clay.*

erratically positioned layers of sand and silt that tend to reconsolidate around piles much more rapidly than the dominant clayey material.

Conclusions

This test program demonstrated that it is possible for pile load to be transferred through low-strength clays to a deep bearing strata during a short-term pile load test. The following conclusions can be derived from this program:

1. Long displacement piles driven into compact sands and sandy silts generated high pore pressures during driving, resulting in highly variable driving resistance at the same pile tip elevation, often within short horizontal distances. Load tests have demonstrated that the design load can be supported by piles driven into this strata to low final resistance.

2. Residual stresses following pile driving were low within the bearing strata based on electrical strain gage readings. These residual stresses were neglected during analysis of loads applied to these test piles. The decrease in residual stress within the bearing strata measured in Test Piles Nos. 1 and 2 may have affected results slightly by indicating very little load transferred to the pile tip. The magnitude of these residual stresses were too small, however, to cancel out all of the indicated pile tension at the lowest telltales. The erroneous results must have been caused by binding between the telltale rods and the oil-filled polyethylene protective tubing, even though high pressures were maintained on the oil during grouting of the anular space around the tubing. These readings also suggest that the load applied at the top of the bearing strata is resisted primarily by the upper very compact sand layer. Higher loading may have demonstrated substantial reserve load capacity at the pile tip. The applied load was limited by the structural capacity of the pile.

3. Even with applied pile loads at 3.5 times design load, little load reached the bearing strata 3 days after the pile was driven, as indicated by the best instrumented test pile results.

4. The earlier the load test began after the pile was driven, the lower the adhesion of clay on the pile sides, permitting increased load to reach the bearing strata. This proved to be the most effective means of transferring larger loads to the bearing strata.

5. Holding the maximum test load for 3 days resulted in a moderate increase in load at the bearing strata, presumably because of creep in the adhesion resistance.

6. Cycling the applied load also shifted pile resistance to lower levels accompanied by additional tip deflection.

7. Use of multiple telltales produced much smaller estimates of load at the bearing strata than would have been estimated from a single telltale at the pile tip due to variation in pile adhesion with depth.

8. An accurate measurement of the elastic modulus of the pile was essential to make a reasonable estimate of load at the bearing strata. Use of the measured variation in modulus with pile strain resulted in higher estimated loads at lower elevations than if an average modulus had been used.

9. Pile adhesion increased with depth for all test piles, as shown in Fig. 10. When pile load tests were performed after a longer interval, Load Test Piles Nos. 2 and 3 indicate that pile adhesion had increased most in the lower part of the clay stratum.

10. Mobilized pile adhesion shortly after driving varied from 0.07 to 0.33 times the shear strength at a 21 m (20 ft) depth. These values fall within the range anticipated for deep clay deposits but are still somewhat below values measured by others [*13*], presumably longer after the piles were driven.

11. It is essential in planning an instrumented pile load test program to provide backup components, so that if one part of the system does not function properly, useful information can still be obtained from the test.

Acknowledgments

The writer appreciates the cooperation and assistance of Onondaga County Department of Public Works; O'Brien and Gere Engineers, Inc., the designer and construction manager; Slattery Associates and Underpinning and Foundation Constructors, Inc., the contractors; M. T. Davisson and William Gamble, consultants to the contractors; Goldberg, Zoino and Dunnicliffe, designers and installers of most of the pile instrumentation; G. G. Goble and Associates, consultants on dynamic instrumentation and several of the partners and staff of the writer's firm. This test program was carried out only through the diligent effort of many people in these organizations too numerous to name.

References

[*1*] Vesić, A. S., North American Conference on Deep Foundations, Mexico City, Soil Mechanics Laboratory, Publication No. 3, Dec. 1964.

[*2*] Vesić, A. S., *Journal of the Soil Mechanics and Foundations Division,* American Society of Civil Engineers, Vol. 96, No. SM2, Mar. 1970, pp. 561-584.

[*3*] Cummings, A. E., Korkhoff, G. O., and Peck, R. B., *Journal of the Soil Mechanics and Foundations Division,* American Society of Civil Engineers, Vol. 74, Dec. 1948, pp. 1553-1563.

[*4*] Orrje, O. and Broms, B. B., *Journal of the Soil Mechanics and Foundations Division,* American Society of Civil Engineers, Vol. 93, No. SM5, Sept. 1967, pp. 59-73.

[*5*] Seed, H. B. and Reese, L. C., *Transactions,* American Society of Civil Engineers, Vol. 122, 1957, pp. 731-754.

[*6*] Lambe, T. W. and Horn, H. M., *Proceedings, 6th International Conference on Soil Mechanics and Foundation Engineering,* Montreal, Vol. II, 1965, pp. 285-289.

[*7*] Lo, K. Y., and Stermac, A. G., *Canadian Geotechnical Journal,* Vol. 1, No. 2, March 1964, p. 63.

[*8*] Bjerrum, L., Brinch Hansen, J., and Sevaldson, R., Publication No. 28, Norwegian Geotechnical Institute, Oslo, 1958, pp. 1-17.
[*9*] Bjerrum, L., and Johannessen, I., *Proceedings, Conference on Pore Pressure and Suction in Soils,* London, March 1960, pp. 108-111.
[*10*] Broms, B. B., and Hellman, L., *Journal of Soil Mechanics and Foundations Division,* American Society of Civil Engineers, Vol. 94, No. SM2, March 1968, pp. 421-429.
[*11*] Reese, L. C., *Journal of Soil Mechanics and Foundation Division,* American Society of Civil Engineers, Vol. 104, No. GT1, Jan. 1978, pp. 91-116.
[*12*] "Design Manual—Soil Mechanics, Foundations and Earth Structures" NAVFAC DM-7, March 1971, Department of the Navy.
[*13*] McClelland, B., *Terzaghi Lectures 1963-1973,* American Society of Civil Engineers, Ninth Terzaghi Lecture, Oct. 1972, pp. 379-421.

G. E. Lamb[1]

Special Requirements for Testing Auger-Placed Grout Piles

REFERENCE: Lamb, G. E., "**Special Requirements for Testing Auger-Placed Grout Piles,**" *Behavior of Deep Foundations, ASTM STP 670*, Raymond Lundgren, Ed., American Society for Testing and Materials, 1979, pp. 381-387.

ABSTRACT: Experience in testing auger-placed grout piles using ASTM Testing Piles Under Axial Compressive Load (D 1143) indicates that modification of the testing methods for this type of pile can improve interpretation of test data and provide greater confidence in the test results. Much of the load testing specified for auger-placed piles is intended to evaluate soil-pile interaction as with driven piles. Other load testing is often specified for auger-placed piles mainly to ensure continuity of the piles. This paper compares pile test procedures and results in a number of auger-placed pile installations and recommends modifications to the testing methods to take the special nature of these piles into account.

KEY WORDS: auger-placed grout piles, concrete piles, load testing

Auger-placed grout piles have been used in the United States for a number of years. The installation technique consists of drilling a hole with a hollow-stem flight auger to approximately the planned tip depth. At that point, the auger is withdrawn a few inches and the plug that seals the discharge port near the tip of the auger is expelled. Grout is then injected while the auger is again advanced to tip depth and is turned for a time without being withdrawn. Once a sufficient head of grout has been developed on the outside of the auger, the auger is slowly and evenly withdrawn. The auger is turned throughout the process. Grout pump pressure is maintained at a level that will keep pressure in the system and will sustain a suitable head of grout above the tip of the auger.

The auger-placed procedure has many advantages, including drastic reduction in vibrations during pile installation, freedom from corrosion or decay, and economy. In recent years, the cost and other advantages of this type of piling has led to its increasing use.

[1]Associate, Dames and Moore, Seattle, Washington 98125; presently president, Fugro Northwest, Inc., Seattle, Wash. 98115.

Unfamiliarity with the technique and uncertainties in the evaluation of piles as installed, have led many designers to specify a much larger number of pile load tests for auger-placed piles than would be specified for driven piles in a similar situation. These load tests are normally performed using ASTM Testing Piles Under Axial Compressive Load (D 1143) or some variation of this method. Experience with this technique leads me to conclude that simply requiring additional standard load tests is not the most effective or the most certain method of ensuring quality piles.

Proper testing for piles, whether augered or driven, should evaluate both the pile-soil interaction and the quality of the pile itself. In the case of driven piles, both these conditions are covered by ASTM standards and industry tests and standards. The materials and fabrication of the pile are subject to strength and performance tests, and the configuration of the piles are set either by relatively standard design practices or by industry and ASTM standards. In the case of auger-placed piles, however, only the pile-soil interaction is covered by a suitable ASTM standard. This, combined with the uncertainties in evaluating augered pile performance outlined above, has contributed to a situation where an economical, often-desirable pile type is either subjected to an inappropriate number of load tests or in many cases is not specified at all due to the difficulty of ensuring proper performance. This paper recommends additional standards to provide for improved testing of this type of piles.

Background—Present Pile-Testing Techniques

Driven Piles

ASTM Method D 1143 was developed for the load testing of driven piles. It has provided good and consistent results in the evaluation of the performance of driven piles. Some of the reasons for the good performance of the test method in the evaluation of driven piles are:

1. It is generally possible to provide test piles for load testing that are quite similar to planned production piles. The characteristics of the test pile are well known and can be taken into account in evaluating the test results.

2. The other aspects of quality control for driven piles (other than load testing) are well established, well understood, and are a routine part of any well-controlled pile-driving operation. These include careful inspection of the piles before driving; comparison of these piles with established standards of quality; limitations on the handling of piles; specifications for the types of hammers, hammer cushions, and pile cushions to be used; tip or butt reinforcing under specified circumstances; and standardized and well-understood methods for inspecting the driving of the piling.

3. Problems during the installation of driven piles are usually visible to careful inspection. These include shaking and splintering in the case of wood piles, head damage and spalling in the case of concrete piles, and irregular or unusual behavior during driving.

From the above, it can be seen that driven piles are "tested" at each pile location for structural soundness and performance under dynamic loads. This is in addition to the physical inspection of the pile prior to driving and any certifications or test results testifying to the quality and integrity of the pile itself.

The purpose, therefore, of load tests on driven piles is almost entirely to evaluate the static soil-pile interaction. It was for this purpose that ASTM Method D 1143 was developed, and it is this purpose that that test method serves very well.

Auger-Placed Grout Piles

Auger-placed grout piles are more difficult to evaluate in nearly every respect than driven piles:

1. Even with very careful inspection, it is difficult to ensure that test piles are typical of production piles. In any case, some of the pile characteristics (for example, Young's modulus) may be varying, even over the period of a load test.
2. The auger-placed piles are created in the field and are not visible or accessible for direct inspection.
3. Problems during installation of auger-placed piles are not so self-evident and easily observed as is the case with driven piles.

Pile-Soil Interaction

In the evaluation of the performance of the pile-soil system for auger-placed piles, several difficulties present themselves. The value of end bearing support is difficult to estimate because of possible disturbance in the tip area during early stages of casting and because the soil beneath the tip has not been consolidated by driving. Skin friction along the pile is also difficult to estimate, largely because of the difficulty in evaluating the friction factor at the concrete-soil contact and because of low and variable lateral pressures as compared with displacement-type driven piles. An additional problem in the evaluation of the soil-pile system is the lack of driving record at each pile location. In the case of driven piles, this provides a crude penetration test to guard against wide variations in soil conditions, while no such check exists for auger-placed piles.

In addition to the problems associated with the pile-soil system indicated above, some difficulties are present in the evaluation of the auger-placed pile itself. These include the possible poor quality of concrete, since no visual inspection of the finished product is possible and the lack of a loading of the finished, cured product during installation. A problem unique to cast-in-place piles is the possibility of "necking" or even discontinuity in the pile as cast. This is most common in weak soils below the water table and is generally caused either by failure to maintain proper pressure on the grout during withdrawal of the auger or by overrapid withdrawal of the auger. Another problem associated with the pile itself is the difficulty in assuring the centering and serviceability of reinforcing, which is inserted into the green concrete grout immediately after installation.

Typical Results

Tests run using ASTM Method D 1143-69T (and modifications) at two locations in the Seattle area illustrate the generally good performance of ASTM Method D 1143 in evaluating pile–soil interaction and the need for supplemental methods and techniques to ensure high-quality augered piles.

East Plaza Parking Garage

On this project, 0.4-m (16-in.) diameter augered piles were designed to support loads of 73 and 82 metric tons (80 and 90 short tons). Soil conditions across the site were quite variable, ranging from glacially consolidated sand and clay at pile-cap depth (where the piles were omitted and the caps redesigned to act as footings) to locations where up to 8 m (26 ft) of fill, debris, and disturbed clay were encountered above the hard, glacially consolidated, clay supporting layer. Because of the irregular soil profile, extensive soil explorations by test borings and auger probes were undertaken before and during installation of the piles. This represents some of the worst conditions under which augered piles would normally be used.

A load test was performed on a test pile installed to a total depth of 15 m (48 ft). Seven meters (23 ft) were in loose fill, 1 m (3 ft) in disturbed clay, and the lower 7 m (22 ft) in hard, undisturbed clay. Reaction forces for the test were provided by eight augered reaction piles. A 0.9 m (36-in.) wide-flange beam transferred the load to the reaction piles. Loads were applied by three identical manifolded hydraulic jacks. Deflection readings were taken from two dial gages. A taut-wire system was used to check early readings and was maintained as a backup system for use in case of problems with the gage system.

The load test was performed in accordance with ASTM Method D 1143-69T and was carried to 200 percent of the design load for a maximum load of 145 metric tons (160 short tons). Evaluation of this test by several techniques led to the conclusion that 73 metric tons (80 short tons) was a suitable design load, and further evaluation permitted estimation of appropriate depths of penetration into the bearing strata for other locations on the site. Extrapolation permitted the estimation of appropriate penetrations for the piles with higher capacities (82 metric tons, 90 short tons), which later became necessary due to design changes.

The wealth of information that was derived from the test was made somewhat less reliable by the large number of variables present. In this case, these included:

1. Uncertainties as to the true Young's modulus in the concrete at the time of the test.
2. Associated uncertainties in estimation of elastic behavior of the pile under load.
3. Difficulties in evaluating the time-deflection performance of the upper fill. (There was evidence that recycling the load caused a significant transfer of load from the upper fill to the underlying soil.)
4. Indications that end-bearing actually plays an important but difficult to quantify role in the support of the pile.

In this case, difficult and variable soil conditions were compensated for by additional testing and careful evaluation of load test data. Building loads have never reached full design loadings due to low utilization of the garage. However, no distress has been observed in the completed structure.

Here the load test provided a suitable guide for design modifications and permitted a good, although limited, evaluation of pile-soil interaction. It did not, however, provide such good guidance as to the quality or configuration of the piles installed or even of the test pile itself.

Sears Roebuck Catalog Order Plant Addition

On this project, 0.3-m (12-in.) diameter augered piles 11 m (36 ft) long were selected to support design loads of 32 metric tons (35 short tons). Two pile loads tests were run to evaluate the accuracy of assumptions as to pile-soil interaction and to investigate the possibilities of using shorter piles for the same design load or of increasing the design load on the 11-m (36-ft) piles. One pull-out test was also conducted to aid in the evaluation of the end-bearing contribution to pile capacity.

Soil conditions at the site were quite uniform. The upper 1.2 to 1.5 m (4 to 5 ft) of soil consist of fill. This is underlain by loose to moderately dense fine to medium sands. The water table was at 2.4 to 3.4 m (8 to 11 ft) below the ground surface.

Load tests were performed on piles at opposite sides of the planned building area. Reaction loads were provided by four anchor piles at each test site. Loads were applied using a single hydraulic jack with a spherical bearing. Deflection readings were made using three dial gages supported on two independent register beams. The test areas were shaded by large awnings to avoid distortions caused by heating by the summer sun.

The load tests were performed in accordance with ASTM Method D 1143-69T, except that the final loading to failure was performed using the constant rate of penetration (CRP) technique, which was not a part of that edition of the method. The technique used in the final loading was very similar to that described in ASTM Method D 1143-74 as the CRP or quick test.

Good agreement was found between the two load tests, and it was concluded that the originally planned depths and design loads should be continued in use. In this case, uniform soil conditions permitted considerable confidence in the soil profile, frequent test breaks of grout cubes provided a reasonably good method for estimating Young's modulus during the test, and unobstructed site conditions permitted construction of production piles in a manner very similar to that used on the test piles. Conditions on this project were very advantageous to the economic, confident use of augered piles. Full design loads will be applied to the piles within the next few months, when the building is extended to its full planned height of nine stories from its present level of seven stories. To date, no distress has been observed in the structure.

Discussion

From my experience with the design, installation, and testing of augered piles, I am convinced that modifications to ASTM Method D 1143 and additional standards are desirable to provide for the proper testing of these useful piles. These include both items that are appropriate for consideration by ASTM and items that I believe are best left to the individual professional judgment of designers and specifiers. A number of considerations of both types are indicated below.

Recommendations

A new ASTM standard should be developed to cover the testing of augered-in-place piles to ensure their structural integrity.

1. The standard should encourage more extensive use of constant rate of penetration (CRP) tests or quick tests as described in ASTM Method D 1143-74 in cases where the test is intended only to evaluate pile continuity. Load tests should be run to failure if at all possible where continuity is an issue. In these cases, the CRP or quick test is quite suitable and much more convenient.

2. Other techniques for ensuring the continuity of the pile should be

evaluated. These would include reflected-impulse techniques or other methods of ensuring pile continuity. If a reliable technique for checking continuity is developed and proven, it should be included in ASTM Method D 1143 or in a separate standard.

3. Procedures for pull-out tests should be included in ASTM Method D 1143. Because of the uncertainties discussed above, detailed pull-out tests are often very desirable to evaluate pile–soil behavior of augered piles.

The following items, while possibly not appropriate for development into ASTM standards, should be considered for inclusion in standards for the proper construction of augered piles.

4. Provisions for a more intensive program of soil testing or other assurance of uniform soil conditions to minimize the uncertainties associated with discontinuous or irregular soil profiles.

5. Techniques for ensuring that an appropriate volume of grout is injected into each pile. This should include appropriate and approved methods of calibration of pumps, etc.

6. Techniques for ensuring that an adequate head is present outside the auger during its withdrawal.

7. Provisions for evaluation of pressure at the pump and effective pressure at the discharge end of the grout system.

8. Strength testing of grout to ensure that the in-place concrete quality is of the specified standards. It should also provide for frequent cube breaks or cylinder tests to assist in the estimation of Young's modulus during load testing.

9. Provisions for ensuring that reinforcing cages, if inserted into the pile, will be properly centered and will provide the expected structural performance.

G. A. Leonards[1] *and D. Lovell*[1]

Interpretation of Load Tests on High-Capacity Driven Piles

REFERENCE: Leonards, G. A. and Lovell, D., **"Interpretation of Load Tests on High-Capacity Driven Piles,"** *Behavior of Deep Foundations, ASTM STP 670,* Raymond Lundgren, Ed., American Society for Testing and Materials, 1979, pp. 388–415.

ABSTRACT: Pipe piles and H-sections about 35 m (115 ft) long were driven through 24 m (80 ft) of alluvial materials to bearing in a stratum of glacial till about 15 m (50 ft) thick overlying bedrock. Load tests were conducted to establish the equipment and driving resistance needed to develop a capacity of 3560 kN (400 tons) solely in the underlying tills. Interpretation of these tests proved to be a challenge that led to the development of a new technique to analyze load transfer using only data obtained from conventional pile load tests. Among the conclusions reached were (1) ultimate shaft friction of steel piles driven into very stiff to hard tills did not exceed 120 kPa (1.25 tsf); (2) for piles as large as 460 mm (18 in.) in diameter, ultimate shaft friction is not necessarily mobilized before the pile capacity is reached; and (3) loading and unloading a pile can cause irreversible changes in the existing as well as on the ultimate shaft friction—methods of interpreting pile load tests to determine load transfer or residual stresses that rely on these changes being negligible, should be used with caution.

KEY WORDS: piles, load tests, ultimate capacity, shaft friction, load transfer, residual stresses

Nomenclature

- A Cross-sectional area of pile (L^2)
- b Arbitrary constant
- c Ratio of pile compression due to a butt load supported entirely by shaft friction to the compression caused by the same butt load supported entirely by point bearing {also, the ratio of the average stress to the maximum stress in the pile for this loading condition}
- c' Ratio of pile compression to the compression caused by the same butt load supported entirely by point bearing (also, ratio of average stress to maximum stress in the pile)

[1] Professor and graduate assistant, respectively, School of Civil Engineering, Purdue University, Lafayette, Ind. 47906.

E Young's modulus (F/L^2)
e Base of natural logarithms
F_s Load supported by shaft friction (F)
F_{s_m} Ultimate shaft friction force (F)
f_s Unit shaft friction (F/L^2)
k Arbitrary constant
L Pile length (L)
m Arbitrary constant
n Slope of butt load versus pile compression diagram times AE/L
P Load on pile butt (F)
P_m Load on pile butt at which either shaft friction or point bearing is fully mobilized (F)
P_{ult} Ultimate pile capacity (F)
P_T Load on pile tip (F)
$P_{T\,ult}$ Ultimate point bearing load (F)
α Ratio of tip load to butt load
κ Constant of integration
λ Constant of integration
δ Displacement of pile butt (L)
δ_T Displacement of pile tip (L)
Δ_e Elastic compression of pile $= (\delta - \delta_T)$ (L)
Δ_{RT} (Upward) displacement of pile tip when the butt is completely unloaded (L)
Δ_s Elastic (recoverable) compression of soil beneath the pile tip $= (\Delta_{RT})$ (L)
Δ_{sp} Plastic (irrecoverable) compression of soil beneath the pile tip (L)
σ Normal stress (F/L^2)
ω Arbitrary constant

Tests on axially-loaded piles are performed either to check the adequacy of production piles or, in the design stage, as an aid in selecting the most efficient pile design, including type, size, length, and method of installation. In the latter case, attention is given to three main considerations:

1. *Ultimate capacity of the pile*, which is used in establishing the embedded lengths and number of piles required in order to achieve a desired factor of safety with respect to bearing capacity. Changes in pile capacity with time after driving may result from dissipation of excess pore water pressures induced during driving and from thixotropic regain in strength of clay soils after remolding. As driving resistance is often the primary means of controlling installation, it is important to assess the difference between the capacity during driving and in the longer term.

2. *Settlement of the pile.* Depending on the preference of the designer, an estimate of pile group settlements may be made and compared with

those considered tolerable for the structure, or settlements may be controlled by increasing the factor of safety with respect to bearing capacity in accordance with prior experience.

3. *Load transfer*, that is, the portion of the butt load carried in point bearing and shaft friction, respectively. The mechanism of load transfer is an important factor influencing the settlement of a single pile and the ratio of the settlement of a pile group to that of a single pile, at the same load per pile [*1*].[2] *Changes in load transfer* under sustained loading generally involve a transfer of shaft friction to point bearing due to creep in the soil adjacent to the shaft or to ground subsidence induced by placing fill, lowering the groundwater table, or remolding the clay around the piles. For these reasons, it may be of primary importance to establish the ultimate pile capacity in point bearing alone.

This paper is concerned with the interpretation of load tests conducted on high-capacity driven piles. The site conditions consisted of 1.5 to 3 m (5 to 10 ft) of slag and cinder fill; 11 to 12 m (35 to 40 ft) of stratified sands, clayey sands, and gravelly sands with seams and pockets of soft (organic) clays; about 12 m (40 ft) of medium stiff silty clay containing thin softer seams; 12 to 15 m (40 to 50 ft) of very stiff to hard glacial tills with an intertill sheet of stiff tough clay. All are underlain by hard dolomite bedrock. Thin remnants of sandstone, limestone conglomerate, or extremely dense sandy silt with cobbles were occasionally encountered between the lowest till sheet and the dolomite. It was desired to develop a design load of 1780 kN/pile (200 tons/pile), with a factor of safety (FS) = 2 against ultimate pile capacity, and a load test program was initiated to

1. verify that the proposed methods for installing the piles could achieve the necessary pile penetrations,
2. choose between two proposed types of piles—an H-section or a pipe pile, and
3. establish the driving resistance (penetration per blow) to be specified so that the desired 3560 kN (400 tons) could be supported entirely by the underlying tills.

The load tests were carried out under the terms of a conventional construction contract at a predetermined price and time to completion. Routine tests were to be conducted with measurement of butt load, butt displacement, and tip displacement (with a telltale). The piles were to be loaded to 3560 kN (400 tons) following ASTM Testing Piles Under Axial Compressive Load (D 1143); after unloading, they would be reloaded to failure, or to 5340 kN (600 tons), which was the maximum kentledge provided. Interpretation of these tests proved to be a challenge that led to the development of an improved technique for analyzing load transfer using only data from

[2] The italic numbers in brackets refer to the list of references appended to this paper.

conventional load tests. The results obtained stimulated the authors to review previous measurements on instrumented piles and caused them to abandon some cherished notions concerning load transfer in driven piles.

Ultimate Pile Capacity

Pile capacity may be viewed either in terms of "failure," that is, rapid progressive settlement at constant load, or as a load satisfying a particular settlement criterion, such as a total settlement of one inch; or twice the load at which the net settlement is 6.35 or 12.7 mm (0.25 or 0.50 in.); or the load at which the total settlement per unit load is 0.03 mm/kN (0.01 in./ton), or when the incremental settlement per unit load is 0.086 mm/kN (0.03 or 0.05 in./ton); or some similar combination of load and component of settlement. Each approach has its uses, especially when correlated with local experience; however, the fraction of the failure load obtained from any settlement criterion depends greatly on the pile type, diameter, length, and method of installation and on the soil conditions at the site. Accordingly, it would be useful if foundation engineers could agree to restrict use of the term "ultimate pile capacity," or "ultimate load," to the maximum load the pile can support before plunging; if the maximum pile load is controlled by limiting the settlement, it could be called the "limiting load."

If it is intended to assess long-term capacity, it is axiomatic that tests should be performed after transient effects due to installation have dissipated. In granular soils, this is a matter of a few days to a week. For friction piles in soft to medium clays, a period of 3 to 6 months may be required. Few data are available in the case of piles driven into stiff to hard clays. Sherman [*2*] adopted "a minimum waiting period" of 2 weeks; Roth [*3*] stated there were no additional effects after 5 to 6 weeks. In the present case, the contract permitted testing piles 7 days after driving.

Many pile tests are performed in which the ultimate load (as previously defined) is not reached. The test may be stopped at twice the design load, or it may be found that the pile continues to sustain increasing loads even at (relatively) large deflections. In granular soils, this may be due to continued dilation and/or crushing of sand grains beneath the tip of the pile. In clay soils, it is commonly caused by consolidation and creep that occurs during the test, which can be minimized if a constant rate of penetration test [*4*] or a Texas quick-load test [*5*] is performed. In any case, a reliable and consistent method of determining the ultimate load is required.

van der Veen [*6*] was probably the first to deal with this problem in a rational manner. He postulated that the relation between the tip load P_T and the tip deflection δ_T could be expressed by

$$P_T = P_{T\,ult}(1 - e^{-\omega\delta_T}) \tag{1}$$

where

$P_{T\,ult}$ = ultimate pile capacity at the tip,
e = base of natural logarithms, and
ω = arbitrary constant.

If it is assumed that a relation of this form is also applicable to the butt load P and butt displacement δ, then

$$\ln (1 - P/P_{ult}) = -\omega\delta \tag{2}$$

where P_{ult} = ultimate capacity of the pile. Accordingly, a value of P_{ult} is selected and $\ln (1 - P/P_{ult})$ is plotted versus δ. The correct value of P_{ult} has been assumed when a straight line plot is obtained.

Davisson [7] assumed that the ultimate pile load is reached when the butt deflection

$$\delta = \Delta_e + \Delta_s + \Delta_{sp} \tag{3}$$

where

Δ_e = elastic compression of the pile,
Δ_s = elastic compression of soil at the pile tip, and
Δ_{sp} = limiting plastic compression of soil at the pile tip.

Davisson proposed that

Δ_e = calculated compression assuming point bearing only
Δ_s = 2.54 mm (0.10 in.) ("typical" quake observed in driving piles about 0.3 m (1 ft) in diameter)
Δ_{sp} = 3.81 mm (0.15 in.)

In a manner analogous to Kondner's [8] pioneering work, Chin [9] proposed that, as an approximation, the butt load versus butt displacement could be expressed by a hyperbola of the form

$$P = \frac{\delta}{m\delta + b} \tag{4}$$

where the asymptote $P_{ult} = 1/m$ and b is a constant.

Transposing terms,

$$\delta/P = m\delta + b \tag{5}$$

hence, a plot of δ/P versus δ is a straight line whose slope $m = 1/P_{ult}$.

The three methods for estimating P_{ult} will now be applied to the results obtained from the pile load tests.

Test Pile No. 1

A pipe pile 356 mm (14 in.) outside diameter and 8.74 mm (0.344 in.) wall thickness was driven with a Vulcan 016 single-acting hammer [66.1 kJ (48 750 ft-lb) rated energy] using a 324 mm (12¾ in.) diameter steel mandrel with 17.5 mm (0.688 in.) wall thickness and a slidable external boot (Fig. 1*a*) at the pile tip. Prior to driving, a 330 mm (13 in. diameter slurry-stabilized hole was prebored to a depth of 21.3 m (70 ft). A cap block and helmet, a striker plate, and a laminated aluminum-micarta cushion were used. The pipe was filled with concrete and a load test performed 7 days after driving. The test length of the pile was 34.7 m (113 ft 8 in.) and the embedded depth was 34.3 m (112.5 ft); driving was stopped at a final set of 0.43 blows/mm (11 blows/in.). Deflections at the butt and tip were measured directly with dial gages. The unconfined compressive strength of the concrete after 7 days was 35 200 kPa (5 100 psi), from which the composite modulus of the concrete-filled steel pipe was calculated to be 4.45×10^7 kPa (6.45×10^6 psi) [*10*]. Figure 2*a* shows the log of a boring taken adjacent to the pile, the driving record, and the load-settlement diagrams obtained from the test.

Figure 3*a* shows van der Veen's plot from which P_{ult} = 1780 kN (200

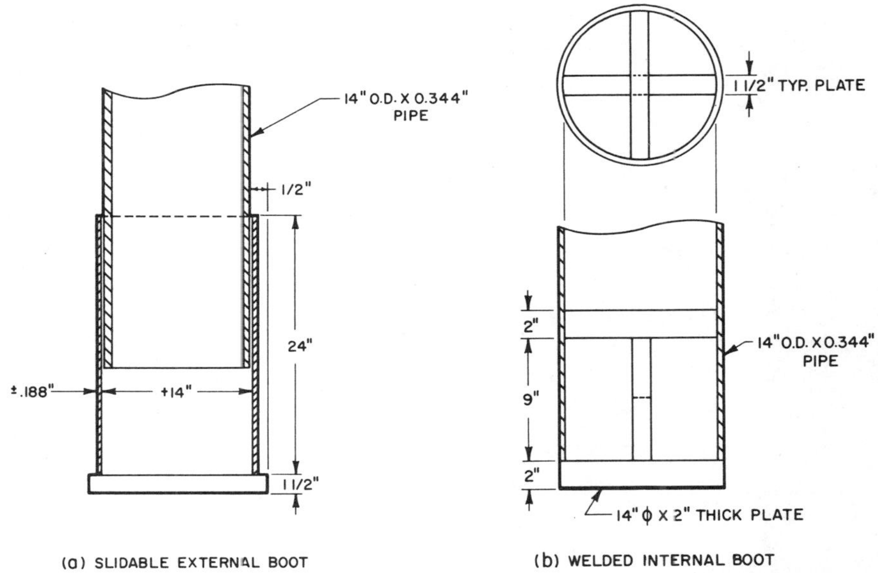

FIG. 1—*Boots used for mandrel-driven pipe piles.*

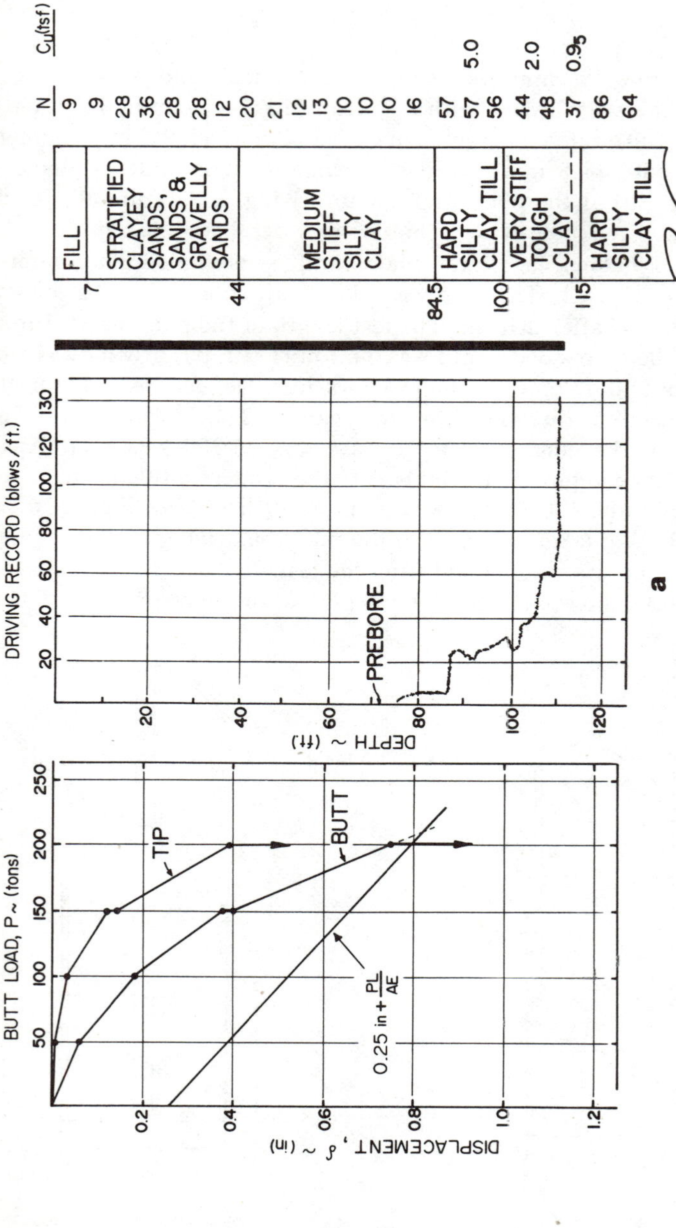
BUTT LOAD, P ~ (tons)
DISPLACEMENT, ρ ~ (in)
TIP
BUTT
0.25 in + PL/AE
DRIVING RECORD (blows/ft.)
DEPTH ~ (ft.)
PREBORE
FILL
STRATIFIED CLAYEY SANDS, SANDS & GRAVELLY SANDS
MEDIUM STIFF SILTY CLAY
HARD SILTY CLAY TILL
VERY STIFF TOUGH CLAY
HARD SILTY CLAY TILL
N
C_u(tsf)
a

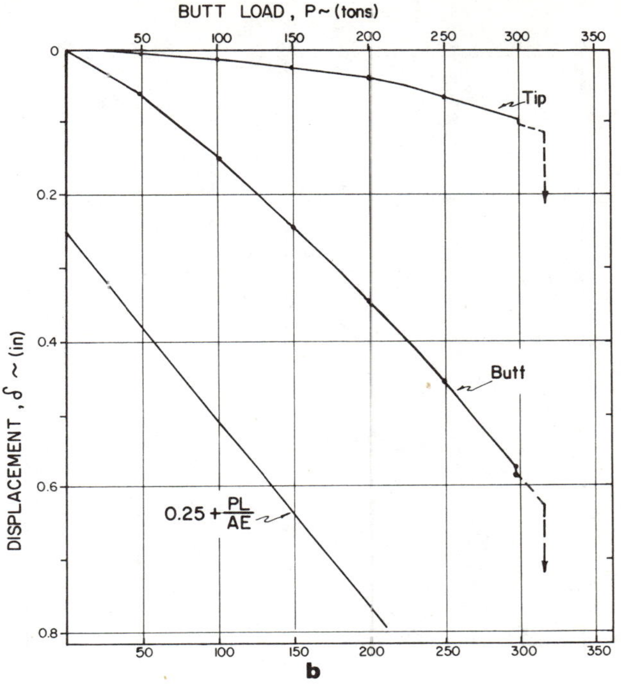

FIG. 2—*Test pile no. 1: 14-in. pipe pile;* (a) *initial loading,* (b) *retest.*

tons) was obtained. Applying Davisson's method (Fig. 2*a*) with $\Delta_s + \Delta_{sp} =$ 6.35 mm (0.25 in.) gave P_{ult} = 1870 kN (210 tons). Figure 4*a* shows Chin's plot from which it was deduced that P_{ult} = 2220 kN (250 tons). When the load increment from 1330 to 1780 kN (150 to 200 tons) was first applied, the deflections stabilized after about 30 min, but the jack load had dropped slightly below 1780 kN (200 tons). The jack pressure was brought back up to 1780 kN (200 tons), and while it was possible to maintain the load at 1780 kN (200 tons), the pile continued to penetrate several inches into the ground. By definition, the ultimate pile capacity was 1780 kN (200 tons).

Retest—Pile No. 1 was retested 29 days after the initial load test (36 days after driving),[3] with the results shown in Fig. 2*b*. When an attempt was made to apply the 2670 to 2890-kN (300 to 325-ton) load increment, the pile started to plunge; it penetrated 152 mm (6 in.) into the ground, and

[3] With f_c' now equal to 44 200 kPa (6400 psi), the composite modulus was calculated to be 4.76×10^7 kPa (6.9×10^6 psi).

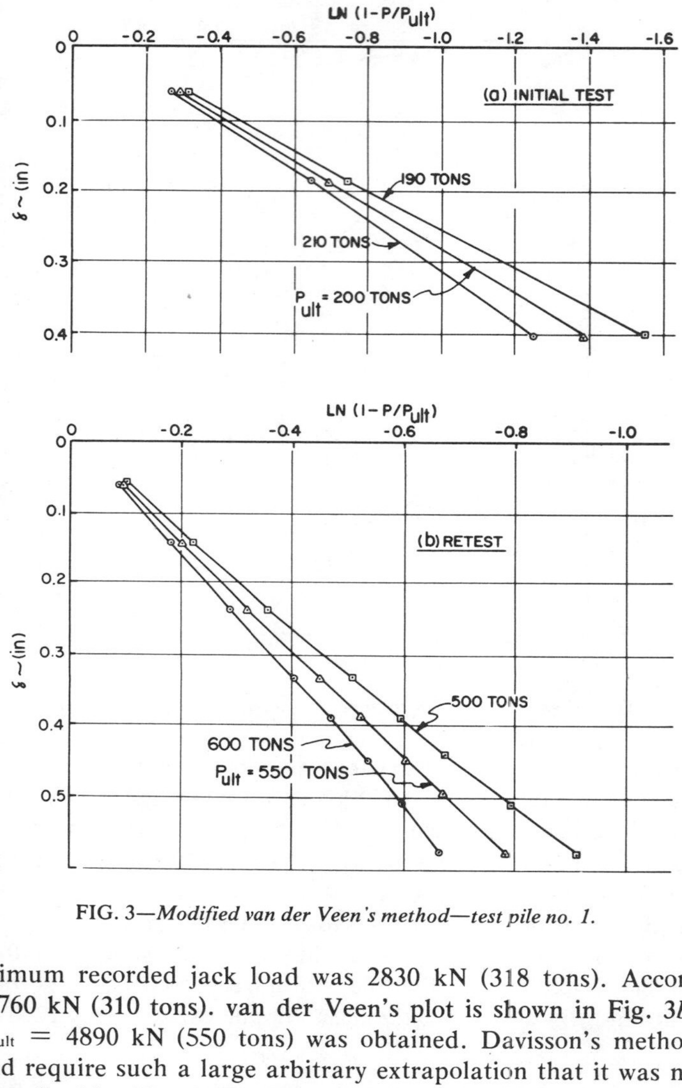

FIG. 3—*Modified van der Veen's method—test pile no. 1.*

the maximum recorded jack load was 2830 kN (318 tons). Accordingly, $P_{ult} \simeq 2760$ kN (310 tons). van der Veen's plot is shown in Fig. 3*b*, from which $P_{ult} = 4890$ kN (550 tons) was obtained. Davisson's method (Fig. 2*b*) would require such a large arbitrary extrapolation that it was not considered applicable. Figure 4*b* shows Chin's plot, which gave $P_{ult} = 8900$ kN (1000 tons).

A comparison of the three methods used to estimate pile capacity is given in Table 1. In the case of the retest, if the loading had been stopped at $P = 2670$ kN (300 tons) (97 percent of P_{ult}), none of the methods would have made a reasonable prediction of the ultimate pile capacity.

The results obtained from test pile no. 1 posed a number of questions:

1. How much of P_{ult} was provided by the till?

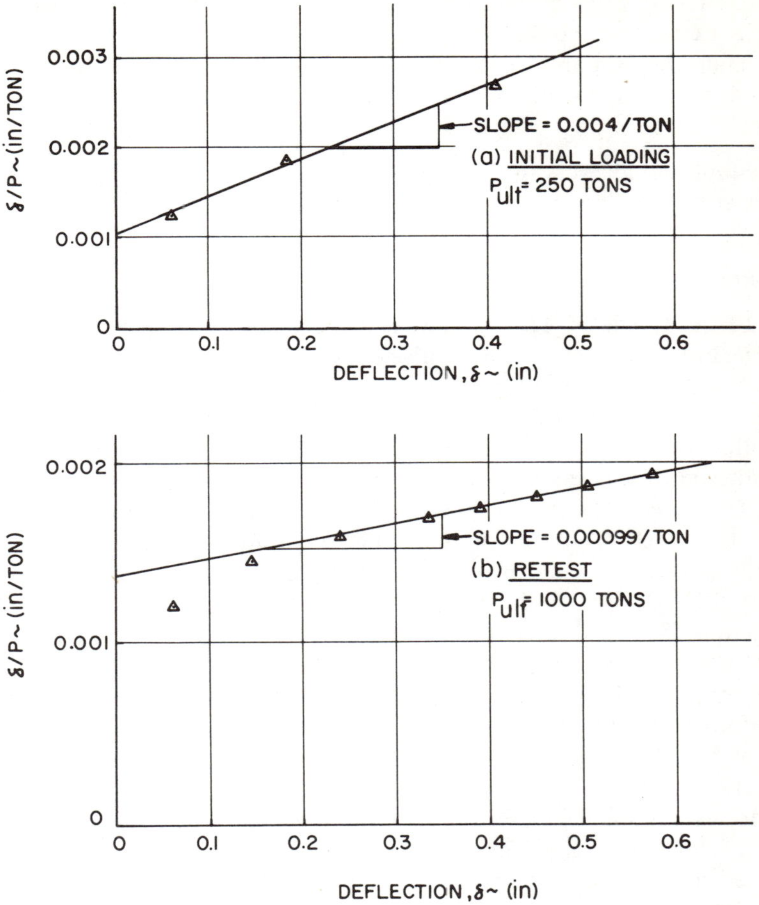

FIG. 4—*Chin's method—Test pile no. 1: 14-in. pipe.*

TABLE 1—*Comparison of methods for estimating pile capacity—Pile No. 1.*

Method	Initial Test (7 Days), tons	Retest (36 Days), tons
van der Veen (modified)	200	550
Chin	250	1000
Davisson	210	...
Load test	200	310

Conversion factor—1 ton = 8.9 kN.

2. How long after driving should load testing be delayed to ensure that transient effects in the till were not a factor?
3. Did the slidable external boot contribute materially to time-dependent effects, and did its use involve a substantial loss in driving energy?

To help obtain answers to these questions, it was decided to instrument the next test pile and measure load transfer directly.

Test Pile No. 2

A 14 in. H-pile at 117 lb/ft was instrumented with pairs of weldable electrical resistance strain gages, five gage pairs on 3-m (10-ft) centers and five on 4.6-m (15-ft) centers on the lower and upper portions of the pile, respectively. The gages were nulled prior to driving with the intent of recording the driving stresses as well as the residual stresses in the pile after driving, as it was felt that the latter could be significant in this case [*11*]. To protect the gages 130 mm (5 in.) channels were welded on both sides of the web. Telltale slots reaching to within 610 mm (2 ft) from the pile tip were provided. The pile was driven without a prebore using the Vulcan 016 hammer; when the driving resistance reached 0.2 to 0.24 blows/mm (5 to 6 blows/in.) no response was received from the lower gages, and shortly thereafter it was observed that the protective channel was free of the H-pile web above the ground surface. Subsequently, no response was obtained from any of the gages. Driving was continued until a set of 1.3 blows/mm (33 blows/in.) was reached at a pile penetration of 37 m (121 ft).

Load testing was begun 8 days after driving. Figure 5 shows the log of an adjacent boring, the driving record, and the load-settlement diagrams obtained from the test. At a load of 5340 kN (600 tons), a lateral shift of the pile head was noticed [at this load the nominal stress in the steel was nearly 241 000 kPa (35 000 psi)]. No attempt was made to estimate P_{ult}, but from the standpoint of soil support, it exceeded 5340 kN (600 tons).

It was desired to instrument another pile using more rugged protection for the strain gages, but this was not possible within the framework of the contract. Instead, an 26.4 m (80 ft) long H-section identical to pile no. 2 was driven just short of the hard glacial till and load tested 15 days later; this pile started to plunge at a load of 2180 kN (245 tons) ($= P_{ult}$).

Test Pile No. 3

Pile no. 3 was identical in every respect to pile no. 1, except that a welded internal boot was used (Fig. 1*b*). At a set of 0.43 blows/mm (11 blows/in.) the pile had penetrated 36.9 m (121 ft) but driving was continued until the set was 2.4 blows/mm (60 blows/in.) at a penetration

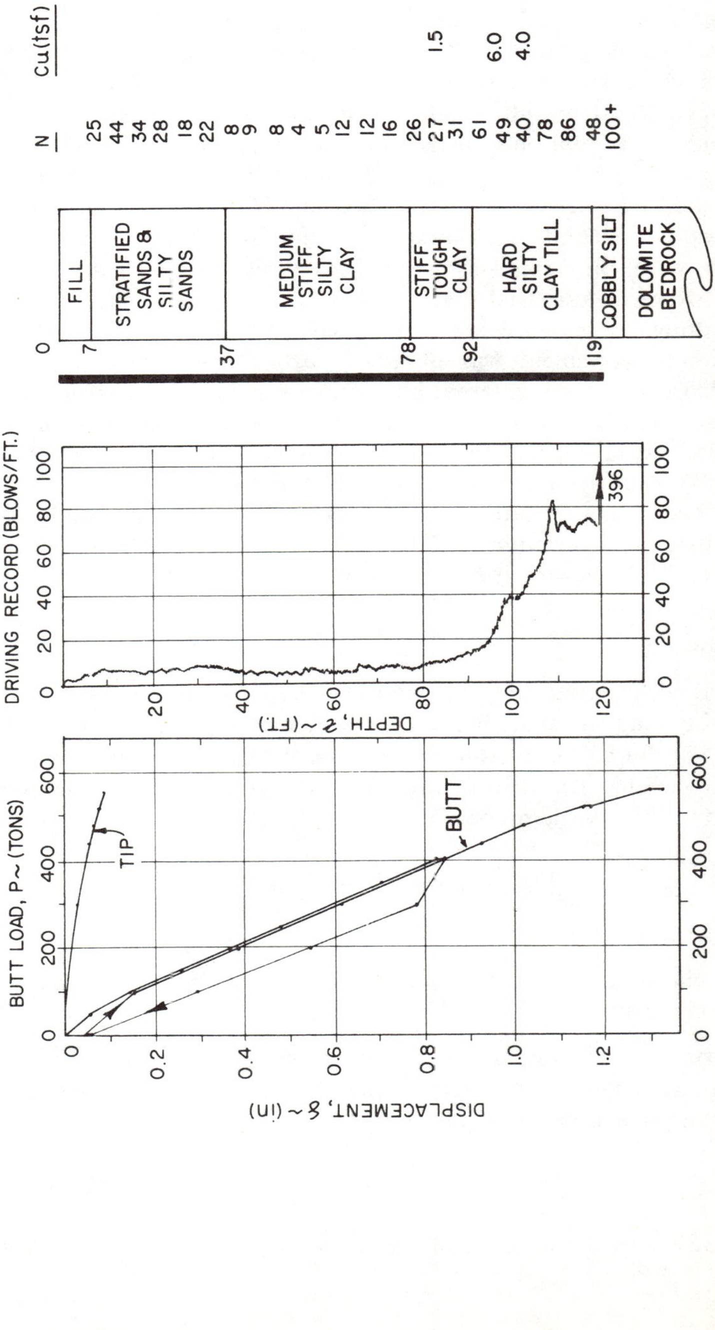

FIG. 5—*Test pile no. 2: 14-in. H-section.*

of 37.5 m (123 ft). Figure 6 shows the log of a boring taken adjacent to the pile, the driving record and the load-settlement diagrams obtained from a load test conducted 18 days after driving. Again, from the standpoint of soil support, P_{ult} was >5340 kN (600 tons).

During the course of the load-testing program, attempts were made to calculate the developed load transfer. Samples of the clay tills were obtained with a Dennison sampler, and the undrained shear strengths were determined from carefully-conducted unconfined compression tests (Figs. 2, 5, and 6). Tomlinson's correlations, as extended by Vesić [*12*] and by Vijayvergiya and Focht [*13*], were duly considered, but the well-documented experiences of Lo and Stermac [*14*] and of Sherman [*2*]—in which values of unit shaft friction essentially equal to the undrained strength were measured—suggested that this approach be used with caution. Nevertheless, trial shaft friction distributions were postulated and the resulting pile compressions compared with the measured values until approximate agreement was obtained. From these calculations it was concluded (*a*) that the external boot initially reduced shaft friction in the till substantially and that at least 5 weeks were required to approach the long-term resistance, and (*b*) that the average ultimate unit shaft friction in the till was of the order of 153 kPa (1.6 tons/ft^2).

Subsequently, the authors made a more intensive study of the possibilities for determining load transfer from the results of conventional load tests. The following is a summary of their findings.

Load Transfer

A method for separating the point-bearing force from the butt load and hence obtaining the shaft friction force F_s was originally proposed by van Weele [*15*]. Based on measurements of tip load P_T, and tip displacement δ_T, on concrete piles driven through weak soils to bearing in relatively dense sand, van Weele proposed that

$$P_T = k\Delta_s \qquad (6)$$

where

Δ_s = elastic (recoverable) compression of soil at the pile tip, and
k = constant

To determine Δ_s the pile is unloaded completely at successive stages in the load test, and if it is assumed that the residual stresses in the pile due to loading and unloading can be neglected,[4] then

$$\Delta_s = \Delta_{RT} \qquad (7)$$

[4]By measuring the pile tip load with a transducer, van Weele showed that this was the case for the conditions of his test piles in Amsterdam.

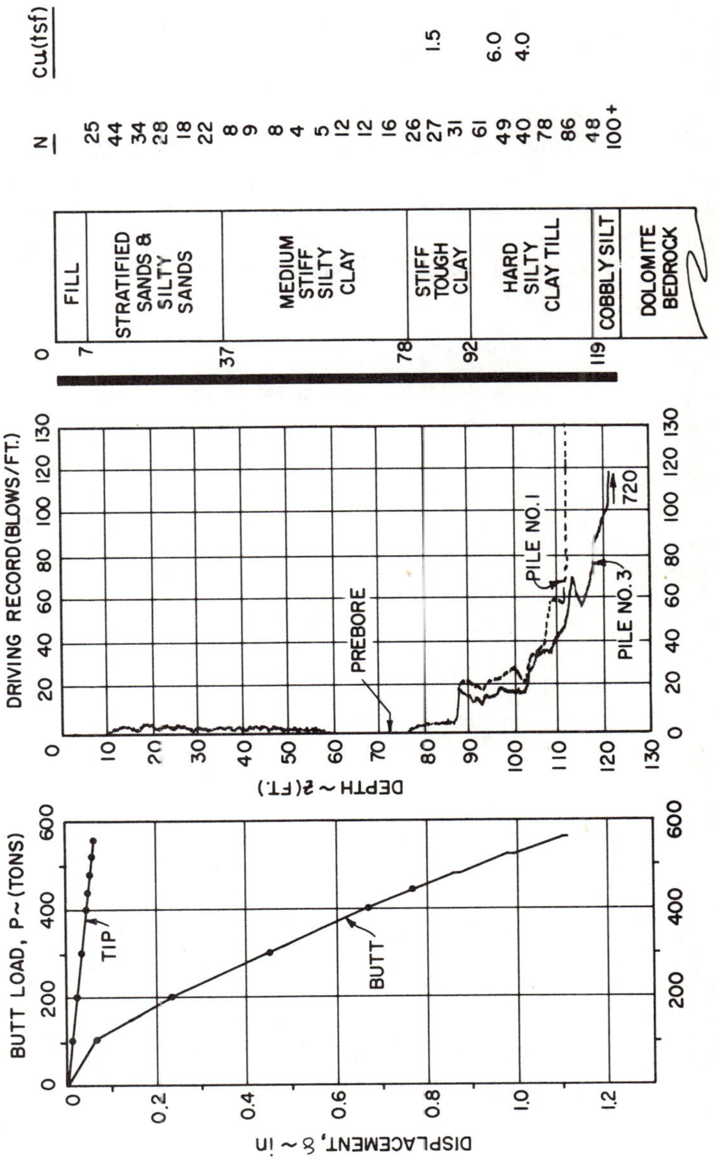

FIG. 6—Test pile no. 3: 14-in. pipe.

where Δ_{RT} is the measured recovery of the pile tip after each unloading. Van Weele recognized that if the shaft friction force F_s was mobilized before the ultimate pile capacity was reached, the increment in butt load would equal the increment in tip load if F_s remained constant. Therefore, a plot of butt load P versus Δ_{RT} (Fig. 7) would become a straight line after F_s was fully mobilized. From Eq 6, a parallel straight line through the origin establishes P_T (and hence F_s) at all values of P. Unfortunately, in 1957, it was not common to use telltales when conducting routine pile load tests, and van Weele felt obliged to recommend that Δ_{RT} (and hence Δ_s) be obtained by subtracting the calculated elastic compression of the pile from the measured butt rebound. This required knowledge of the magnitude and distribution of the unit shaft friction, which van Weele proposed be obtained by correlation with the results of Dutch cone penetration tests.[5]

Mechanics of Pile Compression

At any butt load P, tip load P_T, and shaft friction force F_s, the elastic compression of a pile Δ_e can be expressed as

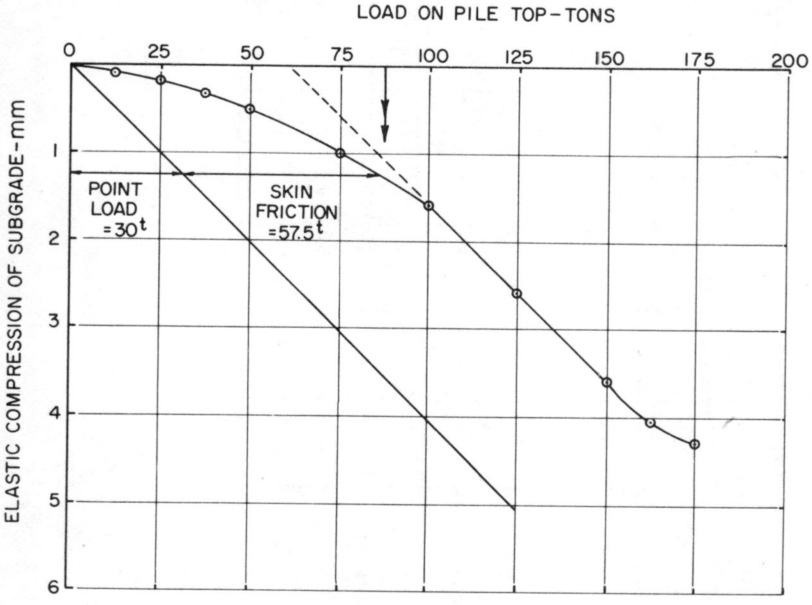

FIG. 7—*Method of separating point load from shaft friction (after van Weele, 1957).*

[5]Rutledge [*16*] suggested that telltale measurements give Δ_{RT} directly, thereby obviating the need to calculate Δ_e, but the utility of his remarks seems not to have been realized.

$$\Delta_e = c' \frac{PL}{AE} = \frac{P_T L}{AE} + cF_s \frac{L}{AE}$$

or

$$\frac{AE}{L} \Delta_e = c'P = P_T + cF_s \tag{8}$$

where

A = cross-sectional area of the pile (to conserve space, the principles are derived assuming the pile has a constant cross-section),
L = pile length,
E = modulus of elasticity of the pile,
c' = ratio of elastic compression of the pile at a butt load P to the elastic compression if all the load P were supported by point bearing only, and
c = ratio of elastic compression of the pile due to a butt load F_s supported entirely by shaft friction to the elastic compression if F_s were supported by point bearing only.

Since

$$P = P_T + F_s \tag{9}$$

substituting Eq 9 into Eq 8,

$$P(1 - c') = F_s(1 - c) \tag{10}$$

Letting $P_T = \alpha P$, then from Eq 9, $F_s = (1 - \alpha)P$, and substituting for F_s in Eq 10,

$$\alpha = \frac{c' - c}{1 - c} \quad \text{or} \quad c = \frac{c' - \alpha}{1 - \alpha} \tag{11}$$

If the *distribution* of unit shaft friction f_s can be estimated, the value of c can be calculated without assigning any magnitudes to f_s (c is also the ratio of the average stress to the maximum stress in the pile due only to F_s). A convenient chart that simplifies this task is given in Fig. 8. With c known, α is obtained from Eq 11, whence P_T and hence F_s are determined.

Differentiating Eq 8 with respect to P,

$$\frac{AE}{L}\frac{d\Delta_e}{dP} = n = c' + P\frac{dc'}{dP} \tag{12a}$$

$$= \frac{dP_T}{dP} + c\,\frac{dF_s}{dP} + F_s\,\frac{dc}{dP} \tag{12b}$$

Advantage can be taken of the fact that segments of the P versus Δ_e plot are often linear. When this is the case, n = constant and Eq 12 can be integrated. For example, regardless of how the pile is transferring load to the soil, for n = constant, Eq 12a gives

$$\frac{dc'}{dP} = \frac{n - c'}{P}$$

which is readily integrated to give

$$c' = n - \kappa\frac{1}{P} \tag{13}$$

where κ = constant of integration. Thus, for any range in P over which n = constant, a plot of c' versus $1/P$ is a straight line with intercept n and slope = κ.

Consider the case when shaft friction is fully mobilized at $P = P_m$ less than P_{ult} so that for $P > P_m$, F_s = constant = F_{s_m}. Thus, $dF_s/dP = 0$, and from Eq 9, $dP_T/dP = 1$. Substituting into Eq 12*b*

$$n = 1 + F_{s_m}\,\frac{dc}{dP}$$

Since n and F_{s_m} are constants, dc/dP = constant. While it is possible for the distribution of unit shaft friction to change in such a way that both F_{s_m} and dc/dP remain constant, the probability is so low, that if F_{s_m} is constant, dc/dP may be taken to be 0. Therefore, n will equal 1, and the P versus Δ_e plot will be parallel to PL/AE.[6]

It is of interest also to consider the case when P_T is mobilized at $P = P_m$, so that for $P > P_m$, P_T = constant = P_{T_m}. Thus, $dP_T/dP = 0$, and from Eq 9, $dF_s/dP = 1$. Substituting into Eq 12*b*,

$$n = c + F_s\,\frac{dc}{dP}$$

Setting $F_s = P - P_{T_m}$ and integrating,

$$c = n - \frac{\lambda}{P - P_{T_m}} = n - \lambda\frac{1}{F_s} \tag{14}$$

[6] If n is found to be greater than 1, the pile is either bowing or yielding or, after full mobilization, there is a reduction in shaft friction. Values of $n < 1$ indicate that shaft friction is not yet fully mobilized.

where λ is a constant of integration. In this case, c is linear with $1/F_s$ and has an intercept n. If $P_{T_m} = 0$ (purely friction pile), $F_s = P$, $c = c'$, and Eqs 14 and 13 are identical. These concepts will now be applied to reanalyze test pile no. 1.

Reanalysis of Test Pile No. 1

Plots of P versus Δ_e for test pile no. 1 are shown in Fig. 9 with corresponding values of n given directly on the figures. In the case of the retest, $n = 0.700$ for four successive load increments right up to P_{ult}. It must be concluded that shaft friction was not fully mobilized until failure occurred. Therefore, either P_T was mobilized first, P_T and F_s both increased right up to failure, or $P_T \simeq 0$, that is, the pile was purely frictional. From bearing capacity considerations,

$$P_{T\,ult} = A(cN_c + \gamma D_f) = 1.07\left[(0.95)10 + \frac{120(112.5)}{2000}\right] \simeq 150 \text{ kN (17 tons)}$$

$$\alpha_{ult} = 17/300 = 0.05667$$

From Eq 11,

$$c_{ult} = \frac{c' - \alpha}{1 - \alpha} = \frac{0.62289 - 0.05667}{0.94333} = 0.600$$

If it is assumed that the distribution of unit shaft friction is as given in Fig. 8*a*, then with $l/L = 84.5/112.5 = 0.75$ and $c = 0.600$,

$$f_{s_1}/f_{s_2} = 0.40$$

therefore, $\pi\, 14/12[84.4(0.4) + 28]f_{s_2} = 300 - 17$ when

$$f_{s_2} = 120 \text{ kPa (1.25 tsf)}; \qquad f_{s_1} = 48 \text{ kPa (0.50 tsf)}$$

A similar analysis for the initial test gave

$$f_{s_2} = 100 \text{ kPa (1.05 tsf)}; \qquad f_{s_1} = 28 \text{ kPa (0.29 tsf)}$$

from which it is seen that although the shaft friction in the till did increase, only 20 of the additional 890 kN (100 tons) carried in the retest was supported in the till; a conclusion that is opposite to the one reached initially.

A plot of c' versus P for the retest is shown in Fig. 10. The discontinuity in curvature corresponds to the condition when pile shortening begins to

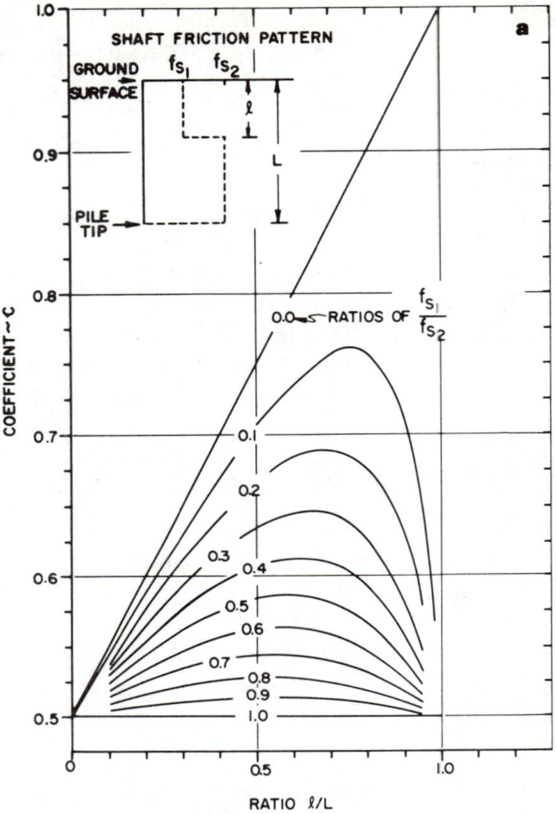

FIG. 8—*Coefficient for various distributions of unit shaft friction.*

occur due to shaft friction being transferred into the underlying till, which can be identified at $P \simeq 1780$ kN (200 tons) and corresponds to

$$f_{s_1} \simeq 57 \text{ kPa } (0.6 \text{ tsf})$$

This is in reasonable agreement with $f_{s_1} = 48$ kPa (0.5 tsf) previously calculated.

Discussion of Results

Ultimate Pile Capacity

Some interesting comparisons between a variety of methods to determine P_{ult} from load test data were made by Fellenius [*17*], which showed a difference of the order of 40 percent between the smallest and largest

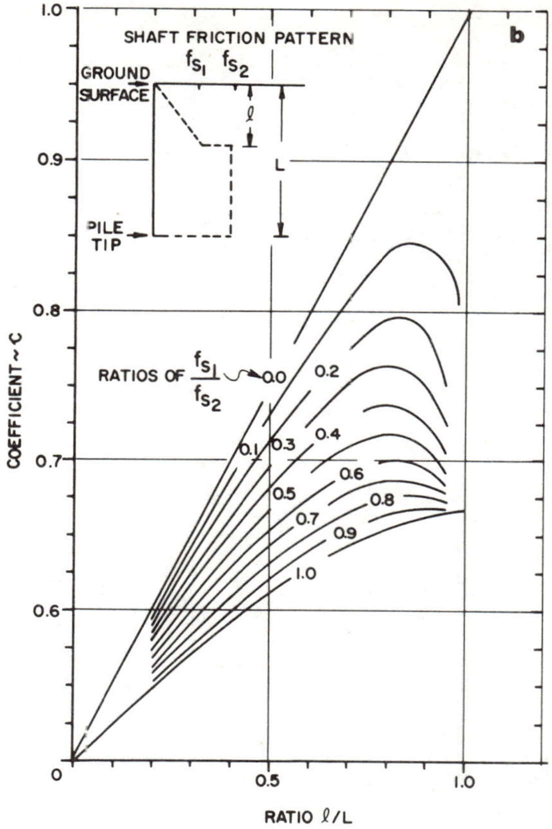

FIG. 8—*Continued.*

interpreted failure values, but no recommendations were given regarding which method is to be preferred. In this connection, the results given in Table 1 (retest) are instructive. With data available up to a load of 2670 kN (300 tons) (97 percent of P_{ult}), van der Veen's method is subject to interpretation (the latter part of the curve was used) and overpredicted P_{ult} by 75 percent; Chin's method gave an outlandish result; and Davisson's method required so extensive an extrapolation of the P versus δ curve, it was not considered applicable. The reason Chin's method gave such a poor result in this case may be deduced as follows:

Assume that $\delta = cP^{\omega}$ when $P \rightarrow P_{ult}$. Differentiating δ/P with respect to δ,

$$\frac{d(\delta/P)}{d\delta} = \frac{P - \delta(dP/d\delta)}{P^2} = \frac{(1 - 1/\omega)}{P}$$

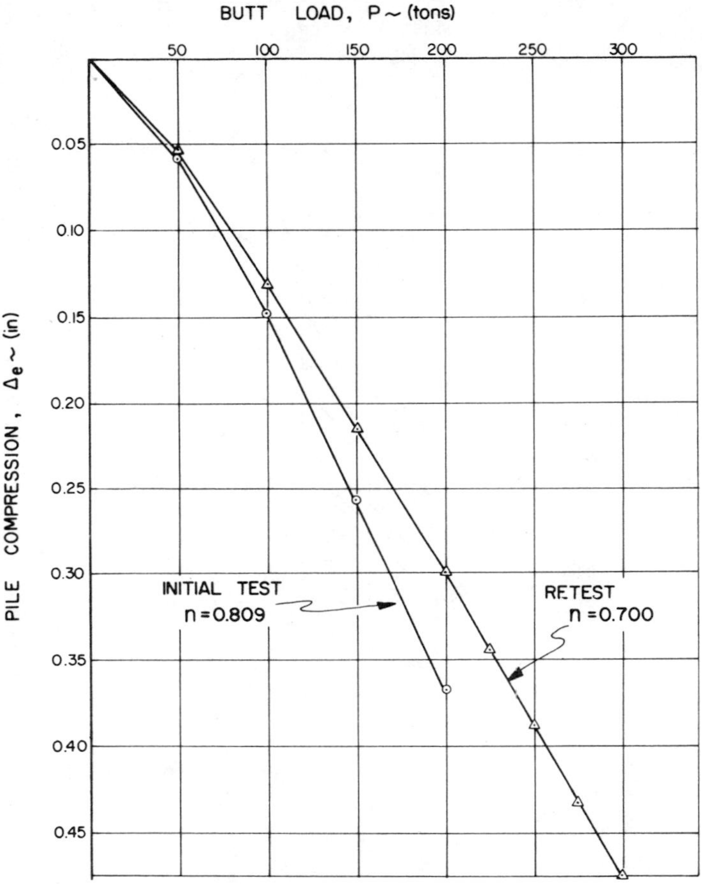

FIG. 9—*Test pile no. 1: 14-in. pipe.*

If ω is large as $P \rightarrow P_{ult}$, the inverse slope of the δ/P versus δ curve approaches P_{ult}. Thus, if the P versus δ plot can be fitted to a polynomial with a large exponent, all methods that extrapolate to an asymptote (hyperbolic, exponential, parabolic, etc.) will give a good estimate of P_{ult}. When the objective of the load test is to obtain P_{ult}, provision should be made to continue loading until the above condition is satisfied. Otherwise, a potentially large error may result no matter which method of extrapolation is adopted.

Load Transfer

A methodology for intepreting load transfer from conventional load tests

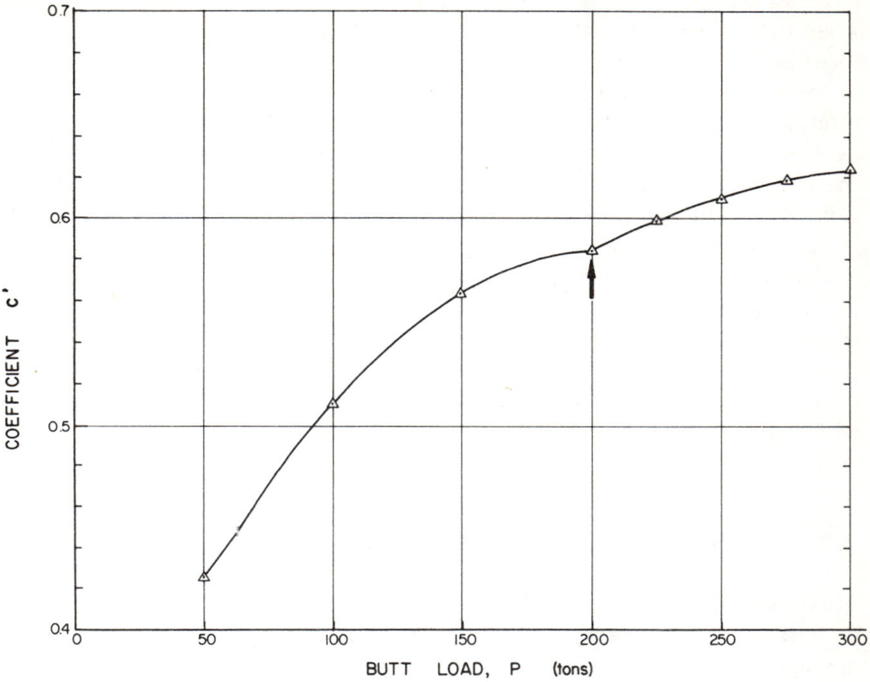

FIG. 10—*Coefficient* c′ *versus butt load—test pile no. 1, retest.*

was presented. It utilizes the fact that a plot of butt load P versus pile compression Δ_e often results in straight line segments; for this condition it is possible to estimate more reliably the separate contributions to pile compression of point load and shaft friction, and hence to estimate their respective values. In a matter analogous to the use of stress-path concepts, with a little practice, considerable insight is gained into the mechanism of load transfer in piles, as illustrated by the following examples.

Test Pile No. 1—It has already been demonstrated that the method led to a different interpretation of load transfer than that deduced initially. Additional shaft friction continued to develop with time between the pile and the prebore as well as in the underlying hard tills. The former effect dominated in this case because only 25 percent of the pile length was embedded in the till in which the increase in average shaft friction was only 18 percent, compared to 70 percent above the till. Thus, with time, the prebore lost much of its effectiveness in reducing shaft friction.

Because the loads applied on test pile no. 3 stressed the concrete well above $\frac{1}{2}f_c'$, it was not possible to interpret quantitatively the effect of the external boot on the shaft friction versus time relationship. However, it is believed that the type of boot had an influence on the effective lateral stresses, especially in the short term, a possibility that was not appreciated when the load-test program was begun. Moreover, inspection of the driving records (Fig. 3) indicates that the external boot resulted in larger energy losses during driving than was the case for the internal boot. Thus, the interpreted ratio of average shaft friction to undrained shear strength that develops in stiff clays (Tomlinson's α value) may depend on the details at the pile tip, as well as on how long after driving the load test is performed. Considering the difficulties in sampling and testing very stiff to hard clays, it also depends on the quality of the effort expended in this connection.

Test Pile No. 2—Plots of P versus Δ_e and c' versus P and $1/P$ for test pile no. 2, reload, are shown in Figs. 11 and 12, respectively. For the load increment 4270 to 4630 kN (480 to 520 tons), $n = 1.00$ (Fig. 11), which indicates that the increment was transferred fully to point bearing. This condition can also be identified in Fig. 12 (arrow). Combining this result with that from the short H-pile [$P_{ult} = 2180$ kN (245 tons)]

$$
\begin{aligned}
F_s \text{ maximum} \quad & 4270 \text{ kN } (480 \text{ tons}) \\
F_s \text{ above till} \quad & \underline{2220} \text{ kN } (250 \text{ tons}) \\
F_s \text{ in till (after 8 days)} = \; & 2050 \text{ kN } (230 \text{ tons}), \text{ or } f_s = 102 \text{ kPa } (1.07 \text{ tsf})
\end{aligned}
$$

This is in good agreement with $f_{s_2} = 100$ kPa (1.05 tsf) previously calculated for test pile no. 1 (initial test at 7 days). Thus, both pile types developed an average ultimate unit shaft friction of 96 to 120 kPa (1.0 to 1.25 tsf) in the very stiff to hard till.

A plot of c' versus P for the initial loading on test pile no. 2 is shown in Fig. 13. Of special interest are the values of c' on unloading. After loading to 3560 kN (400 tons) and unloading to 890 kN (100 tons) a value of $c' = 0.85$ indicates that most of the 890 kN (100 tons) is being supported in point bearing; on the other hand, on first loading, the first 890 kN (100 tons) was supported by shaft friction in the upper part of the pile. Similar results were obtained by Vey [*18*] and by Kerisel and Adam [*19*] using instrumented piles driven into clay soils.

van Weele's Method—As pointed out earlier, the basic assumptions in van Weele's method are: (1) the residual stresses after each unloading are negligible, and (2) the point load is proportional to the elastic compression of the soil beneath the pile tip. It is clear from the previous discussion that, certainly for clays, the first assumption will not generally be satisfied. It is also clear from Fig. 2*b* that at a load of 1780 kN (200 tons), half the ultimate tip displacement had occurred, yet virtually no load was as yet being transferred to the pile tip. It appears that significant soil displacements at the pile tip can occur due to stresses transmitted to the underlying clay by shaft friction forces. Thus, van Weele's assumptions are not valid, in general, and his method should be used with caution—particularly in stiff clay soils.

Mobilization of Shaft Friction—From tests on 360 to 610 mm (14 to 24

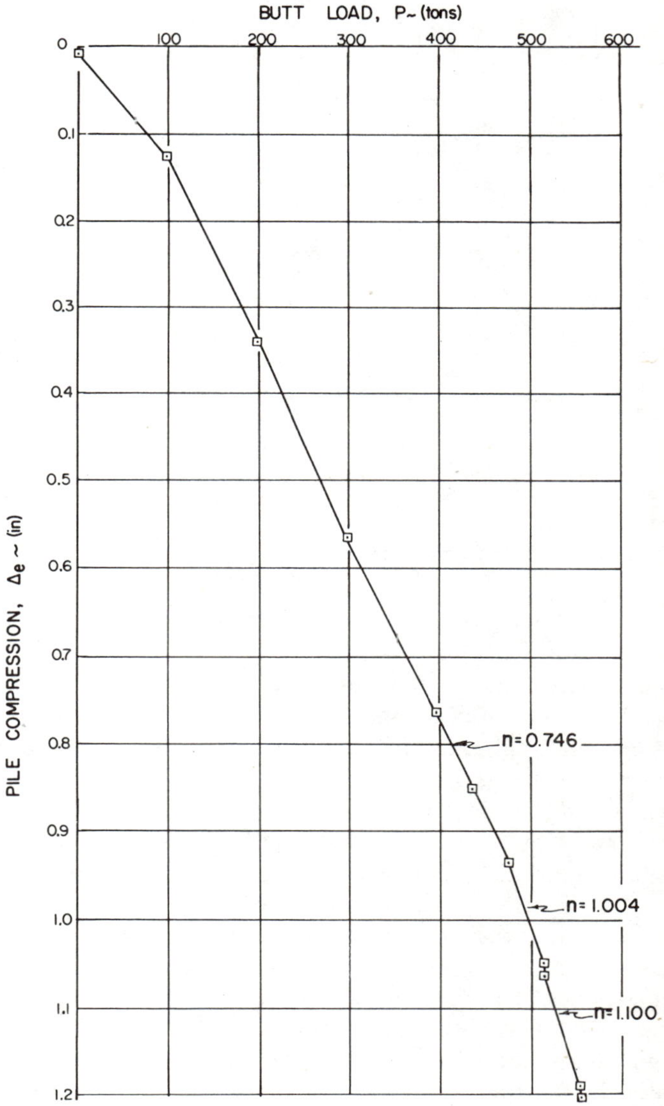

FIG. 11—*Pile compression versus butt load—test pile no. 2, reload.*

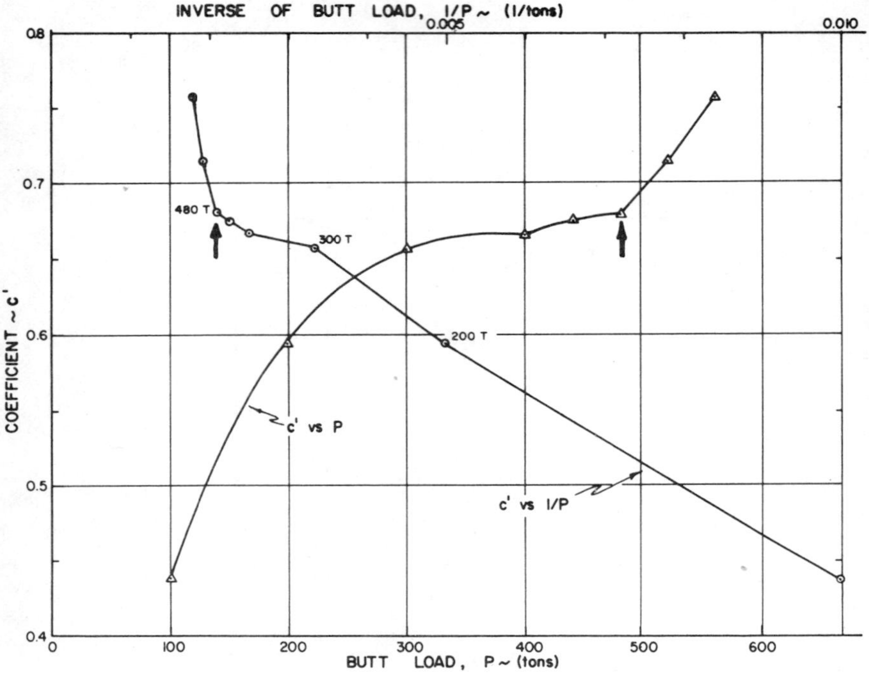

FIG. 12—*Coefficient* c′ *versus butt load—test pile no. 2, reload.*

in.) diameter bored piles in London clay, Skempton [*20*] concluded that the tip displacement at ultimate load was proportional to the pile diameter and that the "shaft adhesion was mobilized at smaller settlements". Much corraborating evidence (for example, O'Neill and Reese [*21*] for bored piles in stiff clays; and Vesić [*10*] for driven piles in sand) conditioned the authors to expect that, for 360 mm (14 in.) diameter piles, shaft friction would be fully mobilized before P_{ult} was reached. This means that *n* should be observed to equal one before failure occurs. Examination of the load test data indicated that this was not always the case, a result that stimulated a careful reexamination of previous measurements on instrumented piles. The outcome was surprising (at least to the authors). For example, 300 to 410 mm (12 to 16 in.) diameter piles driven 15 m (50 ft) through poorly graded sands with clay layers on the Arkansas River Project [*22*] showed that in all cases point bearing and shaft friction continued to increase right up to P_{ult}. Vesić [*23*] reported a test on an 460 mm (18 in.) diameter pile driven 12 m (40 ft) into sands at the Ogeechee River test site in which a butt displacement of 46 mm (1.8 in.) was obtained at a load of 3110 kN (350 tons). Load was added until 3560 kN (400 tons) was applied, with the butt displacement now 130 mm (5 in.), yet strain gages

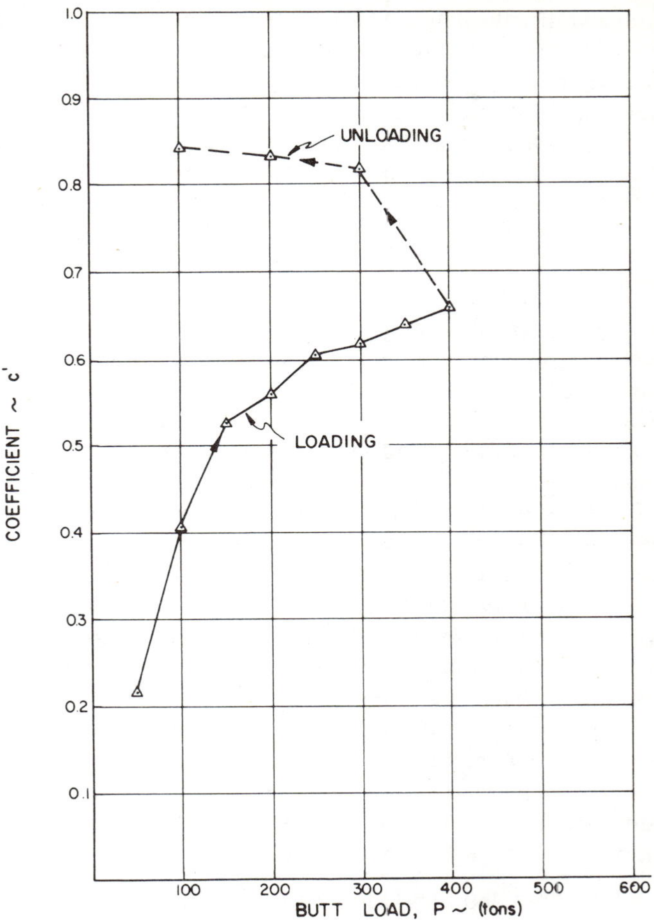

FIG. 13—*Coefficient* c' *versus butt load—test pile no. 2, initial loading.*

showed that the additional 440 kN (50 tons) was supported solely by shaft friction. The increase in skin friction was probably due to a redistribution of effective lateral pressure on the pile shaft. Reluctantly, the authors abandoned their notions regarding early mobilization of ultimate shaft friction. Of course, there are cases when shaft friction is mobilized first, but this cannot be assumed a priori.

Residual Stresses—As is the case when strain gages are nulled after driving, pile compressions measured with telltales reflect the changes in load distribution in the pile due to the applied loads. Residual stresses due to driving are not accounted for. It is clear from the foregoing that loading and unloading in compression followed by loading and unloading in tension can result in considerable changes from the initial conditions, especially in the distribution of shaft friction. Consequently, the procedure proposed by Hunter and Davisson [*11*], and by Mansur and Hunter [*22*], to measure residual shaft friction cannot, in general, be valid. If the residual load in point bearing only is desired, it can be determined directly by conducting a tension test on an instrumented pile. The measured "tension" at the tip of the pile equals the residual compression after driving.

Conclusions

The term "ultimate pile capacity" or "ultimate load" should be used only to connote failure in bearing capacity, that is, when rapid progressive settlement occurs at constant load. If the maximum load on the pile is established by limiting the settlement, it could be called the "limiting load"; the criterion used to establish the limiting load should always be stated.

Load test results should not be extrapolated to determine ultimate capacity if the load-deflection plot is not curving rapidly toward an asymptote; when it is, any of the methods described herein will give a satisfactory estimate of the ultimate load.

A methodology for interpreting load transfer was presented and applied to the interpretation of load tests on high-capacity driven piles. It was determined that, with time, the prebore lost its effectiveness in reducing shaft friction, and that the average ultimate skin friction in the very stiff to hard glacial till was about 120 kPa (1.25 tsf); it is likely that this value is dependent on the type of shoe used at the tip, especially in the short term after driving. The method is not exact, but with practice, it affords the user considerable insight into the mechanics of load transfer. Through its use the authors became aware that:

1. For piles 300 to 460 mm (12 to 18 in.) in diameter, driven into either sands or clays, the ultimate shaft friction is not necessarily mobilized before ultimate load is reached.
2. Loading and unloading a pile (in tension or in compression) can cause irreversible changes in the distribution of effective lateral pressures, and hence on the existing as well as on the ultimate shaft friction. Methods of interpreting pile tests that rely on these changes being negligible (such as Van Weele's procedure to separate shaft friction and point bearing, or Hunter and Davisson's procedure to determine residual shaft friction) should be used with caution. The residual tip load can be determined on an instrumented pile by performing a tension test and measuring the indicated "tension" when the strain at the pile tip becomes constant.
3. Deflection of the pile tip can occur without any load being applied at the point due to strains induced in the soil below the tip by shaft friction forces.

4. Load transfer in piles is sensitive to small changes in soil strain or pile compression. The strains induced by driving and loading pile groups are different from those developed in a single pile. Accordingly, care must be exercised in extrapolating the results of tests on single piles to interpret the behavior of pile foundations.

References

[1] Leonards, G. A., *Proceedings*, American Society of Civil Engineers Specialty Conference on Performance of Earth and Earth Supported Structures, Vol. 1, Part 2, p. 1169.

[2] Sherman, W. C., *Proceedings*, 7th International Conference on Soil Mechanics and Foundation Engineering, Vol. 2, 1969, p. 227.

[3] Roth, C. H. *Foundation Facts*, Raymond Concrete Pile Co., Vol. 8, No. 1, 1972, p. 13.

[4] Whitaker, T. and Cooke, R. W., *Proceedings*, 5th International Conference on Soil Mechanics and Foundation Engineering, Vol. 2, 1961, p. 171.

[5] Butler, H. D. and Hoy, H. E., "Users Manual for the Texas Quick-Load Method for Foundation Load Testing," Federal Highway Administration Offices of Research and Development, 1976.

[6] van der Veen, C., *Proceedings*, 3rd International Conference on Soil Mechanics and Foundation Engineering, Vol. 2, 1953, p. 84.

[7] Davisson, M. T., "High Capacity Piles," Soil Mechanics Lecture Series—Innovations in Foundation Construction, Soil Mechanics and Foundations Division, Illinois Section of the American Society of Civil Engineers, 1973, p. 81.

[8] Kondner, R. L., *Journal of Soil Mechanics and Foundations Division*, American Society of Civil Engineers, Vol. 89, No. SM1, Feb. 1963, p. 115.

[9] Chin, F. K., *Journal of the Soil Mechanics and Foundations Division*, American Society of Civil Engineers, Vol. 97, No. SM6, June 1971, p. 930.

[10] Bozozuk, M. and Labrecque, A. in *Performance of Deep Foundations, ASTM STP 444*, American Society for Testing and Materials, 1969, p. 33.

[11] Hunter, A. H. and Davisson, M. T. in *Performance of Deep Foundations, ASTM STP 444*, American Society for Testing and Materials, 1969, p. 106.

[12] Vesić, A., "A Study of Bearing Capacity of Deep Foundations," Final Report, Project B-189, Georgia Institute of Technology, 1967, pp. 233-5, 238.

[13] Vijayvergiya, V. N. and Focht, J. A., *Proceedings*, 4th Annual Offshore Technology Conference, Vol. 2, 1972, pp. 865-874.

[14] Lo, K. Y. and Stermac, A. G., *Canadian Geotechnical Journal*, Vol. 1, No. 2, March 1964, p. 63.

[15] van Weele, A. F., *Proceedings*, 4th International Conference on Soil Mechanics and Foundation Engineering, Vol. 2, 1957, p. 76.

[16] Rutledge, P. C., *Proceedings*, 4th International Conference on Soil Mechanics and Foundation Engineering, Vol. 2, 1957, p. 449.

[17] Fellenius, B. H., *Journal of the Geotechnical Engineering Division*, American Society of Civil Engineers, Vol. 101, No. GT9, Sept. 1975, p. 855.

[18] Vey, E., *Journal of the Soil Mechanics and Foundations Division*, American Society of Civil Engineers, Vol. 83, No. SM1, Jan. 1957, p. 1160.

[19] Kerisel, J. and Adam, M., *Proceedings*, 7th International Conference on Soil Mechanics and Foundation Engineering, Vol. 2, 1969, p. 134.

[20] Skempton, A. W., *Geotechnique*, Vol. 9, No. 4, Dec. 1959, p. 171.

[21] O'Neill, M. W. and Reese, L. C., *Journal of the Soil Mechanics and Foundations Division*, American Society of Civil Engineers, Vol. 98, No. SM2, Feb. 1972, p. 203.

[22] Mansur, C. I. and Hunter, A. H., *Journal of the Soil Mechanics and Foundations Division*, American Society of Civil Engineers, Vol. 96, No. SM5, Sept. 1970, pp. 1563-1564, 1570.

[23] Vesić, A. S., *Journal of the Soil Mechanics and Foundations Division*, American Society of Civil Engineers, Vol. 96, No. SM2, March 1970, p. 570.

T. D. Lu,[1] *J. A. Fischer,*[1] *and V. G. Miller*[1]

Static and Cyclic Axial Load Tests on a Fully Instrumented Pile

REFERENCE: Lu, T. D., Fischer, J. A., and Miller, V. G., **"Static and Cyclic Axial Load Tests on a Fully Instrumented Pile,"** *Behavior of Deep Foundations, ASTM STP 670*, Raymond Lundgren, Ed., American Society for Testing and Materials, 1979, pp. 416–434.

ABSTRACT: The results of static and cyclic axial pile load tests performed on a fully instrumented steel H-pile driven into a glacial till deposit are presented.

Cyclic loads were applied through a hydraulic system interfaced with an automated electronic–electromechanical closed loop servo system to achieve a controlled load intensity to the pile in a specific shape (sinusoidal) at specific frequency (0.1 Hz).

The load–deflection, load transfer, and configuration of the pile during pile driving and after each load increment was measured through an instrumentation system that consisted of load cells, deflectometers, strain gages, accelerometers, and inclinometers.

Data obtained in this investigation indicated that:

1. Axial deformation of the pile remained essentially constant prior to and after cyclic axial load application. The behavior of the pile was essentially elastic during cyclic load application.
2. Within the scope of testing performed, cyclic vertical load increments do not effect the load carrying capacity or load transfer characteristics of the pile.
3. Load transfer characteristics varied with the magnitude of applied axial load. The ratio of point bearing resistance to the applied load increased with the applied load. However, the rate of increase generally decreases with increasing applied load.

KEY WORDS: pile load testing; instrumented H-pile; instrumentation, static, cyclic, and vertical loads; earthquake simulation; load transfer

A series of pile load tests was performed to obtain representative pile-soil parameters for seismic evaluation of a pile-supported facility originally built in 1963. The data obtained in the pile load test program were to be used in a structural analysis of the existing facility as well as in a structural analysis and design of future pile-supported facilities subjected to earthquake loading conditions.

[1] Senior engineer, partner, and senior engineer, respectively, Dames and Moore, Cranford, N.J. 07016.

Pile-soil parameters in both vertical and lateral directions are required in a soil-pile-structure interaction analysis of a structure subjected to earthquake loading. Thus, a field load test program, which consisted of a static vertical, a cyclic lateral, and a cyclic vertical pile load test (in sequential order) on a fully instrumented pile, was developed and successfully carried out.

This paper presents the results and findings of static and cyclic axial load tests on this test pile. The details of the loading system and instrumentation program are presented elsewhere.[2]

Subsurface Conditions

The subsoils at the site are of glacial origin. In general, these glacial deposits can be characterized as hetrogeneous mixtures of various soil types—cobbles, gravels, sands, silts, and clays. In general, the shale bedrock is interspersed with thin seams of dolomite limestone overlain by three soil layers. Throughout the site, the groundwater level generally ranges from 0 to 5 m (0 to 15 ft) below ground surface.

The upper layer consists of a 4.5 to 7.5 m (15 to 25 ft) thick layer of medium dense to dense mixture of gravels and coarse to fine sands with clayey silts. The natural water content of this layer ranges from 10 to 20 percent.

The middle layer is a gray, medium stiff, silty clay with variable gravel and sand content. The thickness of this layer varies approximately from 9 to 12 m (30 to 40 ft). The natural water content ranges from 12 to 30 percent. The liquid limit of this layer varies from 23 to 40 percent, with plasticity indices ranging from 10 to 20 percent. The undrained shear strength of this layer varies from about 48 to 144 kN/m^2 (1000 to 3000 psf). Consolidation test results indicate that this clay layer is slightly overconsolidated.

The lower layer extends to bedrock and is composed of dense to very dense gravel and clayey silt with occasional weathered shale fragments. Due to the slope of the bedrock across the site, this layer thickness varied from 3 to about 18 m (10 to 60 ft). At the test site, the thickness was about 7.6 m (25 ft) or about 21 m (70 ft) to bedrock. Within the site, this stratum provides bearing support for the piles underlying the existing building.

The results of standard penetration tests performed during the site investigation study are shown in Fig. 1. The correlation between relative density and standard penetration resistance developed by Gibbs and Holtz[3] is

[2] See p. 435.

[3] Gibbs, H. J. and Holtz, W. G., *Proceedings*, 4th International Conference on Soil Mechanics and Foundation Engineering, Vol. 1, London, 1957.

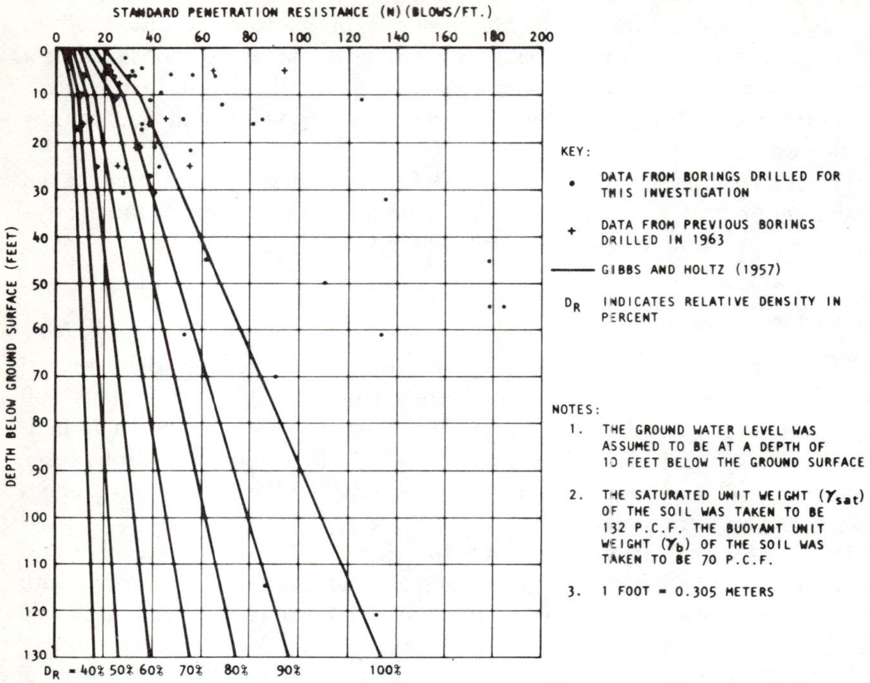

FIG. 1—*Results of standard penetration resistance.*

also shown as part of this figure for demonstration purposes. This correlation cannot be used for indirect determination of relative density for the upper and lower soil stratum due to the presence of cobbles and gravel.

Existing Facilities

The existing facilities were constructed about 15 years ago. They were supported by a total of 508 12BP53 piles. The actual axial load on each pile is about 63 560 kg (70 tons) corresponding to the design loads determined from the results of two pile load tests conducted prior to the construction. The axial load-deflection curves for these two tests are shown in Fig. 2.

The piles under the existing facilities were driven into the soil layer immediately above bedrock. Penetration into this layer generally varied from about 1.5 to 3.0 m (5 to 10 ft). As stated previously, this layer is composed of a dense to very dense gravel and clayey silts with occasional weathered shale fragments. The driving equipment and criteria for the existing piles were:

1. A Raymond 65C differential pile hammer with a rated energy of

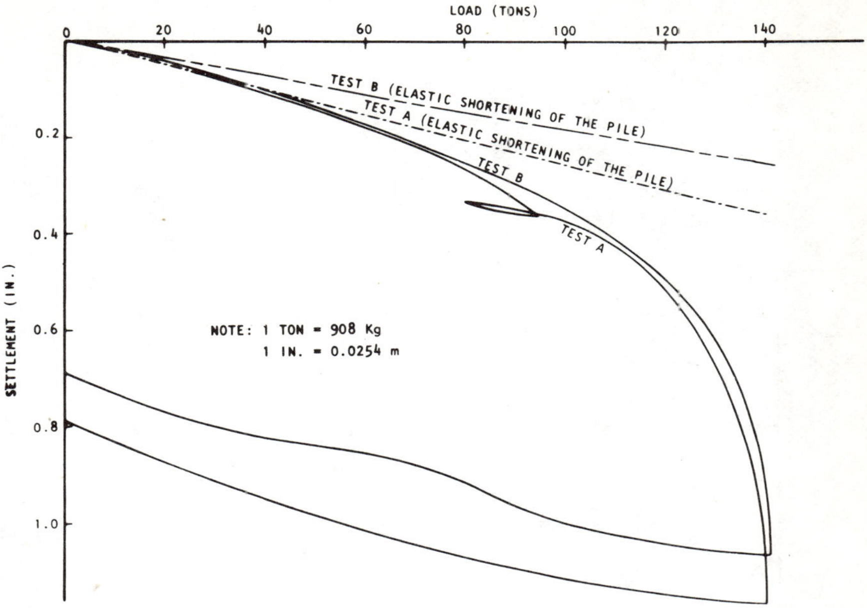

FIG. 2—*Load–settlement curve from pile load test performed prior to construction of existing facilities.*

2700 m·kg (19 500 ft·lb) and a ram weight of 2955 kg (6500 lb) was used in pile driving.

2. Pile was driven to a predetermined depth [if possible, or at least 1.525 m (5 ft) into the dense to very dense gravel and clayey silts].

3. The last 0.305 m (1 ft) of driving had a minimum of 80 hammer blows.

4. The last 0.102 m (4 in.) of driving had a minimum of 10 blows per 0.0254 m (1 in.).

Test Setup

Prior to the installation of the test and the reaction piles, a test pit was excavated. The bottom of the test pit was located at an elevation corresponding to the bottom of the existing pile caps at the site. The configuration of this test pit is shown in Fig. 3. After the excavation of the test pit, the locations of the test and reaction piles were located by survey and staked for subsequent pile driving.

A total of five 12BP53 steel H-piles were installed at the location of the pile load test. The pile load test incorporated the instrumented test pile, the four reaction piles that provided the necessary reaction against uplift

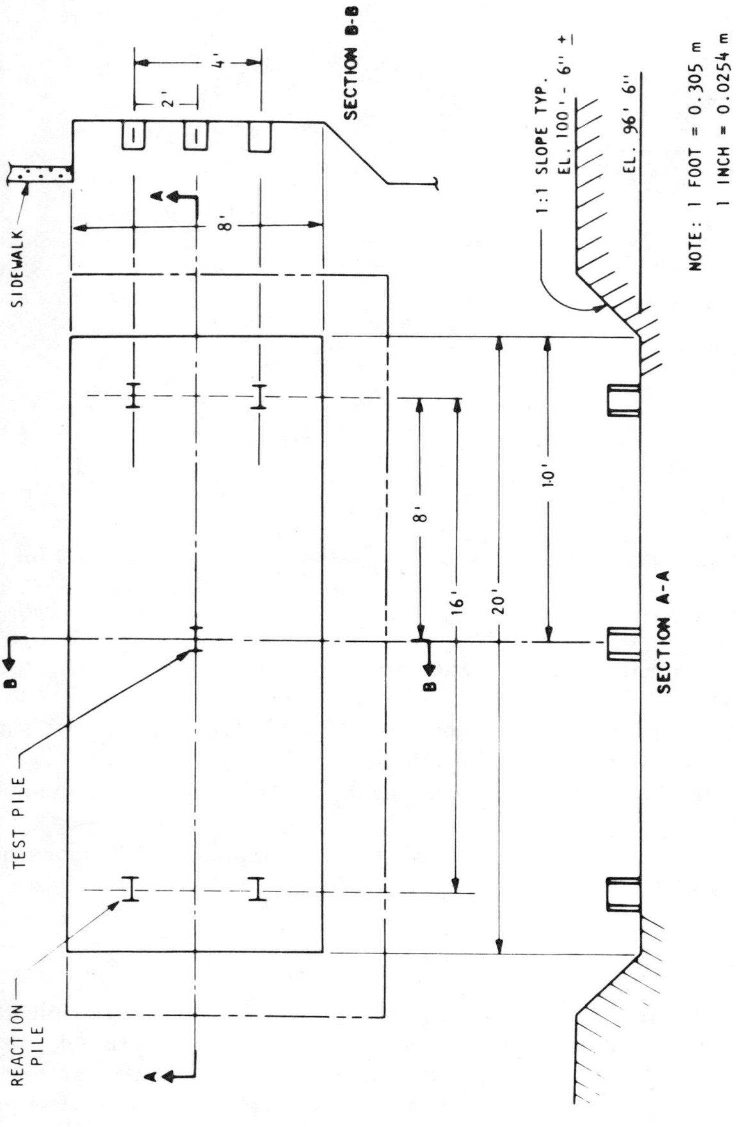

FIG. 3—*Test pit configuration.*

and lateral force, and a test frame to transfer loads from the reaction piles to the test pile. A detailed description of the instrumentation of the test pile has been presented elsewhere (see footnote 2).

Each of the five piles was driven with a Raymond 65C differential acting pile hammer. A micarta-aluminum cap block was used in driving each of these piles. The pile-driving equipment, pile section, and driving criteria used in this study were the same as those used for the 12BP53 H-piles underlying the existing facilities at the site.

The four reaction piles were the first to be driven. These piles were driven without pile points attached. After installation of the reaction piles, the instrumented test pile was then driven with a Pruyn point BP 75600 pile shoe attached. However, during driving of the test pile, it was observed that it had rotated approximately 15 deg with respect to the axis of the proposed load test scheme and the four reaction piles that were previously driven.

The driving resistance encountered in driving each of the five piles were recorded and the results are presented in Table 1. Dynamic driving data (stresses during driving) are presented in Table 2. The driving resistance encountered in installating the test pile is also presented graphically in Fig. 4.

The test pile, during driving, exhibited a somewhat stiffer resistance to penetration (Table 1) than the reaction pile. This higher resistance is of the order of five to six blows per 0.305 m (1 ft), on the average, and is believed to have been caused by one or a combination of three factors. First, during the top 4.5 m (15 ft) of pile driving, the pile hammer was intentionally run at a slower rate to protect the strain gages and lead wires from damage and to ensure verticality of the test pile after pile driving. A slower rate of driving hammer application would be manifested by an apparent stiffer resistance of soil to pile driving. Second, a significant rotation of the pile occurred during the initial driving phase, which could be an indication of the presence of an obstruction, possibly a boulder, in the upper till material. Third, and most probably, the shape and increased projected area of the test pile, due to the addition of protective plates for the strain gages, as well as the slope indicator tubes used to measure the pile verticality and straightness, would increase the penetration resistance of the pile in the upper soils.

The final depth of penetration, however, is entirely consistent with the other (reaction) piles driven and the settlement of the pile due to static loading is comparable both in magnitude and ultimate load capacity with the results of the two load tests performed prior to the construction of the existing facilities (see Fig. 2).

After the installation of the piles was completed, the piles were cut at the required elevation and the reaction frame was fabricated (see Fig. 3 in footnote 2). However, because the test pile had rotated about 15 deg clock-

TABLE 1—*Driving resistance data.*

Depth Below Ground Surface, ft	Blows per Foot of Pile Penetration				
	No. 1 (NE) Pile	No. 2 (NW) Pile	No. 3 (SE)[a] Pile	No. 4 (SW)[a] Pile	No. 5 (Test Pile)[a] Pile
0					
1	2		3	2	9
2	2	2	3	2	9
3	5	4	3	2	9
4	2	5	3	2	9
5	2	2	3	2	9
6	2	3	6	5	12
7	2	3	6	5	12
8	2	6	6	5	12
9	6	6	6	5	12
10	5	6	6	5	12
11	6	7	9	9	20
12	8	8	9	9	20
13	8	7	9	9	20
14	7	8	9	9	20
15	10	11	9	9	20
16	15	11	12	12	19
17	13	13	12	12	13
18	13	12	12	12	13
19	13	13	12	12	15
20	13	13	12	12	15
21	14	13	15	14	15
22	14	12	15	14	16
23	13	13	14	14	17
24	14	13	16	14	18
25	13	13	16	14	19
26	14	13	18	16	20
27	15	15	19	16	20
28	15	15	21	16	22
29	17	15	21	16	22
30	18	17	22	16	20
31	17	17	19	20	22
32	19	17	20	18	23
33	18	17	18	17	23
34	18	18	18	18	22
35	18	18	18	20	24
36	18	17	18	19	23
37	18	18	17	21	26
38	21	21	20	23	27
39	24	27	21	25	31
40	25	29	27	33	32
41	27	31	31	37	38
42	32	35	35	40	40
43	36	40	37	54	44
44	48	50	48	66	50
45	52	66	70	77	60
46	62	66	80	125	80
47	68	78	75	105/8 in.	89
48	74	97	87		49/5 in.
49	99	153	134/11 in.		
50	92/8 in.		93		

TABLE 1—*Continued.*

Depth Below Ground Surface, ft	Blows per Foot of Pile Penetration				
	No. 1(NE) Pile	No. 2 (NW) Pile	No. 3 (SE)[a] Pile	No. 4 (SW)[a] Pile	No. 5 (Test Pile)[a] Pile
51	...	...	95	...	...
52	...	...	80	...	...
53	...	...	80	...	...
54	...	...	90	...	...
55	...	...	108	...	...

Conversion factors—
1 ft. = 0.305 m, and
1 in. = 0.0254 m.
[a] Average blows per foot for 0 to 5 ft, 5 to 10 ft, and 10 to 15 ft below ground surface.

TABLE 2—*Dynamic driving data peak impact stresses (psi).*

Gage No.	Pile Penetration				
	25 ft.	40 ft.	45 ft.	First Blow of Last 34	Thirty-fourth Blow of Last 34
5cc	...[a]	...[a]	...[a]	...[a]	...[a]
10cc[b]	13 337.1	...[a]	14 346.3	13 113.8	9 079.9
15cc[c]	21 953.0	16 846.1	18 693.4	19 453.2	18 368.6
32cc	16 843.2	17 959.7	17 066.5	17 289.8	15 393.2
47cc	12 072.7	11 629.0	13 290.7	14 508.7	13 400.9

Conversion factors—
1 ft. = 0.305 m, and
1 psi = 6.9 kN/m^2.
[a] Indistinct trace.
[b] Gage 10cc noisy—gage may be defective.
[c] Gage 15cc exhibits slope reversal (double peak) on first blow.

wise, the lateral reaction beam had to be rotated through the same angle, and additional struts were added to permit perpendicular lateral load application and to evenly distribute the load to the reaction frame and piles. The final "as built" plot plan of the test pile and reaction piles has been shown in Fig. 4 of footnote 2.

The rotation of the test pile also necessitated a rotation of the vertical jacking system to place it in a plane coincident with the plane of the horizontal jacking system.

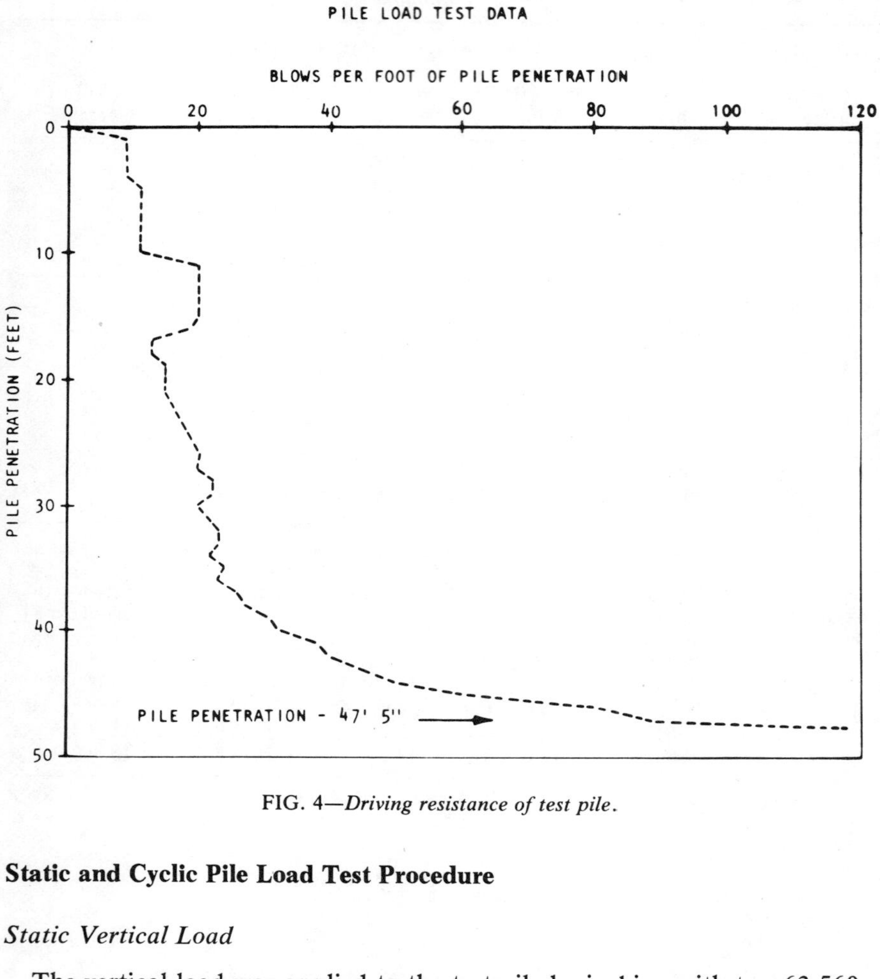

FIG. 4—*Driving resistance of test pile.*

Static and Cyclic Pile Load Test Procedure

Static Vertical Load

The vertical load was applied to the test pile by jacking with two 63 560-kg (70-ton) clevis-ended, calibrated hydraulic jacks with a 0.61-m (24-in.) stroke. One end of each jack was pinned to a 0.0635-m (2½-in.) thick plate that was welded perpendicularly to the top of the test pile. The clevis pin assembly was oriented to provide rotational freedom perpendicular to the direction of loading for the lateral load test. The opposite end of the jacks was pinned to a steel bearing plate 0.0635 m (2½ in.) thick, which in turn was bolted to 318 000-kg (350-ton) capacity calibrated strain gage load cell. The top of the load cell was bolted to a plain milled bearing plate 0.0635-m (2½-in.) thick, which was in turn bolted to the reaction beam. The reaction beam consisted of two wide-flange rolled steel sections connected with stiffeners. The test pile setup has been presented in Fig. 3 in footnote 2.

A seating load of approximately 8172 kg (9 tons) (13 percent of the estimated design load) was first applied to the test pile and held for 15 min. Dial gage readings were taken at intervals of 0, 1, 2, 5, 10, and 15 min, after which the load was released and a new zero established. The load was then applied to the test pile in increments of approximately 25 percent of the design load (70 tons or 63 560 kg) up to and including the design load. Each load increment was sustained until the rate of deflection of the pile was less than 0.025 cm (0.01 in.) per hour. Dial gage readings were recorded for both vertical and lateral deflections at intervals of 0, 1, 2, 5, 10, 15, 30, and 60 min after the application of each load increment. If the rate of deflection criteria had not been satisfied by 60 min, additional readings were taken until the rate was less than 0.0254 cm (0.01 in.) per hour. At this point in time, a cyclic lateral load pile test was performed.

Cyclic Vertical Load

Before the start of the cyclic vertical test, the lateral load was reduced to zero and the load cell and hydraulic jack assembly were removed. The ±22 700-kg (±25-ton) load cell was then repositioned in the vertical assembly, and the vertical load was increased from 63 560 to 77 180 kg (70 to 85 tons). Dial gage readings were recorded for the 77 240-kg (85-ton) load at 0, 1, 2, 5, 10, 15, 30, and 60 min interval. With a zero lateral load and a sustained vertical load of 77 240 kg (85 tons) a cyclic vertical sinusoidally shaped load of ±13 620 kg (±15 tons) was applied to the test pile. The cyclic load was imposed for 20 cycles at a cyclic frequency of 0.1 Hz, and dial gage readings were taken before and after the loading sequence.

The vertical load was then increased to approximately 127 120 kg (140 tons) (200 percent design load), and a vertical cyclic load of ±13 620 kg (±15 tons) was applied. Again, dial gage readings were recorded before and after the cyclic load was applied.

Rebound

The vertical load was held at above 127 120 kg (140 tons) for ½ h. Readings were taken at intervals of 0, 15, and 30 min. The load was then reduced to zero in decrements of 31 780 kg, (35 tons) held for 15 min each. Dial gage readings were recorded at 0 and 15 min.

Results of Static and Cyclic Axial Pile Load Test

Configuration of Pile Before Pile Load Test

Prior to the application of the static vertical load on the test pile, inclinometer data was obtained to measure the variation of the configuration of the pile axis with respect to depth below the top of the pile. The results of the inclinometer measurement are shown in Fig. 5. This figure shows that the top of the test pile was deflected about 0.089 m (3.5 in.) laterally with respect to the bottom of the pile. Thus, the test pile can be considered as essentially vertical prior to the application of the static vertical load.

Static Vertical Load Test

During the static vertical load test, the vertical and lateral movements

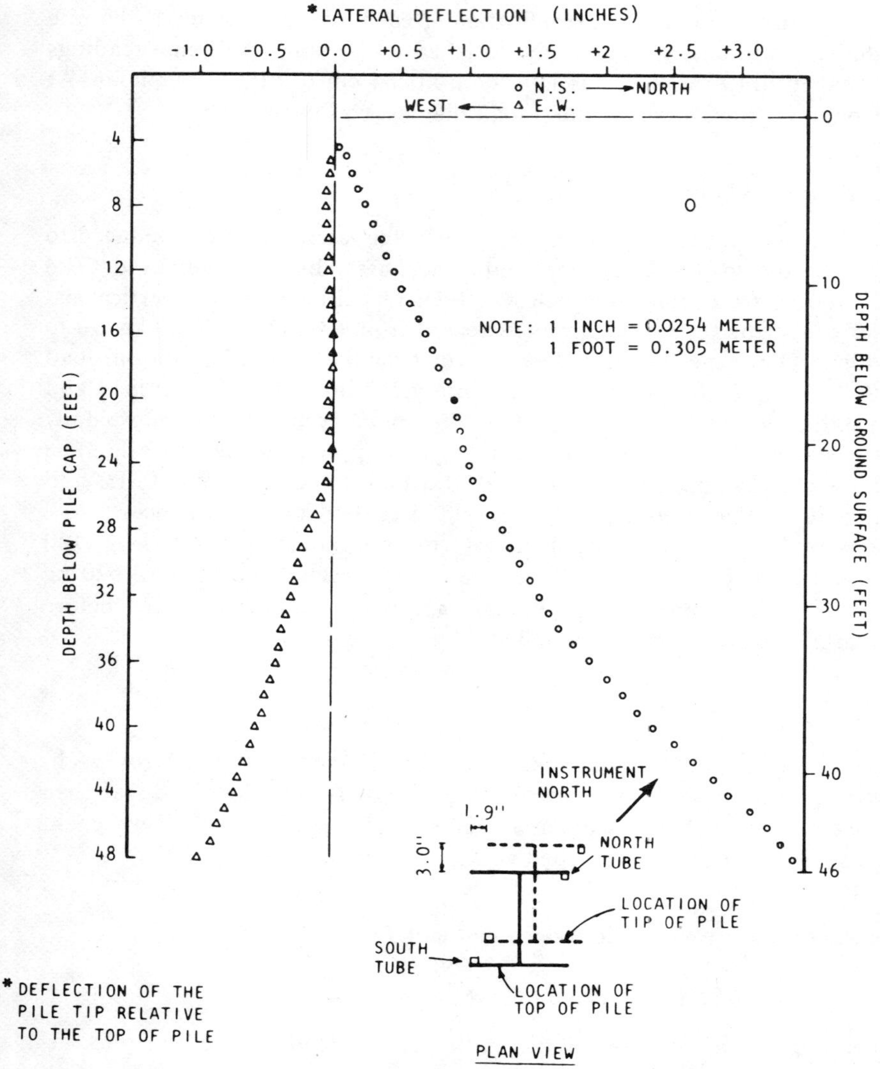

FIG. 5—*Lateral deflection of test pile after pile driving.*

of the top of the test pile were measured by two vertical and two horizontal dial gages, all oriented in the plane of the weak axis. The average vertical deflection of the pile top with respect to the applied load is presented in Fig. 6.

A comparison of Fig. 6 and Fig. 2 indicates the similarity of results of this new pile load test and two previous pile load tests and demonstrates the suitability of the test pile in effectively representing the pile behavior of piles underlying the existing facilities.

The lateral movement of the pile during the static vertical load portion of the test was less than 0.028 cm (0.011 in.) and is considered to be of no consequence.

Strain gage data for the static vertical load tests was also obtained. Typical distribution of the axial compressive load along the pile axis is shown in Fig. 7. This axial load is obtained by multiplying the axial compressive stress by the cross-sectional area of the pile. As can be seen from this figure, the point bearing values of the pile are approximately 60 and 64 percent of the total vertical load for applied axial loads of 63 560 kg (70 tons) and 136 200 kg (150 tons), respectively. Details of the load transfer characteristics are presented later in this paper.

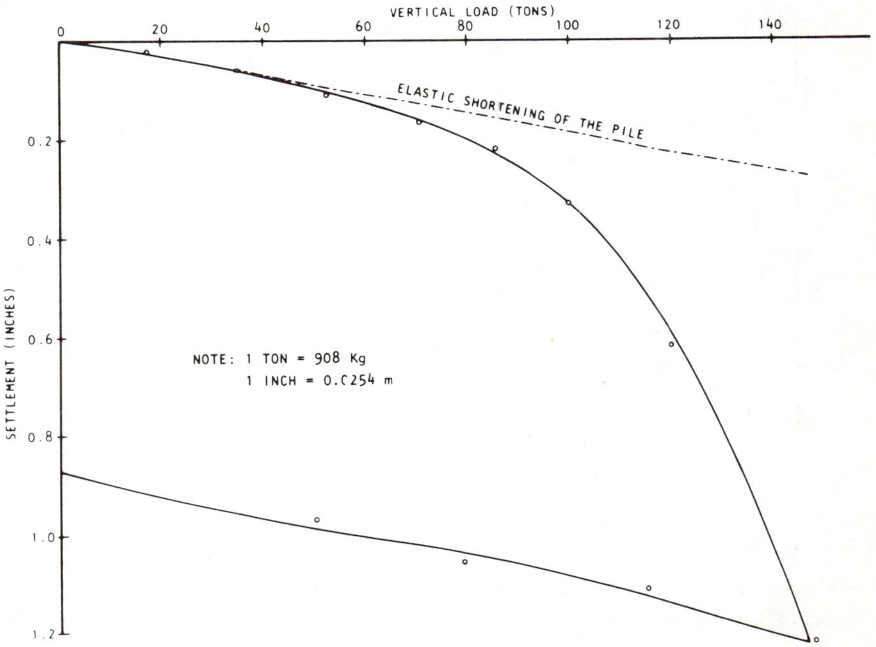

FIG. 6—*Static vertical load–settlement curve from pile load test.*

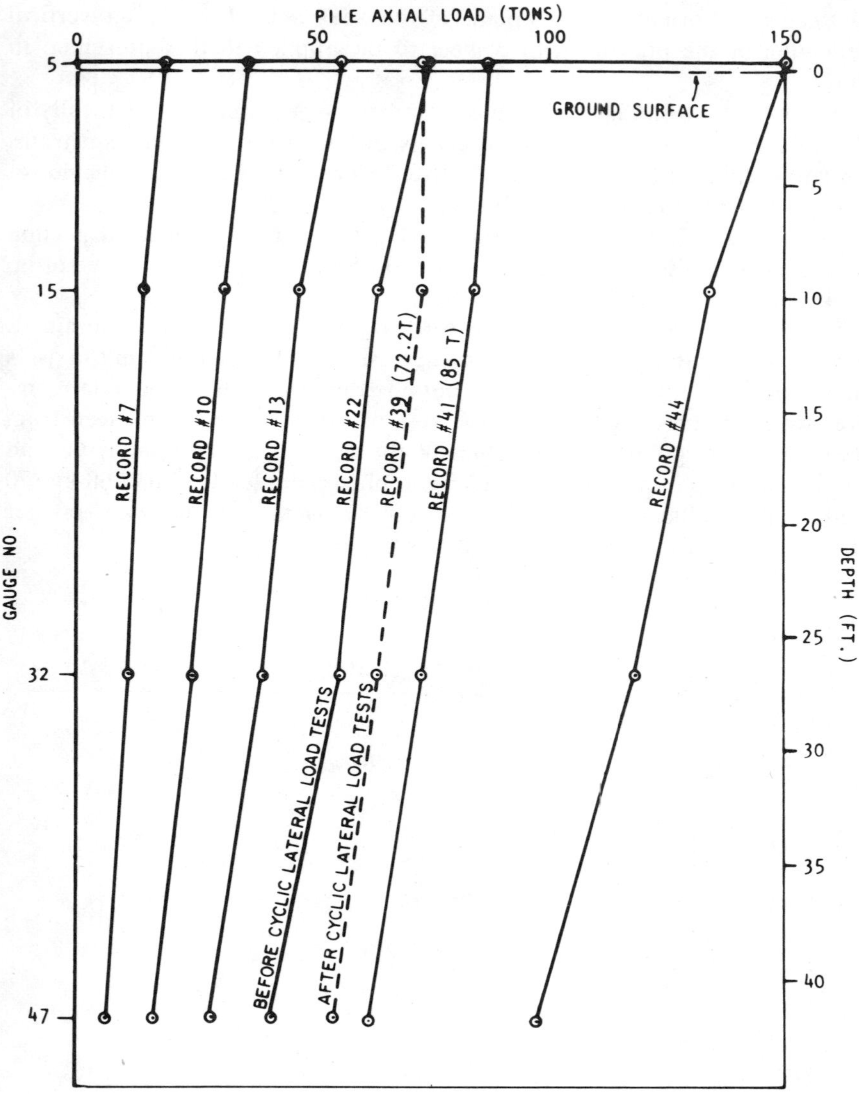

FIG. 7—*Load transfer characteristics of test pile.*

Cyclic Vertical Load Test

Two sequences of vertical cyclic loads were performed. In the first sequence, the static sustained vertical load was increased from 63 560 to 77 180 kg (70 to 85 tons). A cyclic load of ±13 620 kg (±15 tons) was then vertically applied to the test pile. In the second sequence, the test pile was subjected to a sustained static vertical load of approximately 127 120 kg (140 tons) and a cyclic vertical load of ±13 620 kg (±15 tons) superimposed. In all sequences, the applied cyclic load was approximately sinusoidal in shape and was imposed for 20 cycles at a frequency of 0.1 Hz. Dial gage readings were taken before and after the cyclic vertical load test.

The vertical deflection of the test pile during the cyclic vertical load test was monitored by continuous oscillographic records of the deflectometer output.

The results showed that the pile vertical deflection remained practically constant prior and after the application of the cyclic vertical load increments. The net permanent set resulting from 77 240 ± 13 620 kg (85 ± 15 tons) and 77 240 ± 13 620 kg (140 ± 15 tons) test series was found to be 0.05 cm (0.02 in.) and 0.275 cm (0.09 in.), respectively. This indicates that the applied cyclic vertical load had little or no effect on the stiffness of the pile in the vertical direction. The behavior of the pile was essentially elastic during cyclic load application.

The results of the deflectometer data during the cyclic vertical load test were also evaluated. Upon examination of these results, it was found that the cyclic loading essentially followed the same loading and unloading pattern as the static vertical load-deflection curve shown in Fig. 6. The magnitude and number of cycles of the applied cyclic vertical load increment [±13 620 kg (±15 tons)] had not reduced the vertical load resistance capacity of the test pile.

Load Transfer Characteristics of the Test Pile

Compressive forces were monitored during the load test by electric resistance strain gages mounted at five locations on both sides of the web of the pile in a full bridge arrangement. Data were recorded in analog form using oscillographic strip chart recorders as well as in digitized form using a digital voltmeter and, later, using a Vishay strain indicator as a check.

The oscillographic data was converted to digital strain data and then, based on the physical properties of the section, converted to load. After a full examination of the test data, it became apparent that simple conversion of stress to loads, based on the cross-sectional area of the pile, was not adequate. The contributary effects of the protective plates, the slope indicator tube, and the intermittent stitch welding of these accessories had to be evaluated. An upper-lower bound approach solution was then attempted where the items previously described had a contributary effect of 0 and 100 percent, respectively.

With this analysis, it was found that the uppermost gage location (gage no. 5), which was above the ground surface, generally mirrored the applied load for the upper-bound solution as determined by the vertical load cell.

The data obtained from the next lower location—gage Location no. 10, approximately 1.5 m (5 ft) below grade level—was erratic throughout the conduct of the test and was considered suspect and hence was not used.

The data obtained from the third location—gage location no. 15, approximately 3 m (10 ft) below grade level—was consistent, although the upper-bound solution yielded results that were in all cases in excess of the applied loads as determined by the vertical load cell. The maximum deviation of the load measured at this location from the applied load occurred immediately after the cyclic lateral loading program. The output of the gage at this time was assumed to be equal to the applied load, and a data reduction factor was then calculated. The factor that represents approximately 13 percent of the contributary effects of the properties of the protective plate was then applied to the data for this gage location.

Gage 32 and 47—approximately 8 and 12.6 m (25 and 41 ft) below grade level—were in areas where significant continuous welding of the protective plates was specified. For this reason, the upper-bound solution for loading was used. The results of our evaluation of the load transfer characteristics are shown in Fig. 7. The ratio of point bearing resistance to the applied load is shown in Fig. 8.

Evaluation of Test Results

As described previously, the purpose for performing this pile load testing program was to provide the relevant soil-pile parameters for a soil-structure interaction analyses of the existing facilities. There are 508 12BP53 H piles underlying the existing facilities. Although the test pile was of the

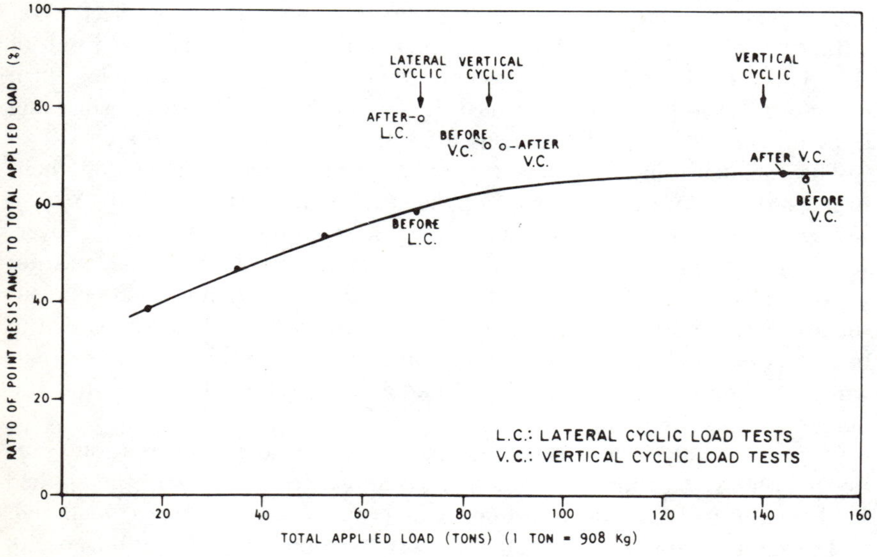

FIG. 8—*Load transfer pattern during load test.*

same 12BP53 H-pile section, the section modulus and moment of inertia of the test pile as well as the cross-sectional area were somewhat different from the existing piles due to the presence of two protective plates to house and protect the strain gages, as well as two square tubes to house the inclinometer to determine the verticality and straightness of the test pile. The characteristics of the test pile and the existing 12BP53 piles are shown in Table 3 for comparison purposes.

Due to the differences between the test pile and the existing 12BP53 piles it was necessary to perform an engineering evaluation based on the pile test results, actual pile characteristics, and configuration to obtain the following design parameters:

1. Vertical design load and ultimate pile capacity.
2. Vertical stiffness (spring constant) of the pile foundation for soil-structure interaction analysis.

Vertical Stiffness (Spring Constant)

The results of this pile load test program indicated that the cyclic load deflection pattern essentially followed the static load-deflection curve, and the cyclic vertical load did not exhibit any softening effect on the vertical stiffness of the pile. The vertical load-deflection curves shown in Figs. 2 and 6 can be used to obtain the vertical spring constant for the 12BP53 pile. The load-deflection relationship is nonlinear in nature and the resultant vertical spring constant for the existing 12BP53 pile expressed as a function of the induced vertical deflection are shown in Fig. 9. A close examination of the results presented in Figs. 2, 6, and 9 shows the vertical spring constants obtained from this pile load test are slightly greater than those obtained from the previous pile load tests performed prior to construction of existing facilities for vertical load values up to 90 720 kg (100 tons). Within this range of vertical load, the induced deflection in the test

TABLE 3—*Characteristics of test pile and existing 12BP53 piles.*

Characteristic	Test Pile	Existing 12BP53 H-Piles
Modulus of elasticity (psi)	29×10^6	29×10^6
Moment of inertia (in.^{4})		
Strong axis	448	394.8
Weak axis	198.3	127.3
Cross-sectional area (in.^{2})	18.91	15.58

Conversion factors—
1 in.^{2} = 6.45 cm^2,
1 in.^{4} = 41.62 cm^4, and
1 psi = 6.9 kN/m^2.

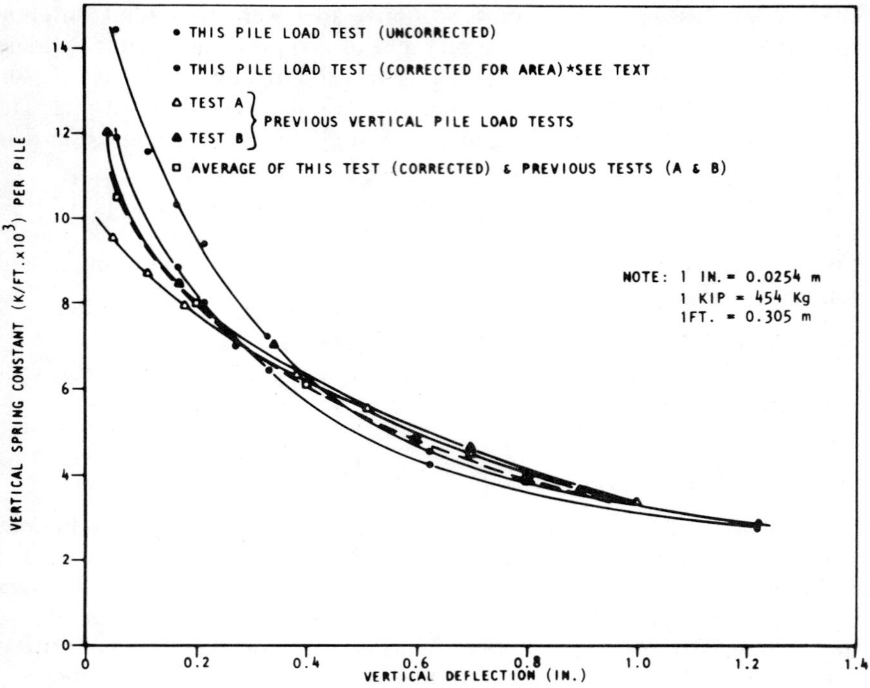

FIG. 9—*Vertical spring constant versus vertical deflection for existing 12BP53 pile group.*

pile was less than the 12BP53 piles tested previously because of the presence of the two protective plates and two slope inclinometer tubes, which increased the cross-sectional area of the pile. In other words, the vertical deflection in the test pile was less than what could have been expected in a regular 12BP53 pile without the protective plates. Correction to the induced vertical deflection can be made to the test pile results to extrapolate the test piles to the regular 12BP53 pile (without the protective plates and slope inclinometer tubes) by the following expression for pile shaft deflection w_s (elastic shortening of pile)

$$w_s = (Q_p + \alpha Q_s)(L/AE_b) \tag{1}$$

where

Q_p = point resistance of the pile,
Q_s = skin resistance of the pile,
A = cross-sectional area of the pile,
E_b = modulus of deflection of pile,
L = length of the pile, and

α = a coefficient depending on the distribution of skin resistance; a value of 0.5 was used in this study in accord with the results of this load test.

This expression was derived by Vesić[4] based on several pile test results and is considered applicable to this study. The results of the load-transfer characteristics of this pile load test are shown in Fig. 8.

The resultant vertical spring constant for the 12BP53 pile by application of the above deflection correction to the test pile data is shown in Fig. 9. The average spring constant results were recommended for use in Fig. 9. The average spring constant results were recommended for use in the structural analysis of the existing facilities.

Summary and Conclusions

In this paper, we presented the results of full-scale static vertical and cyclic vertical load tests on an instrumental pile. These tests were performed to simulate the dynamic behavior of the 12BP53 piles supporting the existing structure under earthquake loading conditions. This paper also provides the soil-pile spring parameters of the existing 12BP53 pile foundation for structural design purpose.

The results of this test, when compared to the previous series of load tests show a remarkable similarity of vertical capacity and demonstrate that the test pile is representative of the vertical pile behavior of the existing piles, both in terms of settlement and load-carrying capacity.

In addition, data derived from compressive gages along the length of the pile has indicated that the piles are primarily point-bearing in nature, with 60 and 64 percent on the applied design and double design test loads, respectively.

Two series of cyclic vertical pile loadings were performed at 77 240 and 127 120 kg (85 and 140 tons), respectively. A cyclic load of 13 620 kg ($\pm$15 tons) was superimposed on the static vertical load increment, and data were obtained.

In conclusion, the results of this static and cyclic pile load test program indicate that:

1. Axial deformation of the pile remained essentially constant prior to and after cyclic axial load application. In other words, the behavior of the pile was essentially elastic during cyclic load application.

2. Within the scope of testing performed, cyclic vertical load increments do not effect the load-carrying capacity or load transfer characteristics of the pile.

[4]Vesić, A. S., "A Study of Bearing Capacity of Deep Foundations," Final Report, Project 189, Georgia Institute of Technology, Atlanta, Ga., May 1967.

3. Load transfer characteristics varied with the magnitude of applied axial load. The ratio of point-bearing resistance to the applied load increased with the applied load. However, the rate of increase generally decreases with increasing applied load.

Based on these findings, the dynamic behavior of the axially loaded pile at the site can be approximated through the results of static axial pile load tests.

T. D. Lu,[1] *V. G. Miller,*[1] *and J. A. Fischer*[1]

Cyclic Pile Load Testing— Loading System and Instrumentation

REFERENCE: Lu, T. D., Miller, V. G., and Fischer, J. A., **"Cyclic Pile Load Testing—Loading System and Instrumentation,"** *Behavior of Deep Foundations, ASTM STP 670,* Raymond Lundgren, Ed., American Society for Testing and Materials, 1979, pp. 435–450.

ABSTRACT: The scope, concepts, methodology, and details of a load application system and instrumentation for the successful operation of a series of cyclic pile load tests are described.

The purposes of the pile tests on an instrumented H-pile were to evaluate the dynamic behavior of piles under cyclic loading conditions for seismic analyses and design of a structure supported on piles. The scope of the pile load tests included a static vertical, cyclic vertical, and cyclic lateral loading, with a sustained constant vertical load.

A major requirement of the cycle testing was to apply a controlled varying load intensity to the pile in a specific wave shape at a specific frequency. To fulfill this requirement, an automated closed loop loading system, which consisted of an electronic-electromechanical control system interfaced with hydraulic loading components, was developed.

An instrumentation program was implemented to measure the static and cyclic response of the test pile to include (*a*) bending moments and axial compression of the pile during driving; (*b*) loads, moments, angular rotation, and deflections of the pile during testing; and (*c*) the configuration of pile after driving and after each increment of load application.

KEY WORDS: pile load testing, instrumental H-pile, instrumentation, cyclic load testing, static and cyclic vertical and lateral load testing, earthquake simulation, load transfer, servo-controlled close loop testing, bonded strain gages

In order to define the available seismic design of an H-pile supported structure, a series of field tests, which consisted of (*a*) static axial, (*b*) cyclic lateral, and (*c*) cyclic axial load tests was performed on a fully instrumented 12BP53 H-pile. These pile load tests were intended to simulate the dynamic

[1] Senior engineer, senior engineer, and partner, respectively, Dames and Moore, Cranford, N.J. 07016.

behavior of the piles under earthquake loading conditions and to provide relevant pile-soil parameters for a seismic evaluation of the existing structure as well as to provide design criteria for a future expansion to be supported on piles of a larger cross section.

This paper describes the scope, concepts, methodology and details of the load application system and instrumentation of these pile load tests. The loading system and instrumentation program described in this paper represents a rather comprehensive attempt to understand the cyclic behavior of piles under various dynamic loading conditions. Although the intended scope of work was successfully carried out through detailed planning and careful execution of the program, refinement is still necessary and possible. It is hoped that the experience learned from this pile load testing program as well as the recommendations provided in this paper will be of benefit to future studies of cyclic pile behavior.

Relevant Considerations

There are various considerations influencing the details of the loading system and instrumentation program for a cyclic pile load test. These considerations include (but are not limited to)

1. Purpose
2. Intended scope of work
3. Subsurface conditions
4. Relevant soil-pile parameters required.

Purposes

As stated previously, the purpose of the cyclic pile load test program was to assess the dynamic behavior of the pile under earthquake loading condition and to provide relevant soil–pile parameters for a seismic evaluation of the existing pile-supported structure. The existing structure is supported by a total of 508 12BP53 H-piles with spacing of the order of about 1.1 m (3.5 ft) center to center. The static load on each pile under the weight of the structure is about 63 560 kg (70 tons). This sustained vertical static load on each pile was simulated in the pile load test program to define cyclic capacity under normal static loading conditions.

In addition, the characteristics of the design earthquake loading was considered. The earthquake ground motion can be characterized as cyclic in nature with erratic magnitude. Thus, cyclic loading in a pile load test should be applied in a specific wave shape and at a specific frequency to simulate the effect of earthquake loading characteristics (magnitude, frequency, and duration) on the structure.

Scope of Work

A loading system and instrumentation program was developed to successfully conduct the intended scope of work for each project as follows:

1. *Static vertical load test.* In this test, a static load of 63 560 kg (70 tons) (design load) was applied in increments of 25 percent of the design load until the full vertical design load was achieved.
2. *Cyclic lateral load test.* In this test, a cyclic lateral load was oscillated in increments of 2270 kg (2.5 tons) on either side of the zero point to the final cyclic load of 11 350 kg (12.5 tons), with a static vertical load of 63 560 kg (70 tons) maintained at all times. The cyclic load was applied for 20 cycles at a frequency of 0.1 Hz (1 cycle per 10 s).
3. *Cyclic vertical load test.* In this test, a variable vertical load cycling between 63 560 to 90 800 kg (70 to 100 tons) and 113 500 to 140 740 kg (125 to 155 tons) was applied for 20 cycles at a frequency of 0.1 Hz.
4. *Static vertical load to failure.* In this test, the static vertical load was increased from the design load of 63 560 kg (70 tons) to 77 240 kg (85 tons) and then to approximately 127 120 kg (140 tons).

Subsurface Conditions

The subsoils at the site are of glacial origin and can be characterized in general as glacial tills. Glacial deposits as a rule are heterogeneous mixtures of various soil types. In general, the bedrock, which is a shale with thin seams of dolomitic limestone, is overlain by three distinct layers of soils. The upper layer consists of a 4½ to 7½ m (15 to 25 ft) thick layer of brown medium dense to dense gravel with fine to coarse sand and clayey silts. The middle layer is a gray medium stiff silty clay with a variable gravel and sand content. The thickness of this layer varies approximately from 9 to 12 m (30 to 40 ft). The lower layer, which extends to bedrock at about 21.3 m (70 ft), is composed of a dense to very dense gravel and clayey silt with occasional weathered shale fragments. Throughout the site, the groundwater level generally ranges from 0 to 5 m (0 to 15 ft) below the ground surface.

Due to the presence of a relatively high gravel content and groundwater, protective measures were necessary to waterproof instrumentation located below grade and to protect them from physical damage due to pile driving. The pile was driven to a depth of 15 m (49 ft) in these soils penetrating about 2.1 m (7 feet) into the bearing stratum, similar to the existing piles in the adjacent structures.

Soil–Pile Parameters

The soil–pile responses required for this pile load test program were

1. Axial load-deflection relations under static and cyclic loading conditions.
2. Lateral load-deflection relations at grade level under cyclic lateral loading.
3. Axial and cyclic lateral load carrying capacity of the pile.
4. The relation between lateral deflection and soil reaction along the length of the pile.
5. Configuration of test pile after pile driving as well as the configuration before and after each static or cyclic load increment.

Items 1 and 2 provide stiffness values of the soil-pile system for seismic evaluation. Item 3 provides design criteria to verify the adequacy of the pile foundation prior, during, and after earthquake loading. It is worthwhile to point out that the lateral pile load testing was performed along the cyclic pile's major axis in this program. Information developed from item 4 is required to evaluate the cyclic lateral behavior of the pile along the minor axis and piles of other cross section for foundation support of future facilities. Data collection in item 5 is intended to check the configuration and adequacy of the test pile before and during the entire test.

Pile Loading System and Instrumentation

Loading System

A major requirement of the cyclic lateral and vertical load testing was to apply a predetermined controlled load intensity to the pile in a specific wave shape at a specific frequency. To fulfill this requirement, an automated loading system, which consisted of an electronic-electromechanical control system interfaced with hydraulic loading components was developed. The electronic and electromechanical control system is a fully automated closed-loop servo system consisting of a function controller, function generator, load cells, and servovalve. The cyclic lateral and vertical load were applied through a hydraulic system consisting of pumps, jacks, and accumulators, interfaced with the closed-loop system.

Electronic and Electromechanical Control System

In earthquake engineering, it is a generally accepted practice to use a frequency ranging from 1 to 2 Hz in testing for seismic response evaluation.[2] Various methods for the application of cyclic loads at the above frequencies were investigated, including manual, semiautomatic, and automatic gas and hydraulic systems. It was found that due to the limitations

[2]Seed, H. B. and Peacock, W. C., *Journal*, American Society of Civil Engineers, Soil Mechanics and Foundation Division, Vol. 97, No. SM 8, Aug. 1971.

imposed by the ability of hydraulic or gas to supply power and due to limitations of the accumulators, cyclic loading at these conventional frequencies would be expensive and impractical. Instead, as a practical and conservative alternate, it was decided that a loading frequency of 0.1 Hz would be utilized in this pile load-testing program.

It is worthwhile to point out that loading frequency has little or no effect on the dynamic response (stiffness and strength) of granular soils.[3] However, for clay soils, lower-frequency cyclic testing will produce less dynamic resistance under dynamic loading conditions.[4] Thus, the use of 0.1 Hz in the cyclic load test will yield conservative results.

With the selection of the cyclic loading frequency of 0.1 Hz, it was concluded that most of the conventional manual and semiautomatic gas and hydraulic systems investigated would be able to develop or deliver the required loads. The major problem, it was felt, would be the shape of the loading curve. Some "tuning" cycles of the loading system would be required to provide a reasonable wave shape approaching the sinusoidal shape desired, before testing, and adjustment of the system would be required during testing to accommodate changes in the pile-soil behavior. Because of these serious drawbacks to the intended operation of the system, a fully automated closed-loop servo system was developed for this test program.

The automated closed-loop servo system was developed through modification of apparatus and techniques used in cyclic triaxial testing of soil samples for dynamic analysis.

The automated closed-loop system consists of a function generator (MTS model 410.31), a controller MTS model 406), strain gage type load cells, and a servovalve. The 410.31 digital function generator is a high-accuracy waveform synthesizer of ramp, sinusoidal, and square wave functions. This generator is specifically designed as a programmer for closed loop servohydraulic testing. The model 406 controller is designed to control one channel of a closed-loop servohydraulic testing system. In general, a closed-loop testing system operates as follows: (1) The digital function generator supplies a signal or program to the controller. (2) The controller compares this to a signal being returned from a transducer (feedback). (3) Any differences between the command and feedback signals is converted into an error signal, which is fed to the servovalve.

More precisely, the command signal is made up of two components, a static level provided by the servocontroller and a dynamic program provided

[3]Lee, K. L. and Fitton, J. A. in *Vibration Effects of Earthquakes on Soils and Foundations, ASTM STP 450*, American Society for Testing and Materials, 1969, p. 71.

[4]Fischer, J. A., Koutsoftas, D. C., and Lu, T. D., "The Behavior of Marine Soils Under Cyclic Loading," Contribution to Proceedings BOSS '76 (International Conference on the Behavior of Offshore Structures), Norwegian Institute of Technology, Trondheim, Norway, Aug. 1976.

by the function generator. Command represents the desired test program, while feedback represents the actual controlled variable—in our case, load. Whenever command and feedback are not equal, an error is present, and the control signal drives the servovalve open in the required direction to bring the controlled variable or feedback signal to the command level. A schematic showing the functioning of the system is shown in Fig. 1.

The cyclic lateral and vertical load were applied through a hydraulic system interfaced with the MTS closed-loop servo system as previously described. The major components of the combined system are shown in Fig. 2.

The system generally was operated in the following manner to produce the required load:

1. The accumulators were precharged to the minimum required pressure to efficiently operate the servo valve [±10350 kN/m² (psi ±1500)] or higher when required for higher lateral load increments.
2. The hydraulic power supply was operated to maintain a uniform oil temperature.
3. The function generator was set for (a) the proper wave shape (sinusoidal) and (b) the frequency was set for 0.1 Hz (1 cycle per 10 s)

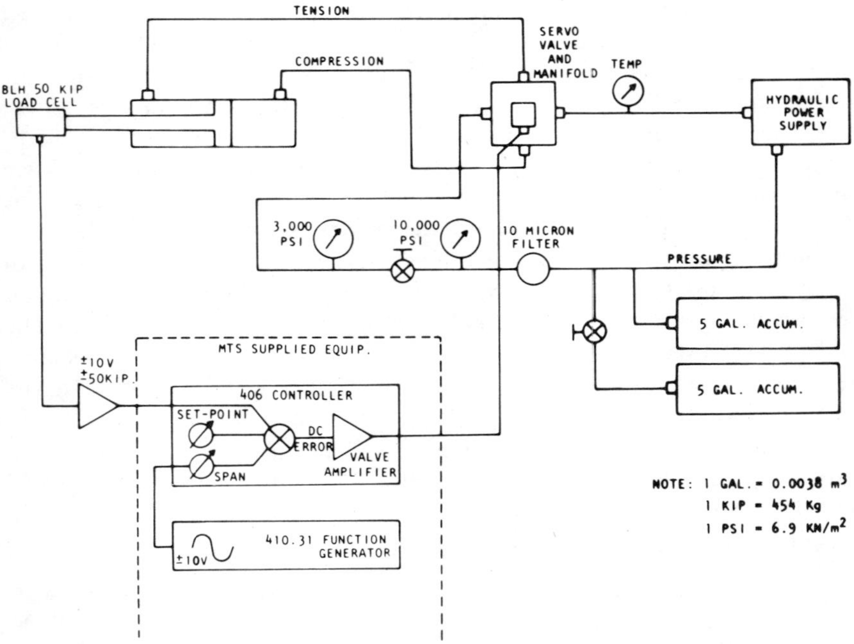

FIG. 1—*Combination of automatic load control and hydraulic system.*

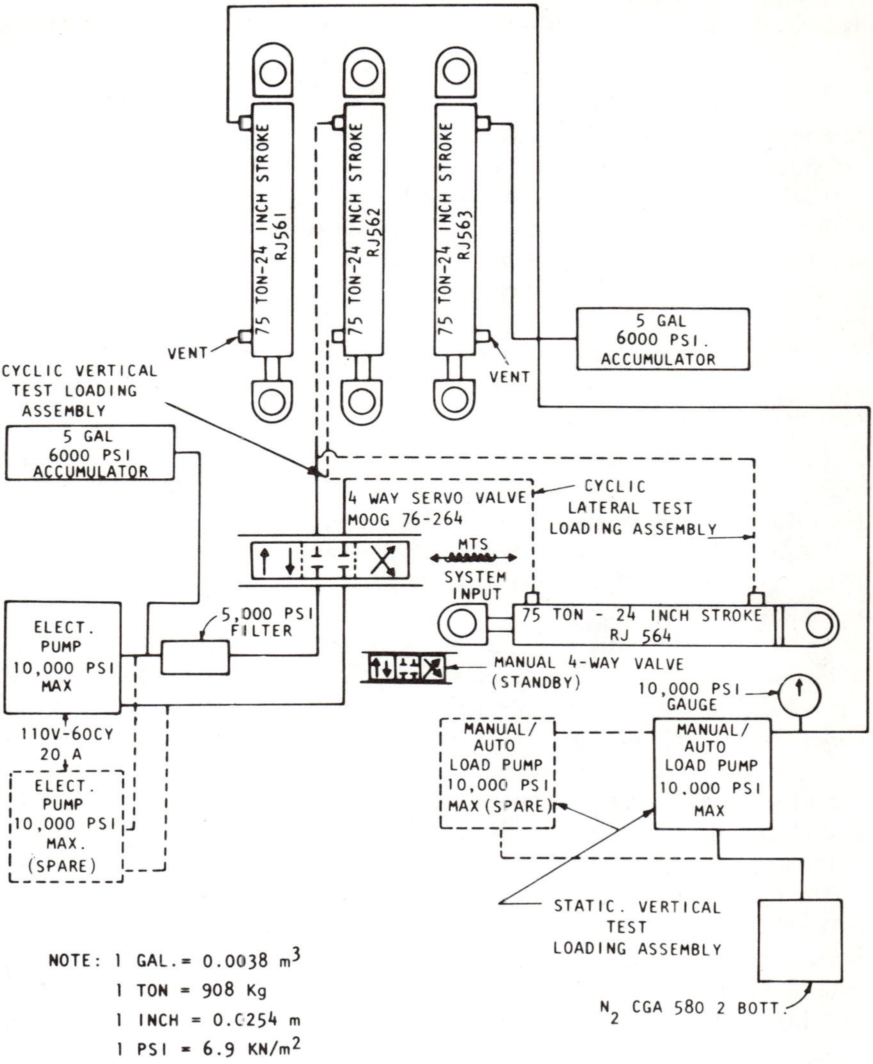

FIG. 2—*Hydraulic system schematic.*

4. The set point control was set to midrange (zero static displacement of the ram).
5. The span control was set for the proper intensity of load.
6. All personnel made final adjustments; static readings were obtained.
7. Recorders were started.
8. Twenty cycles of load were run (on some occasions latter cycles lost

load or less than 20 cycles were run due to physical limitations of the hydraulic loading system utilized in this program).

9. Recorders shut down.
10. Static readings obtained.

The preceeding outlined procedure was followed in the execution of each increment of the cyclic lateral and vertical load tests.

In general, the system functioned well within the limits imposed by the ability of the hydraulic system to supply hydraulic power. The servovalve normally passes a quantity of hydraulic fluid when not in use, and it was necessary to valve it out of the system when not in use or when charging the accumulators.

Hydraulic System

The setup of the hydraulic system is as shown in Fig. 2. All jacks and gages were calibrated prior to the load test. The actual loads applied in both directions, both static and dynamic, were determined by the use of calibrated electrical resistance type strain gage load cells. Data were recorded in analog form on multichannel strip chart recorders and manually converted to digital form.

The loads, both static and dynamic, were applied to the pile by means of a hydraulic loading system. The vertical static loads were applied to the top of the pile by two double-acting, clevis-ended, calibrated hydraulic jacks. The jacks were pressurized by means of a combination hand operated–gas operated hydraulic pump. Loads were determined initially by means of a calibrated pressure gage of test quality and recorded digitally and in analog form electronically by means of a strain gage load cell and multichannel strip chart recorders.

The cyclic lateral load test was performed on the axially loaded test pile [sustained axial load of 63 560 kg (70 tons)] by the application of a variable horizontal thrust in two directions, using a double-acting, clevis ended, calibrated hydraulic jack. The jack was pressurized by means of a 0.00095 m^3/min, 69 000 kN/m^2 (¼ GPM, 10 000 psi) electrically driven hydraulic pump working through a 0.019-m^3 (5-gal) accumulator.

Control of the load to a predetermined intensity as well as shaping of the loading curve to a specified sinusoidal shape and frequency was achieved through the closed-loop servo hydraulic system described previously. A manual control system as backup to the servohydraulic system was available, but not used.

After the performance of the cyclic lateral load test, the test cylinder assembly, including the control system, was disconnected and fitted in the vertical direction between the two 68 000 kg (75-ton) capacity hydraulic jacks that were being used to maintain the static vertical load.

The static vertical load was increased to 77 240 kg (85 tons) and an oscillating vertical load of ±13 620 kg (±15 tons) was superimposed on the static vertical load using the control equipment as previously described. Again, the manual backup system was available, but not used.

Test Setup

A schematic drawing of the reaction frame and test setup for the load system is presented in Fig. 3. This test frame and load setup scheme shows how the pile was loaded as a free end pile in various stages of the pile load test.

A total of five 12BP53 steel H-piles were installed at the location of the pile load test. The pile load test incorporated the instrumented test pile, the four reaction piles that provided the necessary reaction against uplift and lateral forces and a test frame to transfer loads from the reaction piles to the test pile. A detailed description of the instrumentation of the test pile is presented later in this report.

Each of the five piles was driven with a Raymond 65C differential acting pile hammer, with a rated energy of 2700 m-kg (19 500 ft-lbs) per stroke and a ram weight of 2955 kg (6500 lbs). A cap block of micarta–aluminum was used in driving each of the piles. The pile driving equipment, pile section, and driving criteria used in this study were the same as those used for the 12BP53 H-piles underlying the existing facilities at the site.

The four reaction piles were the first to be driven. These piles were driven without pile points attached. The instrumented test pile was driven with a Pruyn point BP 75600 pile shoe attached to prevent the test pile tip from damage and bending during pile driving.

After the installation of the piles was completed, the piles were cut off at the required elevation and the reaction frame was fabricated as shown in Fig. 3. However, because the test pile had rotated about 15 deg clockwise during pile driving, the lateral reaction beam had to be rotated through the same angle, the additional struts were added to permit perpendicular lateral load application and to distribute evenly the load to the reaction frame and piles to provide a frame that will be stiff enough to resist the lateral loads adequately. The final "as-built" plot plan of the test pile and reaction piles is shown in Fig. 4.

The rotation of the test pile also necessitated a rotation of the vertical jacking system to place it in a plane coincident with the plane of the horizontal jacking system.

Instrumentation

Basically, the measured quantities can be separated into below- and above-grade measurements. Below the test pit surface, major and minor

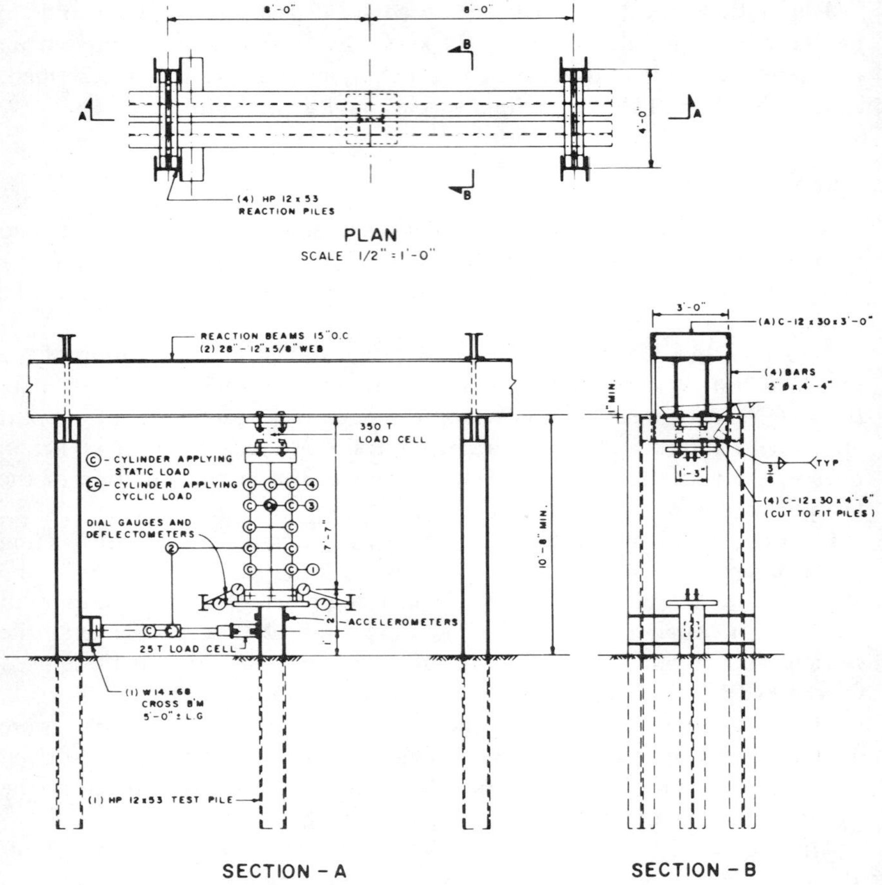

TEST No.	TEST	LOAD SEQUENCE
①	VERTICAL TO 70 T	0 → 17.5 T → 35 T → 52.5 T → 70 T
②	VERTICAL LOAD OF 70 T WITH CYCLIC LATERAL LOAD TO ± 15 T	CYCLIC LATERAL ± 2.5 T ± 5 T ± 7.5 T ± 10 T ± 12.5 T ± 15 T
③	CYCLIC VERTICAL LOAD	CYCLIC BETWEEN 70 T AND 100 T 125 T → 155 T
④	ULTIMATE VERTICAL LOAD	70 T → 105 T → 140 T → 105 T → 70 T → 35 T → 0

NOTE 1 INCH = 0.0254 METER
1 FOOT = 0.305 METER
1 TON = 908 KILOGRAMS

FIG. 3—*Test pile setup—reaction frame, loading system, and above-grade instrumentation.*

bending, compression, and the slope of the pile were measured before and after each increment of load. The above-surface measurements included the vertical and lateral load, the vertical and lateral deflection, and the tilt of the pile top. The instrumentation used to measure these quantities is as follows.

Strain Gages—Test pile fabrication and preparation—As stated previously, the test pile basic section was a 12BP53. As a result of the need to install

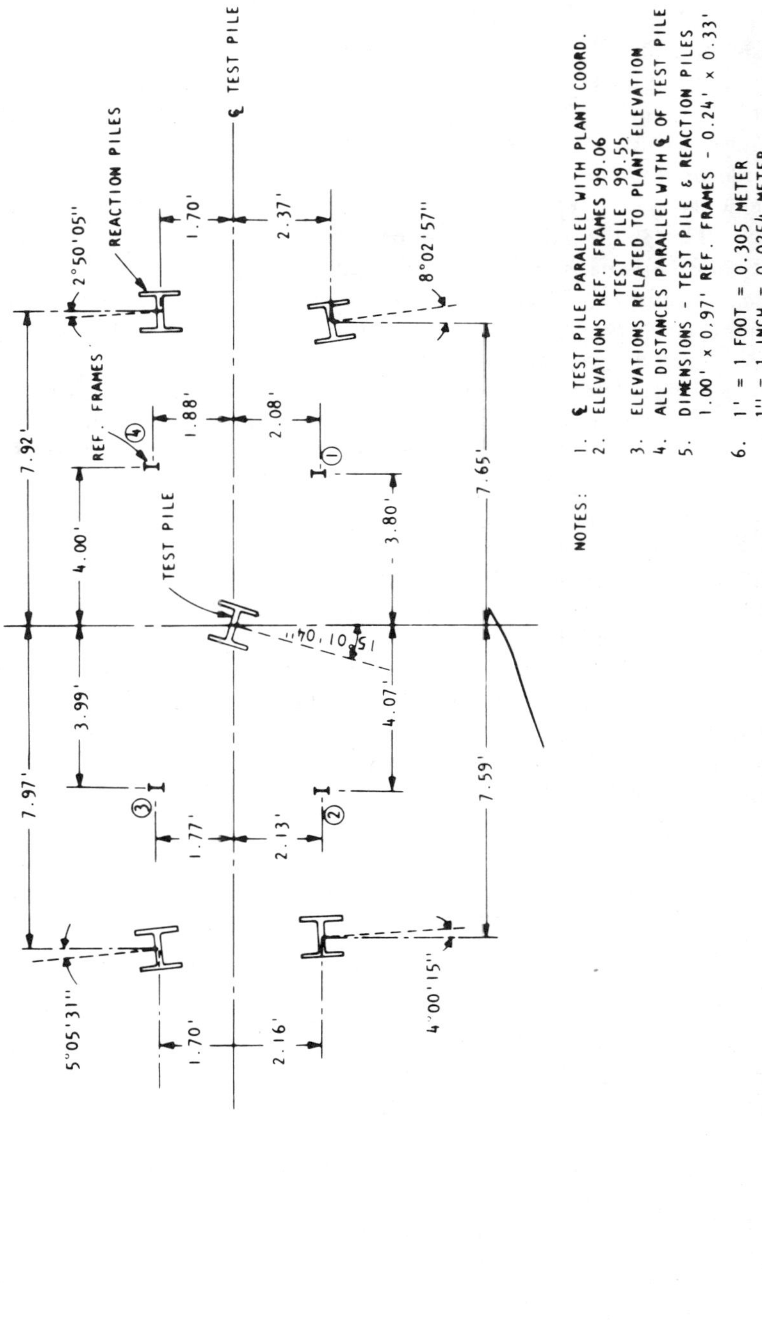

FIG. 4—*As-built configuration of test setup.*

surface strain gages, it was necessary that a fresh-milled structural shape be obtained, free of rust and pitting, to facilitate the application or bonding of this type of strain gages. The location of the 0 and 47 gage stations were established, and the intervening gage locations were established from these. The mill scale was removed by surface grinding from the area directly below and immediately adjacent to the gages, both for gage bonding and for waterproofing purposes.

Strain gaging of the pile—The test pile was instrumented to measure the compressive forces, major bending, and minor bending of the pile during pile driving and during the lateral and vertical load tests. Micro-Measurements type LWK-06-W250B-350 precision weldable strain gages were used. The 350-Ω gages are compensated for steel with a coefficient of linear thermal expansion of 6×10^{-6} cm/cm/°F. The locations of the gages are shown in Figs. 5 and 6.

One hundred forty gages were bonded to the test pile using capacitative discharge stored energy spot welders. After bonding, electrical continuity and electrical resistance to ground were checked. The gages were installed as half bridges, two gages on each side of the pile, and wired using Belden type 8424 cable into full-bridge configuration at the top of the pile. The bridges were wired to measure either major axis bending, minor axis bending, or axial compression.

The Belden 8424 cables were attached to the pile with spot welded shim stock straps and U bolts mounted on the pile within 2 m (8 in.) of each gage location. The cables were protected from damage using fiberglass tape and rubber hose material.

The strain gage installations were waterproofed with a coat of Products Research and Chemical Corporation synthetic rubber. For mechanical protection, this was covered with thin strips of 3-mil shim stock spot welded to the test pile. The shim stock was then overcoated with another layer of the synthetic rubber compound.

These finished installations then underwent 6-h water immersion test and several days of immersion during welding of the protective plates to the pile. At the end of this period, the resistance to ground for all gages was found to be satisfactory. In other words, all gages were found to be functional after the water immersion test.

The major bending strains were measured at 18 specific web locations along the test pile. The gages are in full-bridge (strain electrically additive) arrangement with a bridge constant of about 2.5. The strain gages located between the surface and a depth of about 10.5 m (32 ft) (location of strains of major interest) were assigned oscillographic recording channels of not less than 3.048-cm (1.2-in.) width. The remaining major bending strains were recorded using channel widths of not less than 0.508 cm (0.20 in.).

Minor bending strains were measured at seven specific flange locations along the embedded portion of the test pile. These gages were in full-bridge

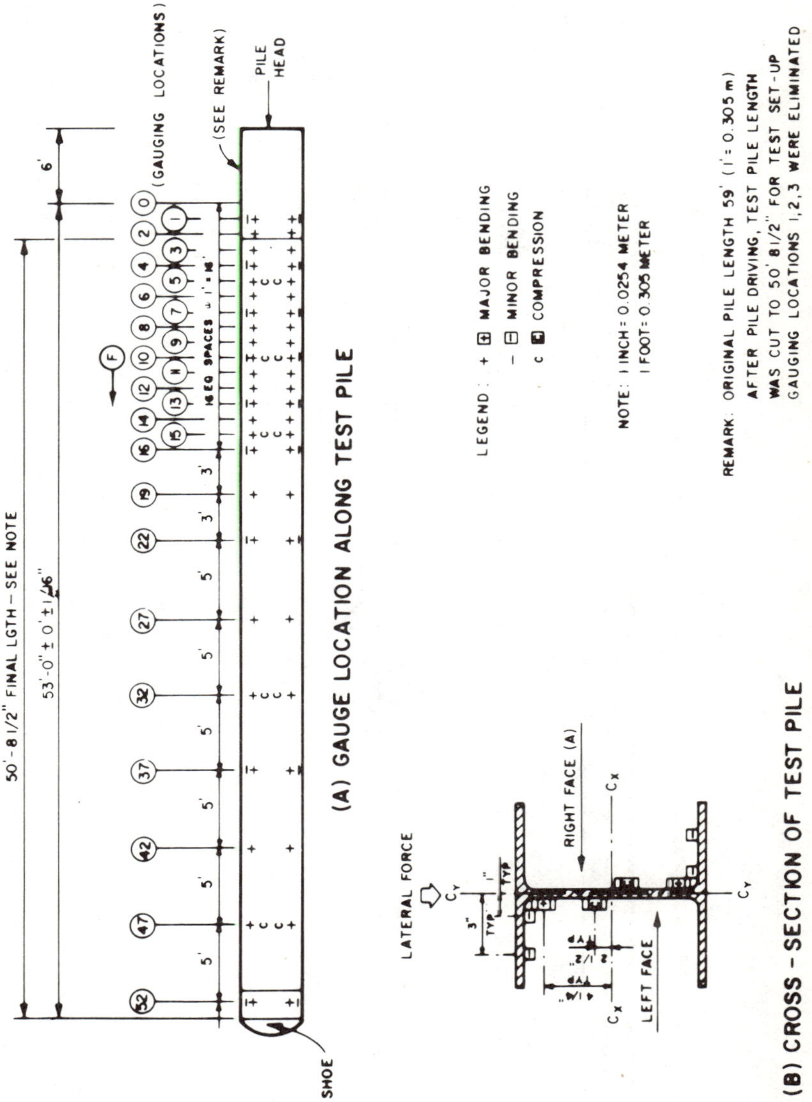

FIG. 5—*Strain gage locations.*

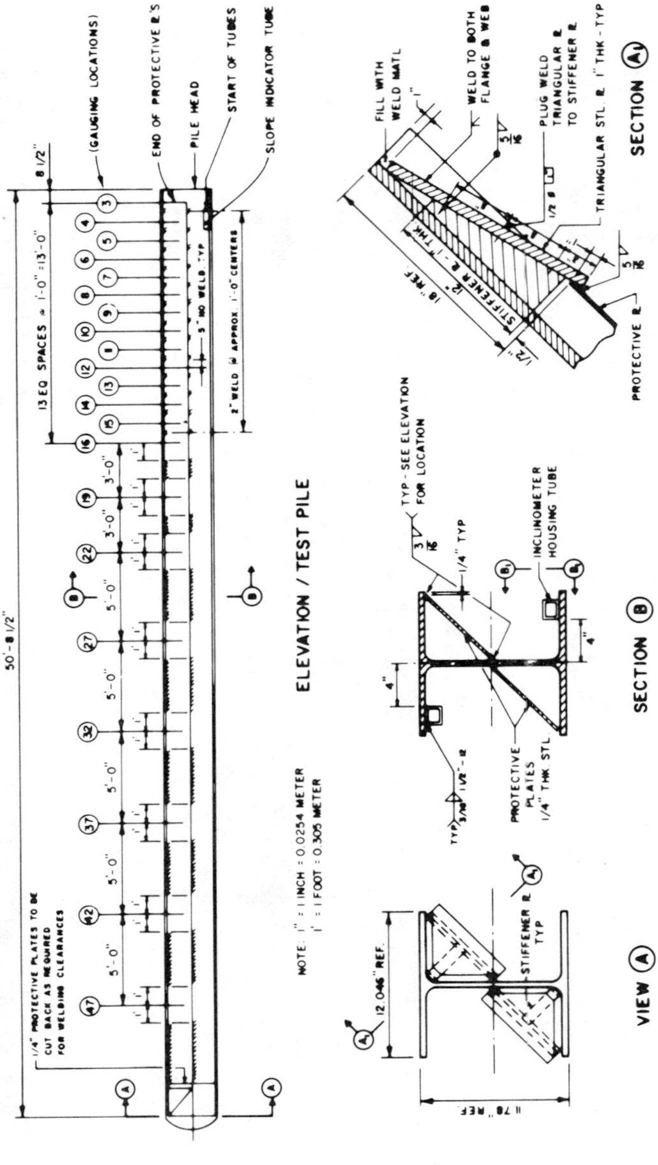

FIG. 6—*Details of protective plates and bottom closure.*

(strain electrically subtractive) arrangement with the bridge constant of approximately 1.3. The bending strains were recorded on channel widths of not less than 0.508 cm (0.20 in.) in width.

The compressive strains were measured at five web locations along the test pile. These gages are in full-bridge (strain electrically additive) arrangement with a bridge constant of about 2.5. The compressive strains were recorded on channel widths of not less than 0.508 cm (0.20 in.).

To protect the gages from damage during pile driving, two 0.64-cm (¼ in.) thick protective plates were spot welded to the web and flanges of the pile along its axial length. Triangular steel plates for bottom closure were also welded at the end of the pile for the same purpose. Details of the protective plates and bottom closure are shown in Fig. 6. After pile driving, all gages were operative except for one compressive strain gage location (location 10cc in Fig. 5), which exhibited erratic behavior.

Inclinometer—In order to measure the position of the tip of the driven pile relative to the top of the pile, a digital inclinometer analysis was used in two square access tubes welded near the edge of diagonally opposite flanges of the test pile.

Data to determine the position of the tip of the pile relative to the top of the pile were acquired at 0.305-m (1-ft) increments by the use of a biaxial accelerometer type inclinometer traveling along the diagonal of a square steel tube, which was welded to the test pile.

These data were later rehandled to provide a basis to determination of the movement of the top of the pile relative to a fixed point below the ground surface [7.625-m (25-ft) depth]. The deflections were resolved into components perpendicular and parallel to the plane of horizontal loading.

The inclinometer used was a Digitilt model 50301 inclinometer manufactured by the Slope Indicator Company of Seattle, Wash. It was fitted to operate within a 3.18-cm (1.25-in.) inside diameter flush-coupled square steel casing.

The Digitilt is a precision inclinometer usually used as an aid in the monitoring of the performance of embankments and the monitoring of the movements of earth slopes. The Digitilt has a claimed system accuracy of one part in 10 000 through a range of ±30 deg from the vertical or ±0.023 m per 100 m of casing.

The sensing elements are two 0.5-g servoaccelerometers mounted at 90 deg to each other within a waterproof casing. A tilt of the inclinometer produces a voltage that is proportional to the sine of the angle of inclination. The output voltage is stable with respect to temperature, input voltage variation, and line resistance. The output of the accelerometers is read on a digital voltmeter located in a control box. The control box provides voltage for energizing the circuits and switching for selection of the proper axis accelerometer. It has a four-digit voltmeter with an automatic polarity indication as well as a BCD output for recording purposes.

Instrumentation above grade—To measure the axial and lateral deflections and the tilt of the pile head during this pile load test program, an above-grade instrumentation package consisting of dial gages, linear variable strain deflectometers, and servoaccelerometers was utilized. A schematic sketch of this instrumentation setup is shown in Fig. 3. All components of this above-grade instrumentation package were calibrated in accordance with standard calibration procedures prior to installation.

Four calibrated dial gages and four deflectometers were positioned on the plate welded to the top of the test pile under various load applications. These gages and deflectometers were mounted on a reference beam, which supported them entirely independent of the test and reaction piles. Two vertical and two lateral dial gages and two deflectometers to measure vertical and lateral motion were positioned at the northeast and the southwest corner of the pile, respectively. Elevations were then obtained on the corners of the reference beams by surveyors, and zero readings were recorded at both locations. Additional readings to provide gross visual control of test operations were also obtained by a surveyor during conduct of the test.

Conclusions and Recommendations

The loading system and instrumentation program for the cyclic pile load tests described in this paper represents a comprehensive attempt to obtain measurements, and understand the behavior of a test pile under cyclic loading conditions. The details and degree of sophistication of the loading system and instrumentation depends on the purpose, scope of work, subsurface condition and required parameters.

The loading system and instrumentation program for the cyclic pile load tests described represents a comprehensive attempt to obtain measurements to understand the physical behavior of the test pile as well as to validate the model used in the analysis. In this respect, the program developed met in all respects, the needs of the project.

Development of programs in the future should be made on the basis of the degree of automation/sophistication required, loading to be modeled and parameters to be measured.

M. W. Montgomery[1]

Pile Load Tests to Evaluate Load Transfer Mechanisms

REFERENCE: Montgomery, M. W., **"Pile Load Tests to Evaluate Load Transfer Mechanisms,"** *Behavior of Deep Foundations, ASTM STP 670,* Raymond Lundgren, Ed., American Society for Testing and Materials, 1979, pp. 451-463.

ABSTRACT: Load transfer mechanisms in 17 test piles are evaluated by means of an extensive monitoring program during full-scale load testing and are compared with behavior computed by a displacement compatibility analysis. H-piles, Raymond step taper piles, auger-cast piles, and steel pipe piles were tested, in many cases to failure. These piles were installed at a power generating station in saprolites of the Piedmont Province of Georgia. Deflection was measured at the butt of each pile and at the tip and one or two intermediate levels in most piles by means of telltales. A boring and laboratory testing program was an integral part of the program. Load transfer mechanisms are evaluated by means of a displacement compatibility analysis using results of the field and laboratory investigations. Good agreement between computed pile behavior and that determined by the monitoring program during load testing is demonstrated. The results of the monitoring program and the enhanced understanding of the load transfer mechanisms involved make a comparison of the effectiveness of the various pile types possible and allow development of rational installation criteria.

KEY WORDS: piles, load tests, shear tests, skin friction, end bearing, load transfer, geotechnical engineering

At Georgia Power Company's Plant Scherer in Monroe County, Ga., two cooling towers and coal-handling facilities required pile support. In order to determine the most economical pile types for the design capacity and to develop rational installation criteria, an extensive field and laboratory investigation was performed. The objective was to assure an appropriate foundation at minimum cost. Consequently, steel H-piles, pipe piles, Raymond step taper piles, and auger-cast piles were installed. In order to account for differences in load transfer by these piles, installation criteria unique to the pile type were developed. These criteria resulted in variable pile lengths at each test area. This paper describes the test program and a

[1] Project Manager, Law Engineering Testing Co., Marietta, Ga. 30067.

displacement-compatability analysis using laboratory data. The test results and the analysis were used to examine load transfer mechanisms for these piles.

Test Program

The test program consisted of three phases—drilling and sampling, laboratory testing of representative soil samples, and installation and testing of the piles. These phases are described in the following sections.

Drilling and Sampling

Thirty-eight borings were made in the areas of interest during subsurface investigations at the site. The borings typically included standard penetration tests (ASTM D 1586) at 1.5- or 3-m (5- or 10-ft) intervals and undisturbed samples of representative soils.

Laboratory Testing

The stiffness of the soils influenced by the piles was determined by direct shear tests and triaxial compression tests. Consolidated undrained direct shear tests were performed on 6-cm (2.36-in.) diameter specimens utilizing a single shear device. The undrained modulus of elasticity of the soils was obtained from saturated consolidated undrained triaxial compression tests performed on 3.5-cm (1.4-in.) and 7-cm (2.8-in.) diameter specimens.

Test Piles

Mandrel-driven cast-in-place concrete (Raymond) piles, steel H-piles, and closed-end steel pipe piles were driven by Raymond International, Inc. The Raymond piles consisted of a lower 25-cm (10-in.) diameter pipe section followed by 2.4- or 3.7-m (8- or 12-ft) step taper segments. The steel piles were HP 14×73 sections and 30-cm (12-in.) diameter 1-cm (3/8-in.) thick closed-end pipes. All piles were driven with a Raymond 00 single-acting steam hammer with a rated energy of 44 100 Nm (32 500 ft-lb) per blow. Step taper and pipe piles were driven after predrilling to various depths. Predrilling was terminated in every case at least 1.8 m (6 ft) above the pile tip elevation.

The Turzillo Corporation installed five auger-cast test piles at the site by grouting through a hollow-stem continuous flight auger as it was withdrawn from the ground. These piles had a nominal diameter of 36 cm (14 in.), except for a single 41-cm (16-in.) pile.

Piles were installed in four test areas (cooling tower 1, cooling tower 2, coal-handling areas 3 and 4). Lengths ranged from 9 to 27 m (30 to 89 ft).

Seventeen piles were tested—five step tapers, five H-piles, five auger-cast, two pipes. Step taper, H-piles, and auger-cast were installed in all four test areas to verify the ability to achieve 623-kN (70-ton) design capacities. Pipe piles were included in test areas 3 and 4. An additional step taper, H-pile, and auger-cast pile were installed in test area 4 to determine whether 890-kN (100-ton) working loads could be achieved.

Load tests were performed essentially in accordance with ASTM Testing Piles Under Axial Compressive Load (D 1143). Loads were applied incrementally by calibrated hydraulic jacks bearing on the pile butts and reacting against steel beams fastened to adjacent reaction piles. Movement of the pile butts was measured by micrometer dial gages and by wire-mirror-scale devices. These measuring devices were supported at a sufficient distance from the piles to be unaffected by them.

Deflections of the pile tips and of intermediate points were measured in many of the piles by telltales. These telltales consisted of a steel rod embedded in the pile concrete at the appropriate depth or affixed to the H-pile and protected by an outer plastic or steel casing. The relative movement between the telltale rod (fixed at its lower end) and the butt of the pile was measured by dial gages. The butt movement less the relative movement represents the deflection of the pile at the point where the telltale was fixed.

Loads were applied in increments of 25 percent of the design capacity up to twice the design value. Each increment was maintained for 1 h, except that at twice the design load, it was maintained for 24 h. During each increment, deflections were measured at standard time intervals. After 24 h at the test load, the piles were unloaded completely. Then the piles were reloaded in relatively short-duration increments to failure or to the limiting jack capacity.

Subsurface Conditions

The site is located within the southern portion of the Piedmont Geologic Province, whose Paleozoic and Precambrian rocks are among the oldest in the United States. The rocks are highly metamorphosed gneisses and schists. These rocks have been deeply weathered in the warm humid climate, forming a residual soil mantle that is generally quite thick. The deeper soils are less weathered and retain the original banding and structure of the parent rock, as evidenced by segregation of minerals into bands. These materials are termed "saprolites." The deeper saprolites may be described as silty sands, with an increasing proportion of silt-size particles with decreasing depth. The upper saprolites may be sandy silts, although generally particles coarser than 0.074 mm predominate. In general, their behavior is typical of that described for sands in the literature when subjected to applied stresses, except for some residual bonds (true cohesion). These bonds have some influence on the soil strength as used in limiting equilibrium analyses, but probably have little effect on the analysis used in this project. Because of their content of mica and other flakey minerals, including kaolinite, they exhibit more compressibility and a lower modulus of elasticity than similarly graded soils consisting entirely of bulky grains.

In the four test areas, soil depths range from 7 to 23 m (23 to 75 ft). These correspond to depths at which saprolites with penetration resistances of 100 blows/ft or greater are encountered. In general, 3 m (10 ft) or more of material with penetration resistances of more than 100 overlies sound rock. This material is commonly termed "partially weathered rock." In general, the groundwater depth in the areas of interest was 3 to 6 m (10 to 20 ft). Figure 1 shows conditions at locations of piles for which results are described in a later section.

Analysis Procedure

The deflection-compatibility analysis utilized is one that is similar to procedures described by Seed and Reese [*1*],[2] Clemence and Brumund [*2*], and Clisby et al [*3*]. All of these procedures differ in some respects, and they all differ somewhat from that utilized in this program.

The analysis used in this project is a numerical solution. The pile is divided into a number of segments. As the number of segments used increases, the accuracy of the approximate numerical approach relative to an exact integration of forces along the pile also increases. As the number of increments approaches infinity, the integration is closely approached. An asymptotic relationship to the closed-form solution is expected.

For each pile length increment, a stress-deflection relationship for the soil adjacent to that segment is assigned, based on direct shear data from undisturbed samples. Adjustment of the data to account for disturbance during pile installation is not considered necessary in this study, as the soils involved are primarily silty sands of medium density, which are not observed to lose significant strength due to disturbance. Straight-line incremental approximations of the actual nonlinear stress-deflection curves are utilized. Figure 2 is an example of the approximations of the actual laboratory-determined stress-deflection curves. There is disagreement in the literature over the location of the shearing surface adjacent to smooth steel piles. For this study, soil-on-soil shear was assumed adjacent to the pile. In the case of H-piles, soil-on-soil shearing along the rectangular surface enclosing the pile was assumed.

The deflection of the soil beneath the pile tip is characterized by the modulus of elasticity. Values of the modulus of the material beneath the pile tip were estimated from static values determined from triaxial shear

[2]The italic numbers in brackets refer to the list of references appended to this paper.

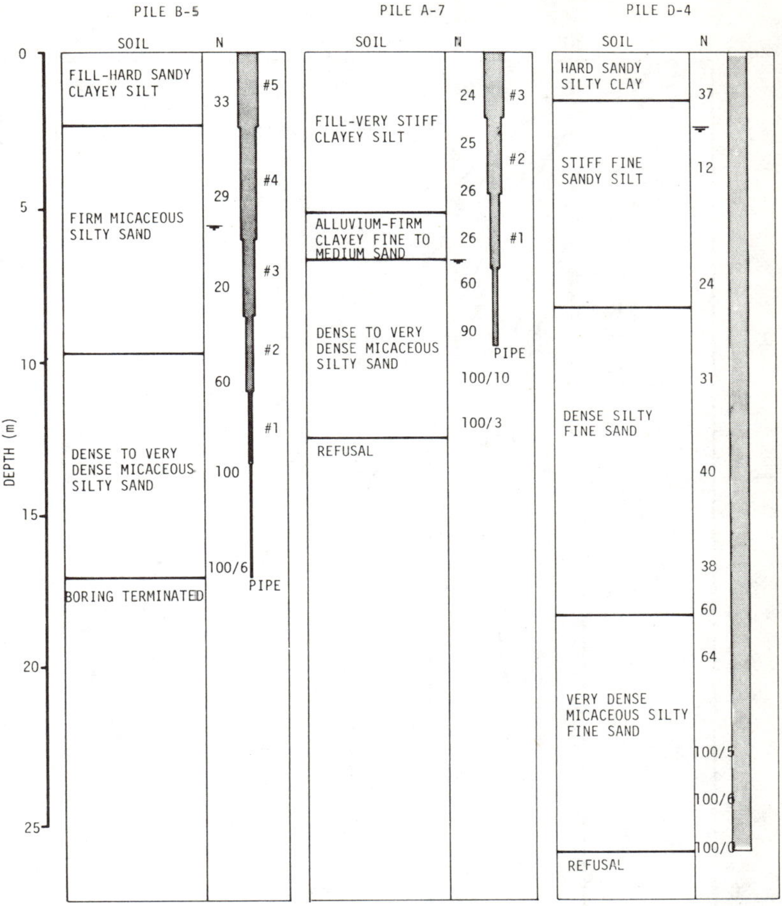

FIG. 1—*Subsurface conditions at pile locations (1 m = 3.28 ft).*

tests (generally consolidated undrained tests cycled a few times within a stress range well below ultimate).

The solution begins with the selection of an arbitrarily assumed tip deflection (see Fig. 3). The relationship between deflection and stress at the pile tip is defined by an equation that characterizes deflection within an elastic medium of the form proposed by Janbu et al [*4*].

Use of the assumed tip deflection in the stress-deflection relationship characterizing the material beneath the tip results in an average contact pressure and force at the tip. The force at the top of the bottom increment is larger because some of the load in the pile is distributed into the soil along this lower segment. An assumed value is selected. The average force

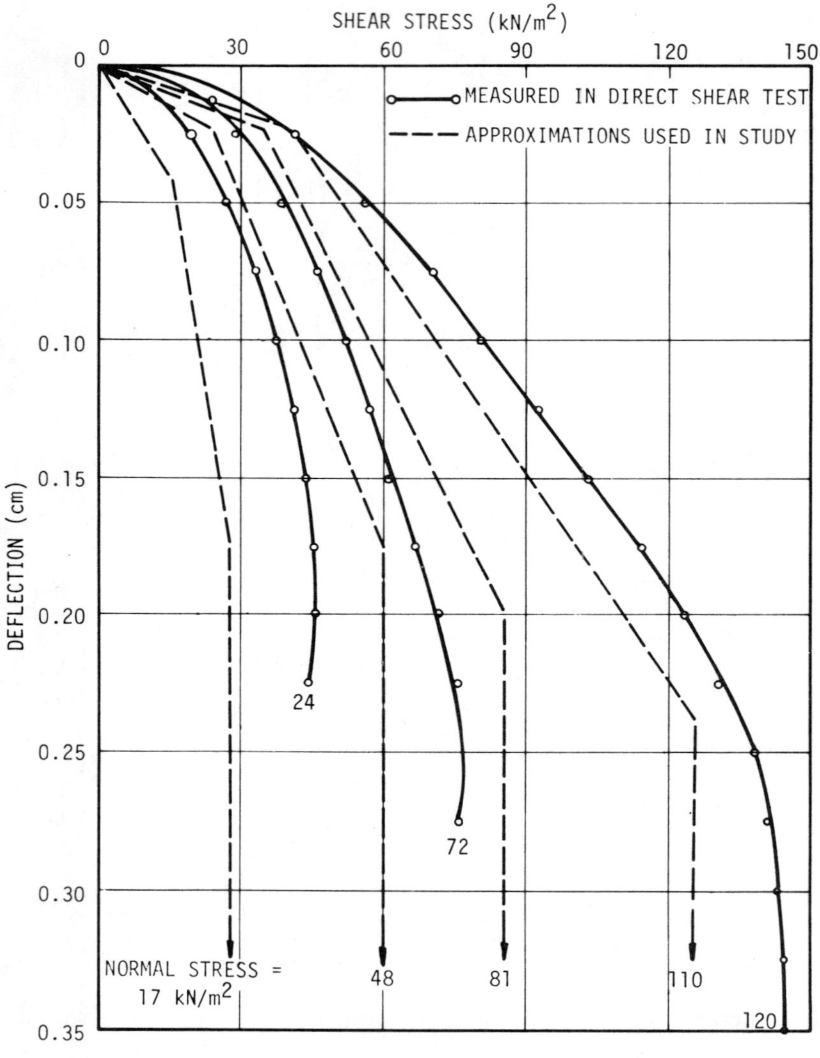

FIG. 2—*Example of direct shear data and approximations used in analysis (1 cm = 0.4 in., 1 kN/m² = 0.14 psi).*

in the lower segment is used to calculate the elastic shortening of this segment. One-half of this calculated elastic shortening plus the assumed tip deflection is taken as the deflection of the midpoint of the bottom section. The relation between shear stress and deflection for the soil around the bottom segment is entered with this midpoint deflection to determine the average shear stress along the bottom segment. The shear stress–deflec-

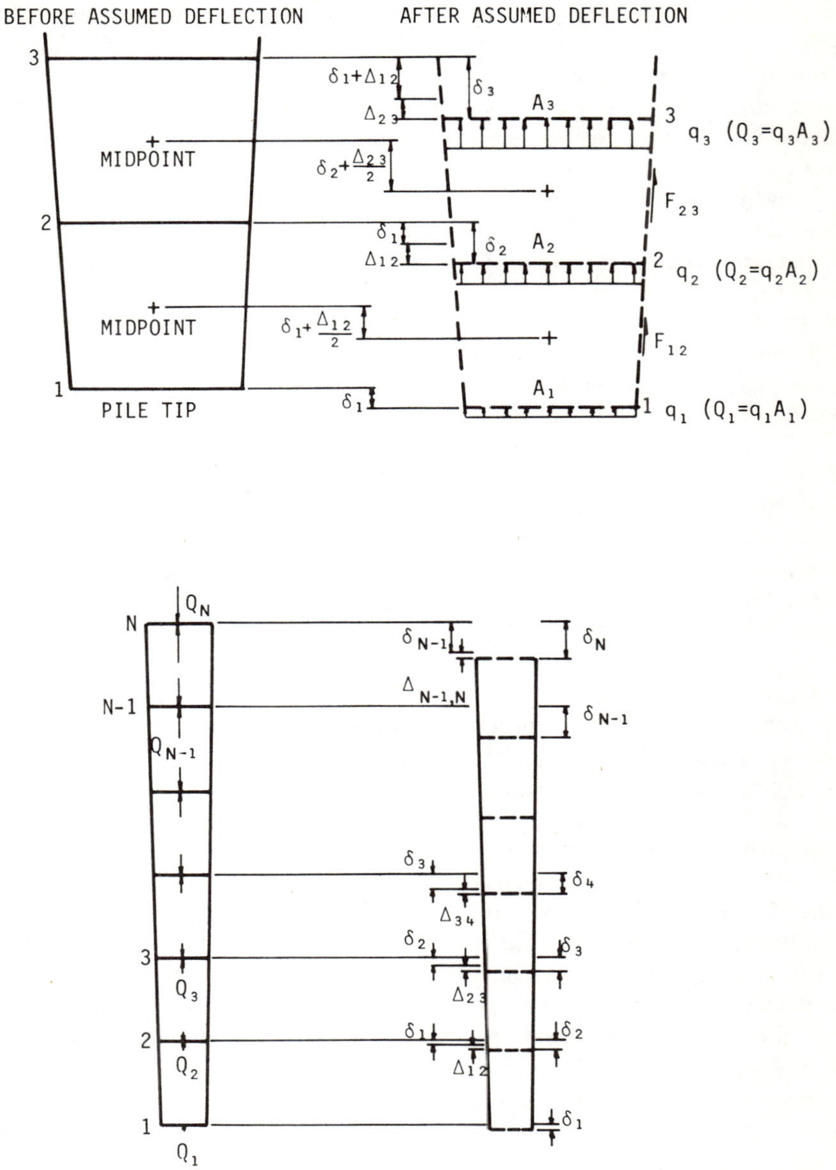

FIG. 3—*Definition of terms in analysis procedure.*

tion relationship is selected based on laboratory test data and the appropriate normal stress. The force resisted by shaft friction in the bottom section is, therefore, the shear stress over the surface area of the pile segment. If a reasonable value of the force at the top of the bottom segment is initially chosen, then that segment will be in equilibrium when top, bottom, and side forces are considered.

Generally, the initial value selected for the force at the top of the segment is not correct, so a new value must be selected and the procedure repeated successively until convergence is achieved. This iterative procedure has now provided the load in the pile at the interface between the two bottom sections and the deflection at that point (assumed tip deflection plus elastic shortening of the lower segment).

With the appropriate value of force at the bottom of the second segment now known (for an assumed tip deflection), the iterative process is repeated for the second segment until convergence is achieved. The process is repeated through each pile segment until the load and deflection at the top of the pile are determined (Fig. 3).

The process described above results in a description of the load-carrying and displacement conditions for a given tip movement. By repeating the process for several values of assumed tip movement, families of curves can be generated showing (1) load in the pile versus depth for a given applied load and (2) short-term deflection of any point on the pile for varying applied loads. Of primary interest in actual pile design problems are the load–deflection curve at the top (butt) of the pile and the load distribution throughout the pile for a given butt load. Obviously, this analysis can be performed in practice only by computer, as manual solution is very time-consuming.

It is emphasized that selection of the normal stress adjacent to a pile is a complex problem that has been extensively studied. For this program, it was assumed that $\sigma'_n = K\sigma'_v$, where $\sigma'_v = \gamma' h$ to the critical depth at 15 pile diameters. σ'_n is the normal effective stress on the pile, K is the ratio of horizontal to vertical stress, σ'_v is the effective vertical stress, γ' is the buoyant soil density, and h is the depth. Below the critical depth, σ'_v was assumed to remain constant. K was selected as 1 for all piles. This does not agree with the literature, where K is generally given as more than 1 for displacement piles [*5,6*]. Predrilling and the micaceous, resilient soil, however, may explain why low K values are reasonable. These values have been confirmed by extensive previous work with this procedure.

Results

Only the driven piles are discussed in this and the following section. Four of the twelve test piles were loaded to failure, at 1670 to 2220 kN (375 to 500 kips). The others sustained loads of up to 3110 kN (700 kips), generally with butt deflections of less than 2.5 cm (1 in.). Figure 4 summarizes all test results during the second loading cycle. Net deflection after the first cycle unloading was less than 0.5 cm. Reasonably good agreement between computed and measured deflections was observed for butts, tips, and intermediate points. Some piles showed better agreement than others, of course. Examples of measured and computed load-deflection curves for two Raymond step taper piles and one H-pile are shown in Fig. 5. Figure 1 shows subsurface conditions at their locations.

The agreement was considered sufficient to lend credence to the computed load transfer distribution determined from the analysis. The computed load transfer distributions for the same step tapers and H-pile are shown in Fig. 6. These figures show that, at this site, most of the applied load in the design load range is taken by shaft friction. Appreciable end bearing is not mobilized until much greater movement takes place, generally at loads in excess of those applied during this test program. The shapes of the load transfer curves are in general agreement with those reported elsewhere [*7–9*].

The shapes of the computed load-versus-depth curves are also indicative of the rate of load transfer throughout the length of the piles. The reciprocal of the slope represents the rate of load transfer. Figure 6 indicates that, for each pile, the slope is steepest near the top throughout the loading range shown, indicating relatively low rates of load transfer at shallow depths. At greater depths, the slopes of the curves become flatter, indicating a higher rate of load transfer, or increasing mobilization of shaft resistance. Most of the load in the range of design capacities is taken in side friction. The curves are generally almost parallel after significant tip resistance is developed. This indicates essentially complete mobilization of side friction before movements are large enough to mobilize appreciable end bearing.

The butt load and tip load-deflection curves clearly indicated the pipe piles to be end-bearing piles, since tip movement is very close to butt movement. The straight-line portion of the butt load-deflection curves prior to plunging failure practically coincides with the elastic shortening of the pipe pile. Therefore, pipe piles at this site are considered to be end-bearing piles, with essentially all of their load transferred to the tips.

Conclusions

The applicability of this analysis is confirmed by the full-scale testing program. Information on load transfer is indirectly determined by simple telltales, rather than by direct stress measurements. Because conditions of deflection and load at any point in the pile are generated, simple deflection measurements can be used to infer information on load transfer rates and mechanisms. Evaluation of load test results in conjunction with this analysis procedure can include more than determination of ultimate or

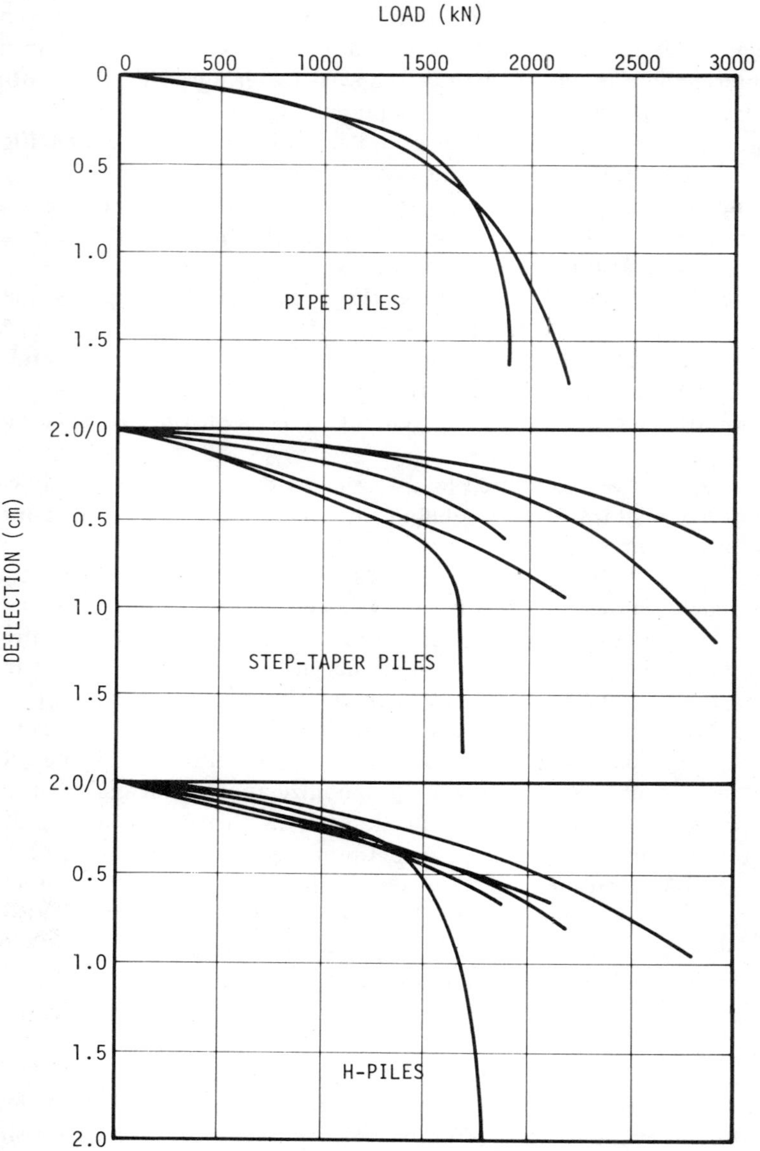

FIG. 4—*Load test results—butt deflection in second loading cycle (1 cm = 0.4 in., 1 kN = 0.11 ton).*

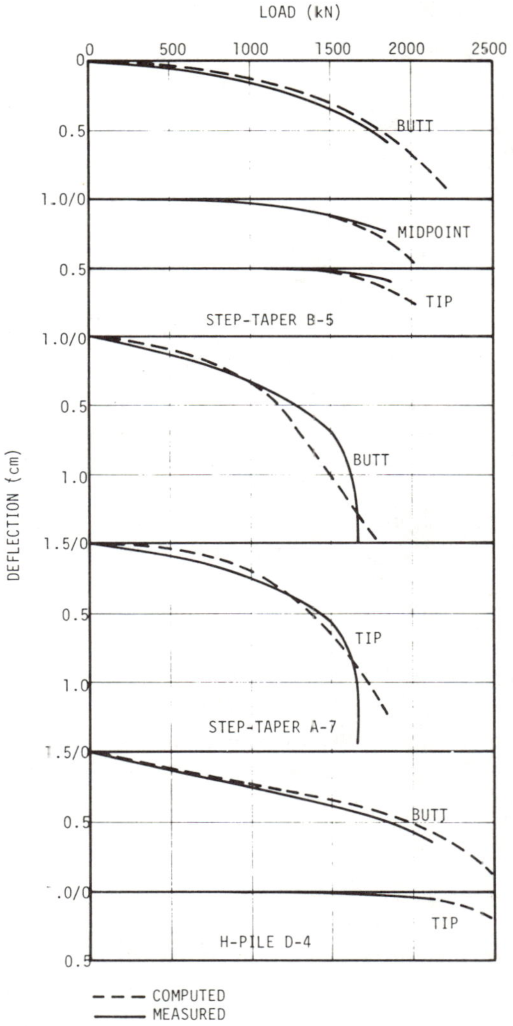

FIG. 5—*Comparison of measured and computed behavior (1 cm = 0.4 in., 1 kN = 0.11 ton).*

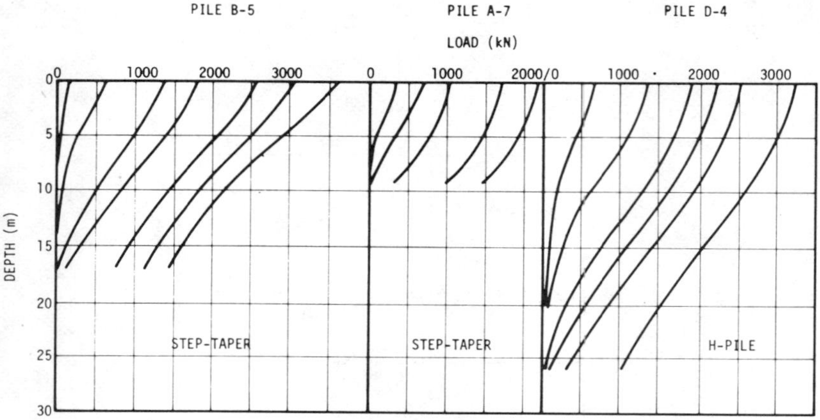

FIG. 6—*Computed load distribution with depth* (*1 m* = *3.28 ft, 1 kN* = *0.11 ton*).

safe working loads. Piles of different types, which transfer load to the soil with different efficiencies, can also be compared, so that realistic length estimates can be made for a given capacity.

Despite the imperfections of this analysis (requires knowledge of the state of stress along pile shaft; failure surface in direct shear test does not simulate exactly the failure zone around a pile), it is a procedure that can describe short-term single-pile behavior with some confidence. It has been utilized for design purposes by the author, over load ranges well below impending failure. Modeling conditions under which tip plunging occurs has not yet been attempted. This analysis procedure requires no more data than conventional limit equilibrium static analyses, but it provides much more insight into pile behavior, particularly when skin friction is significant.

Acknowledgments

The pile test program described herein was carried out for the Georgia Power Company under the direction of their Civil and Mechanical Engineering Department and Generating Plant Construction Department. C. P. Stinespring, structural engineer, administered and directed the program.

References

[1] Seed, H. B. and Reese, L. C., *Transactions*, American Society of Civil Engineers, Vol. 122, 1957, pp. 731-764.

[2] Clemence, S. P. and Brumund, W. F., *Journal of the Geotechnical Engineering Division*, American Society of Civil Engineers, Vol. 101, No. GT6, Proceedings Paper 11369, June 1975, pp. 537-550.

[3] Clisby, M. B., Mattox, R. M., and Webb, J. D., "An Evaluation of Pile Bearing Capacities," *Interim Report II,* Mississippi State University and Mississippi State Highway Department in cooperation with U.S. Department of Transportation Federal Highway Administration, Oct. 1972.
[4] Janbu, N., Bjerrum, L., and Kjaernsli, B., *Norwegian Geotechnical Institute Publication 16,* Oslo, 1956, pp. 30–32.
[5] Sowers, G. B., and Sowers, G. F., *Introductory Soil Mechanics and Foundations,* 3rd ed., Macmillan, New York, 1970, pp. 459–460.
[6] Hirst, T. J., in *Design and Installation of Pile Foundations and Cellular Structures,* H. Y. Fang and T. D. Dismuke, Eds., Envo Publishing, Lehigh Valley, Pa., 1970, p. 46.
[7] Mohan, D., Jain, G. S., and Kumar, V., *Geotechnique,* Vol. 13, No. 1, Mar. 1963, pp. 76–86.
[8] Vesić, A. S., *Journal of the Soil Mechanics and Foundations Division,* American Society of Civil Engineers, Vol 96, No. SM2, Proceedings Paper 7170, March 1970, pp. 561–583.
[9] Touma, F. T., and Reese, L. C., "Behavior of Bored Piles in Sand," presented at the 16–22 Oct. 1972, American Society of Civil Engineers Annual and National Environmental Engineering Meeting, Houston, Tex. (Preprint 1801).

G. Price[1]

Field Tests on Vertical Piles Under Static and Cyclic Horizontal Loading in Overconsolidated Clay

REFERENCE: Price, G., **"Field Tests on Vertical Piles Under Static and Cyclic Horizontal Loading in Overconsolidated Clay,"** *Behavior of Deep Foundations, ASTM STP 670,* Raymond Lundgren, Ed., American Society for Testing and Materials, 1979, pp. 464–483.

ABSTRACT: This paper describes experiments that provided information on the behavior of vertical tubular steel piles under horizontal working loads in London clay. Four aspects of piled foundation behavior were examined:

1. The effect of static and cyclic horizontal working loads on the behavior of a single pile.
2. The magnitude of the induced movements in unloaded piles at three and six pile diameter spacings from a horizontally or vertically loaded pile.
3. The stiffness of a single pile loaded horizontally compared with the stiffness of a row of three piles loaded horizontally at right angles to the row.
4. The effect of a cap on the behavior of a row of three piles loaded vertically and horizontally (at right angles to the row).

The main conclusions drawn are:

1. Low levels of cyclic horizontal loading cause small adverse changes in the behavior of vertical piles.
2. The interaction between piles under horizontal load is considerably less than under vertical load.
3. Conclusion 2 is substantiated by the observations that the stiffness of three piles in a row is approximately three times the stiffness of a single pile under horizontal loads.
4. A cap significantly reduces the horizontal movements under load but has little effect on the vertical behavior of the piles.
5. The experimental techniques used enabled small changes in the pile-soil behavior to be successfully monitored.

KEY WORDS: piles, clay, horizontal and vertical loading, instrumented, pile cap, static and cyclic loading, seasonal effects, group effect, soil displacements

[1]Senior scientific officer, Building Research Station, Garston, Watford, Herts WD2 7JR, England.

At the present time, information is being sought on the ability of vertical piles to carry horizontal loads. Raking piles widely used to resist horizontal loads are difficult and costly to install and can be subjected to large induced bending stresses from vertical loads. This report deals with tests to determine the behavior of vertical piles under horizontal and vertical working loads. Three cases are considered: a single pile, a row of three piles, and a row of three piles with a cap. Static vertical and horizontal loads were applied in all three cases, and cyclic horizontal tests were carried out on the single pile.

The tests were carried out on tubular steel piles that had been hydraulically jacked 4.6 m into the brown oxidized layers of London clay. The piles were 5.1 m long and 168 mm in diameter. Three piles were installed in a row at three pile diameter spacings. The piles had been initially installed to examine their interrelationships under static vertical loading. This work is described by Cooke et al [*1*].[2] For these earlier tests, the loads had been limited to keep the behavior of the pile-soil system elastic. The final vertical load test was carried out with the three piles joined rigidly together by a cast-in-place reinforced concrete cap 500 mm wide by 1500 mm long and 300 mm deep. The soil around the side of the cap was removed so that it only made contact on its underside.

Because the piles had been jacked into the ground, the ultimate vertical loads immediately following installation were known. The vertical working load was taken as one-third of this. To establish the horizontal working load was less straightforward, and a value was chosen that would limit the head movement to within the elastic range of the pile. This was determined from the first horizontal loading test. The purpose of the tests was to investigate:

1. The effect of static and cyclic horizontal working loads on the behavior of a single pile.
2. The magnitude of the induced movements in unloaded piles at three and six pile diameter spacings from a horizontally or vertically loaded pile.
3. The stiffness of a single pile loaded horizontally compared with the stiffness of a row of three piles loaded horizontally at right angles to the row.
4. The effect of a cap on the behavior of a row of three piles loaded vertically and horizontally (at right angles to the row).
5. The effectiveness of the testing methods used.

Site

A comprehensive site investigation had been carried out by Marsland [*2,3*] employing *in situ* and laboratory tests, but only a few of the results applied to the top 5 m of soil in which the test piles were installed. Supplementary un-

[2]The italic numbers in brackets refer to the list of references appended to this paper.

drained triaxial tests were therefore carried out on clay taken from the top region using thin-walled sample tubes. The results are shown in Fig. 1 alongside the average results obtained by Marsland.

Details of Experimental Equipment

The whole experiment, including piles, reference frame, and trench, was enclosed in a polyethylthene-covered framework with a wood and felt roof.

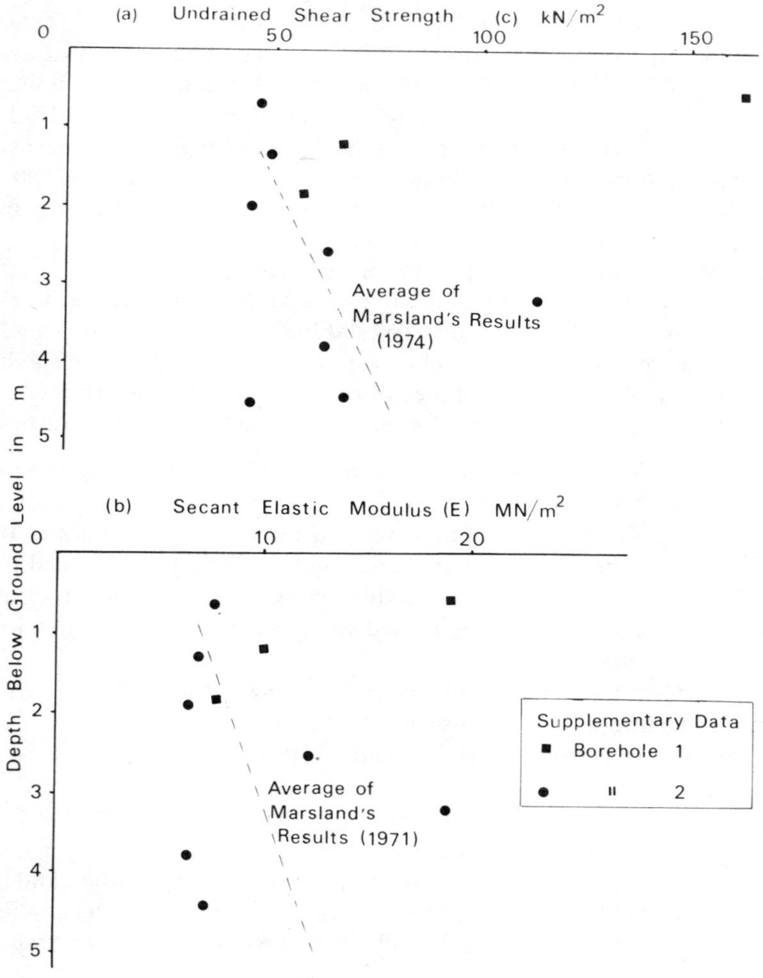

FIG. 1—*Data from supplementary soil samples.*

This enclosure protected both the transducers and reference frame from the effects of wind, sun, and rain; it also provided a degree of comfort for the personnel involved in the tests. All tests were carried out during the night, when temperature fluctuations were small and external vibrations transmitted through the ground from traffic on nearby roads were at a minimum. The internal temperature was kept constant by a heater blower. Figures 2 and 3 show the layout of the experiment.

The trench shown in the figures allowed installation of horizontal inclinometers and horizontal movement probes at various depths into the body of soil around the piles. During construction of the trench, care had been taken to minimize its effect on the surrounding soil, and before the first pile was installed, it was established by observing the readings from two inclinometer trains that the soil adjacent to the trench had stabilized.

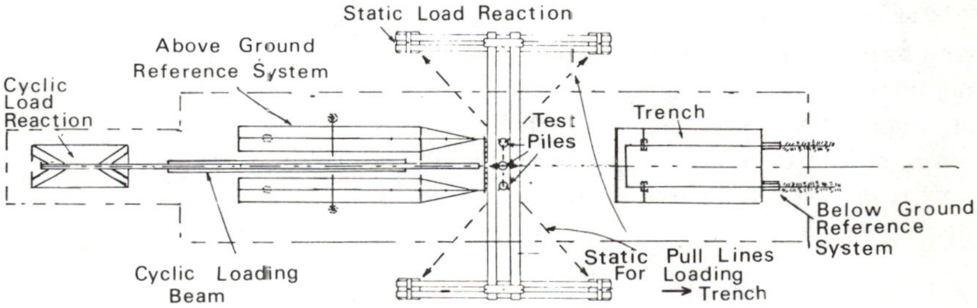

FIG. 2—*Plan of experiment.*

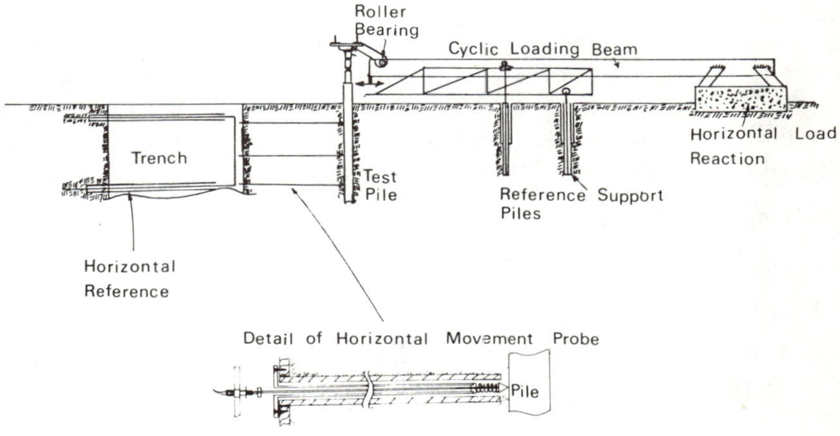

FIG. 3—*Side elevation of experiment and movement probe.*

Measuring System for Pile Movements

The reference system for movements above ground consisted of a light alloy frame cantilevered toward the test piles and supported by five bored piles. The top of each bored pile was sleeved to prevent surface soil movements affecting the reference frame. Below ground level, the horizontal movement transducers were mounted on a reference frame supported on four stainless steel legs fixed 1.5 m into the soil at the far end of the trench. Roller bearings attached to the side of the trench supported the transducer end of the frame at a fixed elevation.

Preliminary tests were carried out to record the degree of cross sensitivity between the vertical and horizontal transducers due to movements at right angles to their respective directions of operation. Only the vertical transducer exhibited some cross sensitivity.

Measuring System for Soil Movements

High-precision horizontal inclinometers consisting of electrolytic levels were used to record vertical movements in the soil. The construction of the inclinometers has been described by Cooke and Price [*4*]. Eight of the inclinometers were connected into trains capable of resolving small vertical movements in the soil, radially from the pile, relative to the inclinometer furthest from the pile. Three trains were used in the experiments described in this report.

Load Measuring and Application Systems

The distributions of load down the pile shafts were recorded by resistance-strain-gaged load cells built into the piles at four levels. The load cells were calibrated before and after the field tests. There was no significant differences between the values obtained.

Loads were applied to the piles in such a way that movements at right angles to the load were not restricted. Cyclic movements were applied to the pile through an adjustable cam driven by a speed-controlled motor. This allowed the amplitude and frequency of the applied oscillations to be varied.

Pile-Testing Procedure

Order in Which Tests Were Carried Out

Four series of tests were carried out in the following order:

1. Static vertical and horizontal loading tests on three piles with a cap.
2. Static vertical and horizontal loading on three piles with the cap undercut.
3. Static vertical and horizontal loading tests on the central pile.
4. Cyclic horizontal loading tests on the central pile (5000 cycles in total).

The static and cyclic loading procedures used in each series of tests will be briefly described later.

Removal of the Cap

For the tests on three piles only, the soil was carefully cut away from underneath the cap, removing the effect of the cap without changing the stiffness of the interconnection between the piles. For the single-pile tests, the cap was completely removed from around the piles.

Static Vertical and Static Horizontal Loading Tests

The tests were carried out in a similar manner. The static load was applied in increments, and the resulting deflections, load distribution, and vertical displacements of the surrounding soil were recorded. When combined loads were applied, the effects of the vertical increment of load were recorded before the horizontal increment of load was applied, after which its effects were also recorded. Only compressive vertical loads were applied. Horizontal loads were applied either toward the trench or toward the reference frame, as shown in Fig. 3.

These tests were capable of detecting small changes in the pile soil behavior, because they were of short duration and the stability of the instrumentation over the testing period was very good.

Cyclic Loading Tests

Cyclic horizontal controlled displacements were applied to the pile while it was under a static vertical load of 30 kN. Initially, the horizontal displacement was set at 1.0 mm and cycled 700 times at each of 3, 6, and 10 cycles/min. The displacements were then increased to 2.0 mm and subsequently to 2.75 mm. In both cases, the pile was cycled approximately 700 times at each of the above three frequencies.

These tests were carried out to investigate the ability of the pile to withstand low levels of cyclic horizontal loading. Changes in the pile-soil behavior as a result of these tests were established by static loading tests.

Order in Which Tests Are Reported

The first tests to be undertaken were the horizontal loading of three piles with a cap. This was the state of the foundation after completing the earlier work on the interaction between piles under vertical loads, reported by Cooke et al [*1*]. It was also the foundation's strongest condition, and therefore, the initial light horizontal loads were unlikely to overload the soil.

A preliminary horizontal loading test was carried out to establish allowable loading limits. Because of the magnitude of the resulting set during this test, it was decided to limit the horizontal load applied during subsequent tests to half the maximum value applied during the first loading (Fig. 4). For the horizontal loading of a single pile, the value was further reduced to a third.

Although the first tests carried out were on three piles with a cap and the last tests were those on a single pile, it was thought that the results were more easily presented and discussed in reverse order. Thus, the work on a single pile is dealt with first and the three piles with a cap last.

Results and Discussion

Effects of Horizontal Loadings on a Single Pile

Effects of Static Horizontal Load on a Single Pile — During January, three static vertical loading tests were carried out on the center pile. These are shown in Fig. 5. The second test was used to check the uniqueness of the first and the third test was undertaken after a series of static horizontal loading tests had been carried out to see if they had significantly affected the

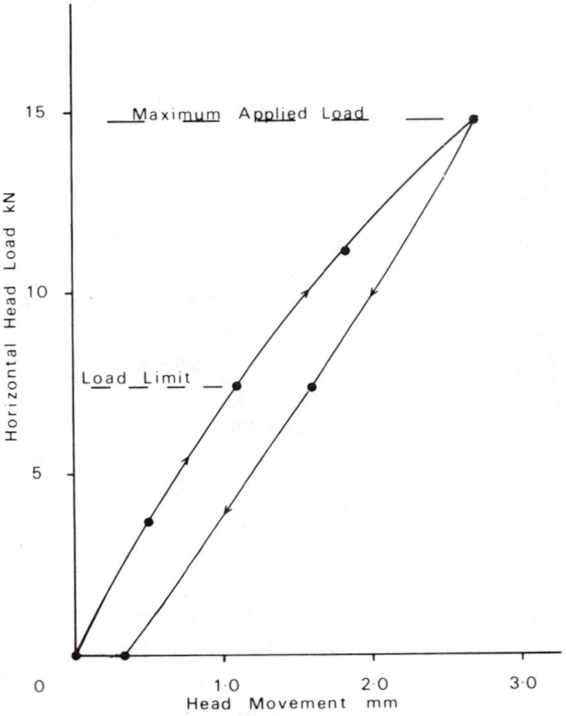

FIG. 4—*Preliminary horizontal loading test for three piles with a cap.*

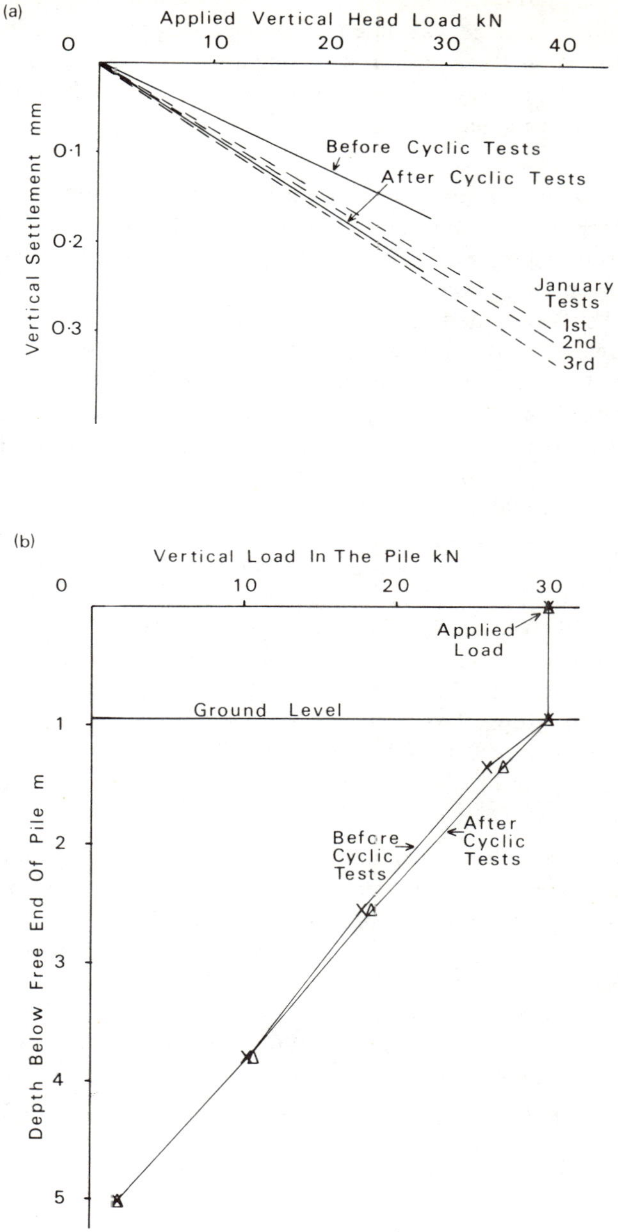

FIG. 5—(a) *Vertical load—settlement for a single pile.* (b) *Vertical load distribution of a single pile.*

behavior of the pile. It can be seen that there was more settlement during the third test, but no change was noted in the load distribution or inclinometer results for any of the three tests. It was concluded from the results of all three instrument types that any changes in pile behavior due to the static horizontal loading tests were very small.

Effects of Cyclic Horizontal Load on a Single Pile—Static vertical loading tests were carried out before and after the cyclic horizontal loading tests in April. The settlement results showed that the pile moved less under vertical load in April (prior to cyclic loading) than in January, indicating that some seasonal soil stiffening had occurred. However, after the cyclic loads, the settlement increased, and Fig. 5*a* shows that at that time the behavior of the pile was almost identical to that observed in the third January test.

The load distributions in the pile before and after the cyclic tests are shown in Fig. 5*b*. Approximately 4 percent more load reached the top load cell after the cyclic loading tests. This suggests that the bond between the soil and the top section of the pile had been slightly reduced.

The results from the topmost horizontal inclinometer train presented in Fig. 6*a* show that in April, before the cyclic tests, the pile settled less than during the January test. Thus they agree with the settlement transducers. The results of the vertical loading test after the cyclic loading show that the vertical movements of the soil were smaller in spite of the fact that the pile settlements were larger (Fig. 6*b*). This could only occur if the bond between the pile and the upper layers of soil had been adversely affected. No significant changes were detected at lower levels by either the pile shaft load cells or the horizontal inclinometers.

Discussion of Results — Horizontal pile deflections under static and cyclic loads— The observations of horizontal pile movement from two of the probes during the static horizontal loading tests on the single pile are given in Fig. 7. No movements were recorded at the lowest level, but a linear relationship between applied loads and displacements was recorded at all higher levels. The recorded deflected shape of the pile under cyclic loads showed that the horizontal restraint provided by the soil was significantly reduced as a result of the cyclic loading. For a given load, the piles horizontal movement at ground level approximately doubled after the cyclic loadings (detailed above). Figure 7 shows that a linear relationship existed between applied head load and horizontal deflection after the cyclic loading.

Effect of seasonal changes in the soil properties—The marked stiffening between the January and April tests, as shown by both the static vertical and horizontal loading results, are explained by seasonal stiffening of the soil. By back analyzing the results of pile tests on the site using a finite element

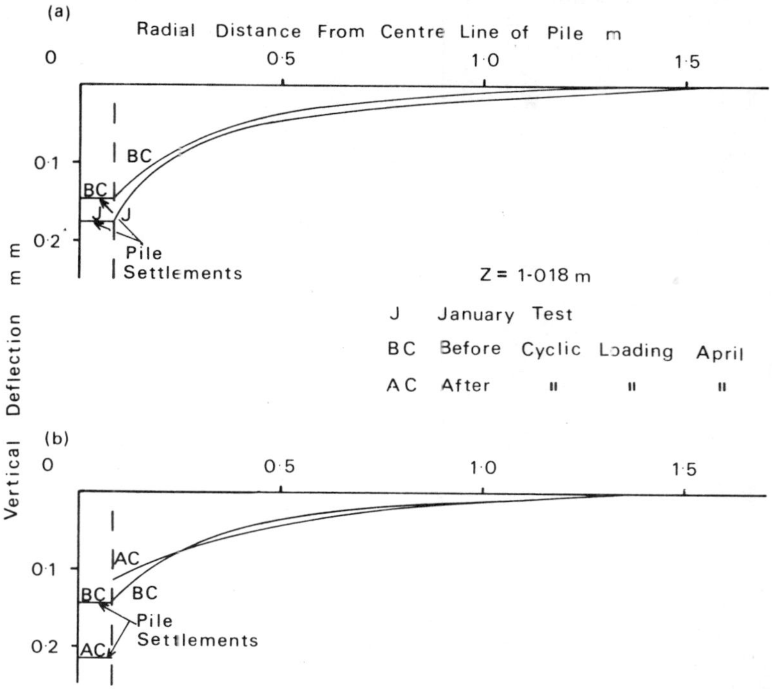

FIG. 6—*Soil deflections around a loaded test pile.*

model, it is estimated that the top metre of soil could have quadrupled in stiffness between the winter and summer months.

Soil softening due to cyclic loading— While seasonal stiffening may be explained by a change in the moisture content in the soil, pile-soil softening due to cyclic loading is a more complex problem, relating to the pile-soil bond and to changes in the soil fabric. Laboratory tests (Sangrey et al [5]) have shown that soil stiffness can be reduced by cyclic loading. The extent of softening depends upon the magnitude of the strains. Cyclic horizontal loading of piles can also affect the stiffness of the surrounding soil, as demonstrated by the tests reported here.

During the static and cyclic loading tests, transducers monitored the movements of the unloaded piles. These movements were expressed as a percentage of the movements of the loaded pile. The results together with those of vertical load interaction tests reported elsewhere [*1*] are shown in Fig. 8.

Effects of Static Horizontal Loading—Figure 8 shows that the magnitudes of induced movements due to horizontal loading are less than those under

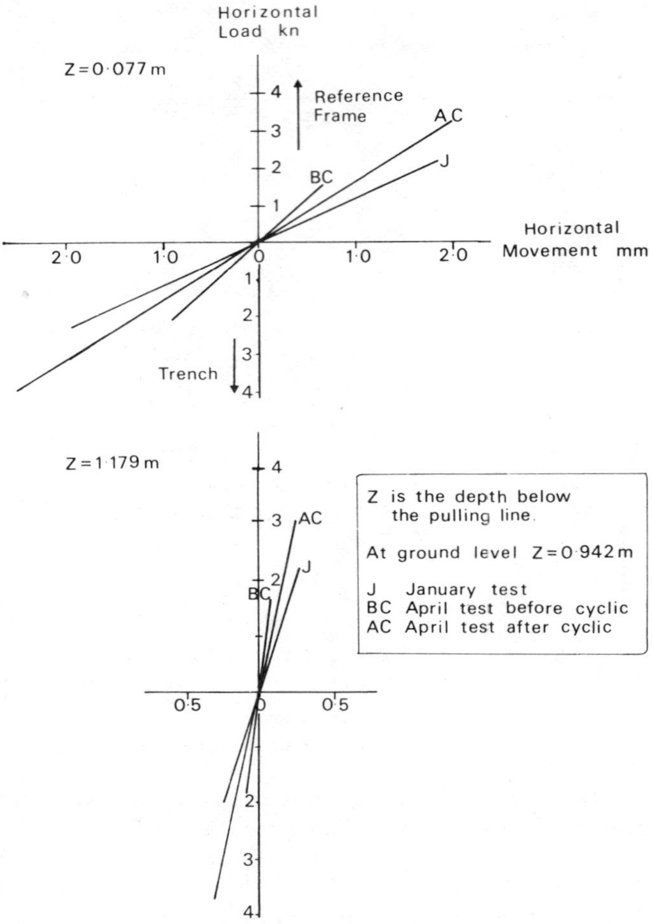

FIG. 7—*Horizontal movement–applied head load for a single pile.*

vertical loading. Figure 8 also shows that the rate of decay of movements decreases with distance from the loaded pile.

Effects of cyclic horizontal loading—The horizontal movements recorded during the first and 5000th cycle of load are shown in Fig. 8. It can be seen that the magnitude of the induced movements dropped after the cyclic loading tests. This could be due to softening of the soil adjacent to the pile.

Comparison Between the Static Loading Behavior of a Single Pile and a Row of Three Piles

Under Static Vertical Load—Interactions between piles under static vertical load have been reported by Cooke et al [*1*].

Diagram of Test Piles

(B) 3pd (A) 3pd (C)

% Movement for
Static Horizontal Loading of

Pile A

9 100 9

Pile B

100 9 4·5

% Movement for
Cyclic Horizontal Loading of
Pile A

Start

16 100 16

After 5000 Cycles

9 100 9

% Movement for
Static Vertical Loading of

Pile B

100 22 10

FIG. 8—*Horizontal and vertical movements induced in adjacent unloaded piles by a loaded pile.*

Under Static Horizontal Load at Right Angles to the Row—Figure 9 shows the deflected shape of a single pile under horizontal loads compared with that of a row of three piles under similar loads per pile. It will be seen that the three piles moved a little less above ground and a little more below than the single pile. Because of the large range of movements involved, the results were plotted on a log scale and therefore could not show the point of contraflecture in the pile.

The movements suggested that the point of contraflecture for the single pile was above that for the three pile row. The data suggest that for approx-

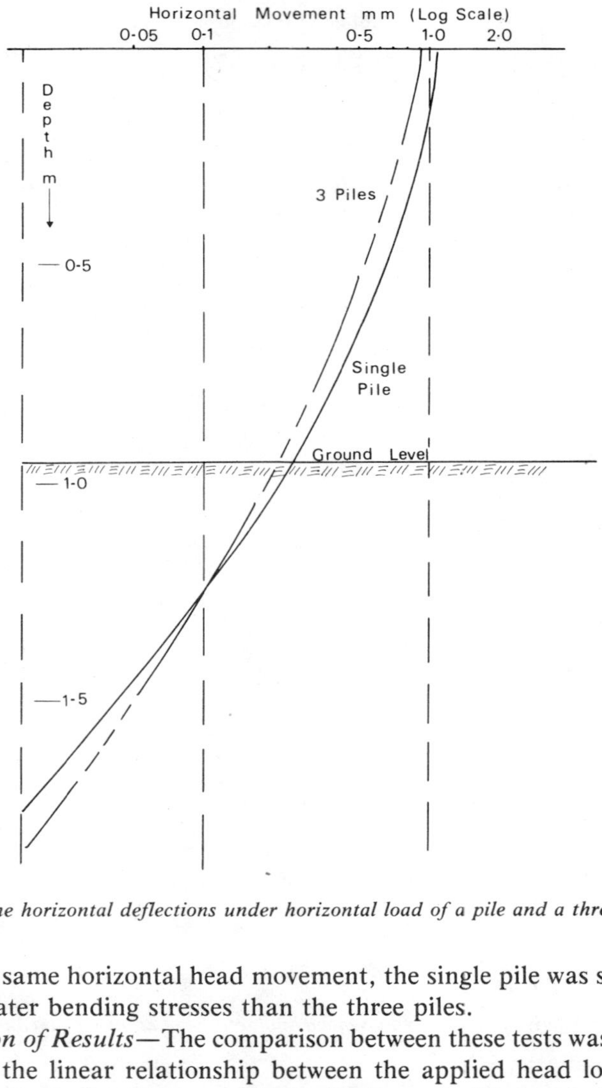

FIG. 9—*The horizontal deflections under horizontal load of a pile and a three-pile row.*

imately the same horizontal head movement, the single pile was subjected to slightly greater bending stresses than the three piles.

Discussion of Results—The comparison between these tests was made easy because of the linear relationship between the applied head load and the horizontal pile shaft movement. Figure 10 shows the linear relationship at a point 0.41 m above ground level. The behavior of the single pile was compared with the row of three piles in tests in which only horizontal loads were applied. Combined load tests were not used, because the vertical loads affected the magnitude of the horizontal movements.

When vertical loads were combined with horizontal loads and the pile pulled toward the trench, the vertical load increased the horizontal movements above ground level and to a depth of 1 m below ground level. At

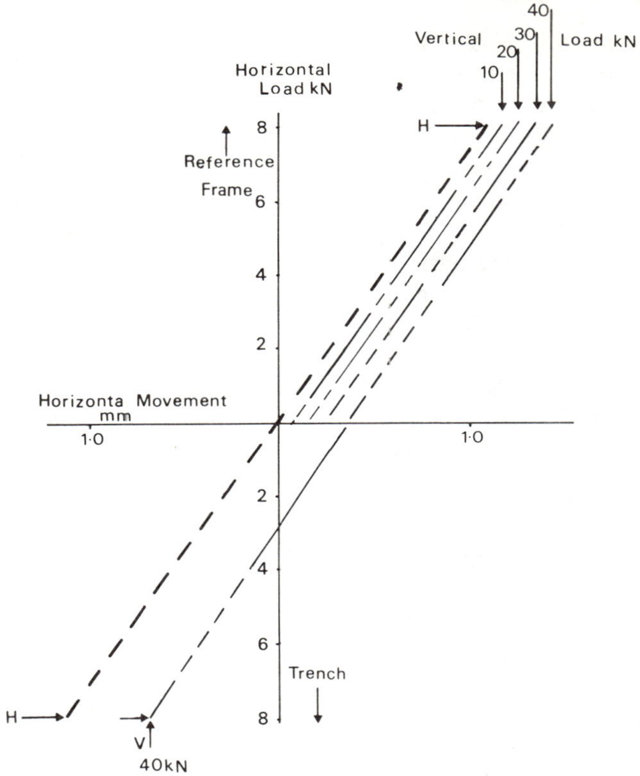

FIG. 10—*Horizontal movements at a point 0.41 m above ground level.*

greater depths, it reduced the movements. When the pull was toward the trench, the effect was reversed. The effects of vertical load on the horizontal movement 0.41 m above ground surface are shown in Fig. 10. Under horizontal load only, the direction of pull did not alter the magnitude of the deflected shape as shown by the chain line. This balanced deflection under horizontal load showed that the trench had no significant effect. It was therefore concluded that despite the piles being installed from a guide frame, they were either bowed or offset slightly from the vertical. Care had been taken to apply the load centrically to the piles.

Since this horizontal movement under vertical load was caused by only slight vertical misalignment or curvature of the pile, it can be seen that raked piles under vertical loads would suffer significant bending stresses. The test results showed that under combined vertical and horizontal loading, the horizontal movements were given by the sum of the movements resulting from the individual loadings.

Effects of a Cap on the Static Loading Behavior of Three Piles

Effects of the Cap Under Vertical Load—Measurements to examine this aspect have been reported by Cooke et al [*1*]. Further tests are reported here because in the earlier tests the cap was added to the three piles, while in these tests it was removed from the three piles by undercutting. The results, therefore, may well be different.

The results of the vertical loading tests on three piles with and without the cap are shown in Fig. 11*a*. The cap appears to play a very small role in reducing the magnitude of the settlement. Figure 11*b* shows that the cap slightly reduced the load transfer from the top 1.5 m of the pile shafts.

The settlement of the piles at the inclinometer levels can be estimated by projecting the deflection curves to the pile (Fig. 12) and assuming no slip between the pile and the soil. The topmost inclinometer train shown in the top half of Fig. 12 shows only a slight reduction in pile settlement under vertical load due to the raft. There are, however, indications that the raft

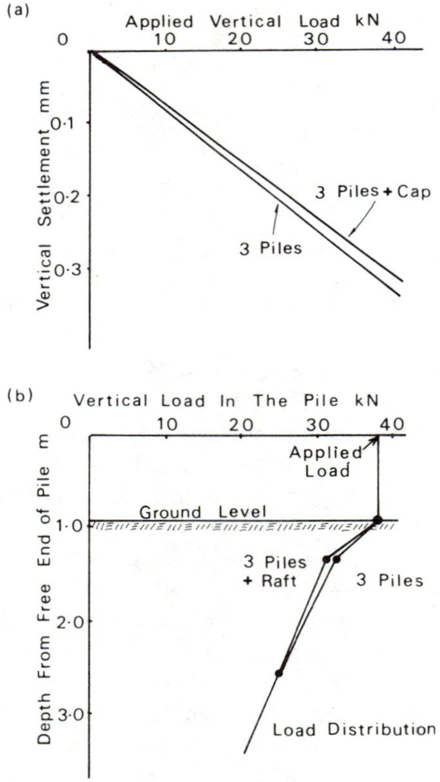

FIG. 11—*Settlement and load distribution for three piles with and without a cap.*

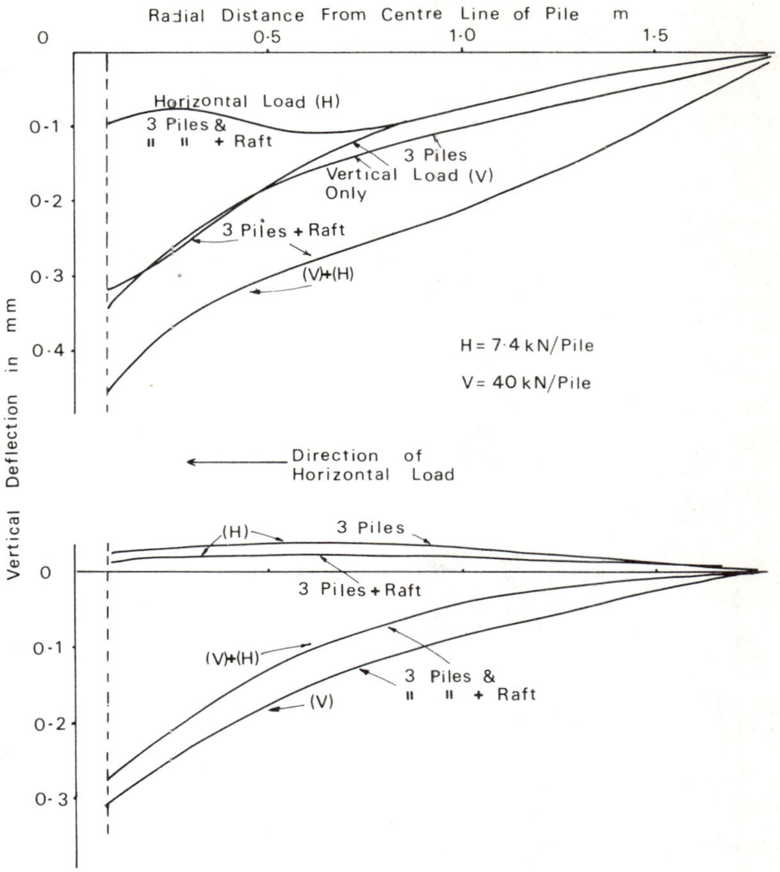

FIG. 12—*Vertical soil deflections around three piles with and without a cap.*

reduced the stress transfer between the pile and the soil at this level. The slopes of the soil deflections at the pile-soil interface with and without the raft under vertical load only are compared in Fig. 12. The shallower slope with the raft indicates a lower level of load transfer between the pile and the soil. Other aspects of Fig. 12 are discussed later.

Effects of the Cap under Horizontal Load—The contribution of the cap to resisting horizontal load is shown in Fig. 13. With the cap in position, the horizontal movements above ground level were halved, irrespective of whether or not vertical load was applied. Below ground (at $Z = 1.493$ m) the cap reduced the movements by a factor of 4 without vertical load and 7 with. These results are shown numerically in Fig. 13, because they cannot be effectively shown graphically.

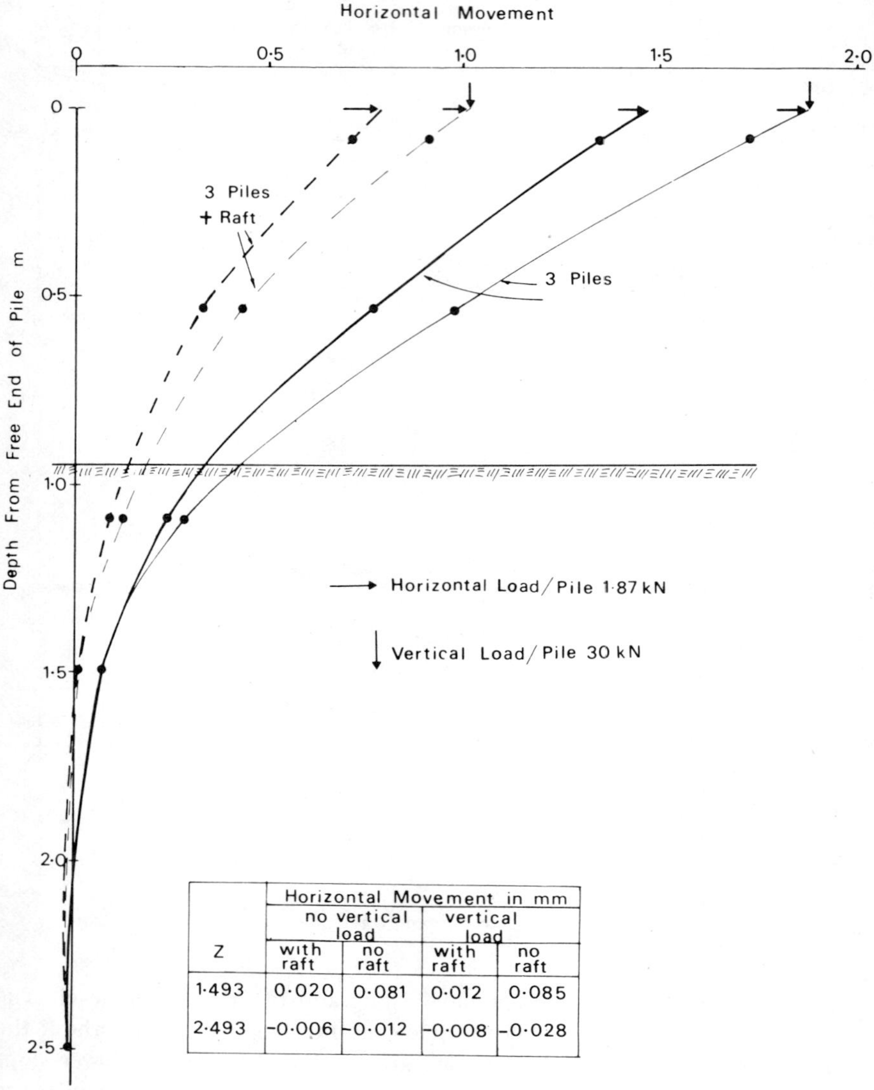

Z	Horizontal Movement in mm			
	no vertical load		vertical load	
	with raft	no raft	with raft	no raft
1·493	0·020	0·081	0·012	0·085
2·493	-0·006	-0·012	-0·008	-0·028

FIG. 13—*Deflected shape of three piles under horizontal loading with and without a cap.*

Discussion of Results—During the tests in which only vertical loads were applied, the magnitude of the horizontal movements were monitored. It was found that the cap reduced these movements by about a half above ground level and to an even larger extent below ground level (Fig. 14) for the above ground effect. It appeared that the effect of vertical load on the horizontal movements of the pile was greater below ground than above. This could be

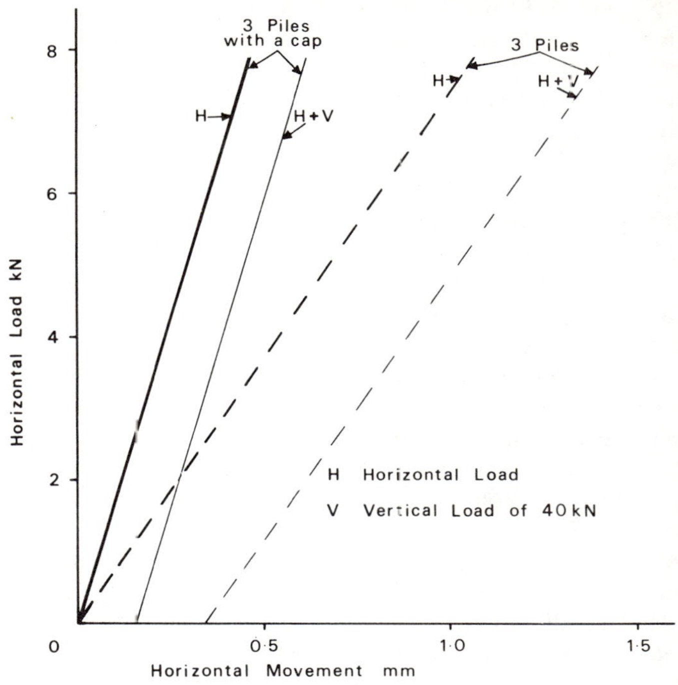

FIG. 14—*Horizontal movement due to vertical loading with and without a cap.*

explained by a bowing mechanism, especially if the pile had a slight bow in it before loads were applied.

Figure 12 shows that under horizontal load (toward the reference frame) the topmost inclinometer train (at Z = 1.0 m) recorded a downward deflection of the soil, as the pile moved away from the inclinometer train. The hump in the curve near the pile was possibly due to the bond between the pile and the soil (in the plane of the inclinometer) deforming the soil upward as it attempts to remain at right angles to the curvature of the pile. The lower inclinometer (at Z = 2.5 m) showed upward movements of the soil, because at its level the pile moved toward the inclinometer (see Fig. 12). The cap slightly reduced this upward movement because it reduced the reverse bending in the pile.

Conclusions

1. The vertical piles were unable to carry low levels of cyclic horizontal load without small adverse changes in performance occurring, that is: (*a*) the horizontal deflections of the pile increased, (*b*) the vertical settlement

of the pile under vertical load increased, and (*c*) the bond between the soil and the upper part of the pile was reduced.

2. The horizontal movements induced in nearby unloaded piles by static horizontal loads are not as great as the induced vertical movements due to vertical static loads. For piles at three pile diameter spacing, the horizontal induced movements were 9 percent, whereas the vertical induced movements were 22 percent. Under cyclic loading, the induced horizontal movements were initially 16 percent, but the value decreased to 9 percent after 5000 cycles.

3. The horizontal resistance of three piles in a row at right angles to the applied load is approximately three times the resistance of a single pile. However, the bending stresses in a single pile are marginally greater than those in a row of three piles when subjected to the same head movement per pile.

4. The pile cap significantly reduced the horizontal movement of the foundation under horizontal load. The cap also reduced the horizontal movements under vertical loads, although it reduced the vertical movements very little.

5. The experimental setup, and particularly the inclinometers and observation trench, enabled the effects of changes in the behavior of the pile under working load conditions to be detected. In particular, the horizontal inclinometers provided information on the elastic modulus of the soil that was useful in finite element analysis.

6. Short-term vertical loading tests effectively demonstrated changes in pile–soil behavior due to cyclic horizontal loading.

7. Small errors in the vertical alignment of the pile or loadings system caused significant horizontal deflection of the pile under vertical load. This highlighted the importance of considering the bending stresses likely to occur when raked piles are subjected to only vertical loads.

8. It is essential to take seasonal changes in the soil into account when evaluating long-term horizontal loading tests on piles. The top 1.5 m of soil can be effected by seasonal changes, and this depth of soil is significant in determining the behavior of a pile under horizontal load.

Acknowledgments

The author wishes to thank R. W. Cooke, I. Wardle, and P. A. Price for their help in preparing this paper. The work described has been carried out as part of the research program of the Building Research Establishment of the Department of the Environment and this paper is published by permission of the Director.

References

[1] Cooke, R. W., Price, G., and Tarr, K. J., "Jacked Piles in London Clay: Interaction and Group Behavior Under Working Conditions," to be published July, 1979.

[2] Marsland, A., *Proceedings*, Symposium on Interaction of Structure and Foundation, Midland Soil Mechanics and Foundation Engineering Society, Department of Civil Engineering, University of Birmingham, July 1971, pp. 7-17. (Also BRE Current Paper CP24/73)

[3] Marsland, A., *Proceedings*, European Symposium on Penetration Testing, Stockholm, Sweden 1974, Vol. 2/2, pp. 245-252. (Also BRE Current Paper CP87/74)

[4] Cooke, R. W. and Price, G., *Proceedings*, Symposium on Field Instrumentation, British Geotechnical Society Institute of Electrical Engineers, London, June 1973. (Also BRE Current Paper CP26/73)

[5] Sangrey, D. A., Henkel, D. J., and Esrig, M. I., *Canadian Geotechnical Journal*, Vol. 6, 1969, pp. 241-252.

M. F. Randolph[1] *and C. P. Wroth*[1]

A Simple Approach to Pile Design and the Evaluation of Pile Tests

REFERENCE: Randolph, M. F. and Wroth, C. P., **"A Simple Approach to Pile Design and the Evaluation of Pile Tests,"** *Behavior of Deep Foundations, ASTM STP 670,* Raymond Lundgren, Ed., American Society for Testing and Materials, 1979, pp. 484–499.

ABSTRACT: In the past, the design of a pile in a given soil against undue settlement has entailed lengthy numerical analysis for a variety of pile geometries. Similarly the back analysis of load tests on piles has usually been accomplished by repeated analysis with differing soil stiffness profiles in order to arrive at the profile which gives best agreement with the measured results. The recent development of an approximate closed form solution for the problem of a vertically loaded pile has enabled these two processes to be considerably simplified.

The present paper takes the solution as developed and extends it to the case of piles bearing on a stiffer stratum of soil. The solution may then be used both for the design of single piles and also for the back analysis of pile tests. For given soil conditions, it is possible to draw up design charts showing how the settlement of a pile may vary with load level and pile geometry. This process is demonstrated for some typical soil conditions. For evaluating load tests, the closed form solution may be manipulated to give an immediate estimate of the average soil stiffness down the pile for the case of tests on uninstrumented piles. If additional information (such as the settlement of the pile base) is available, it is possible to obtain an estimate of the soil stiffness profile down the length of the pile without recourse to a computer. Examples of some actual pile tests are given.

KEY WORDS: analysis, elastic theory, soil mechanics, piles (foundations), design, end-bearing piles, settlement, sheer modulea, pile load tests

In spite of the nonlinear nature of most soil behavior, it is customary, when estimating the likely settlement of a piled foundation, to assume that the soil stiffness may be characterized by some suitably chosen elastic modulus. Even with this simplifying assumption, a rigorous solution to the problem of a pile embedded in an elastic material is only possible with the

[1]Assistant lecturer and reader in soil mechanics, respectively, University Engineering Department, Cambridge University, Cambridge CB2 1PZ, England; Wroth is presently professor of engineering science, Oxford University, Oxford, England.

aid of of a computer, using numerical methods of analysis. Such methods are inefficient and cumbersome to use in the initial stages of design or when back analyzing the results of a pile test. Charts may be prepared showing how the settlement of a pile varies with the pile geometry and stiffness [*1*],[2] but it is not clear how these results should be modified to account for the conditions at a particular site, for example, a softer layer of soil at some depth down the pile or perhaps a partially sleeved pile.

Recently, a simple closed-form solution has been presented [*2,3*] that gives the load-settlement ratio for a pile embedded in a wide range of nonhomogeneous elastic soil. The solution is approximate, but the accuracy of the results has been extensively checked with the aid of more rigorous methods of analysis. At present, the solution is applicable to soils where the stiffness increases linearly with depth. This paper extends the solution to include the case of end-bearing piles where the pile is driven through one soil deposit to bear on a stiffer layer of soil. For design purposes, charts may be prepared rapidly for a particular soil type showing how the settlement of the pile varies with the stiffness and geometry of the pile. Examples of preparation and use of such design charts are given. Finally, it is demonstrated how the solution may be used to back analyze pile tests in order to give an estimate of the stiffness profile of the soil.

Analytical Solution

The solution is derived [*2,3*] by assuming that the load-settlement behavior of the pile shaft may be considered separately from that of the pile base. Accordingly, the soil surrounding the pile may be divided into two layers by a line *AB* drawn at the level of the pile base (see Fig. 1). Initially, it is assumed that the soil above *AB* will be deformed solely by the stresses transferred from the pile shaft and that the soil below *AB* will be deformed solely by the stresses at the pile base. Some modification of this assumption is necessary in order to take account of the interaction between the upper and lower layers of soil; the interaction will serve to limit the deformation of the upper layer of soil, reducing the deformations to negligible size at some radius r_m. The main steps leading to the solution will be outlined below.

Deformation of Pile Shaft

From consideration of vertical equilibrium of elements of soil around the pile, it may be shown that the shear stress in the soil decreases inversely

[2]The italic numbers in brackets refer to the list of references appended to this paper.

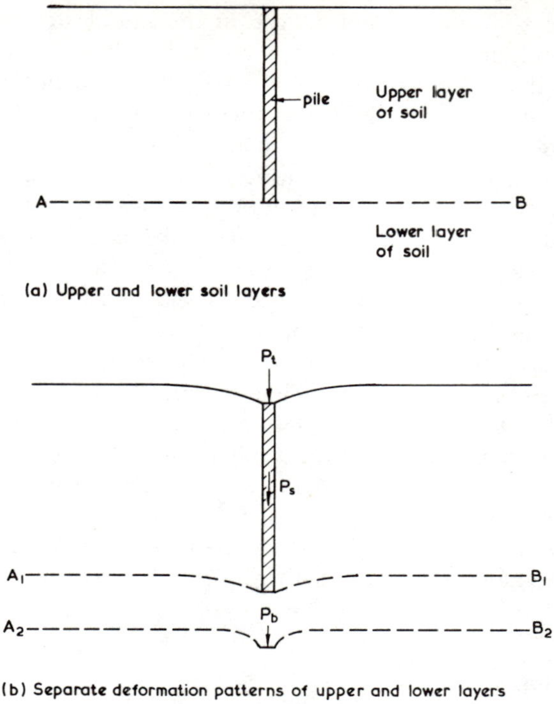

FIG. 1—*Uncoupling of effects due to pile shaft and base.* (a) *Upper and lower soil layers.* (b) *Separate deformation patterns of upper and lower layers.*

with radius from the pile [*4-6*]. Thus, the shear stress at radius r may be written

$$\tau = \tau_0 r_0 / r \tag{1}$$

where τ is the shear stress, r is the radius, and the subscript 0 implies conditions at the pile face.

For soil that is radially homogeneous around the pile, Eq 1 implies that the shear strains in the soil will also vary inversely with the radius. Thus, integration of the shear strains leads to deflections in the soil that vary in a logarithmic manner with the radius:

$$w = w_s - \frac{\tau_0 r_0}{G} \ln r/\bar{r}_0 \tag{2}$$

where w is the vertical deflection of the soil, w_s is the vertical deflection of the pile shaft, and G is the shear modulus of the soil. The logarithmic variation has been confirmed by measurements from pile tests in the field [*4*] and also by finite element analyses [*5,6*]. Assuming that the movement in the soil will be negligible beyond the radius r_m, the settlement of the pile shaft may be written as

$$w_s = \zeta \tau_0 r_0 / G \tag{3}$$

where $\zeta = ln(r_m/r_0)$. The ratio of the total shaft load to the pile displacement is given by

$$\frac{P_s}{Gr_0 w_s} = \frac{2\pi r_0 l \tau_0}{Gr_0 w_s} = \frac{2\pi}{\zeta} \frac{l}{r_0} \tag{4}$$

The load-settlement ratio of the pile shaft has been nondimensionalized by dividing by the shear modulus of the soil and by the pile radius. The values of w_s and τ_0 will be average values down the pile shaft [*2*].

Deformation of Pile Base

The pile base acts as a rigid punch on the surface of the lower layer of soil (Fig. 1*b*). The load-settlement ratio may be obtained from the Boussinesq solution [*7*]

$$\frac{P_b}{Gr_0 w_b} = \frac{4}{(1 - \nu)} \tag{5}$$

where ν is the Poisson's ratio for the soil. It should be noted that the upper layer of soil will not contribute to the stiffness of the pile base, since it is being deformed by the shear stresses acting down the pile shaft [*2,3*].

Combination of Shaft and Base Behavior

For a rigid pile, the settlement of the pile shaft will be uniform and equal to that at the pile base. Thus, the overall load-settlement ratio is given by

$$\frac{P_t}{Gr_0 w_t} = \frac{P_b}{Gr_0 w_b} + \frac{P_s}{Gr_0 w_s} = \frac{4}{(1 - \nu)} + \frac{2\pi}{\zeta} \frac{l}{r_0} \tag{6}$$

This is the approximate solution developed for a rigid pile embedded in a homogeneous elastic half-space. The limiting radius r_m has been found, empirically, to satisfy $r_m = 2.5l(1 - \nu)$ [*2,3*]. For this value of r_m, very good agreement is obtained between results from Eq 6 and corresponding results from integral equation analyses [*8, 9*]. Typical results are shown in Fig. 2.

The solution above has been extended to deal with compressible piles and also soils that are nonhomogeneous [*2,3*]. In general, the stiffness of a soil deposit will tend to increase with depth. The solution may be applied to any soil where the stiffness increases approximately linearly with depth (Fig. 3*a*). The degree of homogeneity may be conveniently expressed by a factor $\rho = G_{l/2}/G_l$ which is the ratio of the shear modulus at the pile middepth to that at the pile base. This factor will vary between 0.5 for a soil whose stiffness is proportional to depth and 1.0 for a homogeneous soil. The derivation of the full solution may be found in Refs *2* and *3*. The complete expression is

$$\frac{P_t}{G_l r_0 w_t} = \left[\frac{4}{\eta(1-\nu)} + \frac{2\pi}{\zeta}\rho\frac{l}{r_0}\frac{\tanh(\mu l)}{\mu l}\right] \times \left[1 + \frac{4}{\eta(1-\nu)}\frac{1}{\pi\lambda}\frac{l}{r_0}\frac{\tanh(\mu l)}{\mu l}\right]^{-1} \quad (7)$$

where

$\lambda = E_p/G_l$,
$\mu l = (2/\zeta\lambda)^{1/2} l/r_0$,
$\rho = G_{l/2}/G_l$, and
$\zeta = \ln(r_m/r_0)$.

Young's modulus for the pile has been written as E_p, and the shear modulus of the soil at the level of the pile base as G_l. A factor η has been introduced to enable the solution to be extended to underreamed piles [*2*]. For a ratio of underream of r_b/r_0, the factor $\eta = r_0/r_b$ accounts for the greater load taken by the pile base. For cases where there is a softened zone around the pile, the constant ζ may be taken as

$$\zeta = G_\infty \int_{r_0}^{r_m} (Gr)^{-1}\, dr \quad (8)$$

where G varies with radius from the pile, tending to G_∞ at large values of r (where the soil has not been disturbed by the pile installation process).

The value of r_m is given by

$$r_m = 2.5\rho l\,(1-\nu) \quad (9)$$

for piles in a deep stratum of soil (of greater depth than about three times

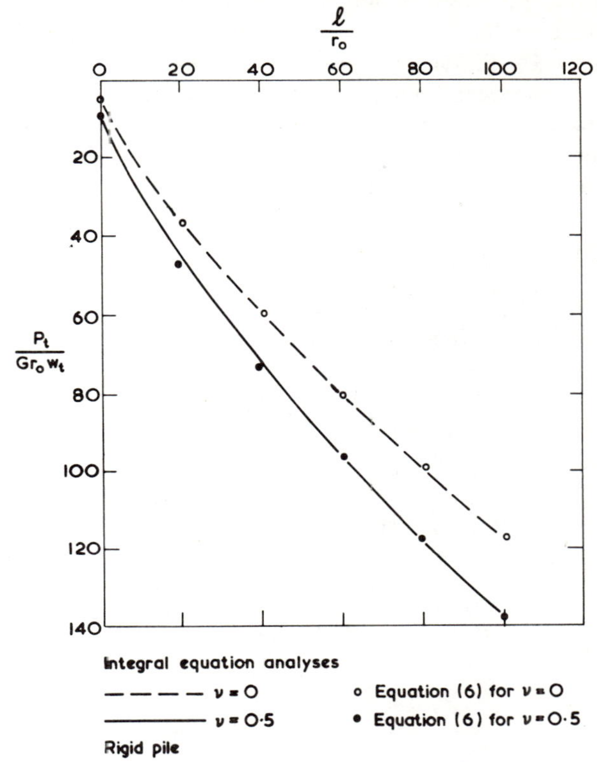

FIG. 2—*Comparison of load-settlement ratios for different slenderness ratios of rigid piles in homogeneous soils.*

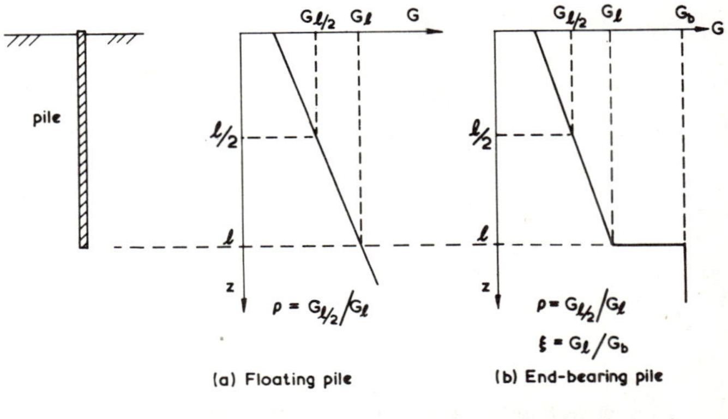

FIG. 3—*Assumed variation of shear modulus of soil with depth.*

the pile length). For most practical cases, however, the soil stiffness increases considerably at depths less than $3l$. It has been found empirically that a value of $r_m = 2\rho l\,(1 - \nu)$ should then be taken [*2,3*].

Figure 4 shows typical curves given by Eq 7 for a variety of pile stiffness ratios and the two extreme soil types of $\rho = 1$ and $\rho = 0.5$. It should be noted that many piles may be regarded as essentially rigid. A useful guideline is that for piles to be classed as rigid (that is, $w_t/w_b < \sim 1.2$),

$$(l/r_0)^2/\lambda < 0.5 \tag{10}$$

For such piles, Eq 7 reduces to

$$\frac{P_t}{Gr_0w_t} = \frac{4}{\eta(1-\nu)} + \frac{2\pi}{\zeta}\,\rho\,\frac{l}{r_0} \tag{11}$$

For compressible piles, the variation of settlement down the pile is approximately given by [*2,3*]

$$w(z) = w_b \cosh[\mu(l - z)] \tag{12}$$

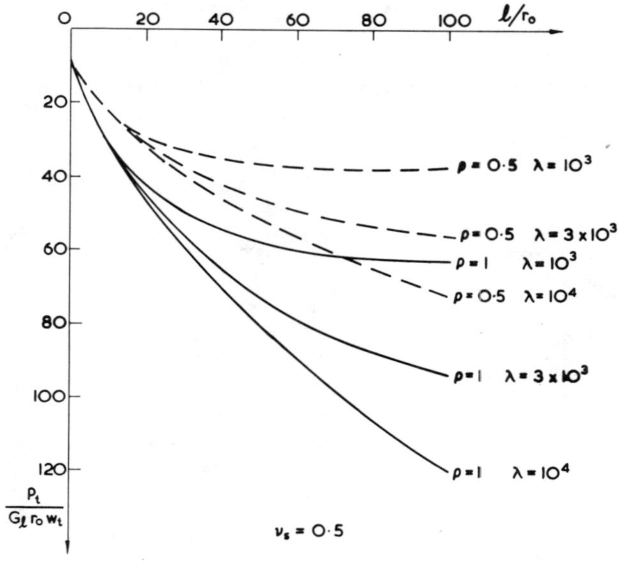

FIG. 4—*Variation of load-settlement ratio with slenderness ratio for different pile stiffness and soil types.*

The ratio of the settlement at the top of the pile to that at the pile base is then

$$w_t/w_b = \cosh(\mu l) \tag{13}$$

Extension of Solution to End-Bearing Piles

In many piled foundations, the piles are driven through a relatively soft layer to bear on a stiffer stratum of soil. The stiffness of the upper layer of soil may also increase gradually with depth, and so a typical (idealized) stiffness profile is shown in Fig 3*b*. There will be two main effects of the sudden increase in the shear modulus of the soil at the level of the pile base:

1. The load at the pile base will be increased (for a given settlement) by a factor G_b/G_l.
2. The value of the maximum radius of influence of the pile will be decreased. It will be assumed that this decrease is linear with the factor G_l/G_b. However, since for a completely rigid substratum some load transfer down to the pile shaft will be expected [*10*], there will be some finite lower limit to r_m.

It has been found empirically that a suitable expression for r_m for end-bearing piles is

$$r_m = l\{\tfrac{1}{4} + [2\rho(1-\nu) - \tfrac{1}{4}]\xi\} \tag{14}$$

For $\xi = 1$, this reduces to that recommended previously; as $\xi \rightarrow 0$, the limiting value of r_m is $l/4$. Using this expression for r_m and a modified factor η given by $\eta = (r_0/r_b)\,\xi$, Eq 7 may be used to calculate the load-settlement ratio of end-bearing piles.

Figure 5, taken from Poulos and Mattes [*10*], shows good agreement between results from integral equation analyses and those given by the above solution. For a very stiff substratum (for example, $\xi < 0.01$) the terms containing $1/\eta$ dominate in Eq 7. Thus, the load-settlement ratio reduces to

$$\frac{P_t}{G_l r_0 w_t} \simeq \frac{\pi\lambda}{l/r_0}\,\frac{\mu l}{\tanh(\mu l)} \tag{15}$$

For still piles, tanh $(\mu l) \simeq \mu l$ and, as found by Poulos and Mattes [*10*], the load-settlement ratio is given by the elastic settlement of the pile, since, from Eq 15,

$$w_t \simeq \frac{P_t}{\lambda G_l \pi r_0}\,\frac{l}{r_0} = \frac{P_t}{E_p A_p}\,l \tag{16}$$

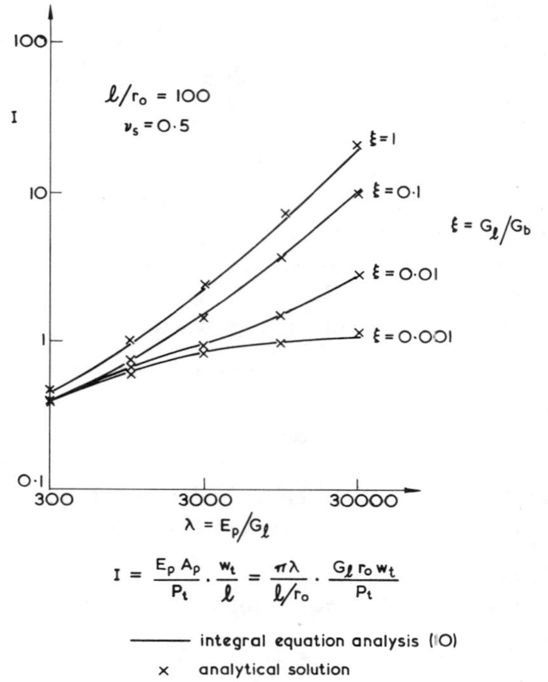

FIG. 5—*Settlement coefficients for end-bearing piles* (*after Poulos and Mattes* [*10*]).

In practice, the stiffness of the soil at the pile base is often of the same order of magnitude as that of the overlying layer (that is, $1 \leq G_b/G_l \leq 10$). In this case, the full solution (Eq 7) must be used with the limiting radius given by Eq 14. Typical values of the load-settlement ratio are given in Table 1 for piles of three different slenderness ratios in the two extreme soil types—homogeneous ($\rho = 1$) and stiffness proportional to depth ($\rho = 0.5$). Results from finite element analyses are also given for comparison; the two sets of results are in good agreement.

Application of Solution to Pile Design

At present, piles are designed on the basis of ensuring overall stability of the foundation. However, the permissible settlement of the foundation is often a more stringent design criterion than the overall stability. In addition, restricting the settlement to a value less than that necessary to mobilize full shaft friction will automatically satisfy the criterion of overall stability. To facilitate the choice of pile type when the load and permitted settlement are known, design charts may be prepared using the

TABLE 1—*Comparison of load-settlement ratios for end-bearing piles.*[a]

		Load-Settlement Ratio ($P_t/G_l r_0 w_t$)					
	Soil Type	$\xi = 1$	$\xi = 0.5$	$\xi = 0.25$	$\xi = 0.1$	$\xi = 0.01$	
$l/r_0 = 20$	$\rho = 1$	41.5	51.7	64.8	88.6	158.5	finite element
		41.6	52.1	66.1	90.3	156.9	Eq 7
	$\rho = 0.5$	25.0	33.3	45.3	69.4	142.4	finite element
		28.5	36.7	48.4	71.6	150.2	Eq 7
$l/r_0 = 40$	$\rho = 1$	53.6	65.1	73.2	84.6	104.2	finite element
		54.5	62.6	71.8	84.3	106.3	Eq 7
	$\rho = 0.5$	34.8	43.3	51.0	62.8	84.1	finite element
		34.5	41.0	49.5	64.5	101.2	Eq 7
$l/r_0 = 80$	$\rho = 1$	65.3	69.3	72.4	75.5	78.8	finite element
		62.1	67.1	71.8	76.9	82.8	Eq 7
	$\rho = 0.5$	35.8	39.0	42.1	45.6	49.8	finite element
		37.2	42.0	48.0	58.0	78.7	Eq 7

[a] $E_p/G_l = 10^3$ for all piles, $\nu_s = 0.4$.

analytical solution outlined above. For any particular profile of soil stiffness, Eq 7 may be used to calculate the ratio P_t/w_t for a range of possible pile geometries and stiffness. Contours of equal values of P_t/w_t may then be drawn to aid design.

Floating Piles

Figure 6*a* shows a design chart for piles in a soft clay deposit. The soil stiffness profile has been obtained from tests at Canvey Island using a self-boring pressure meter [*11*]. These tests gave a mean profile of shear modulus of $G = 0.9z$ MN/m², where z is the depth in metres. The pile stiffness has been taken as $E_p = 2.10^4$ MN/m²—a suitable lower bound for a concrete pile. The chart shows contours of P_t/w_t for a range of pile lengths and radii. The design geometry of the pile may be chosen from known values of the pile load and permitted settlement, together with any additional limitations on either the pile length or pile radius.

End-Bearing Piles

For piles that are to be driven to a specified depth in order to bear on a stiffer stratum of soil, the length of the pile is fixed. The pile radius and pile type (which will affect the stiffness E_p) are then the important variables. A typical soil profile has been taken from a site where 15 m of soft to firm clay overlay laterite. Shear modulus values were obtained using a self-boring pressure meter.[3] These gave an average of $G = 10$ MN/m² for

[3] Clarke, B. G. (1977), private communication.

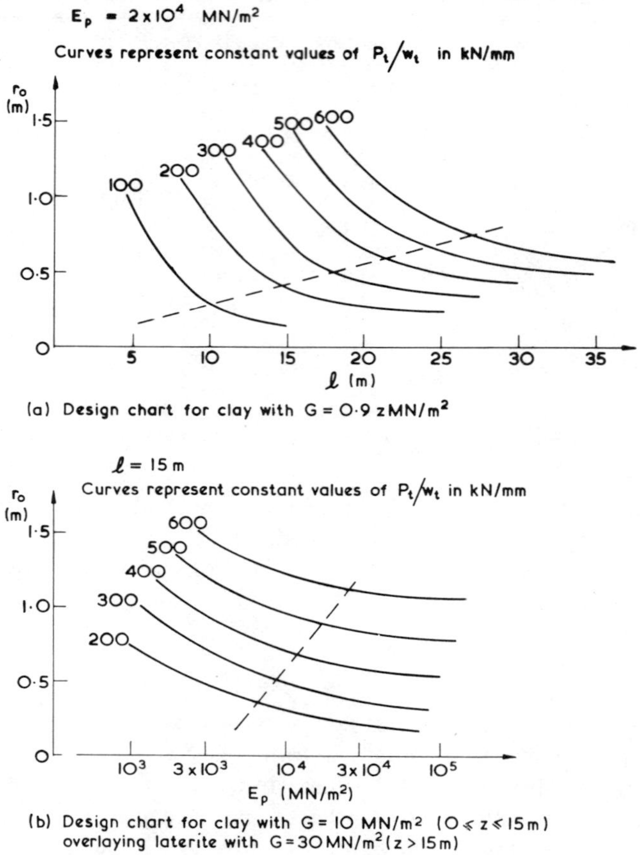

FIG. 6—*Examples of design charts for single piles:* (a) *for clay with* $G = 0.9z$ *MN/m*^{2}*;* $E_p = 2 \times 10^4$ *MN/m*^{2}*;* (b) *for clay with* $G = 10$ *MN/m*^{2} *(*$0 \leq z \leq 15$ *m) overlying latenite with* $G = 30$ *MN/m*^{2} *(*$z > 15$ *m),* $1 = 15$ *m. Curves represent constant values of* P_t/w_t *in kilonewtons per millimetre.*

$0 \leq z \leq 15$ m and $G = 30$ MN/m^2 for $z \geq 15$ m. Figure 6*b* shows a design chart for end-bearing piles in such a soil. Contours of P_t/w_t have been plotted for a range of pile stiffnesses E_p and pile radii r_0. The length of the pile is fixed at 15 m.

It is necessary to ensure that the calculated settlement for the designed pile is below that necessary to mobilize full shaft friction. From Eq 3, the shaft settlement as a fraction of shaft radius is $w_t/r_0 = \zeta\tau_0/G$. Typically, the factor ζ is of the order of 3 (see Ref *5* and *6*), increasing as the pile length increases and decreasing as the degree of homogeneity of the soil decreases. The ultimate shaft friction may be written as $(\tau_0)_f = \alpha c_u$. Thus, the shaft settlement to cause full shaft friction may be written as

$$(w_t/r_0)_f \sim 3\frac{\alpha}{G/c_u} \tag{17}$$

Reasonable limits on the ratio G/c_u are $G/c_u \sim 150$ for stiff, heavily over-consolidated clay and $G/c_u \sim 50$ for normally consolidated clay. The appropriate value of α is more difficult to determine and will depend on the method of installation of the pile. In order to obtain approximate limits on the settlement to mobilize full shaft friction, it will be assumed that α will lie between 0.5 for heavily overconsolidated clay (recommended by Skempton [*12*] for bored piles in London clay) and 1.0 for normally consolidated clay [*13*]. Thus, depending on the past stress history of the soil and also on the method of pile installation, sensible limits on the pile settlement necessary to cause shaft failure are

$$0.01 < (w_t/r_0)_f < 0.06 \tag{18}$$

These figures are only guides, and the appropriate values of $(w_t/r_0)_f$ for a given pile and soil type should always be calculated from Eq 3. For a given permitted settlement of the designed pile, this will lead to a minimum safe value of the pile radius if failure of the pile shaft is to be avoided.

It should be noted that in the design charts, contours of constant P_t/w_t tend to become parallel to the length axis (Fig. 6*a*) at high values of l/r_0 and parallel to the pile stiffness axis (Fig. 6*b*) at high values of E_p. This implies inefficient areas of design. For floating piles, increasing the length of long (compressible) piles shows little reduction in the load-settlement ratio of the pile. For end-bearing piles, beyond a certain pile stiffness, little compression of the pile shaft takes place, and thus nothing is to be gained by increasing the pile stiffness further. Both these limits on efficient design are indicated by dashed lines in the two design charts.

Application of Solution to the Back Analysis of Pile Tests

The analytical solution may be used to interpret the results of load tests on piles. If measurements are taken during the load test of the settlement of the pile base, as well as of the pile head, then it is possible to estimate the profile of soil stiffness from a single pile test [*2,3*]. To illustrate the manner in which the solution may be applied to more standard pile tests, where measurement of the pile head settlement only is made, a series of pile tests in chalk will be considered. Lord [*14*] reports the results of tests on six piles (four different lengths and two different diameters) driven into chalk at Portsmouth Harbour. The piles were driven precast shell piles, as constructed by West's Piling and Construction Co. Ltd, with an estimated average stiffness of $E_p = 2.8 \times 10^4$ MN/m².

The chalk varies in quality from grade VI in the top few metres, to grade I at depth (see Ward et al [*15*] for definition of chalk grades). Typical profiles of the chalk grade from borehole logs and the positions of the six test piles are shown in Fig. 7. In order to analyze the pile tests, it will be assumed that the soil stiffness increases proportionally with depth (see Burland et al [*16*]) until the grade I chalk is encountered, whereupon the soil stiffness will increase suddenly (as in Fig. *3b* with $\rho = 0.5$). Poisson's ratio will be taken as 0.24 [*16*].

The three shorter piles (W3, W7, and W8 in Table 2) will behave as essentially rigid piles, and so the nondimensional load-settlement ratio may be obtained from Eq 11 with $\eta = 1$ and $\rho = 0.5$. The values of the quantity $P_t/(G_l r_0 w_t)$ for the three pile tests are 40.3, 48.4, and 40.0, respectively. For the measured ratios of P_t/w_t (at the design settlement of 6 mm) given in Table 2, these values lead to separate estimates of the rate of increase of G with depth of 0.97, 0.80, and 1.00 MN/m² per metre, respectively. The more accurate expression, which takes account of the

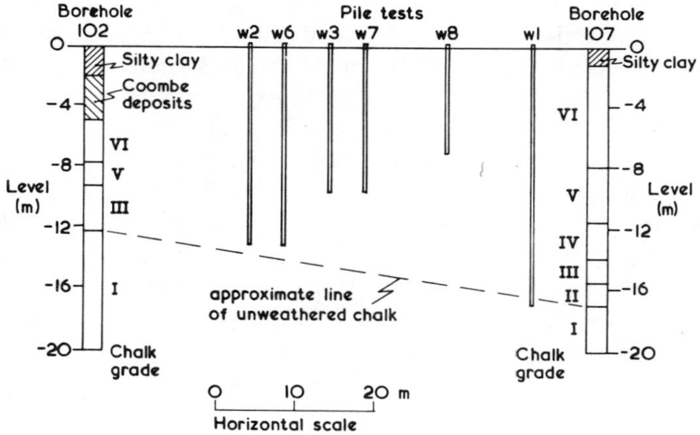

FIG. 7—*Profiles of chalk grades and positions of pile tests* [*14*].

TABLE 2—*Data for pile tests.*[a]

Pile number	W1	W2	W3	W4	W5	W6
Diameter, m	0.51	0.51	0.51	0.38	0.38	0.38
Length, m	17	13	9.5	13	9.5	7
P_t/w_t (at w_t = 6 mm), kN/mm	200	167	95	140	70	53

[a] From Lord [*14*]. E_p = 28 000 MN/m².

small amount of pile compression, yields corresponding values of 1.08, 0.92, and 1.07 MN/m² per metre. Thus, it seems reasonable to assume that the shear modulus profile in the weathered chalk may be approximated as $G = 1.0z$ MN/m², where z is the depth in metres.

The test piles W2 and W6 were 13 m long and, being quite close to borehole 102, appear to be end-bearing on the unweathered (grade I) chalk. Test pile W1 reached the unweathered chalk at a penetration of 17 m. If it is assumed that the stiffness of the unweathered chalk is the same ($= G_b$) at each of the three pile positions and that the stiffness of the weathered chalk is given by $G = 1.0z$ MN/m², then Eq. 7 may be used to estimate ξ and thus G_b. This leads to independent estimates for G_b of 31.6, 33.1, and 43.7 MN/m², respectively (tests W1, W2, and W6). Thus, it seems that a sensible estimate for the shear modulus at the top of the unweathered chalk at this site is about 36 MN/m². This figure is in reasonable agreement with values at the pile bases estimated by Lord [*14*] from the part of the load-settlement curves after full mobilization of shaft friction. He gives values varying from 18 to 50 MN/m², with an average of 32 MN/m².

Figure 8 shows a comparison between the measured load-settlement curves for the six pile tests and the (linear) curves predicted assuming that the modulus of the weathered chalk is given by $G = 1.0z$ MN/m² and that of the unweathered chalk is $G = 36$ MN/m². There is good agreement in the early stages of the tests for all six piles.

It should be noted that the analysis above may be performed with a

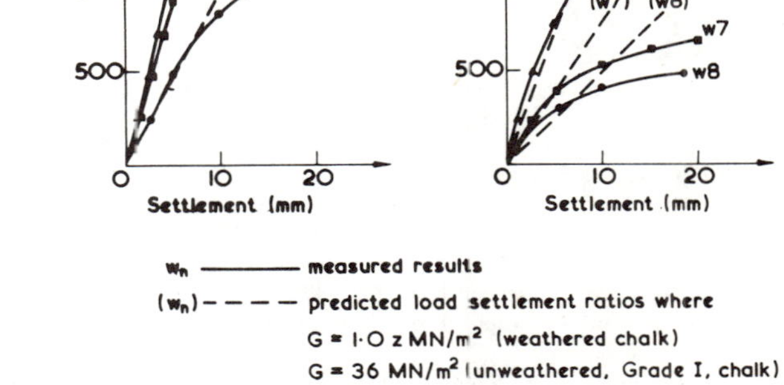

FIG. 8—*Measured and calculated load-settlement curves* (*14*).

hand calculator in a matter of minutes. There is thus a considerable saving of effort compared to traditional methods of back analysis, which involve reading load-settlement ratios and correction factors off a number of different charts (Poulos [*1*]) or perhaps performing a succession of finite element analyses (Cooke and Price [*17*]).

Conclusions

If the behavior of the soil is to be idealized by assuming that it may be characterized by suitably chosen elastic constants, then it is essential that simple solutions be available that show how a foundation will interact with the soil. This paper has outlined the use of such a solution for a single vertically loaded pile. The solution, originally developed to analyze fully floating friction piles, has been extended to deal with the case of piles driven through a soft stratum to bear on a stiffer stratum of soil.

Much design of piled foundations is concerned with the behavior of groups of piles. The analytical model may be used for the analysis and design of pile groups [*2*]. However, it is often necessary to design single piles capable of withstanding some specified load without undergoing excessive settlement. In such circumstances, the solution presented may be used to advantage in the preparation of suitable design charts.

Detailed information on the shear modulus profile of the soil may be obtained from instrumented pile tests where the distribution of load or settlement down the pile is measured. The testing of uninstrumented piles of different embedded lengths may also provide this information. The analytical solution provides an efficient means whereby such tests may be back analyzed to give estimates of the shear modulus profile of the soil for subsequent use in design.

References

[*1*] Poulos, H. G., *Journal of Soil Mechanics and Foundation Engineering*, Vol. 98, No. sm 9, Sept. 1972.
[*2*] Randolph, M. F., "A Theoretical Study of the Performance of Piles," Ph.D. Thesis, Cambridge University, Cambridge, England, 1977.
[*3*] Randolph, M. F. and Wroth, C. P., *Journal of Geotechnical Sciences*, Vol. 104, No. GT12, Dec. 1978, pp. 1465–1488.
[*4*] Cooke, R. W., The Settlement of Friction Pile Foundations, *Proceedings*, Conference on Tall Buildings, Kuala Lumpur, 1974.
[*5*] Baguelin, F., Bustamante, M., Frank, R., and Jezequel, J. F., *Annals de l'Institut Technique du Bâtiment et des Travaux Publics*, Suppl 330, Serie SF/116, 1975.
[*6*] Frank, R. "Étude Théorique du Compartement des Pieux sous Charge Verticale. Introduction de la Dilatance," Docteur Ingenieur Thesis, University of Paris VI, Nov. 1974.
[*7*] Timoshenko, S. P. and Goodier, J. N., *Theory of Elasticity*, 3rd ed., McGraw-Hill, New York, 1970.
[*8*] Poulos, H. G. and Davis, E. H., *Geotechnique*, Vol. 18, No. 3, 1968, pp. 449–471.
[*9*] Banerjee, P. K., "A Contribution to the Study of Axially Loaded Pile Foundations," PH.D. Thesis, Southampton University, Southampton, England, 1970.

[10] Poulos, H. G. and Mattes, N., *Geotechnique,* Vol. 19, No. 2, 1969, pp. 285–300.
[11] Pender, M. J., "A Soft Clay Deposit Under Embankment Loading," Internal Report, Cambridge University, Cambridge, England, 1974.
[12] Skempton, A. W., *Geotechnique,* Vol. 9.
[13] Tomlinson, M. J., "Some Effects of Pile Driving on Skin Friction," Institution of Civil Engineers, 1970.
[14] Lord, J. A., *Geotechnique,* Vol. 26, No. 1, pp. 73–93.
[15] Ward, W. H., Burland, J. B., and Gallois, R. W., *Geotechnique,* Vol. 18, No. 4, 1968, pp. 399–431.
[16] Burland, J. B., Sills, G. C., and Gibson, R. E., "A Field and Theoretical Study of the Influence of nonhomogeneity on Settlement," *Proceedings,* 8th International Conference SM & FE, Moscow.
[17] Cooke, R. W. and Price, G., "Strains and Displacements around Friction Piles," *Proceedings,* 8th International Conference SM & FE, Moscow.

F. Rausche[1] *and G. G. Goble*[2]

Determination of Pile Damage by Top Measurements

REFERENCE: Rausche, F. and Goble, G. G., "**Determination of Pile Damage by Top Measurements,**" *Behavior of Deep Foundations, ASTM STP 670,* Raymond Lundgren, Ed., American Society for Testing and Materials, 1979, pp. 500–506.

ABSTRACT: Measurements of force and acceleration at the pile top during driving can be used to detect the presence of damage at points along the pile below the ground surface. Analytical considerations based on one-dimensional wave mechanics are discussed, and a method is derived to place a quantitative measure on the degree of damage. Actual field measurements on damaged piles are presented.

KEY WORDS: field tests, foundations, inspection, pile driving

In the past few years, the capability has been developed to measure routinely force and acceleration at the top of a pile during driving. The primary reason for making these measurements has been to determine pile static capacity using the Case method [*1*][3] to verify hammer performance parameters, or to determine soil resistance characteristics by analysis with the Case Pile Wave Analysis Program (CAPWAP) [*2*]. With both force and acceleration time records available from the Case Method measurements, it is possible to detect discontinuities or reductions in cross section of the pile even though the point of disturbance is hidden below the ground surface. By one-dimensional wave propagation considerations, it is possible to reach quantitative conclusions regarding the degree of section reduction at the discontinuity.

In this paper, the basic mechanics will be presented for evaluation of pile damage. Some actual measurements on damaged piles will be shown and evaluated.

[1]President, Goble and Associates, Inc., Cleveland, Ohio 44106

[2]Chairman, Department of Civil, Environmental, and Architectural Engineering, University of Colorado, Boulder, Colo. 80309.

[3]The italic numbers in brackets refer to the list of references appended to this paper.

Fundamental One-Dimensional Wave Mechanics

When a suddenly applied axial load acts on a slender rod whose length is much greater than its diameter, the induced stress propagates away from the point of application. Pile driving is an example of this phenomenon. The variations of stresses in the rod have a wave-type appearance. In the case of the stress wave induced by impact, the more suddenly load application occurs, the better the applicability of stress wave mechanics.

The stress wave will propagate along the rod at a velocity c, and this velocity is a material property. Wave speeds in piles will be greater than 3000 m/s (10 000 ft/s) for typical pile materials. The application of the suddenly applied force will induce velocities at the end of the rod that propagate with the stress wave. These particle velocities will have the same general magnitude as the velocity of the impacting ram and for pile driving typically range from 1.8 to 4.6 m/s (6 to 15 ft/s).

In addition to the forces induced by impact, there are also forces caused by passive effects such as soil resistance. When the stress wave propagates along the pile below the ground surface, the pile displaces downward. Due to the displacement, upward-acting soil resistance forces are generated that in turn induce stress waves in the pile that propagate away from the point of action and are superimposed on the impact wave. Other forces caused by passive effects are changes in cross section and reflections from the pile tip.

Development of Damage Detection Relations

It can be shown [*3*] that a proportionality exists between force and velocity at the pile top during and after impact so long as no return waves have reached the top. Thus, for a uniform cross section pile with no damage or soil resistance and having a length L, the measured force and velocity for a time after impact less than $2L/c$ are related by

$$F = v\,EA/c \tag{1}$$

where F is the pile force, v is the particle velocity, E is the modulus of elasticity of the pile material, A is the pile cross-sectional area, and c is the velocity of stress wave propagation. The quantity EA/c is generally known as the impedance and will hereafter be designated as I.

Figure 1 shows measurements that were made on a very long steel pile. The measured acceleration has been integrated to obtain velocity, and that quantity multiplied by I is plotted together with the force. This pile was of uniform cross section and was driven inside an empty casing for all but the last 40 ft of its length. Thus, reflections from soil resistance or cross section changes were impossible until slightly before the first reflection from the

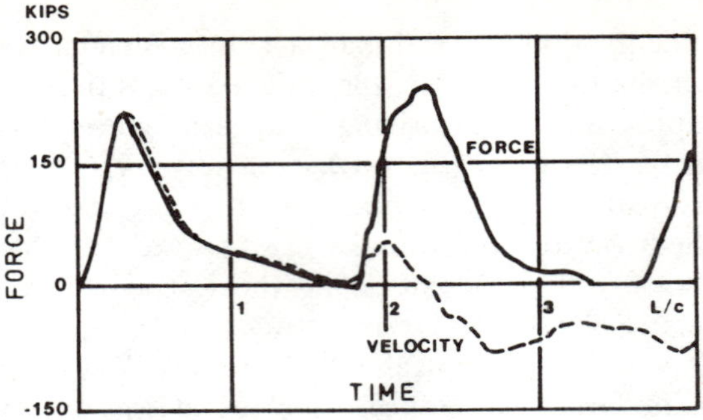

FIG. 1—*Measured force and velocity times* EA/C *for a pile with no side resistance except on the lower 40 ft.* L = *393 ft,* c = *16 800 ft/s,* EA/c = *44.7 kips/ft/s (1 ft = 0.3048 m, 1 ft/s = 0.3048 m/s, kips/ft/s = 14.59 kN/m/s).*

pile tip at $2L/c$ after impact. The records start to diverge when soil resistance forces have their effect at the pile top. Note that both force and velocity turn upward together at about $1.8L/c$. Since the pile is of uniform cross section, it can be concluded that this increase is due to an input force from the driving system. This input came when the hammer assembly impacted with the helmet, a fact that can be verified by examination of the acceleration record (not given here). Very shortly after assembly impact, the reflection from the tip soil resistance arrived causing force and velocity records to diverge. The difference

$$R_\Delta = F - vI \tag{2}$$

at a time t_x after impact is caused by resistance forces that at a distance

$$x = ct_x/2 \tag{3}$$

from the pile top. The resistance forces R_Δ may be either concentrated or distributed.

If a pile changes cross section at a depth x, then at the time $2x/c$ after impact, a wave effect can be observed at the pile top in both force and velocity records. If the upper pile section has an impedance I_1 and the lower one an impedance I_2, then the pile top force will change by

$$F_\Delta = -2F_i \frac{I_1 - I_2}{I_1 + I_2} \tag{4}$$

if the pile top is fixed [*4*].

In Eq 4, F_i is the force at impact. Similarly, a change of velocity can be observed at a free pile top

$$v_\Delta = 2v_i \frac{I_1 - I_2}{I_1 + I_2} \tag{5}$$

with v_i being the velocity at impact. Thus, the change between force and velocity from either consideration at the pile top due to a cross-sectional change is

$$u_\Delta = 2F_i \frac{I_1 - I_2}{I_1 + I_2} \tag{6}$$

This is true as long as the total force at the pile top does not reach zero (separation) and for times less than $2L/c$ after impact.

In general, the impact wave is reduced by the effect of soil resistance at the time it gets to the point of cross-section change. In fact, it will have decreased by $R/2$ for only elastoplastic resistance forces. More realistically, however, it should be assumed that F_i has decreased by R_Δ, since damping wave effects are not additive. Thus, Eq 6 becomes

$$u_\Delta = 2(F_i - R_\Delta) \frac{I_1 - I_2}{I_1 + I_2} \tag{7}$$

A relative lower cross section

$$B = \frac{I_2}{I_1} = \frac{A_2}{A_1} \tag{8}$$

will be introduced that is valid for piles of only one material. Thus, subsituting Eq 8 in 7 and solving for B leads to

$$B = \frac{2 - \mu_\Delta/(F_i - R_\Delta)}{2 + \mu_\Delta/(F_i - R_\Delta)} \tag{9}$$

which can be further reduced when defining

$$\alpha = \frac{u_\Delta}{2(F_i - R_\Delta)} \tag{10}$$

which is a relative pile top wave effect. Thus,

$$B = \frac{1 - \alpha}{1 + \alpha} \tag{11}$$

Example of Damage Determination

Consider Fig. 2, which is a record taken on a broken concrete pile. The early force and velocity behavior shows the typical force-velocity proportionality. At a time of 8.6 ms after impact, the velocity shows an increase relative to the force. This increase can only be explained by a small pile mass and stiffness approximately 16.5 m (54 ft) below the pile top [$x = \frac{1}{2} c t_x$ = (54 ft)]. Before the relative velocity increase became noticeable, the proportional velocity had already dropped by an amount R_Δ = 925 kN (208 kips) relative to the force. The relative increase of this velocity at the point of damage effect was u_Δ = 1352 kN (304 kips) and the impact force was F_i = 2304 kN (518 kips). Therefore

$$\alpha = \frac{u_\Delta}{2(F_i - R_\Delta)} = \frac{304}{2(581 - 208)} = 0.49$$

and

$$B = \frac{1 - \alpha}{1 + \alpha} = 0.34$$

Damage Classification

There is no experimental proof available justifying the following classi-

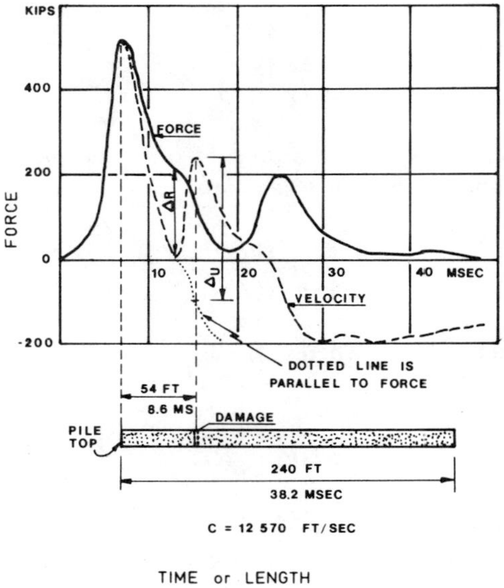

FIG. 2—*Example of a force and velocity measurement on a broken pile.*

fication. It was set up under the presumption that B actually indicates how much pile cross section is left. The following classification assumes that the soil resistance immediately below the point of breakage is negligible.

B	Severity of Damage
1.0	undamaged
0.8–1.0	slight damage
0.6–0.8	damage
Below 0.6	broken

Cracks and Slacks

Slacks, as they often occur between spliced pile sections, can well be discovered in easy driving. One important distinction between a crack/slack and a damage is that the latter becomes worse while a crack diminishes as driving becomes harder.

Once it is established that a reflection is due to a slack rather than to damage, its gap width can be determined in an approximate manner. This determination is based on an integration of the relative velocity change (measured from the force as a zero line). Thus,

$$\delta = \frac{1}{2} \int_{t1}^{t2} \left(v - \frac{F - R_{\Delta}}{I} \right) dt \qquad (12)$$

where t_1 and t_2 refer to the beginning and end of the slack effect and R_{Δ} is the difference between force and proportional velocity at time t_1. Figure 3 shows a sample calculation in which the shaded area corresponds to the integral of Eq 12.

The slack of 1.12 mm (0.044 in.) determined in Fig. 3 is probably a lower bound, since precompression effects had already closed the gap to a certain degree. It is also possible that the gap was not uniform (gap closed on one side, open on the other side). In this case, it would take a certain force to reduce the slack distance. Note that the slack reduced the speed of wave transmission to the lower section and the pile tip reaction occurs later than expected. The time of the wave return indicates a slack at 17 m (56 ft) below the top, a point that is very close to a pile splice. A greater accuracy in distance determination cannot be expected. The splice was of a mechanical type and had a design slack of about 0.79 mm (0.032 in.).

References

[1] Goble, G. G., Rausche, F. and Likins, G. "Bearing Capacity of Piles from Dynamic Measurements, Final Report," Report No. OHIO-DOT-05-75, Ohio Department of

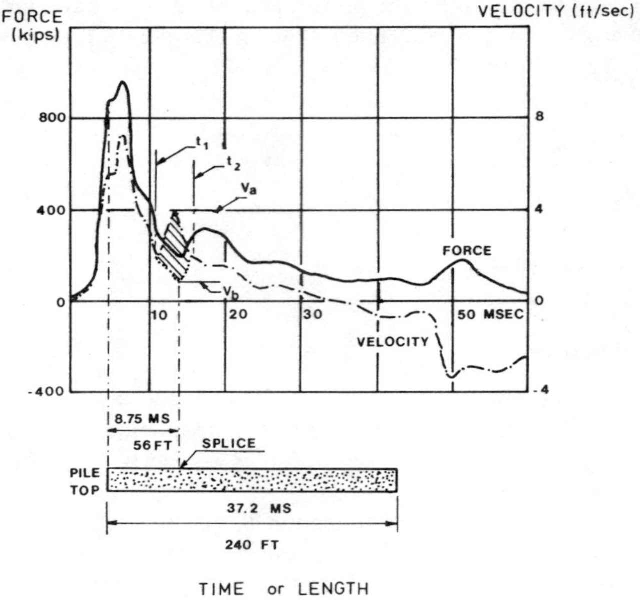

FIG. 3—*Example of force and velocity records for a pile containing a slack.* c= *12 900 ft/s,* V_a = *4 ft/s,* V_b = *0.9 ft/s,* t_1 = *11.3 ms,* t_2 = *16.0 ms, slack* = *½ (4.0 − 0.9) (16.0 − 11.3) (12/1000) = 0.044 in. (1 ft/s = 0.3048 m/s, 1 in. = 25.4 mm).*

Transportation, Department of Solid Mechanics, Structures and Mechanical Design, Case Western Reserve University, March, 1975.

[*2*] Rausche, F., Moses, F., and Goble, G. G., *Journal of the Soil Mechanics and Foundations Division, American Society of Civil Engineers,* Vol. 98, No. SM 9, Proceedings Paper 9220, September, 1972.

[*3*] Goble, G. G., Rausche, F., and Moses, F., "Dynamic Studies on the Bearing Capacity of Piles, Phase III," Report No. 48, Division of Solid Mechanics, Structure and Mechanical Design, Case Western Reserve University, 1970.

[*4*] Rausche, F., "Pile Capacities from Dynamic Measurements," Ph.D. thesis, Case Western Reserve University, Cleveland, Ohio, 1969.

D. M. Rempe[1]

Building Code Requirements for Maximum Design Stresses in Piles

REFERENCE: Rempe, D. M., **"Building Code Requirements for Maximum Design Stresses in Piles,"** *Behavior of Deep Foundations, ASTM STP 670,* Raymond Lundgren, Ed., American Society for Testing and Materials, 1979, pp. 507–519.

ABSTRACT: The maximum pile design stresses specified in building codes exert a strong influence on the design of pile foundations and have a major impact on the efficiency of utilization of pile materials. This paper presents an overview of code requirements in the United States, with emphasis on model codes and the recommendations of technical societies and trade organizations, but including a sampling of requirements in local codes. This is followed by a discussion of the current trend toward higher design stresses, potential problems resulting from this trend, and means by which codes can be structured so as to minimize or avoid these problems.

KEY WORDS: building codes, pile foundations, steel piles, concrete piles, wooden piles

Building codes typically specify maximum allowable axial compressive stresses for steel, concrete, and timber in foundation piles. In the past ten years, several code-writing agencies have increased the allowable stresses; other agencies are currently considering such increases. The increased code-allowable stresses have been adopted principally on the basis of theoretical structural capacity, with inadequate consideration given to problems of installation, quality control, and durability.

Code allowable stresses have a strong impact on pile design, due principally to the common practice of using the building codes as design guides. Although building codes are not intended to replace engineering design, there is a tendency among many engineers to use a pile design stress equal to the limit allowed in the code without further analysis. By this process, code stress limits become, in effect, design standards. This practice becomes more dangerous as code stress limits increase, leading to smaller pile cross-sectional area for a given load. The reduced cross-sectional area results in in-

[1]Geotechnical consultant, Champaign, Illinois 61820.

creased difficulty in developing bearing capacity, diminished margin for error in design and quality control, and greater possibility of failure due to deterioration. Thus, in many foundations, a design stress that is within code limits and that provides an apparently adequate theoretical factor of safety with respect to structural failure may in fact be unsafe.

The first part of this paper is an overview of code specifications for axial compressive stresses for piles in the United States, with emphasis on model codes and the recommendations of technical societies and trade organizations, but including a small sampling of local codes. This is followed by a discussion of the justification presented for recent increases in code-allowable stresses, the problems associated with these increases, and means by which the problems can be minimized or avoided.

Code Requirements

Types of Code

A complex array of building codes exists throughout the United States. Foundation design on a given project may be governed by a local code, a state code, a federal agency code, or no code at all. The governing code may incorporate all or part of a model code, or code recommendations by technical societies and trade organizations.

Four model codes are available for adoption by state or local governments: (1) the Basic Building Code (BBC), issued by the Building Officials and Code Administrators International, Inc.; (2) the National Building Code (NBC), issued by the American Insurance Association; (3) the Standard Building Code (SBC), issued by the Southern Building Code Congress International, Inc.; and (4) the Uniform Building Code (UBC), issued by the International Conference of Building Officials [*1–4*].[2]

Technical societies such as the American Society for Testing and Materials (ASTM), American Society of Civil Engineers (ASCE), American Association of State Highway and Transportation Officials (AASHTO), Prestressed Concrete Institute (PCI), and American Concrete Institute (ACI) periodically issue standards, manuals, and technical reports that may be incorporated by reference into building codes. Trade organizations created by materials suppliers, such as the American Iron and Steel Institute (AISI), Portland Cement Association (PCA), and National Forest Products Association (NFPA) also issue technical reports, manuals, and code recommendations. Government agencies, such as the General Services Administration, Corps of Engineers, Naval Facilities Engineering Command, and many others, issue manuals or codes.

There is considerable interaction among the model, technical society and

[2]The italic numbers in brackets refer to the list of references appended to this paper.

trade organization codes with respect to pile design stresses. This is due at least in part to the code-writers' reliance on the technical societies and trade organizations for code provisions, and the trade organizations' aggressive advocacy of code provisions favorable to their product.

Code Allowable Stresses

In current practice, piles are designed by the working stress method. Thus, the codes specify allowable working stresses for various pile materials. Typically, these stresses are selected by the code writers on the basis of the structural capacity of the pile. There is usually no requirement for consideration of attainable bearing capacity or durability within the allowable stress provision of the code; these factors are usually mentioned elsewhere in the code (see Additional Limitations, below).

Tables 1 through 4 contain allowable stress provisions of several selected model, technical society, trade organization, and local codes. Table 1 pertains to steel in H and pipe piles. To facilitate comparison of code provisions that are expressed in varying formats, the equivalent allowable stresses for two commonly-available steel strengths are shown on the table. Table 2 pertains to cast-in-place concrete in uncased (cast *in-situ*) piles, cased piles (thin shell), and concreted pipe. Table 3 pertains to precast reinforced concrete

TABLE 1—*Allowable stresses for steel in pipe and H-piles.*

	Allowable Compressive Stress[a]			
Source [Reference]	Pipe[b]	H Piles	f_y = 36 ksi[c]	f_y = 50 ksi[c]
UBC[d] [*4*]	$0.35f_y$ (max 12.6 ksi)	$0.35f_y$ (max 12.6 ksi)	12.6 ksi	12.6 ksi
NBC [*2*]	$0.50f_y$	...	...	...
SBC[d] [*3*]	12.6 ksi	12.6 ksi	12.6 ksi	12.6 ksi
BBC[e] [*1*]	$0.45f_y$ (max 20 ksi)	16.0 ksi	16.0 ksi	16.0 ksi
AISI [*5*]	$0.50f_y$[f]	$0.50f_y$	18.0 ksi	25.0 ksi
Chicago [*13*]	12.0 ksi	12.0 ksi	12.0 ksi	12.0 ksi
New Orleans[g] [*14*]	$0.50f_y$ (max 25 ksi)	$0.50f_y$ (max 25 ksi)	18.0 ksi	25.0 ksi
New York City [*15*]	$0.35f_y$ (max 12.6 ksi)	$0.35f_y$ (max 12.6 ksi)	12.6 ksi	12.6 ksi

[a] 1 ksi = 6.9 MPa; f_y = minimum yield stress; additional restrictions on total pile load, minimum metal thickness and other items are not listed.
[b] See Table 2 for allowable stress on concrete.
[c] Values are calculated on basis of code requirements for H-piles.
[d] Higher stresses permitted if substantiated.
[e] Provisions are subject to interpretation.
[f] Reduce by 20 percent for open-end piles on rock.
[g] Allows 33 percent increase for wind load.

TABLE 2—*Allowable stresses for cast-in-place concrete.*

Source [Reference]	Allowable Compressive Stress[a]		
	Uncased	Cased (Shell)	Cased (Pipe)[b]
UBC[c], NBC, SBC[c] [*4,2,3*]	$0.33f_c'$	$0.33f_c'$ to $0.40f_c'$[d]	$0.33f_c'$
BBC[e] [*1*]	...	...	...
ACI [*10*]	$0.33f_c'$	$0.33f_c'$ to $0.40f_c'$[d]	$0.33f_c'$
PCA [*6*]	$0.33f_c'$	$0.33f_c'$ to $0.40f_c'$[d]	...
Chicago [*13*]	...	$0.40f_c'$	$0.40f_c'$
Los Angeles [*16*]	$0.225f_c'$	$0.225f_c'$	$0.25f_c'$
New Orleans[f] [*14*]	$0.225f_c'$	$0.25f_c'$	$0.275f_c'$
New York City[g] [*15*]	...	$0.25f_c'$	$0.25f_c'$

[a] f_c' = concrete strength; additional restrictions on total pile load, pile size, and other items are not listed here; allowable stress on reinforcing steel not listed.

[b] See Table 1 for allowable stress in pipe.

[c] Higher stresses are permitted if substantiated.

[d] The higher value is permitted if requirements for confinement of concrete are met.

[e] Provisions are subject to interpretation.

[f] Allows 33 percent stress increase for wind load.

[g] Code allows determination of capacity by either working stress or ultimate strength methods of calculation. Values shown are for working stress method.

TABLE 3—*Allowable stress for precast and prestressed concrete piles.*

Source [Reference]	Allowable Compressive Stress[a]	
	Precast	Prestressed
UBC[b], NBC, SBC[b] [*4,2,3*]	$0.33f_c'$	$0.33f_c' - 0.27f_{pe}$
BBC[c] [*1*]	...	...
ACI [*10*]	$0.33f_c'$	$0.33f_c' - 0.27f_{pe}$
PCI [*17*]	...	$0.33f_c' - 0.27f_{pe}$
PCA [*6*]	$0.33f_c'$	$0.33f_c' - 0.27f_{pe}$
Los Angeles [*16*]	$0.225f_c'$	$0.33f_c' - 0.27f_{pe}$
New Orleans[d] [*14*]	$0.33f_c'$	$0.33f_c' - 0.27f_{pe}$
New York City[e] [*15*]	$0.25f_c'$	$0.25f_c'$

[a] f_c' = concrete strength; f_{pe} = effective prestress; additional restrictions on pile load, pile size, and other items are not listed; allowable stress on reinforcing steel is not listed.

[b] Higher stresses permitted if substantiated.

[c] Provisions are subject to interpretation.

[d] Allows 33 percent stress increase for wind load.

[e] Code allows determination of capacity by either working stress or ultimate strength methods of calculation. Values shown are for working stress method.

TABLE 4—*Allowable stress for treated timber piles.*

	Allowable Compressive Stress, ksi[a]			
Source [Reference]	Southern Pine[b]	Red Oak[c]	Douglas Fir[d]	Notes
UBC[e] [*4*]	1.20	1.10	1.20	...
NBC, BBC [*2, 1*]	1.09 to 1.37	1.02 to 1.16	1.25	ASTM D 2899[f]
SBC[e] [*3*]	1.20	1.10	1.25	ASTM D 2899[f,g]
NFPA [*8*]	1.09 to 1.37	1.02 to 1.16	1.25	ASTM D 2899[f]
New Orleans[h] [*14*]	1.20	1.02 to 1.16	1.25	ASTM D 2899[f,g]
New York City [*15*]	1.20	1.20	1.20	...

[a] 1 ksi = 6.9 MPa; additional restrictions on pile load, pile size, and other items not listed; stresses are range for species listed (see note *b*).

[b] Longleaf, shortleaf, loblolly, and slash.

[c] North and southern.

[d] Coast.

[e] Higher stresses permitted if substantiated.

[f] Code specifies ASTM Method 2899-74, which gives method of calculation of maximum stress. ASTM Method D 2899 permits increased stress with distance from tip.

[g] Stresses tabulated for selected species; for others, use ASTM Method D 2899.

[h] Allows 33 percent stress increase for wind load.

and prestressed concrete piles. Table 4 pertains to three common species of treated timber piles. For simplicity, many supplementary code provisions related to pile size, allowable load, and allowable stress have been omitted from the tables.

The code-allowable stresses are expressed in one of the following formats:

Type 1: $f_d = Kf_u$

Type 2: $f_d = Kf_u$ $\quad f_d$ not to exceed f_m

Type 3: $f_d = f_m$

where

f_d = maximum allowable stress under service loads,

K = reduction factor,

f_u = minimum failure or yield stress of pile material, and

f_m = maximum design stress, regardless of f_u.

For steel, all three types of format are used in the various codes. Type 1 format is used for concrete, with modification for prestress and steel reinforcement where appropriate. For timber, the type 3 format is used; in many cases, f_m is required to be determined as recommended in ASTM Establishing Design Stresses for Round Timber Piles (D 2899-74). Some codes, such as UBC and SBC, allow stresses higher than the specified limits if substantiation in the form of analysis, load tests, etc. is provided.

The greatest variations in allowable stress occur in the provisions for steel piles (Table 1). For steel with a yield stress of 248 MPa (36 ksi), the allowable stress varies from 83 to 124 MPa (12 to 18 ksi). With a yield stress of 345

MPa (50 ksi), the range is from 83 to 173 MPa (12 to 25 ksi). The wide variation reflects acceptance by some codes of higher stresses recently proposed by AISI [*5*].

There is considerable uniformity among the model, technical society and trade organization codes with respect to allowable stresses in concrete (Tables 2 and 3). The uniformity results from widespread acceptance of PCA recommendations [*6*]. However, many local codes contain lower allowable stresses than do the model codes.

Many codes have adopted timber stresses determined according to ASTM Method D 2899-74 or have incorporated D 2899-74 by reference. This standard specifies a method for calculating maximum design stresses on the basis of the crushing strength of small clear specimens as per ASTM Establishing Clear-Wood Strength Values (D 2555). It is important to note that ASTM Method D 2899 does not include a factor of safety in the recommended design stresses. The recommended stresses are the measured average failure stresses of small clear specimens with reduction for specimen size, variability of specimens, load duration, treatment, and other factors. A formal factor of safety of 1.25 is specified as optional. Stresss calculated as per ASTM Method D 2899 can be as high or higher than the long-term failure stress. ASTM Method D 2899 has been challenged on the grounds that the recommended design stresses are unsafe [*7*].

Additional Limitations

In addition to the maximum stress provisions, codes typically specify that total pile load shall be limited to the supporting capacity of the soil or rock, as determined by static analysis or by an energy formula, such as the *Engineering News* formula. For loads exceeding a nominal value, a static load test is usually required. In some cases, total pile load limits are specified, depending upon the type of pile (not the cross-sectional area) and the nature of the soil or rock support.

Most codes also require protection of piles against deleterious environmental conditions where such conditions exist. In some cases, this can be accomplished by increasing the pile cross-sectional area.

Trend to Higher Stresses

The recent code-allowable stress increases do not reflect any change in the character of the pile materials, but rather an attempt to use these materials more efficiently. Much of the impetus for change has come from manufacturers and suppliers of the various pile materials, who are interested in maintaining or improving the competitive position of their product. Toward this end, they have published reports and code recommendations advocating or supporting increased stresses [*5,6,8,9*]. Arguments in favor of increased allowable stresses include the following:

1. The factor of safety against pile overstress need not be larger than the factor of safety against pile bearing capacity failure, which is nominally 2.0 in most codes. On this basis, the design stress for steel, for instance, can be raised to one-half of yield.
2. Some soils exhibit "freeze," that is, an increase in pile capacity with time after driving. In these conditions, piles can be driven with very little soil resistance and correspondingly low dynamic stress levels, and subsequently loaded to high stress levels. Higher allowable stresses permit the designer to take advantage of soil freeze. Load tests in areas where soil freeze occurs are cited as evidence for the safe use of higher design stresses (less favorable load test results from areas where soil freeze does not occur are not considered).
3. Code allowable stresses that are appropriate to unfavorable construction conditions are wasteful as applied to projects with favorable conditions. It is argued that engineers should be permitted to take advantage of freeze and other favorable conditions, and should be relied upon to apply reduced stresses where conditions are unfavorable.
4. Pile capacity can be checked by means of static load test. If the application of higher allowable stresses results in pile loads greater than the soil/rock support, the error will be detected by load testing prior to construction.

All the reasons cited above are valid with respect to selected pile foundation projects. A decision as to their general validity, however, requires consideration of several problems associated with increased allowable stresses that have heretofore been neglected.

Problems Associated with Higher Stresses

As pile stresses increase, the engineer must deal rationally with certain problems that heretofore have largely been avoided by use of a generous factor of safety against pile structural failure. These problems will be described in the context of a brief examination of current practice with respect to factors of safety.

There are three basic modes of failure for a pile foundation: (1) bearing capacity failure of the pile relative to the soil/rock support, (2) structural failure of the pile, and (3) unacceptable settlement without pile failure. The designer should estimate a margin of safety against each mode of failure. Each mode will be discussed with respect to pile stress.

Factor of Safety with Respect to Bearing Capacity Failure

The nominal factor of safety with respect to bearing capacity failure is equal to the specified minimum ultimate pile capacity, as determined by analysis and/or load test, divided by the design working load per pile. The ultimate capacity is modified if necessary to account for group effects. The factor of safety is commonly specified as minimum 2.0, with 1.5 often permitted for transient loads such as wind or earthquake. In some cases, the transient load reduction is incorporated into the design load by means of load factoring.

The actual factor can be less than the nominal factor under the following conditions:

1. Error in prediction of ultimate capacity, not adequately checked by load test.
2. Load test not representative of service installation and loading conditions due to errors in test procedure, variations in installation methods, and nonuniform soil profile. The last is particularly critical when the design relies on soil freeze.
3. Failure to account properly for group loading effects and negative skin friction.
4. Error in calculation of design load.
5. Occurrence of transient loads that were reduced in the calculation of design load.
6. Failure to allow for possible load increase due to pile misalignment within specified tolerances.
7. Differential pile settlement.
8. Failure of quality control during installation.

Factor of Safety with Respect to Structural Failure

The nominal factor of safety with respect to structural failure is equal to the specified material stress at pile failure, divided by the design stress on the pile material at working load. The codes, by specifying allowable stresses, dictate the minimum value of this factor for each pile material.

For steels with a yield stress of 248 MPa (36 ksi), values of the nominal factor ranging from 2.0 to 3.0 are indicated by the allowable stress provisions listed in Table 1. For concrete piles designed according to ACI allowable stresses (Table 2 and 3), ACI states a minimum nominal factor of 2.2 based on ultimate strength [*10*]. Timber piles designed as specified in ASTM Method D 2899-74, without the optional formal factor of safety, may have a nominal factor that can be less than 1.0 under long-term loading (see Code Allowable Stresses, above).

The actual factor of safety with respect to structural failure may be less than the nominal value under several conditions, as follows:

1. Failure of quality control, leading to substandard material or reduction in effective pile cross section as installed. For driven piles, this includes damage during driving due to dynamic stresses.

2. Deterioration of pile material with time or load repetition, leading to change in strength characteristics or reduction in pile cross section.

3. Error in calculation of design load.

4. Occurrence of transient loads that had been reduced in the calculation of design load.

5. Failure to allow for load increase due to pile misalignment within specified tolerances.

6. Differential pile settlement.

7. Bent pile.

8. Unequal bearing on boulder or bedrock.

Margin of Safety with Respect to Settlement

In current practice, the margin of safety with respect to settlement is not expressed as a numerical factor. Rather, estimates are made of the probable maximum total and differential settlements. The margin of safety lies in the competence with which the settlement estimates are made.

Settlement of the top of the pile consists of two components: (1) shortening of the pile and (2) settlement of the pile tip. In general, the increase in pile allowable stresses has a direct effect on pile shortening and only a secondary effect on tip settlement. Elastic shortening is directly proportional to the applied stress and can be significant for rigid structures founded on long piles. Settlement of the pile tip is not directly proportional to pile stress, because in order to develop the increased bearing capacity associated with the higher pile stress, the piles must be founded in stiffer soil or rock materials. Some pile and soil materials exhibit an increased rate of creep at higher stress levels. Deep-seated settlements, due to compression well below pile tip elevation are a function of total weight of structure and are not affected by pile stress.

The actual settlements can exceed predicted values due to failure to account for stress transfer, negative skin friction, and other factors which are beyond the scope of this discussion. Higher-than-predicted settlements can result also from the conditions leading to reductions in structural and bearing capacity, as listed above. Decreased pile stiffness may coincide with decreased structural capacity, resulting in greater elastic shortening. Decreased soil stiffness may coincide with reduced bearing capacity, causing greater tip settlement.

Effects of Defects on Factors of Safety

As summarized above, there are many sources of design and quality control error that can lead to defects in piles. The defects may take the form of overloads, imperfections in materials, and inadequate bearing capacity. Increasing the design stress on the piles reduces the margin for error and thus multiplies the importance of each defect. With a nominal factor of safety against structural failure of 3.0, for instance, considerable imperfection can be tolerated while maintaining an actual factor of 2.0 or greater.

Considering the high probability of defects, the specification of a nominal factor of safety equal to 2.0 implies acceptance of an actual factor equal to less than 2.0. In the opinion of the writer, this is unsafe unless accompanied by rational analysis of the probability for defects, indicating that the actual factor of safety is acceptable.

Recommendations for Allowable Stress Provisions

Allowance for Defects

The conditions under which the actual factors of safety are less than the nominal factors, as listed above, all imply failure of design engineering or job quality control and thus are theoretically avoidable. However, considering the inevitability of defects, code-allowable stresses should include an allowance for defects. The proposed allowance is analogous to the strength reduction factor ϕ applied to reinforced concrete construction in the ACI code [*11*].

Ideally, the effect of the defect allowance would be to vary the allowable stress according to the potential for defects in the completed piles. For example, allowable stresses would be lower for H-piles driven to tip bearing in a boulder field than for the same type pile driven in uniform sand, where the potential for damage is less. Similarly, a lower stress would be allowed for encased, than for cased, cast-in-place concrete piles in soil conditions wherein intrusion of soil into the pile is a potential problem.

Considerable study will be required in order to determine defect allowances appropriate to the various pile types and driving conditions.

Drivability

One of the common defects in pile design practice is the failure to account for pile drivability. There are two aspects of drivability: (1) the pile must have sufficient stiffness to transmit driving forces large enough to overcome soil resistance, and (2) the pile must have sufficient strength to withstand the driving forces without damage. For a given soil condition and pile length, the strength and stiffness of a pile, taken together, determine its drivability.

For a given soil condition and pile length, drivability manifests itself as the maximum capacity, at the time of driving, to which the pile can be driven regardless of the hammer used or the final blow count. The only exception to this maximum is a pile penetrating soft soil to bear evenly on sound unweathered rock, with no transition zone of weathered rock between the soft soil and sound rock. In this unusual case, the full structural strength of the pile can be developed. Pile capacity gained or lost after driving, due to soil freeze or relaxation, is a function of soil properties, and pile configuration and is not related to drivability.

Davisson [12] suggests the following drivability limits applicable to most pile lengths and soil conditions, and normal driving equipment:

Material	Design Stress Limit (F_{ss} = 2.0)	Notes
Steel	83 MPa (12 ksi)	...
Concrete, precast	11.0 MPa (1.6 ksi)	minimum f_c' = 35 MPa (5.0 ksi) (f_c' = compressive strength)
Timber	8.3 MPa (1.2 ksi) on tip area	friction pile
Timber	5.5 MPa (0.8 ksi) on tip area	point-bearing

The stresses are in general limited by pile stiffness. Thus, increases in strength yield little benefit in most cases. Davisson noted that these limits represent stress levels that can consistently be developed with normal methods. The limits can be exceeded, for example, in the case of short point-bearing piles and other special cases, and with optimized driving equipment.

Drivability limits are real, if approximate, physical limits that should be recognized in the building codes. The alternative is a high incidence of failed load tests on projects where such tests are required and unanticipated reductions in factors of safety on projects where tests are not required.

There are two alternate methods of recognizing drivability limits within the code stress provisions: (1) incorporate drivability limits into allowable stress provisions and (2) require a check on drivability as part of the substantiation for a pile design.

Generalized drivability limits on allowable stresses, as described above, can be developed for application to a wide variety of driving conditions. Such limits are appropriate for incorporation into code stress provisions intended for general use under reasonably unfavorable driving conditions.

The generalized limits, however, may be overly conservative when applied to especially favorable conditions. Therefore, the codes should allow stresses in excess of the generalized limits when properly substantiated by load tests and analysis. Load tests are a useful and indispensable means for checking drivability. They can be misleading, however, particularly in projects wherein soil conditions are variable. Therefore, the code should require an analytical check on drivability by means of wave equation analysis as a supplement to load tests.

Wave equation analysis accounts for pile stiffness and predicts driving stresses as well as pile capacity at the time of driving; thus, it provides a check on drivability. Although the method is by no means perfect, it represents a great improvement over obsolete energy formulas such as the

Engineering News and Hiley formulas. A requirement for drivability analysis by wave equation would permit the elimination of current code provisions requiring predictions of capacity by the energy formulas.

Conclusions

1. As a result of the trend toward higher code allowable stresses for pile foundations, the engineer bears increased responsibility to analyze thoroughly each pile design and not to rely on code allowable stresses as design guides.
2. Codes should be structured in such a way that allowable stress levels are adjusted to the size and difficulty of the project. For small projects, the engineer should have the option of designing at a low stress level with reduced substantiation. For large projects or in soil freeze locations, he should have the opportunity to invest appropriate engineering effort and to take advantage of higher design stresses where possible.
3. Code requirements should recognize the drivability limits of pile materials for driven piles. A design stress based solely on the structural capacity of the pile section may result in a pile that cannot be driven to the desired capacity due to inadequate stiffness or strength. Drivability limits may be recognized in codes either by restrictions on allowable stress or by requirements for drivability analysis and/or load tests.
4. Code allowable stresses should include an allowance for defects. The allowance should vary according to the probability of defects normally associated with the type of pile and installation conditions.
5. Existing code-allowable stresses that correspond to a nominal factor of safety against structural failure of 2.0 or less should be examined critically, particularly in light of the several conditions under which the actual factor of safety can be less than the nominal factor.

References

[1] *BOCA Basic Building Code/1978*, 7th ed., Building Officials and Code Administrators International, Chicago, Ill., 1978.

[2] *National Building Code*, 1976 ed., American Insurance Association, New York, N.Y., 1976.

[3] *Standard Building Code*, 1976 ed., Southern Building Code Congress International, Birmingham, Ala., 1976.

[4] *Uniform Building Code*, 1976 ed., Vol. 1, International Conference of Building Officials, Whittier, Calif., 1976.

[5] *Pile Foundations*, 5th ed., American Iron and Steel Institute, New York, N.Y., April, 1973.

[6] "Report on Allowable Stresses in Concrete Piles," Portland Cement Association, Skokie, Ill., June, 1971.

[7] Armstrong, R. M., "Structural Properties of Timber Piles," this volume.

[8] *National Design Specification for Stress-Grade Lumber and Its Fastenings*, 1973 ed., National Forest Products Association, Washington, D.C., 1973, with supplement through Nov. 1974.

[9] Blessey, W. E., *High Capacity Long Steel Piles*, American Iron and Steel Institute, New York, N.Y., 1969.

[10] ACI Committee 543, *ACI Journal, Proceedings*, Vol. 71, No. 10, Oct. 1974.

[11] "ACI Standard Building Code Requirements for Reinforced Concrete (ACI 318-77)," American Concrete Institute, Detroit, Mich., 1977.

[12] Davisson, M. T., "Pile Load Capacity," *Proceedings: Design Construction and Performance of Deep Foundations*, American Society of Civil Engineers, University of California, Berkeley, Calif., Aug. 1975.

[13] *Chicago City Building Code*, Chicago, Ill., amendments through 1976.

[14] *New Orleans Building Code*, New Orleans, La., amendments through 1976.

[15] *Building Code of the City of New York*, New York, N.Y., amendments through 1976.

[16] *Los Angeles City Building Code*, Los Angeles, Calif., amendments through 1976.

[17] PCI Committee on Prestressed Concrete Piling, "Recommended Practice for Design, Manufacture and Installation of Prestressed Concrete Piling," *Journal*, Prestressed Concrete Institute, March-April, 1977.

[18] *Baltimore City Building Code*, Baltimore, Md., amendments through 1976.

[19] *Building Codes—City of Atlanta*, Atlanta, Ga., amendments through 1976.

K. E. Robinson[1]

Horizontal Subgrade Reaction Estimated from Lateral Loading Tests on Timber Piles

REFERENCE: Robinson, K. E., "**Horizontal Subgrade Reaction Estimated from Lateral Loading Tests on Timber Piles,**" *Behavior of Deep Foundations, ASTM STP 670,* Raymond Lundgren, Ed., American Society for Testing and Materials, 1979, pp. 520–536.

ABSTRACT: Lateral loading tests on timber piles in a variety of soil conditions are reported. Measured horizontal deflections and slopes of the piles are compared with predicted movements based on elastic theories and published values of horizontal subgrade reaction. It was found that the actual deflections and slopes of the piles were consistently less than the computed values, particularly for cohesionless soils. Based on the test results presented in this paper, suggested relationships between horizontal subgrade reaction and soil consistency are given.

The horizontal subgrade reaction is a function of the magnitude and position of load application (it decreases with increasing soil stress). For cohesionless soils, increases in soil density caused by driving displacement piles apparently increase the horizontal subgrade reaction.

KEY WORDS: lateral loads, piles, timber, subgrade modulus, full-scale tests, soil consistency, elastic theory

The design of pile-supported structures occasionally requires an assessment of the allowable lateral loads that may be assigned to the piles. A number of technical papers—Reese and Matlock [*1*],[2] Davisson and Prakash [*2*], Davisson and Gill [*3*], Broms [*4*], and Davisson and Robinson [*5*]—have presented theoretical analyses whereby the behavior of laterally loaded piles can be predicted. These analyses, which require an assessment of the horizontal subgrade reaction, are based on elastic behavior of the soil-pile system. Davisson and Prakash have summarized values of horizontal subgrade reaction suggested by various authors for different soils. However,

[1] Partner, Dames and Moore, North Vancouver, British Columbia, V7J 2L1, Canada.
[2] The italic numbers in brackets refer to the list of references appended to this paper.

few data have been published whereby the validity of these suggested values of horizontal subgrade reaction can be checked. This paper presents the results of eleven lateral loading tests on timber piles driven in a variety of soil conditions. Applying the theoretical analysis to the load test results, values of horizontal subgrade reaction values have been computed. These values vary somewhat from the values summarized by Davisson and Prakash. Based on the test data and a review of tests by others, relationships between horizontal subgrade reaction and soil consistency are suggested for specific loading conditions on displacement piles.

Lee [*6*] and Alizadeh [*7*] have discussed the application of the theoretical derivations to the results of both model and full-scale tests. Although these authors have determined values of horizontal subgrade reaction on the basis of the measured deflections, their values have not been related to other soil parameters. In this paper, the horizontal subgrade reaction is tentatively related to the standard penetration test for cohesionless soils and to undrained shear strength for overconsolidated cohesive soils.

Theoretical Considerations

For practical purposes, a pile subjected to lateral load can be considered as a beam supported on a series of elastic springs with stiffness k_h as a reasonable approximation. The governing equation, as presented by Hetenyi [*8*], is

$$\mathrm{EI}\frac{d^4y}{dx^4} = p = k_h(y) \tag{1}$$

where

EI = the flexural stiffness of the pile,
p = the load produced by the springs,
x and y = coordinate axes (x is the vertical coordinate, y is the horizontal coordinate or deflection, and
k_h = the coefficient of horizontal subgrade reaction.

According to Terzaghi [*9*], k_h is nearly constant with depth for cohesive soils and increases linearly with depth ($k_h = xn_h$) for a pile of width 30 cm (1 ft), where n_h equals the constant of horizontal subgrade reaction for cohesionless soil (Fig. 1*a*).

The author's definitions of k_h and n_h are the same as those used by Terzaghi. However, Terzaghi relates to the pressure on the pile face p (kg/cm^2) (lb/in.^{2}) so that for a pile of width 30 cm (1 ft), $k_h = p/y$ (kg/cm^2, lb/in.^{2}) and $n_h = K_h/x$ (kg/cm^3, lb/in.^{3}). In this paper k_h and n_h are related to the load per unit of depth on the assumption that they are independent of the pile width. This assumption is considered to be reason-

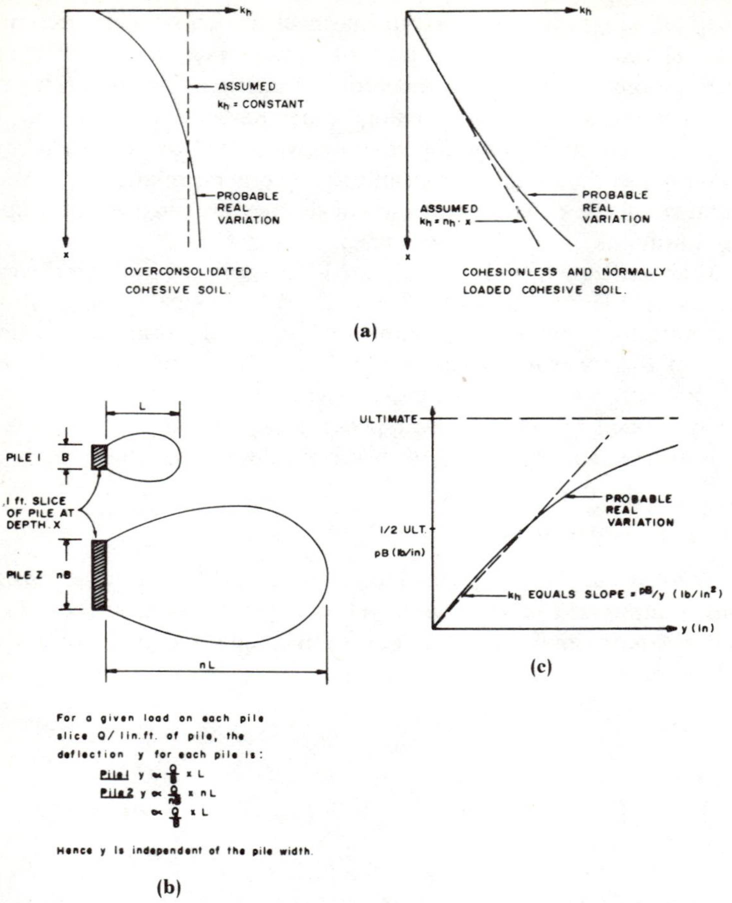

FIG. 1—(a) *Variation in horizontal subgrade reaction with depth.* (b) *Influence of pile width on dimensions of bulb of pressure.* (c) *Relationship between soil* pB (*pounds per inch*) *and deflections* v (*inches*).

able for the range of pile widths used in most engineering problems, that is, 20 to 90 cm (8 in. to 3 ft).

Consider the bulb of pressure behind a unit length of a pile at depth x below the ground surface as shown in Fig. 1*b*. If the pile width B is increased to nB, then the length of the pressure bulb is also increased from L to nL (Terzaghi). The deflection is approximately proportional to the length of influence of the pressure bulb and the average pressure. For a given load Q per unit length of pile, the pressure for width B is Q/B, and for width nB, the pressure is Q/nB. Then the deflection $y \propto Q/B \times L$ for the first case and $Q/nB \times nL = Q/B \times L$ for the second case, giving approximately the same net effect.

Once the values of k_h and n_h for a specific pile width of approximately 30 cm (1 ft) have been determined in terms of the force per unit depth and unit deflection, it will be approximately the same for any width of pile. It is also assumed that the stress-strain ratio of the soil-pile system is constant, that is, behaves elastically within the stress range considered. For this paper, the units of k_h and n_h are kg/cm^2 (lb/in.^{2}) and kg/cm^3 (lb/in.^{3}), respectively.

The coefficient of horizontal subgrade reaction varies with the magnitude of load as well as the depth below ground surfaces. At soil reactions below about half of the ultimate passive pressure of the soil, it is assumed that k_h is constant and approximately equal to the tangent modulus as shown on Fig. 1*c*. Where the soil reaction is greater than one-half the ultimate value, a secant modulus must be assumed.

Davisson and Gill present derivations for the case where k_h is a constant with depth and for the case of a two-layered cohesive soil (the layers having different values of k_h). For their derivations, the relative stiffness factor is defined as

$$R = 4(\mathrm{EI}/k_h)^{1/2} \tag{2}$$

Where a pile is driven into uniform soil (namely, a one-layer system), the pile length below the ground surface is greater than $4R$, and the load is applied at the ground surface, the deflection and pile slope at the ground surface can be determined theoretically from the following equations

$$Y_g = 1.4\, QR^3/\mathrm{EI} \tag{3}$$

$$\theta_g = Q/k_h R^2 \tag{4}$$

where

Y_g = the horizontal deflection of the pile at the ground surface,
θ_g = pile slope at the ground surface in radians relative to its slope prior to load application, and
Q = horizontal load.

Nondimensional parameters for pile deflection and moment at any location for a two-layered cohesive soil and for any pile length can be estimated using procedures presented by Davisson and Gill.

For cohesionless soils, Reese and Matlock have presented solutions using nondimensional parameters for pile deflection and change in pile slope. For their derivations, the relative stiffness factor is defined as

$$T = 5(\mathrm{EI}/n_h)^{1/2} \tag{5}$$

Where the pile length is greater than $4T$, the deflection and change in pile slope at the ground surface can be estimated from the following equations

$$Y_g = \frac{2.4QT^3}{EI} + \frac{1.62MT^2}{EI} \tag{6}$$

$$\theta_g = \frac{1.63QT^2}{EI} + \frac{1.75MT}{EI} \tag{7}$$

where M = the moment at the ground surface.

Nondimensional parameters for pile deflection, change in slope, shear, and moment for any pile length or loading condition can be estimated from the paper by Reese and Matlock.

Davisson and Robinson have shown that a long pile subjected to bending can be analyzed as though it were a free-standing pile with a fixed base at some depth below the ground surface. Their solutions for deflections where the load is applied at the ground surface are identical to those given above.

Table 1 summarizes values k_h and n_h suggested or computed by a number of authors [*1,2,5,9–11*].

Field Tests

Lateral loading tests on timber piles were conducted at eleven sites in the lower Fraser Valley of British Columbia. The test locations are shown in Fig. 2. All piles were driven with a drop hammer weighing approximately 1360 kg (3 000 lb) and falling between 1.5 and 3 m (5 and 10 ft). Vertical loading tests were carried out about 1 week after driving (except for test 10, where the lateral load test was conducted 9 months after driving). The piles were load tested vertically 1 day prior to the lateral loading. Vertical loading tests were continued until either the test pile or the reaction piles failed, whichever occurred first. However, horizontal load test results are presented herein only for one of the piles that did not fail during the vertical load test. In general, it was found that the lateral deflections for the pile that failed during the vertical load test were greater than for the pile that did not fail (in some cases, twice as much).

The lateral load tests consisted of jacking apart two adjacent piles, as shown on Figs. 3 and 4. The lateral deflections of the piles were recorded as the loads were increased. With the exceptions of tests 1 and 11, loads were applied in approximately 907-kg (1-ton) increments up to 2720-kg (3-tons) and then cycled from one to five times. The load was then increased to approximately 5440 kg (6 tons), or until the lateral deflection of the test and reaction piles exceeded the extension of the jack, whichever occurred first. Test 1 in peat was cycled at 1360 kg (1.5 tons) after excessive deflections

TABLE 1—*Suggested values of* k_h *and* n_h.

a. *Overconsolidated Cohesive Soils*

Source	k_h (lb/in.^{2})
Davisson and Robinson [5]	67 S_u [a]

b. *Cohesionless and Normally Loaded Cohesive Soils*

Sand

	n_h (lb/in.^{3})							
		Loose		Medium		Dense		
Source	Very Loose Repetitive Loading	Dry	Submerged	Dry	Submerged	Dry	Submerged	Very Dense
Terzaghi [9]		8	5	24	17	65	40	
Rowe [10]		3				93		
Davisson and Prakash [2]	1.5							>100

Other Soils

Source	n_h (lb/in.^{3})	Soil Description
Davisson and Prakash [2]	1.0[b] and 2.0	soft normally loaded clay
Peck and Davisson [11]	0.4–1.0	very soft organic silt
Reese and Matlock [1]	0.6–12.7	soft clay

[a] S_u = undrained shear strength (lb/in.^{2}).
[b] Repetitive Loading

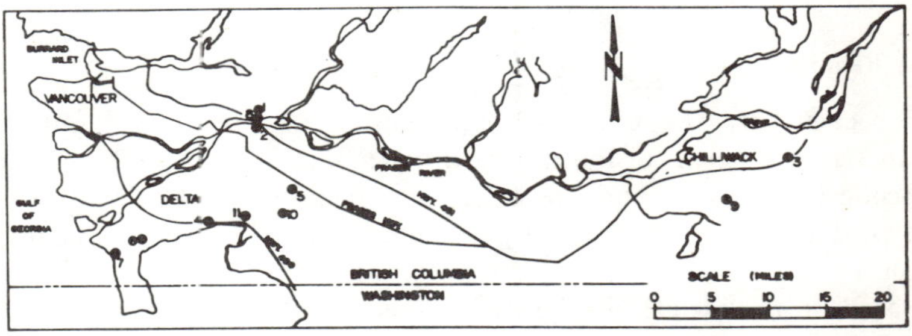

FIG. 2—*Key plan.*

were recorded at a load of 1815 kg (2 tons). The pile was unloaded and then reloaded at a constant rate to about 10.9 tonnes (12 tons).

At Tests 1 and 2, the piles were driven into peat. At Test 2, the peat was covered by a 90-cm (3-ft) surface layer of compact sand. At Tests 3 to 5, the piles were driven into cohesive soils with stiff surface layers present in Tests 4 and 5. At Tests 6 to 11, the piles were driven into cohesionless

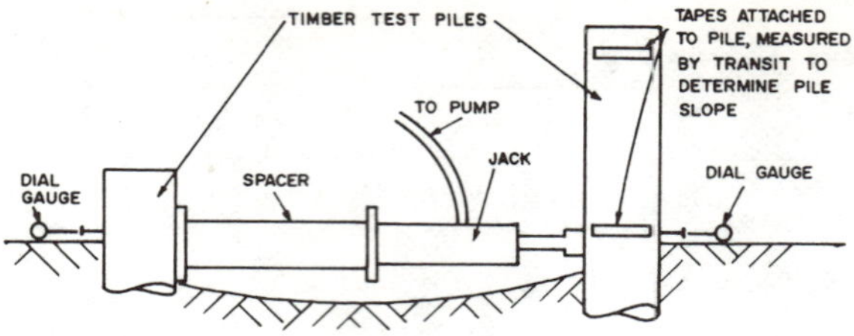

FIG. 3—*Typical pile test setup details.*

FIG. 4—*Typical test setup.*

soils, although some were quite silty. It was assumed that $k_h = n_h x$ would be applicable for Tests 6 through 11. Drill-hole logs and load deflection plots are provided in Figs. 5 to 10.

Test Results

For each site, the soil conditions within the upper 3 m (10 ft) were summarized in terms of the standard penetration resistance blow count (N) for cohesionless soils and the undrained shear strength for cohesive soils and peat. From these, average parameters k_h and n_h were estimated from Table 1.

Based on the measured pile deflections, k_h and n_h were computed using Eqs 2 through 7 for each test site. The average soil parameters, deflections, and estimated and computed k_h and n_h at a load of 2720 kg (3 tons) are summarized in Table 2. The measured deflections for tests 6 and 11 were adjusted for bending moments as loads were applied about 60 cm (2 ft) above the ground surface. The recorded deflections for these tests are at the point of load application, not at the ground surface. The n_h values computed at loads other than 2720 kg (3 tons) for cohesionless soils are plotted relative to the N value in Fig. 11.

For all the tests, values of k_h and n_h, estimated from Table 1, were well

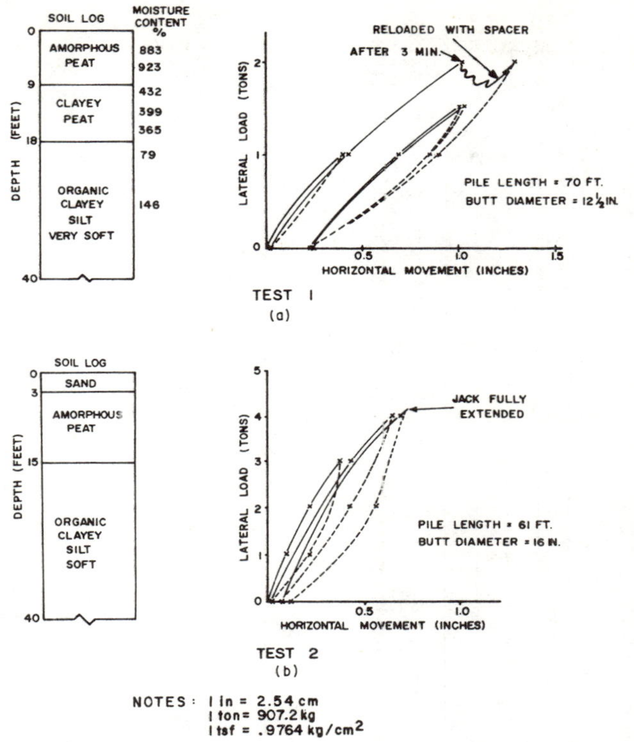

FIG. 5—(a) *Test 1.* (b) *Test 2.*

below the values computed using the recorded deflections. For the two tests in peat, it was found that a reasonable value of k_h, where water contents are of the order of 500 to 1000 percent, is 7 kg/cm^2 (100 lb/in.^{2}). Typical field vane shear strengths of this peat in similar areas of the lower Fraser Valley average about 0.15 kg/cm^2 (300 lb/ft^2). Assuming a constant of horizontal subgrade reaction, a computed n_h of 0.12 kg/cm^3 (4.2 lb/in.^{3}) was found for a load of 1815 kg (2 tons) for Test 1. Only two cycles of loading were applied, and it is difficult to estimate the ultimate increase in deflection for continued cycling.

For tests in cohesive soils, k_h, determined from the test results, was about twice that estimated from assumed and observed shear strength parameters using Table 1 for both one- and two-layered soils. For these materials, some consolidation may have occurred after driving, which could account for the increase in k_h. Four additional loading cycles produced horizontal deflections approximately 20 percent larger than the first cycle deflection. However, the additional deflection decreased with each successive cycle of loading.

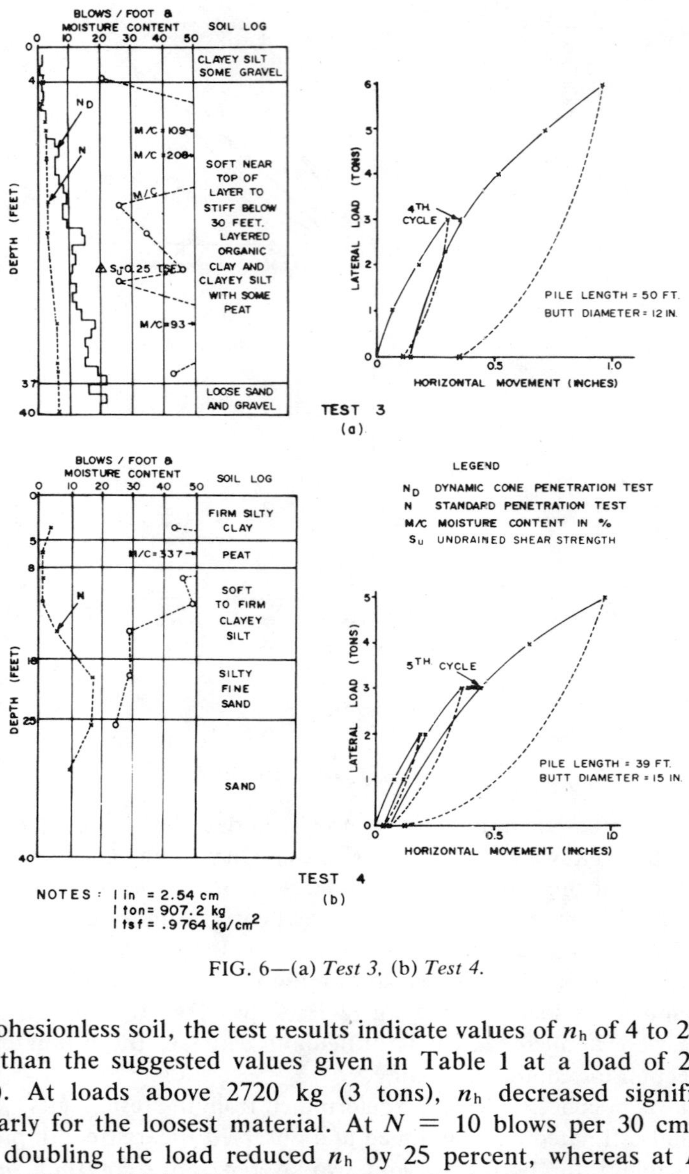

FIG. 6—(a) *Test 3*, (b) *Test 4*.

For cohesionless soil, the test results indicate values of n_h of 4 to 20 times greater than the suggested values given in Table 1 at a load of 2720 kg (3 tons). At loads above 2720 kg (3 tons), n_h decreased significantly, particularly for the loosest material. At $N = 10$ blows per 30 cm (ft) in Test 8, doubling the load reduced n_h by 25 percent, whereas at $N = 2$ blows per 30 cm (ft) in Test 10, doubling the load reduced n_h by about 66 percent. It should also be pointed out that the relative soil consistency, as determined by the standard penetration resistance blow count, is not an exact value but forms an easy basis for comparison with other sites. However, inaccuracies because of the test could not account for the great differences between values of n_h computed from observed movements and

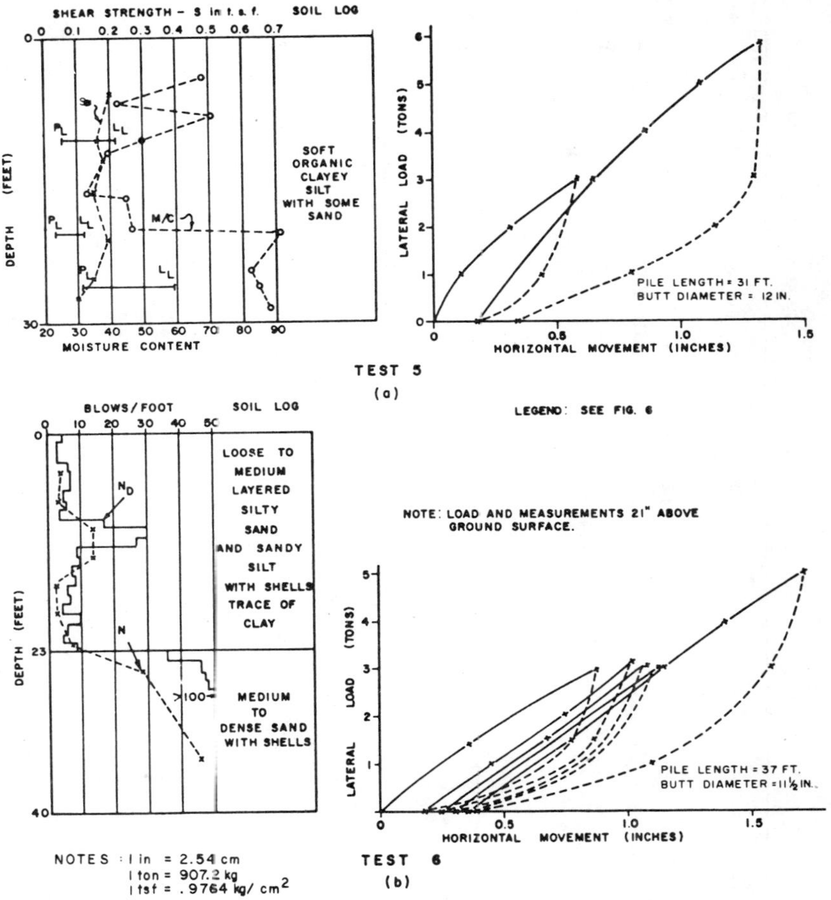

FIG. 7—(a) *Test 5*, (b) *Test 6*.

values predicted from Table 1. Part of this discrepancy could be the result of an increased density of the material caused by driving the tapered displacement piles.

Cyclical loading produced increases in deflection at a decreasing rate with successive cycles. The increase in deflection at the design load of 2720 kg (3 tons) varied with consistency of the soil. For the looser soils at Tests 6 and 10, the deflection increased by about 30 percent after five cycles. For the denser materials at Tests 7, 8, and 9, deflections increased by less than 10 percent during cyclical loading.

In general, k_h and n_h, computed using the slope of the pile at the ground surface, compared favorably with the value computed using deflection. The exceptions were Tests 7 and 9 performed in cohesionless soils, where

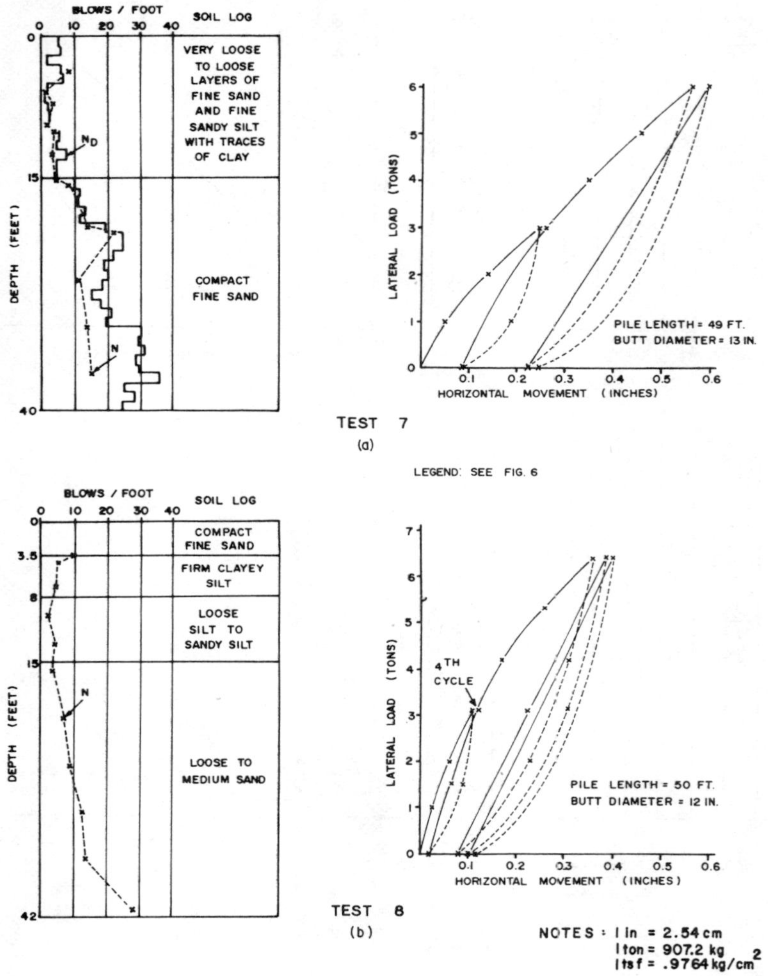

FIG. 8—(a) *Test 7.* (b) *Test 8.*

the computed n_h from the pile slope was significantly higher than that predicted from deflections. However, this is on the conservative side, and n_h is very sensitive to small changes in slope angle.

For most tests, the maximum applied lateral load was 4535 or 5440 kg (5 or 6 tons). The lower maximum loads in the softest soils were governed by the extension length of the jack, not by the soil-pile capacity. Test 11, in relatively loose cohesionless soil, was loaded at a constant rate to 12 tons without structural failure. Test results indicate that for normal vertical design loads of 20 to 30 tons for timber piles, the ultimate lateral load capacity would be more than 20 percent of the vertical design load for all

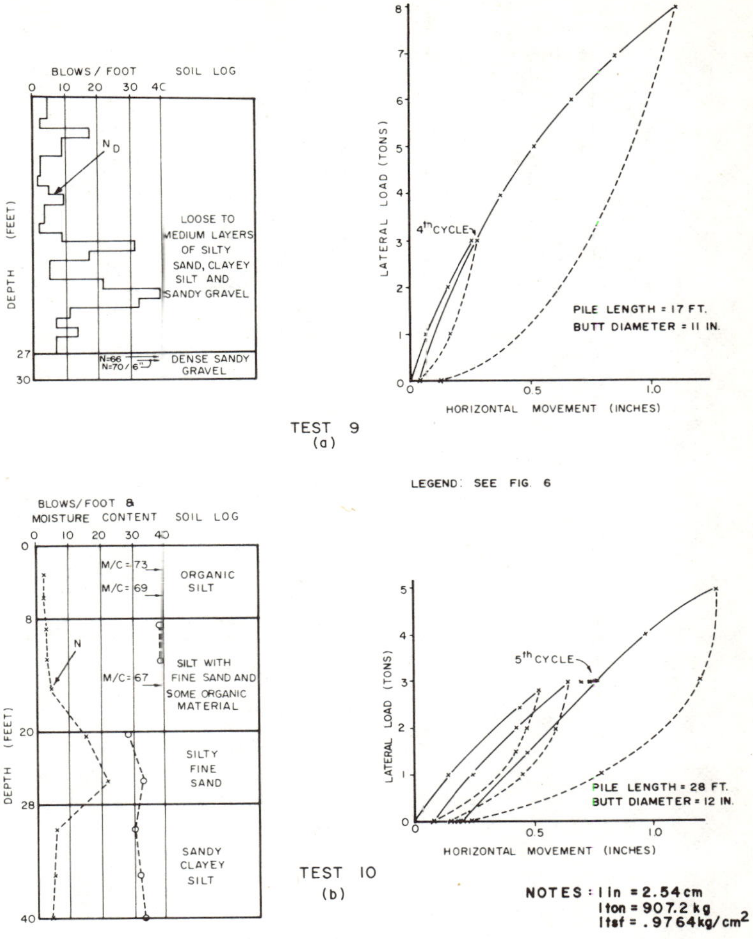

FIG. 9—(a) *Test 9.* (b) *Test 10.*

soils except very soft peat and organic deposits. However, for most structures, the lateral load will not exceed 10 percent of the vertical design load. If it does, the allowable horizontal displacement will probably govern the design. Some fixity of the top of the pile within the concrete cap will reduce deflections, as will the passive earth pressure acting on the vertical face of the pile cap.

Tests by Others

Few papers have been written on lateral-load test results, and of those

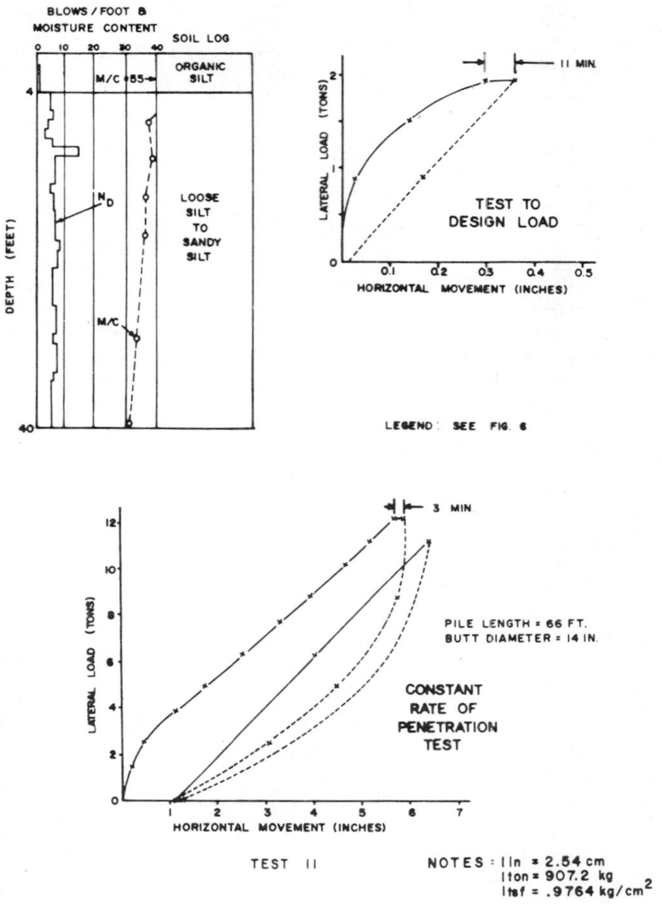

FIG. 10—*Test 11.*

that have, many do not provide sufficient soil data to give a reliable comparison with the results presented in this paper.

For timber piles in cohesionless soils, Gleser [*12*] reports the results of lateral loading tests on piles driven into sand compacted in the upper 1 m (3 ft). His test results indicate $n_h = 1.5$ kg/cm^3 (55 lb/in.^{3}); however, no data were provided on soil density. Wagner [*13*] provided test results in loose and in dense sandy soils, and Alizadeh and Davisson [*14*] presented test results on a timber pile driven in Arkansas River sand. The values of n_h indicated by these test results are plotted in Fig. 11 and agree well with the present test results.

Broms [*15*] has summarized lateral loading tests completed by others in cohesionless soils. Using the same theoretical derivations as presented

TABLE 2—*Comparison between suggested and observed values of* k_h *and* n_h *at load of 3 tons applied at the ground surface against displacement piles.*

Test No.	Soil Description	N [a]	S_u [b]	Horizontal[c] Movement, in.	Type[d,e]	Suggested from Table 1	Computed[f] From Defl.	Computed[f] From Slope
	Soil Conditions				**Horizontal Subgrade Reaction**			
1	Amorphous peat	<1	...	1.04	k_h	70	100	140
					n_h	<1	4.2	
2	3 ft sand	...	...	0.38	k_{h1}	400	500	...
	Over amorphous peat				k_{h2}	70	100	...
3	4 ft gravelly clay	...	800[g]	0.31	k_{h1}	370	740	...
	Over clayey silt	1.5	400[g]		k_{h2}	185	370	...
4	5 ft stiff clay	3	1200[g]	0.37	k_{h1}	270	500	...
	Over silt and peat	1	400[g]		k_{h2}	90	180	...
5	Organic clayey silt	<1	300	0.60	k_h	140	300	300
					n_h	1	16.5	...
6	Layered silty sand and sandy silt	3	...	0.90[h]	n_h	5.6	30	30
7	Layered sand and sandy silt	5	...	0.25	n_h	8	62	205
8	3.5 ft sand	10	...	0.11	n_h	13	256	250
	Over clayey silt	4						
9	Silty sand	5	...	0.24	n_h	8	100	290
10	Slightly organic silt	2	...	0.64	n_h	3.8	15	...
11	3 ft organic silt	1	...	0.68[h]	n_h	5.6	34	...
	Over sandy silt	3	...					

[a] Average standard penetration blow count upper 10 ft.
[b] Average undrained shear strength upper 10 ft in lb/ft^2.
[c] Movement at load of 3 tons, except test 1 at 2 tons.
[d] k_h coefficient of horizontal subgrade reaction in lb/in.^{2}; k_{h1} coefficient of horizontal subgrade reaction in lb/in.^{2} for surface layer; k_{h2} coefficient of horizontal subgrade reaction in lb/in.^{2} for lower layer.
[e] n_h constant of horizontal subgrade reaction in lb/in.^{3}
[f] Computed from movements at 3 tons, except test 1 at 2 tons.
[g] Undrained shear strength estimated from N value and tests on similar soil types at nearby locations.
[h] Movements measured about 2 ft above ground surface.

herein, his data indicated that predicted deflections for driven piles, using k_h and n_h from Table 1, were consistently higher than the measured values. He used average soil parameters within a depth of $4R$ or $4T$, which is generally greater than 10 ft. For analyses of the tests reported herein, the writer used the average parameters within the upper 10 ft of the soil profile. Hence, Broms should have found closer correlations between measured and calculated deflections, because N normally increases with depth (i.e., n_h would be higher if it is based on N averaged over a depth greater than 10 ft). Broms also found that deflections for driven piles were much less than for jetted piles.

Tests by Hoadley et al [*16*] provided sufficient soil data for comparison

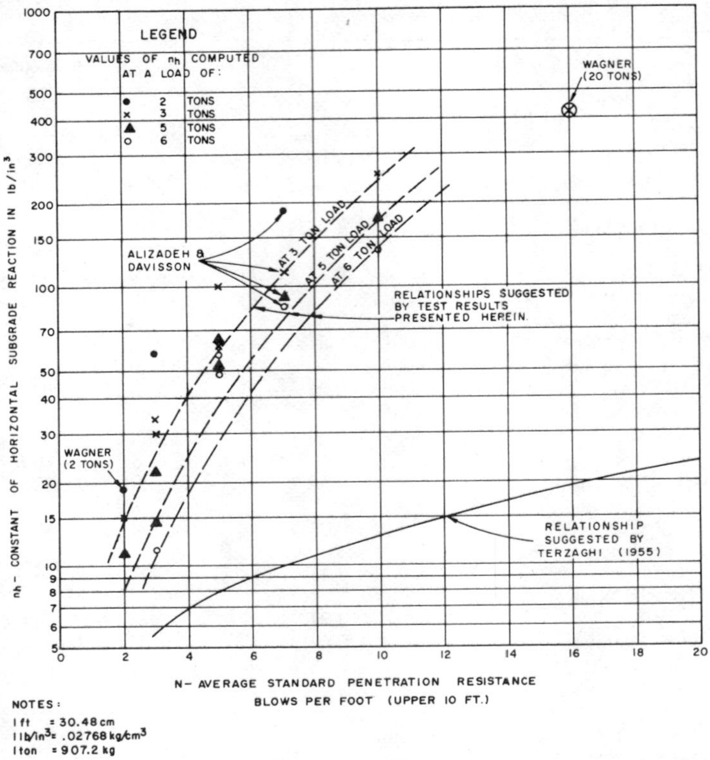

FIG. 11—N *versus* n_h *for cohesionless soil where loads are applied at the ground surface against displacement piles.*

of results in cohesive soils. They present results of tests on hollow metal piles and found the relationship $k_h = 73$ Su, which is lower than the results of the present tests. However, they did not specify the lateral load at which k_h was computed. It is likely that the load would be greater than 2720 kg (3 tons), as design load for pipe piles are usually higher than for timber piles.

Tests in peat by Wagner provided values of $k_h = 13.4$ kg/cm^2 (190 lb/in.^{2}) or $n_h = 0.22$ kg/cm^3 (8 lb/in.^{3}) at a load of 1815 kg (2 tons).

Suggested Values of Horizontal Subgrade Reaction

Results of the limited number of tests reported herein and the tests by others suggest that values of k_h and n_h published by various authors and presented in Table 1 are conservative for the loads and positions of load application used in the present series of tests. The densification effect of driving tapered displacement piles in cohesionless soils should be taken into account, as should the magnitude and position of lateral load. For loads up to 5440 kg (6 tons), the few tests reported indicate the relationship between N in the upper 3 m (10 ft) and n_h is approximately as shown in Fig. 11. Figure 11 is based on static loading conditions. Based on the present test results and results provided by Broms [*15*], Alizadeh [*7*], and Prakash [*17*], repetitive loading on very loose soil could double the deflection, while for N of 10 or more, a 25 percent increase over the first cycle deflection would be a conservative assumption.

For cohesive soils, the computed values of k_h were approximately twice the values given in Table 1. However, due to the increase in strain with repeated loading and considering test results by Hoadley et al, it is suggested that the relationship $k_h = 67$ Su be used for loads up to 2720 kg (3 tons) under cyclical loading.

For higher lateral loads, deflections will increase at an increasing rate with load (that is, the suggested values of k_h and n_h will decrease at higher loads).

Based on Tests 1 and 2, an empirical value of $k_h = 7$ kg/cm^2 (100 lb/in.^{2}) or $n_h = 0.1$ kg/cm^3 (4 lb/in.^{3}) is indicated for amorphous peat having moisture contents between 500 and 1000 percent (tests by Wagner gave higher values). Deflections will probably double under repetitive loading conditions. Where there is no stiff surface crust, the horizontal subgrade reaction will be significantly lower for lateral loads greater than 1815 kg (2 tons).

For partially embedded piles where loads are applied well above the ground surface, the induced moments will produce additional stress in the soil and result in lower values of horizontal subgrade reaction. Hence, computed values of n_h for very soft silt of 0.01 to 0.025 kg/cm^3 (0.4 to 1.0 lb/in.^{3}), as found by Peck and Davisson for load application 10 m (30 ft) or more above the ground surface, are significantly lower than would have been found if the load had been applied at the ground surface.

Conclusions

The existing theoretical expressions for determining horizontal pile movements at the ground surface are suitable, provided realistic values of k_h and n_h are used. The test results suggest that the values proposed by various authors and given in Table 1, may be conservative for tapered timber piles, particularly in sands. Based on the limited number of tests reported herein and the tests reported by Wagner, possible relationships between N and n_h for various lateral loads applied at the ground surface are given on Fig. 11.

The results of Test 11 and extrapolation of the other test data indicate that structural failure will not occur for lateral loads up to 12 tons applied at the ground surface, unless the soil is very soft or very loose. The failure load for the soil-pile system would probably be reached when the structural capacity of the pile in bending is exceeded. The soil will continue yielding and redistributing stresses as the loads increase.

Acknowledgment

This paper is published with the permission of British Columbia Hydro and Power Authority.

References

[*1*] Reese, L. C. and Matlock, H., "Non-Dimensional Solutions for Laterally Loaded Piles with Soil Modulus Assumed Proportional to Depth," 8th Texas Conference on Soil Mechanics and Foundation Engineering, 1956.

[*2*] Davisson, M. T. and Prakash, S., A Review of Soil-Pile Behaviour, Highway Research Board Record No. 39, 1963.

[*3*] Davisson, M. T. and Gill, H. L., *Journal of the Soil Mechanics and Foundations Division, Proceedings,* American Society of Civil Engineers, Vol. 89, No. SM 3, May, 1963.

[*4*] Broms, B. B., *Journal of the Soil Mechanics and Foundations Division, Proceedings,* American Society of Civil Engineers, Vol. 91, No. SM 3, May 1965.

[5] Davisson, M. T. and Robinson, K. E., "Bending and Buckling of Partially Embedded Piles," 6th International Conference on Soil Mechanics and Foundation Engineering, Vol. 2, 1965.

[*6*] Lee, K. L., *Journal of the Soil Mechanics and Foundations Division, Proceedings,* American Society of Civil Engineers, Vol. 94, No. SM 1, Jan. 1968.

[*7*] Alizadeh, M. in *Performance of Deep Foundations, ASTM STP 444,* American Society for Testing and Materials, 1969, p. 379.

[*8*] Hetenyi, M., *Beams on Elastic Foundations,* University of Michigan Press, Ann Arbor, Mich., 1946.

[*9*] Terzaghi, K., *Geotechnique,* Vol. 5, 1955.

[*10*] Rowe, P. W., Discussion of "Evaluation of Coefficients of Subgrade Reaction" by K. Terzaghi, *Geotechnique,* Vol. 6, 1956.

[*11*] Peck, R. B. and Davisson, M. T., Discussion of "Design and Stability Considerations for Unique Pier," by J. Michalos and D. P. Billington, *Transactions,* American Society of Civil Engineers, Vol. 127, Part IV, p. 413, 1962.

[*12*] Gleser, Sol. M. in *Symposium on Lateral Load Tests on Piles, ASTM STP 154,* American Society for Testing and Materials, 1953, p. 75.

[*13*] Wagner, A. A. in *Symposium on Lateral Load Tests on Piles, ASTM STP 154,* American Society for Testing and Materials, 1953, p. 59.

[*14*] Alizadeh, M. and Davisson, M. T., *Journal of the Soil Mechanics and Foundations Division,* American Society of Civil Engineers, Vol. 96, No. SM 5, Sept. 1970.

[*15*] Broms, B. B., *Journal of the Soil Mechanics and Foundation Division, Proceedings,* American Society of Civil Engineers, Vol. 90, No. SM 3, May 1964.

[*16*] Hoadley, P. J., Francis, A. J., and Stevens, L. K., "Load Testing of Slender Steel Piles in Soft Clays," 7th International Conference on Soil Mechanics and Foundation Engineering, Vol. 2, 1969.

[*17*] Prakash, S., "Behaviour of Pile Groups Subjected to Lateral Loads," Ph.D. thesis, University of Illinois, 1962.

M. Spanovich[1] *and R. G. Garvin*[1]

Field Evaluation of Caisson–Shale Interaction*

REFERENCE: Spanovich, M. and Garvin, R. G., **"Field Evaluation of Caisson–Shale Interaction,"** *Behavior of Deep Foundations, ASTM STP 670,* Raymond Lundgren, Ed., American Society for Testing and Materials, 1979, pp. 537–557.

ABSTRACT: Load tests were conducted on ten full-sized cast-in-place drilled caissons (bored piles, drilled shafts) of varying diameter. Three caissons were tested to evaluate end-bearing on the surface of fissile shale; five caissons were tested to evaluate side shear of shale sockets; one caisson was tested to evaluate end-bearing below weathered shale; and one caisson was tested to evaluate side shear in soil. Nine of the shale-bearing caissons tested exhibited subgrade failure, observed as excessive settlement, while the remaining shale-bearing caisson, drilled to end-bearing below the weathered shale, failed by shearing of the concrete shaft. The end-bearing test values exceed commonly employed empirical design values by factors ranging from 3 to 23. The shale sockets supported loads of at least one-third of and in some cases exceeding the ultimate end-bearing load.

KEY WORDS: axial load, bored piles, construction, drilled piers, field tests, foundations, load testing, shafts, shale

Assigned nominal contact bearing values for cast-in-place drilled caissons (bored piles, drilled piers, drilled shafts) bearing on shales and claystones based on empirical or semiempirical criteria are so conservative that they often render such foundations either prohibitively costly or unreasonably overdesigned.

Unlike piles, rock-bearing caisson designs suffer from inadequate field test data that would contribute to the understanding of caisson–subgrade interaction and load transfer mechanisms. To provide such prototype test data, load tests were conducted on full-scale cast-in-place drilled caissons founded on shale bedrock. As caissons support axial loads by both end-bearing and side friction, the test program was designed to isolate rock

*Original experimental data were measured in customary inch-pound units.

[1] Principal and project engineer, respectively, Engineering Mechanics, Inc., Pittsburgh, Pa. 15205.

socket side shear from end bearing as much as practicable in a group of caissons of 46-cm (18-in.) diameter, 61-cm (24-in.) diameter, and 76-cm (30-in.) diameter bearing on the surface of and socketed into shale. Field and laboratory tests were conducted to isolate and evaluate the friction along the side walls of the test caissons through the soil overburden. Soil and rock tests were also performed to determine physical properties of the soil overburden and shale bedrock.

Site Investigation

The caisson test layout and the locations of two test borings are shown on Fig. 1. Standard penetration tests (SPT), in compliance with the ASTM Penetration Test and Split-Barrel Sampling of Soils (D 1586-67), were conducted at 3-ft vertical intervals in the soil overburden of both borings. The SPT provides an indication of the density or consistency or both of the soils encountered as well as soil specimens for laboratory evaluation. The SPT were continued to refusal on bedrock, from which level 3 m (10 ft) of continuous NX 54-mm (2.12-in.) diameter rock core were obtained from each boring in compliance with the ASTM Diamond Core Drilling for Site Investigation (D 2113-70). Undisturbed 76-mm (3-in.) diameter thin-walled steel Shelby tube samples were also obtained from the soil overburden for laboratory evaluation in accordance with the ASTM Thin-Walled Tube Sampling of Soils (D 1587-74). The undisturbed samples were subjected to unconsolidated-undrained direct shear tests, unconfined compression tests, natural density determinations, and plastic limit and plasticity index (Atterberg limits) tests. Upon completion of the caisson testing program, NX concrete cores were obtained from test caissons 5 and 11 and the shale bedrock below these caissons.

The test site was selected because it represented a subsurface situation where, because of poor soil overburden and relatively shallow depth to bedrock, caissons would provide a feasible foundation for a settlement-sensitive structure. The soil mantle is composed of 4.6 m (15 ft) to 6.4 m (21 ft) of randomly placed fill, mostly clayey silt with some rock fragments and occasional organic inclusions, underlain by residual soil, the product of in-place decomposition of bedrock. The residual soil, averaging 0.3 m (1 ft) in thickness, is basically decomposed silty shale grading into silty shale bedrock. The generalized subsurface profile and laboratory test results are shown in Fig. 2.

Using Terzaghi and Peck's [*1*][2] correlation of SPT blow counts N with the consistency of cohesive soils, the fill was medium to stiff with soft zones. The water contents of the SPT fill samples ranged between 17 and 30 percent, averaging about 25 percent. The liquid limit of the fill varied

[2] The italic numbers in brackets refer to the list of references appended to this paper.

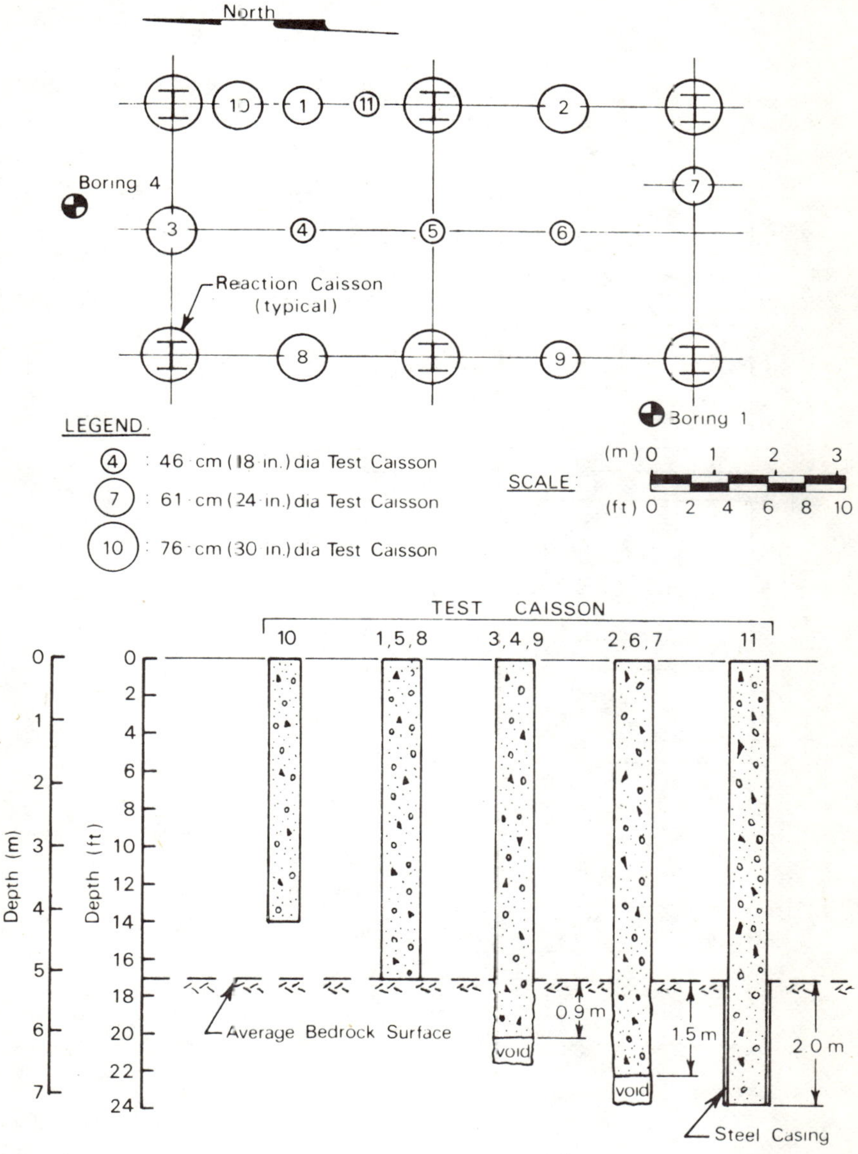

FIG. 1—*Test caisson layout and caisson profiles.*

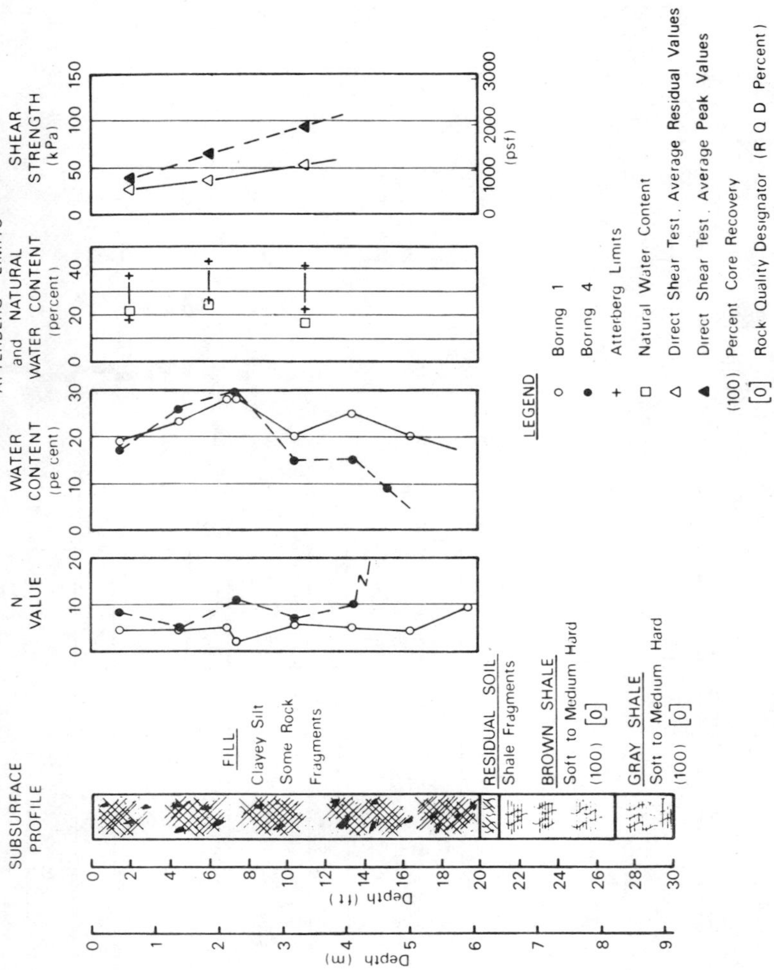

FIG. 2—*Subsurface profile and laboratory test results.*

from 37.2 to 42.7 percent, while the plastic limit varied between 18.3 and 26.6 percent. The water content of the undisturbed fill samples varied from 16.5 to 25.0 percent. The residual soil was of stiff consistency where cohesive and was very dense on the relative density scale at lower reaches where shale fragments dominated its structure.

Again, using Terzaghi and Peck's correlation of N values with consistency, the shear strength of the fill ranged from 24 to 96 kPa (500 to 2000 psf). The average values of the peak shear strength from the direct shear tests ranged from about 36 kPa (750 psf) at a depth of 0.6 m (2 ft) to about 91 kPa (1900 psf) at a depth of 3.4 m (11 ft), increasing linearly with depth as shown in Fig. 2. The undisturbed samples of fill at depths of 0.6 and 1.8 m (2 and 6 ft) exhibited shear strengths (one-half of the unconfined compressive strength) of 25 and 73 kPa (525 and 1525 psf), respectively. The undisturbed soil sample at a depth of 3.4 m (11 ft) was inadequate for such testing.

Bedrock at the test site in Allegheny County, Pa., is a shale unit in the Little Pittsburgh member of the upper Conemaugh Formation, Pennsylvanian System, approximately 12 m (40 ft) below the Pittsburgh coal horizon. It is a fissile, soft to medium hard silty shale with occasional micaceous sandy seams. Underwood [*2*] lists the following physical properties typical of the Pennsylvanian Cyclothemic shales encountered in Eastern Ohio and Western Pennsylvania:

Compressive strength	1449 kPa (210 psi)
Modulus of elasticity	6.9 GPa (1 000 000 psi)
Cohesion	386 kPa (56 psi)
Angle of internal friction	36 deg
Dry density	221 kg/m^3 (138 lb/ft^3)
Natural moisture	9.1 percent

The upper meter of shale at this site was extremely weathered and variegated and included soft clay seams. The current City of Pittsburgh Building Code assigns a nominal allowable caisson end-bearing design value of 766 kPa (8 tons/ft^2) to "soft rock and shale."

Caisson Testing Procedure

Eleven straight-shaft test caissons were constructed by drilling to the desired level with an auger and filling the holes with fluid concrete. As shown on Fig. 1, test caissons 1, 5, and 8 were terminated at end bearing on the surface of the shale bedrock; test caissons 3, 4, and 9 were socketed 0.9 m (3 ft) into the shale; and test caissons 2, 6, and 7 were socketed 1.5 m (5 ft) into the shale. Test caisson 10 was installed to determine the shear strength being developed along the soil-caisson side wall for a full-scale caisson. It was therefore terminated in fill above the bedrock surface. Test caisson 11, the only end-bearing caisson extended below the bedrock surface, was socketed 2 m (6.5 ft) into the shale to evaluate the bearing capacity of the more competent shale below the extremely weathered zone. The 2-m (6.5-ft) shale socket of test caisson 11 was lined with a steel casing to eliminate bonding within the shale socket.

To isolate side shear from end bearing, it was necessary to create a void below the concrete shaft of the socketed caissons. This was accomplished by installing a 0.6-m (2-ft) thick styrofoam plug at the bottom of the drilled hole prior to placing concrete. A 25-mm (1-in.) diameter tube was attached to the center of the plug extending to the top of the caisson. Through this tube, a dissolving agent (perchlorethylene) was poured into the styrofoam mass, dissolving it to create a void. This procedure was employed for each socket-bearing caisson and for test caisson 10, the 76-cm (30-in.) diameter caisson terminated in soil. All caisson holes were filled with standard 20.7 MPa (3000 psi) concrete without reinforcing steel.

Figure 1 also shows the locations of six caissons installed for the load test reaction platform. The reaction caissons were installed by drilling 91-cm (36-in.) diameter straight shafts socketed at least 3 m (10 ft) in bedrock. HP 14 × 74 steel beam sections were installed vertically in each caisson after which the drilled holes were filled with concrete. Loads were applied to the caissons by jacking against a horizontal beam constructed of double W 36 × 330 sections spanning between the reaction caissons.

The load tests were conducted in compliance with the ASTM Method of Testing Piles Under Axial Compressive Load (D 1143-74) and included at least two loading and unloading cycles. Whenever possible, two load tests were conducted concurrently. The loads were applied to the test caissons generally in 178 and 222-kN (20 and 25-ton) increments. Each load increment during the initial loading cycle was maintained for 1 h before increasing the load. Loads of 890 kN (100 tons) and multiples of 890 kN (100 tons) were maintained for 2 h before unloading the caissons in 222 and 445-kN (25 and 50-ton) decrements; each unloading decrement and the zero load were maintained for 20 min. The loading-unloading cycle was then repeated with reloading increments and unloading decrements each maintained for 20 min; newly applied loads (exceeding the maximum load of the previous cycle) were maintained for 1- and 2-h intervals. The loading-unloading cycles were continued until an apparent failure of the caisson, or its supporting subgrade, occurred. Modifications of the loading sequence or duration of load application or both were made during the tests to evaluate further the ultimate and failure loads.

The settlement of the top of each test caisson was measured by two independently supported dial gages, with an accuracy of 0.0254 mm (0.001 in.), located on diametrically opposite sides of the jack. The dial gage measurements were averaged to compute gross settlement; the gross settlement was also checked with an independently supported horizontal wire passed by a vertical scale mounted on the jack.

Analytical Considerations

The results of each caisson test were analyzed to determine ultimate caisson load capacity. Attempts were also made to isolate the ultimate load supported by the shaft in soil from that portion of the load supported by either the shale socket or by the shale in end bearing.

To determine failure and ultimate support loads from the test results, it was necessary to define failure as related to the test data. The ultimate caisson capacity can be defined as the load beyond which the caisson begins to punch into the ground, observed as an abrupt increase of settlement with respect to the load increase (that is, as the slope of the load-settlement curve approaches a vertical line). However, where failure is not so evident, it is common engineering practice to select the load beyond which the rate of settlement, in comparison with the loading increment, exceeds a certain magnitude. For these tests, the ultimate caisson capacity was defined as the test load beyond which the rate of gross settlement exceeded 0.029 mm/kN (0.01 in./ton) of additional load. Further, it is desirable to limit total gross settlement under any load to 25.4 mm (1.0 in.).

The cyclic loading of the test caissons was conducted for two reasons: to study the load-settlement-rebound characteristics for loading cycles less than failure and to achieve at least one complete loading cycle before reaching the failure load. The loading cycle below failure can be analyzed for the effects of elastic caisson shortening as well as yielding of the supporting subgrade. The rebound curves indicate the amount of net settlement after the caisson has rebounded and after the subgrade stresses have relaxed or rebounded or both. The remaining net settlement represents the amount of tip settlement or permanent strain of the supporting subgrade.

Load Test Results

Soil Side Shear

The load-settlement curve for test caisson 10, the 76-cm (30-in.) diameter caisson terminated in soil above the bedrock surface, is shown on Fig. 3. The limiting rate of gross settlement, 0.029 mm/kN (0.01 in./ton) of additional load, was exceeded at a load of 445 kN (50 tons) during the initial loading cycle. The total gross settlement after 1 h under the 356-kN (40-ton) load was 2.77 mm (0.109 in.), whereas after 30 min under the 445-kN (50-ton) load, the gross settlement was 7.80 mm (0.307 in.). The caisson supported an ultimate load of 356 kN (40 tons), which when divided by the peripheral area of the shaft, yields an average soil shear capacity of 34.5 kPa (5 psi). It appears that the maximum soil resistance was mobilized at a settlement of between 2.54 mm (0.1 in.) and 7.62 mm (0.3 in.).

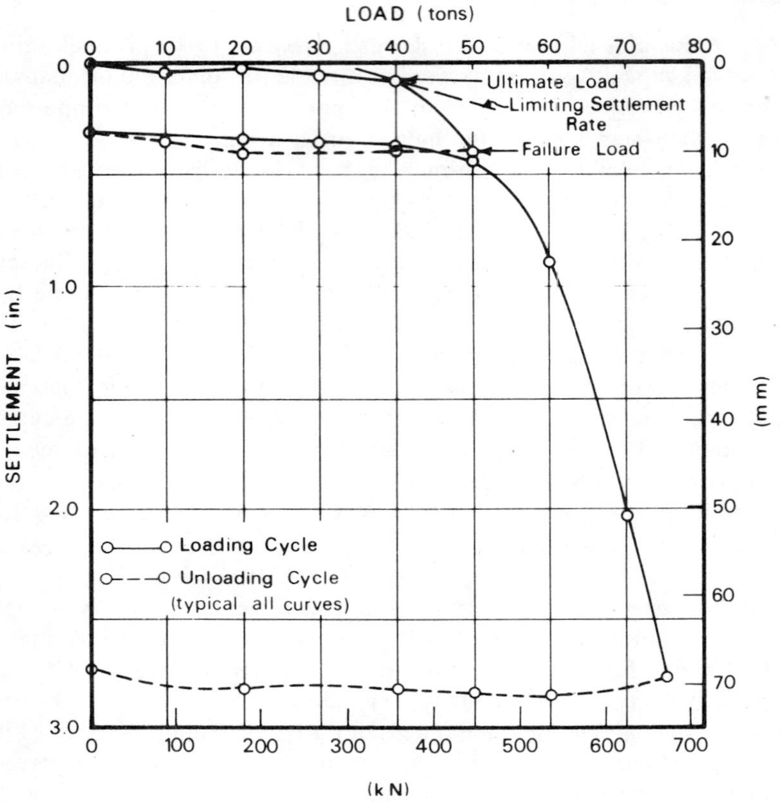

FIG. 3—*Load-settlement curve: test caisson 10*

0.9-m (3-ft) Socketed Caissons

Figure 4 shows the test results for test caisson 4, the 46-cm (18-in.) diameter shaft, which failed at a load of 1246 kN (140 tons) during the second loading cycle; it therefore supported an ultimate load of 1068 kN (120 tons). Using the soil side shear value derived from test caisson 10, the computed ultimate soil side shear for test caisson 4 is 240 kN (27 tons); hence, the shale socket supported a load of 828 kN (93 tons), or 628 kPa (91 psi) of socket area. The computed maximum elastic shortening for test caisson 4 under a load of 1068 kN (120 tons) is 1.73 mm (0.068 in.); the net settlement at the ultimate load is therefore 12.12 mm (0.477 in.).

The failure of test caisson 4 occurred as a subgrade failure, whereby the caisson settled excessively into the ground at loads beyond 1068 kN (120 ton). The caisson settled about 12.7 mm (0.5 in.) under the 1246-kN (140-ton) load and over 30.5 mm (1.2 in.) under a load of 1424 kN (160 tons) applied for only 30 min.

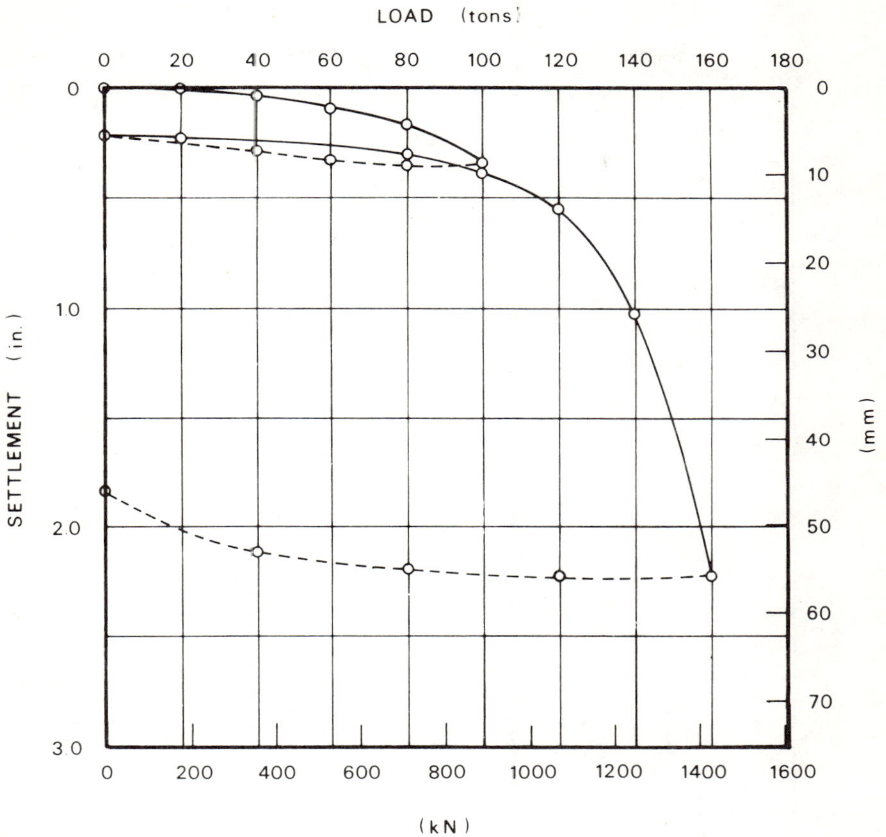

FIG. 4—*Load-settlement curve: test caisson 4*

The load-settlement curve for test caisson 9, the 61-cm (24-in.) diameter shaft is shown in Fig. 5. Failure of this caisson was observed at a load of 1335 kN (150 tons) during the second loading cycle. The ultimate load of 1112 kN (125 tons) produced a gross settlement of 11.53 mm (0.454 in.), of which 1.24 mm (0.049 in.) is the computed maximum elastic shortening of the shaft concrete. Beyond the 1112-kN (125-ton) ultimate load, the caisson began to settle excessively into the ground (subgrade failure). The computed maximum soil side shear capacity for test caisson 9 is 409 kN (46 tons). The rock socket therefore supported 703 kN (79 tons), or 400 kPa (58 psi) of socket area.

The load-settlement curve for test caisson 3, the 76-cm (30-in.) diameter shaft is shown in Fig. 6. The failure load for this caisson, 3560 kN (400 tons) of the sixth loading cycle, caused excessive settlement (subgrade failure) of the caisson. The caisson was able to support an ultimate load of 3338 kN (375 tons), but the gross settlement of 31.37 mm (1.235 in.)

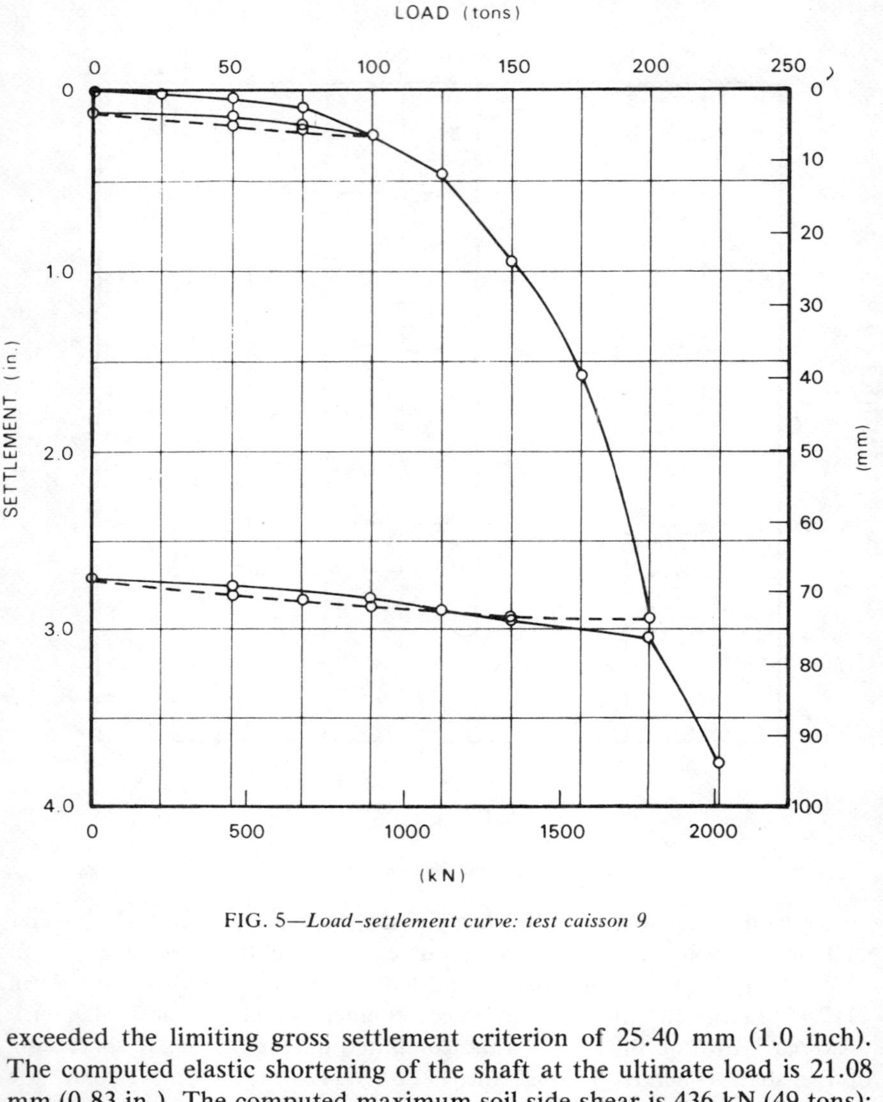

FIG. 5—*Load-settlement curve: test caisson 9*

exceeded the limiting gross settlement criterion of 25.40 mm (1.0 inch). The computed elastic shortening of the shaft at the ultimate load is 21.08 mm (0.83 in.). The computed maximum soil side shear is 436 kN (49 tons); the shale socket therefore supported a load of 2902 kN (326 tons), or 1322 kPa (193 psi) of shale socket area.

An accurate evaluation of the settlements and rebounds for test caisson 3 is difficult because of disruptions during the test program. The hydraulic jack failed during the third and fifth loading cycles, necessitating immediate unloading of the caisson. Also, welds on the reaction platform failed at the 3115-kN (350-ton) load of the sixth cycle and the 3560-kN (400-ton) load of the seventh cycle; in these cases, the loads were maintained until

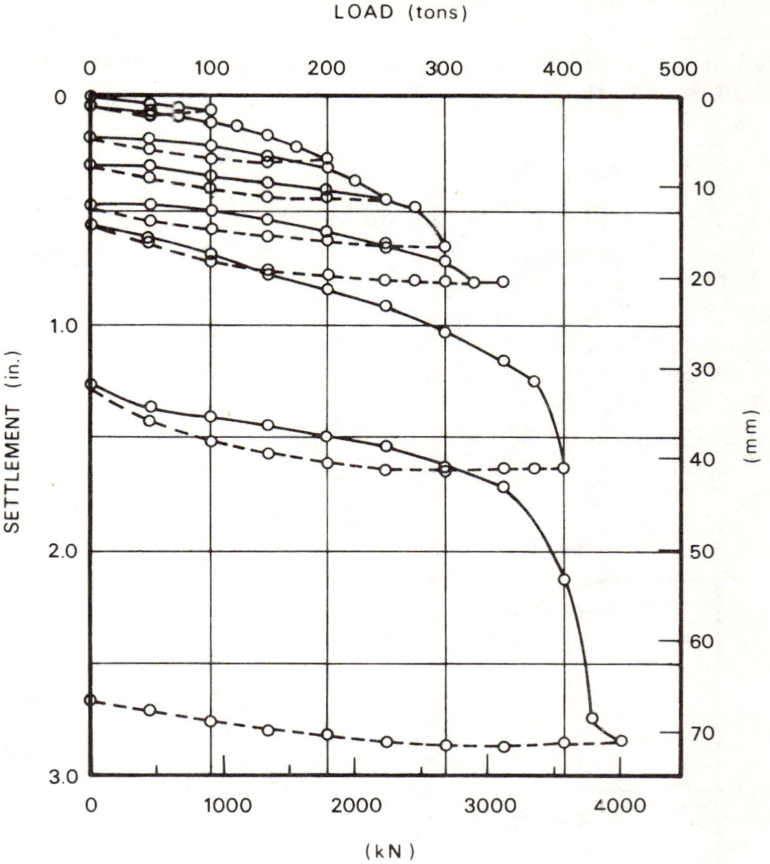

FIG. 6—*Load-settlement curve: test caisson 3*

the welds were repaired and the loading program was resumed. It is notable that test caisson 3 sat without loads for 3½ days after completion of the fifth loading cycle, while a new hydraulic jack was installed. It is felt that the disruptions during the test program, and the sometimes prolonged load applications, caused the excessive gross settlement under the ultimate load. The repeated loading and unloading cycles probably caused a substantial remolding of the subgrade, which reduced its shearing resistance. This is illustrated by the slope of the reloading portion of the sixth cycle, when the caisson settled at an average rate of 0.004 mm/kN (0.0014 in./ton) of load as compared with 0.002 mm/kN (0.0007 in./ton) during the first five loading cycles. At the end of the fifth cycle, the caisson had undergone a permanent settlement of 14.50 mm (0.571 in.). The caisson settled an additional 16.89 mm (0.665 in.) during the 3338-kN (375-ton) load increment of the sixth cycle. The 3560-kN (400-ton) load caused the caisson to settle at the rate of 0.05 mm/kN (0.016 in./ton) of load increment, compared to the maximum limit of 0.029 mm/kN (0.01 in./ton) of additional load adopted for this test.

1.5-m (5-ft) Socketed Caissons

As shown in Fig. 7, the ultimate load for test caisson 7, the 61-cm (24-in.) diameter shaft, was 1335 kN (150 tons) of the second loading cycle, beyond which load the caisson settled excessively. The caisson underwent 11.96 mm (0.471 in.) of gross settlement under the ultimate load, of which 1.50 mm (0.059 in.) represents the maximum computed elastic shortening. The maximum computed soil side shear is 312 kN (35 tons), or 352 kPa (51 psi) of shale socket area.

The ultimate load for test caisson 2 is 2670 kN (300 tons) of the third

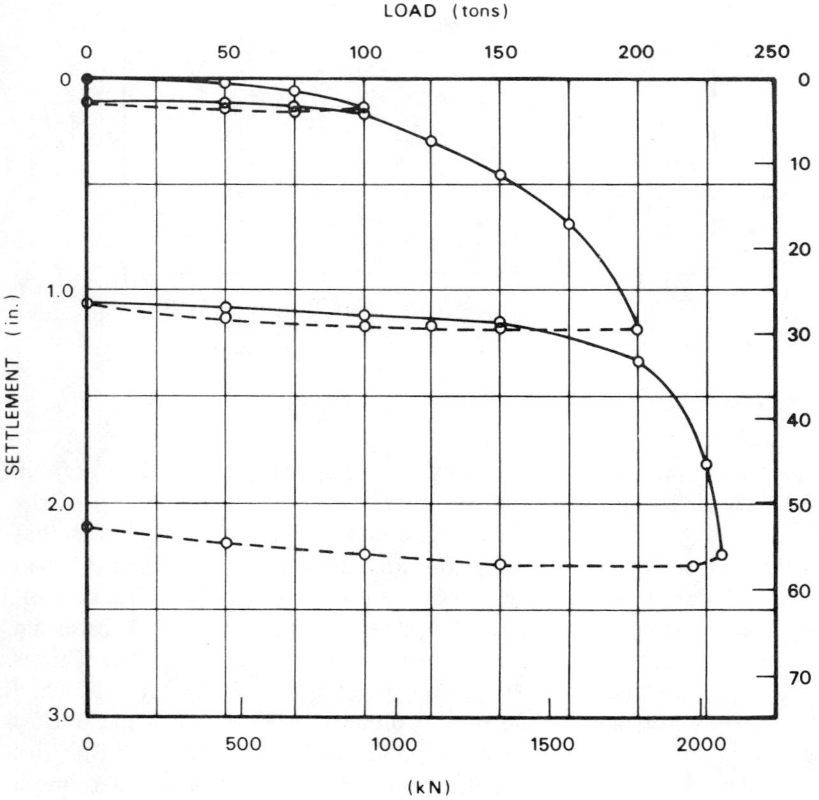

FIG. 7—*Load-settlement curve: test caisson 7*

loading cycle, as shown in Fig. 8. The 2892-kN (325-ton) load of the fourth loading cycle caused excessive settlement of the caisson (subgrade failure). The ultimate load caused a gross settlement of 21.46 mm (0.845 in.), of which the elastic shortening accounts for 1.75 mm (0.069 in.). The computed maximum soil side shear for test caisson 2 is 409 kN (46 tons); therefore, the shale socket supported 2261 kN (254 tons), or 621 kPa (90 psi) of socket area.

Test Caisson 6, the 46-cm (18-in.) diameter shaft was deleted from the test program.

End-Bearing Caissons

As shown in Fig. 9, the 46-cm (18-in.) diameter shaft of test caisson 5 supported an ultimate load of 890 kN (100 tons). The computed maximum

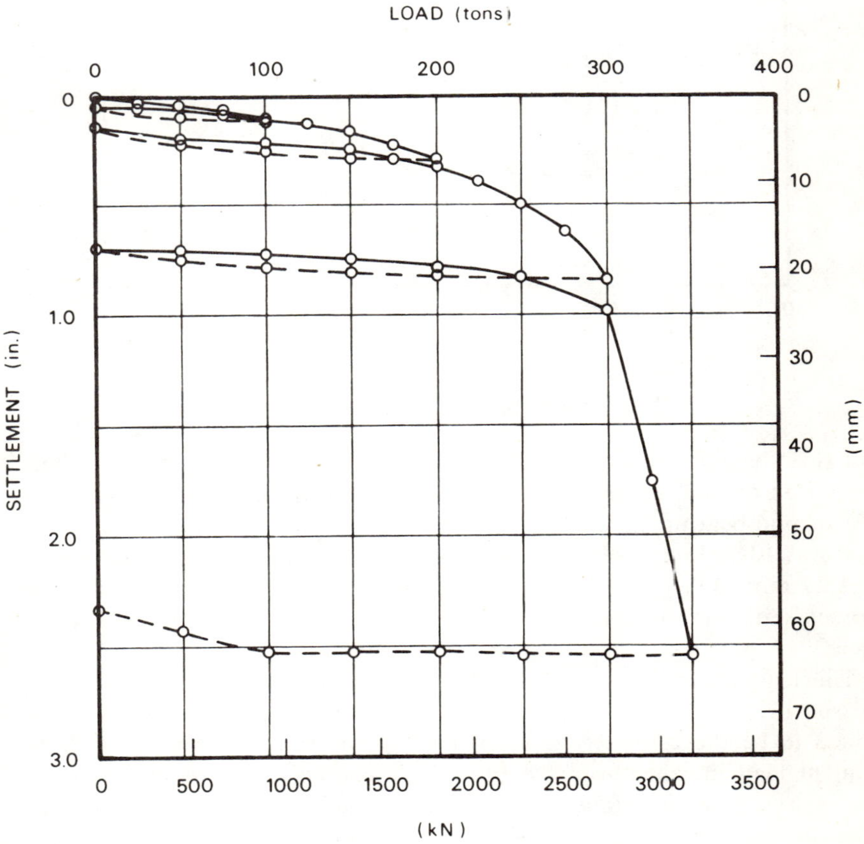

FIG. 8—*Load-settlement curve: test caisson 2*

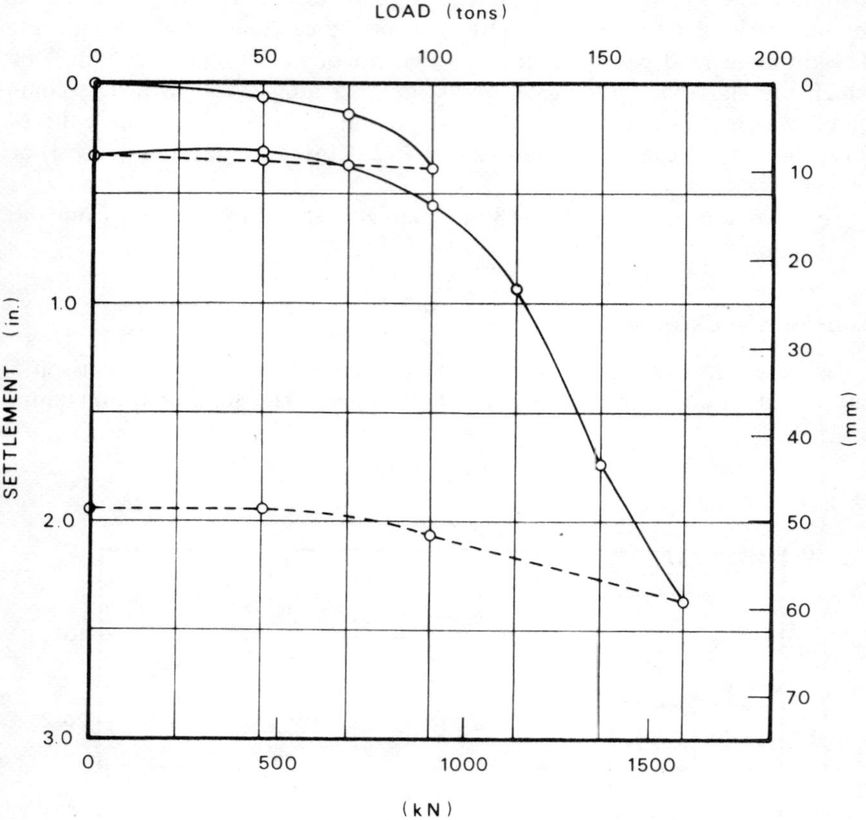

FIG. 9—*Load-settlement curve: test caisson 5*

soil side shear for test caisson 5 is 249 kN (28 tons); therefore, the shale supported an end-bearing load of 641 kN (72 tons), or 3928 kPa (41 tons/ft^2) of end-bearing area. The observed gross settlement at the ultimate load was 10.16 mm (0.400 in.), compared to a computed elastic shortening of 1.27 mm (0.05 in.). The failure of test caisson 5 occurred as excessive caisson settlement under the 1112-kN (125-ton) load of the second loading cycle.

The load-settlement curve for the 61-cm (24-in.) diameter shaft, test caisson 1, is shown in Fig. 10. The ultimate load for this caisson was determined to be the 2448-kN (275-ton) load of the third loading cycle. Subsequent load increments of the third cycle caused excessive settlement of the caisson (subgrade failure). The gross settlement of the caisson at the ultimate load was 25.20 mm (0.992 in.), while the computed elastic shortening of the concrete shaft was 2.03 mm (0.080 in.). The calculated maxi-

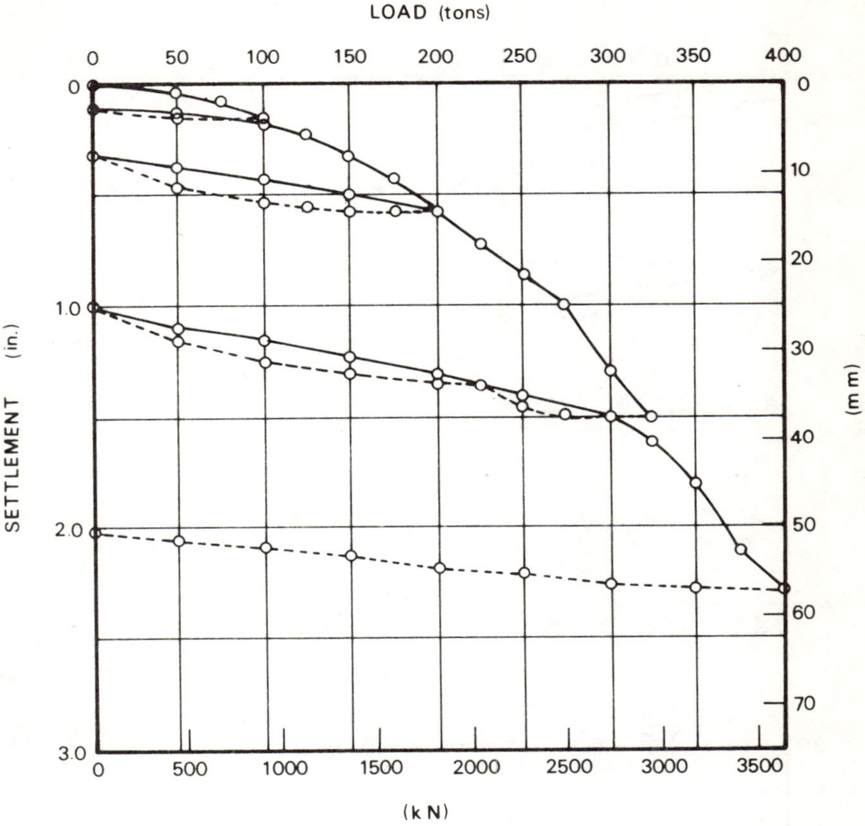

FIG. 10—*Load-settlement curve: test caisson 1*

mum soil side shear for test caisson 1 was 347 kN (39 tons); therefore, the shale supported 2100 kN (236 tons), or 7185 kPa (75 tons/ft^2) of end-bearing area.

Figure 11 shows the load-settlement curve for the 76-cm (30-in.) diameter shaft, test caisson 8, which experienced excessive shaft settlement (subgrade failure) beyond the ultimate load of 1558 kN (175 tons) of the second loading cycle. The computed maximum soil side shear for test caisson 8 is 472 kN (53 tons); therefore, the shale supported 1086 kN (122 tons), or 2395 kPa (25 tons/ft^2) of end-bearing area. The computed elastic shortening of the shaft, 0.89 mm (0.035 in.) at the ultimate load, represents 4.2 percent of the observed gross settlement, 21.16 mm (0.833 in.).

The test results for test caisson 11, the 46-cm (18-in.) diameter shaft bearing 2.0 m (6.5 ft) below the shale surface, are shown in Fig. 12. Failure of test caisson 11 occurred in the concrete shaft, which sheared at a depth

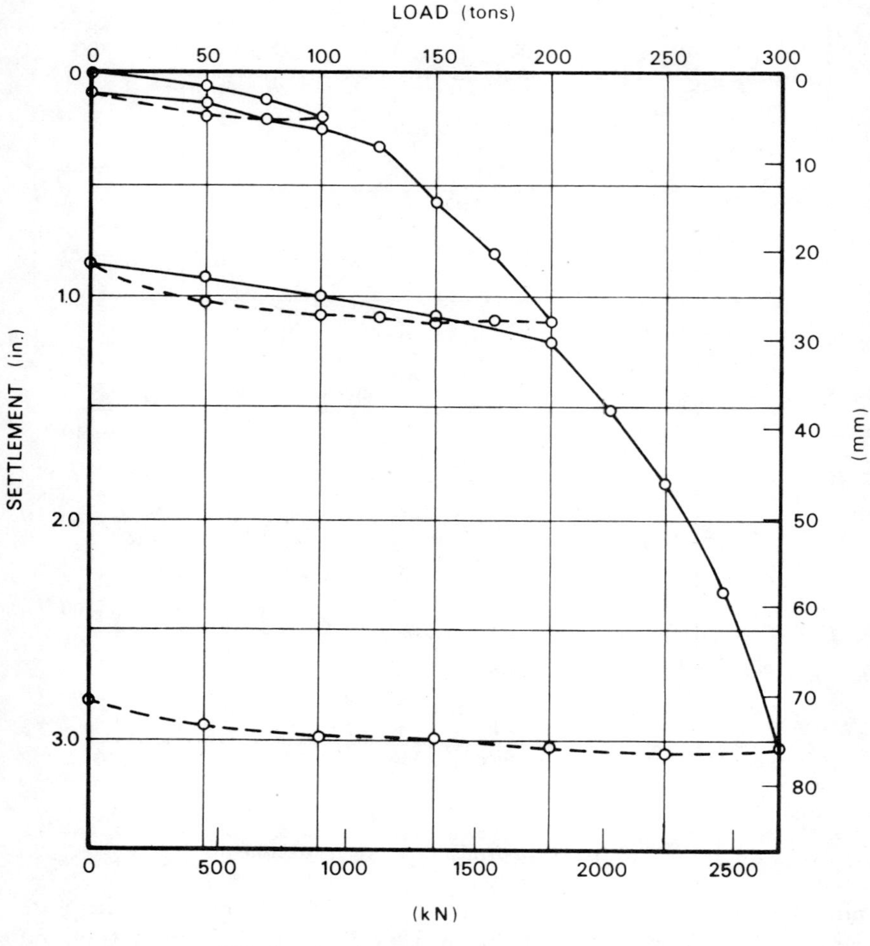

FIG. 11—*Load-settlement curve: test caisson 8*

of 1.74 m (5.7 ft) below ground surface under a load of 3338 kN (375 tons) during the fourth loading cycle. The compressive stress in the concrete shaft at failure was computed at 20.31 MPa (2943 psi). The test data show that the subgrade adequately supported a 3115-kN (350-ton) load at a rate of settlement of 0.006 mm/kN (0.002 in./ton) of load increment and that it also supported the 3338-kN (375-ton) load for 30 min at the same settlement rate prior to failure of the shaft. At the 3115-kN (350-ton) load, the gross settlement of the shaft was 21.06 mm (0.829 in.), of which the computed elastic shortening represented 5.33 mm (0.21 in.). The data show that the shale supported an end-bearing load of at least 2875 kN (323 tons), or 17 436 kPa (182 ton/ft^2) of end-bearing area.

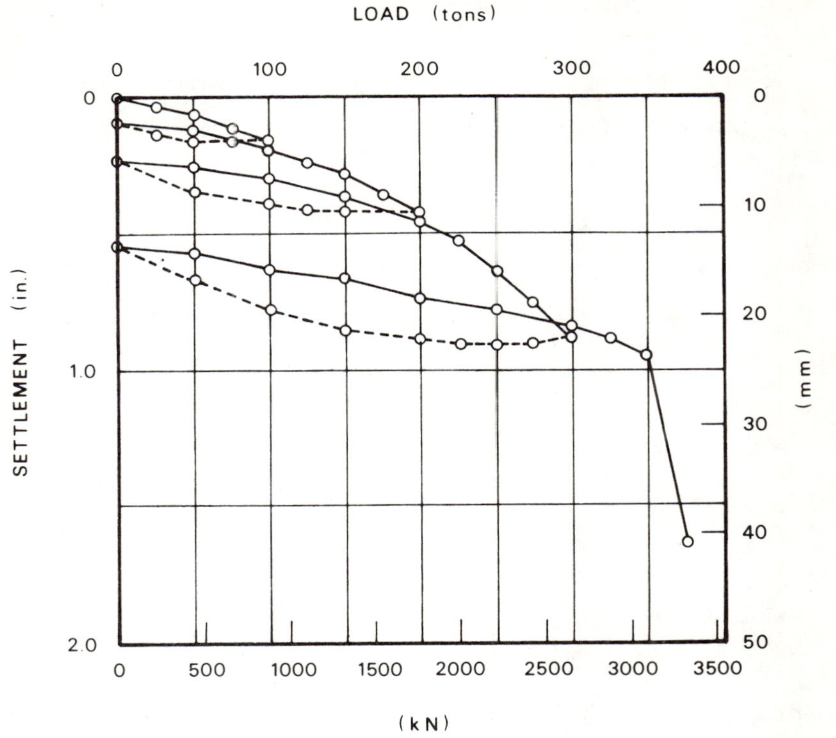

FIG. 12—*Load-settlement curve: test caisson 11*

For the three loading cycles before the shaft failure, the computed elastic shortening represented between 21 and 31 percent of the gross settlement at the maximum cycle load; the percentage decreased as the maximum cycle load increased. The corresponding computed elastic rebound represented between 54 and 73 percent of the total observed rebound; the percentage also decreased as the maximum cycle load increased.

Discussion

Table 1 summarizes the caisson test results. The socket side shear values range from 352 to 1325 kPa (51 to 192 psi), averaging about 656 kPa (95 psi). The end-bearing values range from 2395 to 7185 kPa (25 to 75 tons/ft^2) for caissons bearing on the surface of the weathered shale; at a depth of 2 m (6.5 ft) into the relatively fresh shale, an end-bearing value of at least 17 436 kPa (182 tons/ft^2) was obtained. The wide variation of test results is best explained by the variations in hardness and continuity of the shale throughout the test site. For example, the shale at test caisson

TABLE 1—*Summary of caisson test results.*

Test Caisson Number	Diameter, cm	Socket Length, m	Ultimate Load, kN	Computed Ultimate Soil Load Capacity, kN	Ultimate Socket Load, kN	Ultimate Socket Shear Strength, kPa
			Socketed Caissons			
4	46	0.9	1068	240	828	628
9	61	0.9	1112	409	703	400
3	76	0.9	3338	436	2902	1322
7	61	1.5	1335	312	1023	352
2	76	1.5	2670	409	2261	621
10	76	[a]	356	356	...	34[b]
			End-Bearing Caissons			
					Ultimate End-Bearing Contact Pressure, kPa	
5	46	0	890	249	3 928	
1	61	0	2448	347	7 185	
8	76	0	1558	472	2 395	
11	46	0	3115+	240	17 436	

[a] Caisson terminated in soil.
[b] Side shear in soil.

supported three times the contact pressure supported by the shale at test caisson 8.

The observed ultimate load for test caisson 10, defined by the failure criteria selected for these tests, was 356 kN (40 tons). This load, divided by the peripheral area of the shaft, produces an average side shear resistance of 34.87 kPa (728 psf). The ultimate shaft resistance for test caisson 10, computed by the widely used relationship $P_s = \pi D_s L \alpha S_u$ (where S_u = average peak undrained shear strength), is 367 kN (41.2 tons), or an average side shear resistance of 35.93 kPa (750 psf). Based on the results of field tests in clays and shales [*3*], α for this analysis was assumed to be 0.5, and S_u, the average peak shear strength value of 71.85 kPa (1500 psf), was derived from the direct shear tests. The direct shear tests also produced an average residual shear strength of 38.32 kPa (800 psf) for test caisson 10.

Results of the load test for test caisson 10 indicate that the peak soil shearing resistance was mobilized at an axial settlement of between 2.5 and 10.0 mm (0.1 and 0.4 in.). Similar displacements of load-tested piers in clays were observed by Whitaker and Cooke [*4*], 6.4 mm (0.25 in.), and Reese and O'Neill [*5*], 5 mm to 10 mm (0.2 in. to 0.4 in.). The ultimate base resistances of the end-bearing caissons were achieved at settlements ranging from 10 to 23 mm (0.4 to 0.9 in.). This indicates that, for the soil strains required to develop the ultimate end-bearing caisson capacities, it is likely that the soil is strained beyond its peak stress where the residual shear strength of the soil provides the shaft resistance. Following this rationale, higher values for the ultimate load capacity of the shale would be derived by deducting the lower soil side shear resistance based on the residual shear strength of the soil. This rationale also applies to caissons socketed into shale, since the ultimate socket and end-bearing capacities occurred at essentially the same magnitude of settlement.

The theory states that, initially, the load is supported by the caisson shaft resistance, and the observed settlement represents elastic shortening of the shaft. As the settlement exceeds the elastic shortening, the shaft tip begins to settle and base loading is initiated. The load is then supported by a combination of shaft and base resistances. The load-settlement curves show that, for loading cycles less than the failure load, all caissons experienced permanent tip settlement. Computed maximum elastic shortening of the caissons represents from 8 to 31 percent of the observed gross settlements for loads less than the failure load. This means that the maximum elastic shortening and the corresponding maximum soil shaft resistance were exceeded during the initial load increments. For the end-bearing caissons, base resistance was therefore developed during the initial loading cycles.

The rebound illustrated by the unloading portion of a load-settlement curve represents the combined elastic rebound of the caisson and rebound of the supporting subgrade. The test data indicate that the elastic rebound occurs instantaneously as the caisson is unloaded; the subgrade rebound is more gradual, with the rate of rebound generally decreasing with time. The load-settlement curves show that tip settlement represents from 30 to 60 percent of the observed gross settlement at maximum cycle loads less than the ultimate load. Therefore, as the shafts were unloaded, a maximum of 70 percent of the gross settlement was recovered in the form of elastic caisson rebound and subgrade rebound. The rebound curves for unloading after failure of the caissons also indicate elastic rebound of the caissons and some subgrade rebound. The total rebounds after failure are less than 20 percent of the total gross settlements at the failure loads.

The criteria for establishing the failure and ultimate test loads were selected because they best represent failure indicated by the load-settlement curves and test data. The criteria limit gross settlements to 25 mm (1 in.) and differential settlements (comparing all caissons) to less than 13 mm (½ in.). These are generally accepted tolerances for column-supported structures. The only test that exceeded the gross settlement limit at the ultimate load was test caisson 3. Problems with the jacking apparatus and reaction frame caused prolonged settlements under some of the load applications and possibly detrimental remolding of the soil and shale subgrade. It is felt that lesser settlements would have been observed if the equipment problems had not occurred.

The test results indicate that ultimate socket side shear and end-bearing capacities are mobilized at essentially the same magnitude of caisson settlement. This means that, for an end-bearing caisson socketed into shale, the ultimate total caisson capacity can be derived from a summation of the predicted side shear and end-bearing load capacities.

Conclusions

Even though the test results show substantial variation, the lowest values are at least three times greater than the City of Pittsburgh Building Code allowable design value, which already includes a substantial safety factor. The test results indicate that, for similar shales and subsurface conditions, caissons could be designed for an end-bearing contact pressure of 958 kPa to 1437 kPa (10 to 15 tons/ft^2). Side shear in shale, as a fraction of the end-bearing capacity, could be designed for 95.6 to 191.6 kPa (14 to 28 psi). These values would include safety factors of 1.7 and 2.5 for end-bearing and 2.1 to 4.2 for socket shear, again based on the lowest test results. A substantially greater end-bearing capacity could be achieved by drilling the caissons into the lower, less weathered shale; an end-bearing capacity of 17 436 kPa (182 tons/ft^2) was observed at failure of the concrete shaft of the test caisson bearing 2 m (6.5 ft) below the shale surface.

The calculated shaft resistance developed in soil represents up to 30 percent of the ultimate capacity of the caissons tested. In every case, the soil capacity alone, based on peak shear strength, exceeded the City of Pittsburgh Building Code allowable design load of 766 kPa (8 tons/ft^2). This suggests that, for similar soil and rock conditions, the soil contribution to the design capacity can be very significant and should be evaluated in the caisson design.

The residual shear strength of the randomly placed fill was found to be about one-half of the peak shear strength. If the soil contribution to the load capacity of the caisson is to be considered, then residual shear strength values should be used to compute soil side shear capacity, especially if the caisson is expected to develop end-bearing resistance.

Because any fill will consolidate to some degree under its own weight, dragdown forces from negative skin friction should be a consideration in evaluating the long-term load capacity of caissons. For this reason, it may be desirable, or even necessary, to ignore the soil side shear, since it becomes negative if significant consolidation occurs. Negative friction forces were not considered for these test caissons because of the instantaneous loading nature of the test.

The test program evaluated the worst caisson bearing conditions, whereby all but one of the caissons were terminated in the extremely weathered shale without benefit of stringent manual cleaning operations. Most installations of caissons bearing on similar shales would by-pass the extremely weathered zone, especially considering the ease with which it can be penetrated with an earth auger. Therefore, considering that the building code design value still applies for the relatively unweathered shale tested by test caisson 11, caissons socketed at least 1 m (3 ft) into the shale more often than not incorporate safety factors exceeding 20 based on end-bearing alone. Even higher safety factors are incorporated in the design, as it is common practice to ignore the shale socket support, which these tests prove is quite substantial.

The caisson tests provide useful data for more realistic evaluations of caisson-shale interaction. The variations of the test results emphasize the desirability of full-size contract caisson tests to develop a workable range of design values. By correlating such load test data with pertinent soil and bedrock data, caisson foundations can be used more efficiently.

Acknowledgments

This research program was funded by Eckert Drilling Company of Pittsburgh, Pa. The cooperation and assistance of Walter Eckert, P. E., and Donald Anderson, P. E., and the workmen of Eckert Drilling Company are most appreciated.

References

[*1*] Terzaght, K. and Peck, R. B., *Soil Mechanics in Engineering Practice*, 2nd Ed., Wiley, New York, 1967, p. 347.

[*2*] Underwood, L. B., 'Classification and Identification of Shales," *Journal of the Soil Mechanics and Foundation Division*, American Society of Civil Engineers, Vol. 93, Nov. 1976, pp. 97-116.

[*3*] Woodward, R. J., Gardner, W. S., and Greer, D. M., *Drilled Pier Foundations*, McGraw-Hill, New York, 1972, pp. 264-268.

[*4*] Whitaker, T. and Cooke, R. W., *Proceedings*, Symposium on Large Bored Piles, Institution of Civil Engineers and Reinforced Concrete Association, London, Feb. 1966.

[*5*] O'Neill, M. W. and Reese, L. C., *Journal of the Soil Mechanics and Foundations Division*, American Society of Civil Engineers, Feb. 1972.

W. F. Swiger[1]

General Discussion—Comments on Working Loads for Pile Foundations

During the past several years, there has been a significant increase in working loads for piling. Even higher loads are now being proposed. The rationale for proposing these high working loads and accompanying high stresses is that the allowable stress in a pile should be determined by its structural characteristics. These allowable stresses are determined by applying a factor of safety to either ultimate or yield stresses, assuming the pile to act as a short column. These then would establish maximum pile loadings, which would be modified as considered proper by the designing engineer if soil conditions were such that total pile loads would exceed resistance of the supporting soils or as dictated by driving stresses.

The fallacy in this approach is that it fails to recognize a number of factors that reduce the apparent factor of safety used. First, piles in contrast to caissons or above-grade structural members are installed by driving. The effects of driving and the limitations on safe performance inherent in driving cannot safely be ignored in establishing pile loads. There is no way in which the completed pile or the soil that supports it can be inspected upon completion of the work. Closed-end pipe piles or the shells of cast-in-place concrete piles can be checked for straightness or whether they are plumb before filling with concrete, but these elementary checks are too often omitted. Even these simple inspections are not possible for other types of piling. Evaluation of adequacy of each pile depends on observing its performance while being driven—a boring job usually assigned to the newest and least experienced engineer or technician or sometimes to a man without a technical background.

Even with most careful layout, spotting of piles, and realigning during driving, piles move out of position during driving. They bend, move sideways, and drive out of plumb. If you carefully check a closed-end pipe pile job you will find very few that are straight. I have seen H-piles rotate

[1] Vice president and senior consulting engineer, Stone and Webster Engineering Corporation, Boston, Mass.

90° or more while being driven, and on one of our jobs, an H-pile wandered so severly as to strike and sever a pile previously driven for the footing of an adjacent column. These distortions result in significant bending stresses that must be allowed for.

Experience has shown that the dynamic driving resistance (except possibly for piles driven to bearing on hard rock) as computed, for example, by the wave equation, should be at least twice the static design load and preferably 2.5 or even 3 times static design load for long piles. This is to ensure adequate penetration into the bearing stratum to provide long-term support while allowing for temporary resistance from materials lying above the bearing stratum.

Most piles are driven through soft, compressible soils to bearing in a dense bearing stratum or upon rock. For tapered piles in such conditions, the most critical section usually is near the tip. For working of stresses exceeding 1000 psi in the tip sections of wood piles of fir, pine, or red oak, there is very little margin between driving sufficiently hard to ensure adequate penetration into the bearing stratum and driving so hard as to damage or break piles. A pile hammer is not a precision tool, and meeting desired resistances without breaking piles can be extremely difficult in practice. This was a recognized problem even years ago with the then lighter loads on wood piles than now proposed and piles of better materials as regards knots and crooks than now permitted. It has been discussed by a number of authors, for example, Chellis.[2]

The same problem occurs with H-piles when working stresses approaching one-half the yield strength. Because of the tendency of the piles to bend during driving, camber from rolling, alignment, sloppy fit into pile caps, and similar factors, stresses near the head of the pile are never uniform. Bending of webs and flanges usually starts when computed driving stresses reach about 35 ksi for A 36 steel.

The problem is more acute for pipe piles, since they usually are filled with concrete, which is included in estimating structural strength for design loads. Thus, the pipe as driven may be quite highly stressed.

Parenthetically, H- and pipe piles especially should be carefully checked to be sure that elastic compressions (springing) during driving are not so large as to absorb so much energy that piles cannot be driven to desired dynamic resistances. This should be done using the wave equation analysis. Pile driving formulas such as the *Engineering News* formula provide no information on this point and no dependable information on driving resistance for most cases.

Precast concrete piles may be damaged during driving either near the head or by tensile cracking by the reflected wave unless heavily post-tensioned.

[2] Chellis, R. C. in *Pile Foundations*, 2nd ed., McGraw-Hill, New York, 1961, pp. 62–65.

The strength of cast-in-place piles should be carefully checked, especially for the critical sections near the tip. Concrete for cast-in-place piles should have high slumps, preferably 4 to 6 in., to ensure against arching during placement. Concrete with such slumps tends to bleed weakening the concrete where this excess water accumulates. Complete removal of water from the pile shell is sometimes not practical, and sometimes there is slight seepage at joints. Either can affect strength. Present codes permit ultimate strengths to 5000 psi. Where such strengths are used in design, it is necessary that very careful workmanship and supervision be assured.

Because of the deflection of the soil under load, the piles near the edge and especially at the corners of a pile group carry more load than the central piles. This is especially a consideration for piles into clay.[3] It also occurs in piles driven into sand and gravel, but to lesser degree.[4] Stiffness of the structure may also result in significant redistribution of loads to individual pile groups such that certain groups especially at corners and along exterior walls actually carry larger loads than computed. Some codes include group reduction factors that decrease the allowable pile loads for groups usually with larger reduction for large groups. This is not really a rational solution, since uniform loadings of piles is still assumed although by reducing pile loads it provides additional safety against failure of the more heavily loaded piles. Theoretically, the stress distribution in the soil should be approximated and the piles so located as to be about equally loaded. If uniform pile loading is assumed, the actual load on the corner and exterior piles may significantly exceed design values with consequent reduction in assumed factors of safety. It is my opinion that these several factors and problems must be considered in establishing loads in piles, especially code loading values.

It is also prudent for the engineer to consider code loads from an entirely different aspect. Even though a code loading is defined as a "maximum," it very quickly becomes a "minimum" because of competitive pressures. If a failure occurs, the designing engineer is in an extremely vulnerable position against a claim or law suit if he has designed to such "maximum" values, especially if there is clear wording in the technical literature that such values are maximum and should be used only when the engineer has carefully investigated all relevant factors. But explaining to a client that lower pile loads are appropriate to his building than apparently are allowed in the code or were used just down the street can be difficult, if not completely frustrating. In considering this matter it is wise to keep in mind the fact that, except for fairly large groups of piles, failure of a single pile under a footing can result in serious overloading of the remaining piles or

[3] Masters, F. M., *Transactions*, American Society of Civil Engineers, Paper 2174, Vol. 108, 1943, pp. 115–173.

[4] Swiger, W. F., "Foundation Tests for Los Angeles Steam Plant," *Civil Engineering*, Dec. 1941.

even gross failure, and the higher the loadings used the fewer the piles in each group.

I do not believe it in the interest of either the profession or the public to adopt pile loads using rather minimal factors of safety based on short column strengths. In short, I do not consider certain of the pile loadings now being proposed to be safe or prudent.

C. D. Thompson[1] *and D. E. Thompson*[2]

Influence of Driving Stresses on the Development of High Pile Capacities

REFERENCE: Thompson, C. D. and Thompson, D. E., **"Influence of Driving Stresses on the Development of High Pile Capacities,"** *Behavior of Deep Foundations, ASTM STP 670,* Raymond Lundgren, Ed., American Society for Testing and Materials, 1979, pp. 562–577.

ABSTRACT: The impact stresses have been measured near the top of piles as they are driven to practical refusal in very dense granular soils and shale bedrock. These stresses have been compared with capacities determined from static load tests. The comparisons were carried out on 24 piles of various types and lengths driven with a wide range of hammers and capblocks at different sites. They indicate that the ultimate bearing load divided by the pile shaft cross-sectional area during driving averages 1.2 times the impact stress at practical refusal with a standard deviation of ±20 percent. This is in close agreement with the Case method wave equation analysis. It is concluded that, for conditions similar to those tested, driving criteria to establish a required bearing load may be estimated from impact stress measurements at the top of the pile. Other methods, such as conventional dynamic formulas, are not as suitable, as they cannot measure the effects of hammer and cap block variations.

KEY WORDS: driven piles, dynamic measurements, impact stresses, static bearing capacities, very dense granular soil, shale bedrock, pile hammers, cap blocks, wave equation analysis, pile-driving criteria

Nomenclature

- A Cross-sectional area of pile shaft as driven.
- C Wave speed velocity in pile.
- E Modulus of elasticity of pile.
- F_1 Impact force near the pile top.
- F_2 Force near the pile top at a time interval of $2L/C$ after impact.
- J_c Soil damping factor used in the Case method.
- L Length of driven pile.
- M Mass of pile as driven.

[1] Vice president, The Trow Group Ltd., Rexdale, Ontario M9V 3Y8, Canada.
[2] Manager, Oshawa Office, The Trow Group Ltd., Oshawa, Ontario L1H 5N7, Canada.

R_u Ultimate static bearing capacity.
STAT Static compression force on pile for diesel hammers.
V_1 Impact velocity near pile top.
V_2 Velocity near pile top at a time interval of $2L/C$ after impact.
ρ Density of pile material.
σ_1 Impact stress near pile top.

For many years, engineers have realized that the prediction of the bearing capacity of driven piles by means of dynamic formulas is unreliable and frequently inaccurate. This was clearly demonstrated by the Michigan pile load tests [*1*].[3] Subsequent work by Forehand and Reese [*2*] showed that the wave equation analysis based on Smith's [*3*] simplified discrete element model may be used as a tool for more accurate bearing capacity prediction. This increases the potential for optimizing the working load on piles. Economic conditions have dictated that such optimization is desirable, since, for large expensive and heavy structures, a small increase in the bearing capacity can result in relatively large savings in foundation costs. To this end, there has been considerable progress in refining the practical aspects of wave equation analysis [*4,5*]. However, although wave equations indicate that impact velocity, and therefore impact stress, is a major aspect in developing bearing capacities, little research has been undertaken to examine its significance as a design tool for pile bearing capacities.

This paper presents information on driving stresses at the top of piles and bearing pressures of high capacity piles driven to bearing in very dense granular soil (glacial deposits) and shale bedrock in Ontario, Canada. Its intent is to indicate the significance of the magnitude, variations, and knowledge of driving stresses in relation to the optimization of the working loads on the pile. Data from nine sites has been compiled where driving stresses in addition to normally used parameters, such as set and ultimate bearing capacity, were measured. Additionally, three case histories are presented to illustrate the significance of driving stresses as a control mechanism and the unreliability of more conventional procedures.

Procedure

The experimental work associated with documentation for this paper was undertaken on construction projects over a period of two years from 1975 to 1977. For each site, the following field testing procedure was employed: (a) detailed geotechnical investigation in accordance with ASTM standards, (b) pile load tests carried out to failure in accordance with ASTM Testing Piles Under Axial Compressive Load (D 1143) and (c) measurements of strains and accelerations near the top of piles during driving em-

[3] The italic numbers in brackets refer to the list of references appended to this paper.

ploying the Goble-Case Western equipment. This involved attaching strain gages and accelerometers near the top of piles to obtain the dynamic measurements. For each hammer blow, electrical signals from these instruments were fed into a preprogrammed portable computer (pile-driving analyzer) to be converted into force and velocity parameters. From these, the force (stress), energy, and ultimate bearing capacity predicted by the Case method wave equation analysis were automatically calculated. The force and acceleration data were recorded on magnetic tape with an instrumentation tape recorder. They were subsequently reanalyzed by the laboratory computer to obtain more detailed information about the driving characteristics.

Results

The results of 24 load tests on piles where impact stresses have been measured are presented. Additionally, three case histories are documented to illustrate the conclusions presented by the study. In all cases, the geotechnical investigations were used to establish the maximum bearing capacity of the bearing strata, and the piles were driven close to practical refusal for the hammer and soil conditions. Consequently, the limitations of the hammer-cap block-pile system to develop a bearing load were being measured rather than the ultimate capacity of the bearing stratum.

Case History A

Information on load tests carried out between 1966 and 1975 on similar piles in fairly uniform soil conditions in one area of industrial development in Southern Ontario are presented in Table 1 [*6*]. A typical geotechnical profile of the soil conditions at the site is presented in Fig. 1 [*6*]. This profile is near the location of pile test 8 in Table 1.

All the piles were 32.4-cm outside diameter closed-end pipe, which generally drive to practical refusal in the silt till at depths varying from 21.3 to 30.5 m. Support to the piling is primarily in end bearing, although some frictional capacity is developed in the silty clay till.

A typical load-deflection curve is presented in Fig. 2 [*6*]. This plot is for test 6 in Table 1. It illustrates the definitions of failure used to estimate the ultimate capacity at this site. It should be noted that the Davisson criterion [*7*], which is employed in this documentation, frequently results in more conservative estimates of ultimate bearing capacity than those commonly employed in Ontario [*8*].

Case History B

The driving characteristics and geotechnical profile for case history B are

TABLE 1—*Increase of bearing capacities at site over a period of 10 years* [6].

		Hammer Data				Field Results			
								Ultimate Capacity, kN	
Test No.	Date of Test	Type	Rated Energy, kJ per blow	Cap Block	Pile Data[a] Driven Length, m	Final Driving Resistance, blows per 25 mm	Impact Stress at Pile Top, MPa	After Davisson	Used In Design
1	3/66	single-acting air	33.2	7.6 cm fir-hardwood	21.4	28	. . .	2162	1779
2	12/68	double-acting air	33.2	not known	21.3	42	. . .	1895	1913
3	11/69	single-acting air	33.2	10.2 cm wood packing	32.0	40	. . .	1779	2046
4	2/70	single-acting air	33.2	not known	21.5	40	. . .	1157	1600
5	4/75	open-end diesel	53.9	10.2 cm plywood, 10.2 cm hardwood	24.4	134/19 mm	150	2254	3114
6	4/75	open-end diesel	53.9	3.8 cm plywood, 10.2 cm hardwood	24.4	130	203	2722	3380
7	4/75	open-end diesel	53.9	3.8 cm plywood, 10.2 cm hardwood	24.4	49	164	2304	2580
8	4/75	double-acting air	33.2	steel cable	24.4	44	175	1948	2135

[a] All piles were 324-mm OD closed end steel pipe piles. Tests 1–4 and 8 were on 7.1-mm wall thickness ASTM A252, Grade 2 steel. Tests 5–7 were on 9.4-mm wall thickness grade 3 steel.

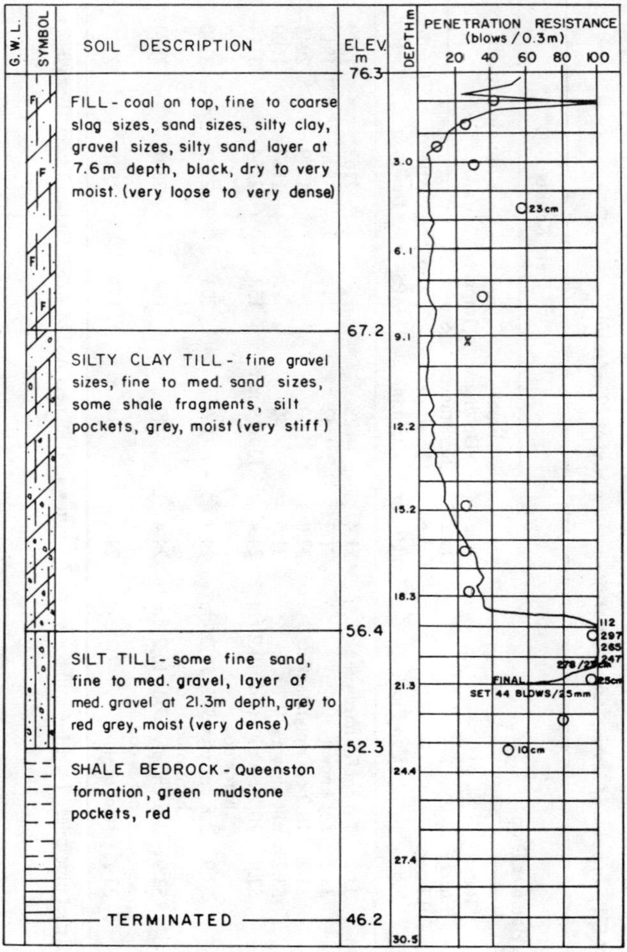

FIG. 1—*Typical geotechnical profile for case history A* [6].

given in Fig. 3 [*9*]. An HP 305 mm by 396 N 12.2-m-long pile was driven to practical refusal at 25 blows per 25 mm, terminating in the very dense silt till, with an open-end diesel hammer having a rated energy of 62.4 kJ per blow. The cap block in the system was a micarta and aluminum sandwich. The same hammer-cap block combination was employed on resumption of driving of the pile on the following day.

Case History C

The pile-driving characteristics and geotechnical profile for case history

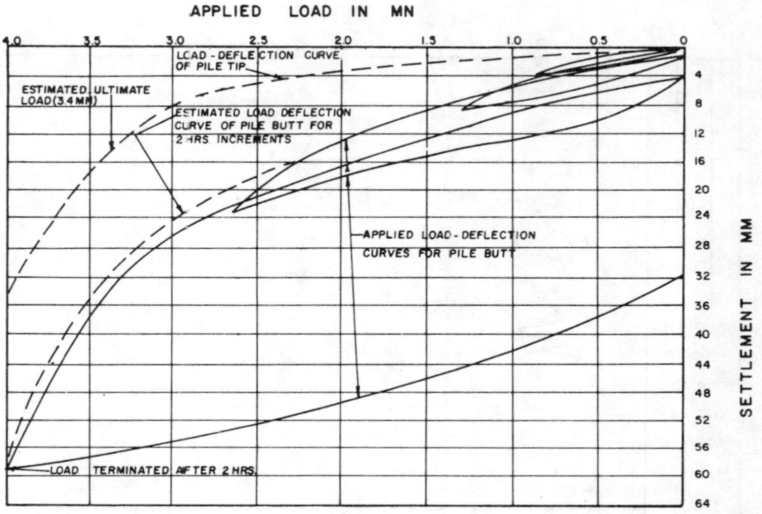

FIG. 2—*Typical load test curve for case history A* [6].

C are presented in Fig. 4. Steel H-piles were used for this project. Figure 4 illustrates the driving characteristics of a 12.2-m-long HP 305 mm by 329 N pile with a minimum yield strength of 248 MPa. It was driven with an open-end diesel hammer with a rated energy of 53.9 kJ per blow, and the cap block consisted of an aluminum and micarta sandwich. The pile was driven to a final driving resistance of eleven blows per 25 mm in very dense silty sand and gravel till.

General Results

The results of 24 pile load tests and the impact driving stresses are given in Table 2. The results cover a wide range of hammers, cap blocks, piles, and soil types. In addition, Fig. 5 presents a plot of the impact driving stress against the load per unit shaft area. The impact stress was obtained from strain measurements near the pile top. The ultimate bearing pressure is defined as the ultimate static bearing load from load tests divided by the cross-sectional shaft area of the piles as they were driven.

Discussion

The driving forces or stresses that can be developed in piles are considered to have a major influence on the achievement of high bearing capacities in very dense granular soils. Wave equation analysis models the hammer, cap block, pile, and soil to estimate the force and velocity param-

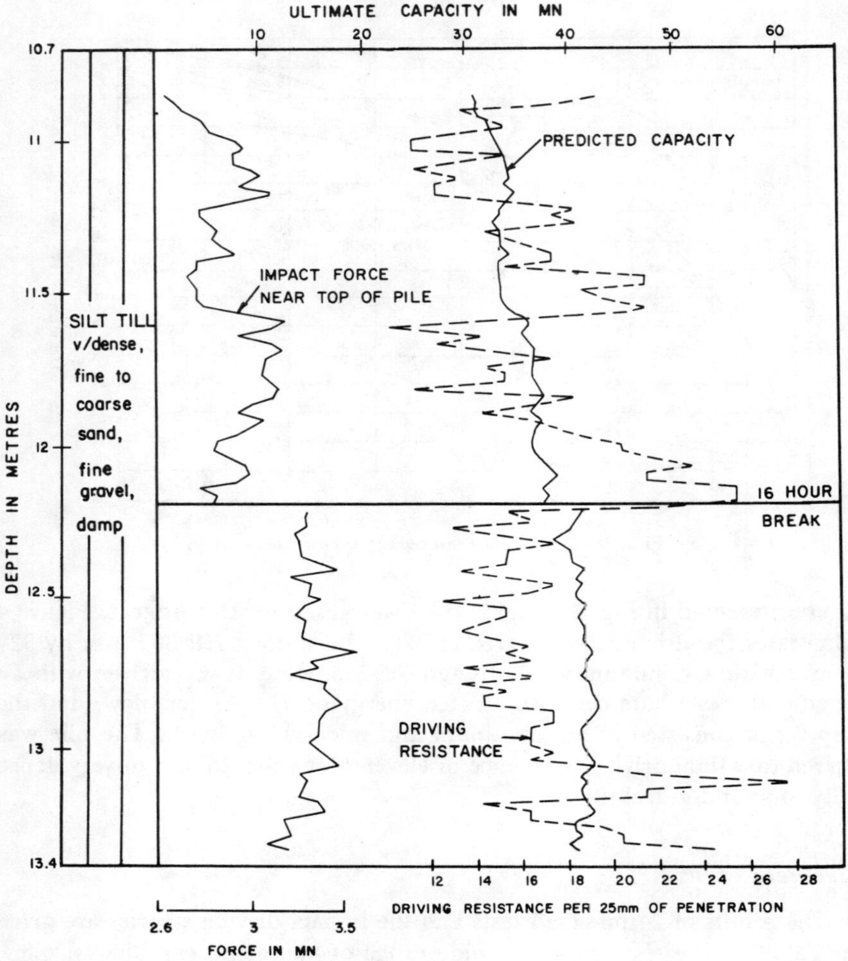

FIG. 3—*Pile-driving record for case history B* [9].

eters in the pile. However, most suffer from difficulties in modeling the hammer, cap block, and soil. The Case method (Eq 1) eliminates errors due to modeling of the hammer and cap block as the strain and acceleration measurements near the top of the pile are the input to the analysis. If some of the parameters illustrated in Fig. 6 are approximated, the ultimate bearing load R_u, is observed to be a function of the impact force F_1 (Eq 9), Also, experience indicates that the changes in F_2, V_2, and J_c are of magnitudes that would usually change the relationship by less than 20 percent for piles driven to practical refusal in very dense granular soil. The impact

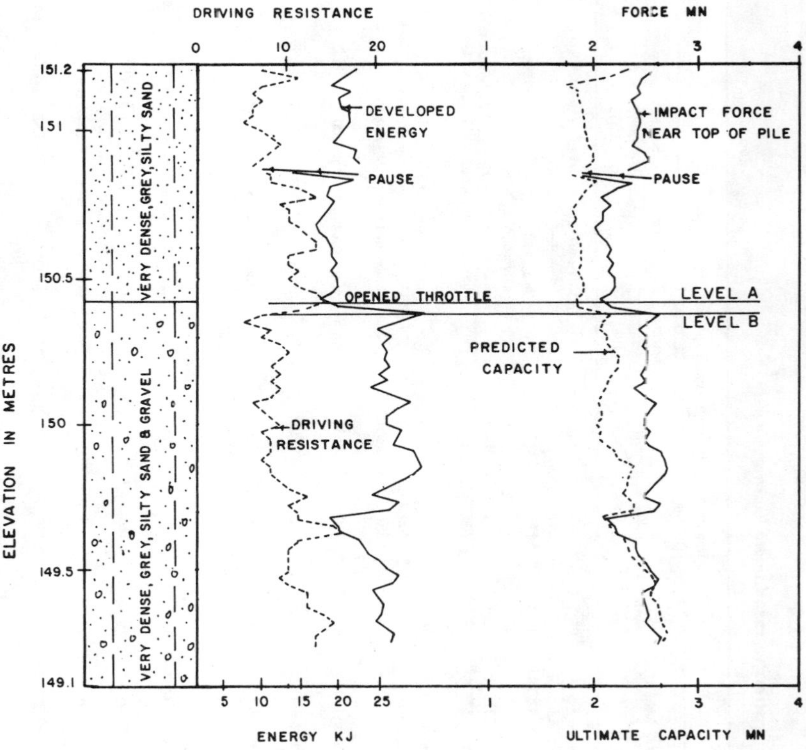

FIG. 4—*Pile-driving record for case history C.*

force or stress at the top of the pile should therefore be a reasonably accurate method of determining ultimate bearing loads in these conditions. The purpose of the analysis for this paper is to check this hypothesis and examine its significance in terms of achieving high bearing loads on driven piles.

Case Method Wave Equation Analysis

$$R_u = (1 + J_c)\frac{F_1 + F_2}{2} + \frac{MC}{2L}(V_1 - V_2) - 2J_cF_1 + J_c\mathrm{STAT} \qquad (1)$$

where

R_u = ultimate static bearing load of pile,
J_c = damping factor for soil,
F_1 = impact force at the top of the pile,
F_2 = force at the top of the pile at a time interval of $2L/c$ after impact,

TABLE 2—*Impact driving stress versus ultimate bearing capacity.*

	Hammer Data			Pile Data			Ultimate Bearing Capacity		
Pile No.	Type	Rated, Energy, kJ	Cap Block	Type and Size	Driven Length, m	Impact Driving Stress at Pile Top, MPa	Load, MN	Bearing Pressure, MPa	Soil Description at Pile Tip
1	double-acting air	33.2	steel cable	HP 305 mm × 329 N	24.5	99	1228	87	silty sand and gravel till, shale fragments and boulders, $N > 100$
2	open-end diesel	53.9	aluminum and micarta	HP 305 mm × 329 N	16.7	195	2758	196	silty sand till, traces of clay and gravel, very dense, $N > 100$
3	open-end diesel	30.5	aluminum and micarta	HP 305 mm × 329 N	6.9	158	1423	101	shale bedrock, weathered limestone layers
4	open-end diesel	53.9	aluminum, micarta, and steel cable	HP 305 mm × 329 N	9.1	206	2402	171	shale bedrock, weathered limestone layers
5	open-end diesel	62.4	aluminum and micarta	HP 305 mm × 329 N	36.9	223	3604	256	silty very fine sand, $N = 138$
6	open-end diesel	30.5	aluminum and micarta	HP 305 mm × 329 N	7.0	200	1778	177	shale bedrock, weathered limestone layers
7	open-end diesel	62.4	aluminum and micarta	pipe—324 mm OD × 8.4 mm	19.4	251	2091	252	sandy silt till, $N = 50$-100
8	open-end diesel	46.0	aluminum and micarta	pipe—324 mm OD × 8.4 mm	18.3	119	1655	176	sandy silt till, $N = 50$-100
9	open-end diesel	46.0	aluminum and micarta	pipe—324 mm OD × 9.5 mm	18.3	192	1708	181	sandy silt till, $N = 50$-100
10	open-end diesel	46.0	aluminum and micarta	pipe—324 mm OD × 9.5 mm	20.1	154	2002	213	sandy silt till, $N = 50$- 100
11	drop—20-kN ram, 1.4-m drop	27.3	20.3 cm oak, grain vertical 5.1 cm Masonite	pipe—324 mm OD × 6.9 mm	15.5	154	1156	169	sand, fine to medium, $N = 31$-52
12	drop—20 kN-ram, 1.4-m drop	27.3	20.3 cm oak, grain vertical 5.1 cm Masonite	pipe—324 mm OD × 6.9 mm	14.9	153	947	150	sand, fine to medium, $N = 31$-52
13	drop—20 kN-ram, 1.4-m drop	27.3	20.3-cm oak, grain vertical 5.1 cm Masonite	pipe—324 mm OD × 6.9 mm	14.6	165	1121	164	sand, fine to medium, $N = 31$-52
14	drop—20-kN ram, 1.4-m drop	27.3	20.3 cm oak, grain vertical, 5.1 cm Masonite	pipe—324 mm OD × 6.9 mm	14.9	149	1085	159	sand, fine to medium, $N = 31$-52
15	open-end diesel	53.9	10.2 cm plywood, 10.2 cm hardwood	pipe—324 mm OD × 9.4 mm	24.4	150	2722	243	silt till, some fine sand and gravel, $N = 80$ to >100
16	open-end diesel	53.9	3.8 cm plywood, 10.2 cm hardwood	pipe—324 mm OD × 9.4 mm	24.4	203	2722	286	silt till, some fine sand and gravel, $N = 80$ to >100

17	open-end diesel	53.9	3.8 cm plywood, 10.2 cm hardwood	pipe—324 mm OD × 9.4 mm	24.4	164	2304	248	silt till, some fine sand and gravel, N = 80 to >100
18	double-acting air	33.2	steel cable	pipe—324 mm OD × 7.1 mm	24.4	175	1948	210	silt till, some fine sand and gravel N = 80 to >100
19	open-end diesel	62.4	aluminum and micarta	pipe—324 mm OD × 6.4 mm	45.6	261	1780	281	silty, very fine sand, N = 201-105
20	open-end diesel	53.9	19.1 cm hardwood	pipe—245 mm OD × 9.8 mm	17.7	205	1566	218	shale bedrock, weathered
21	open-end diesel	30.5	19.1 cm hardwood	pipe—178 mm OD × 7.9 mm	15.2	188	1264	302	shale bedrock, weathered
22	open-end diesel	62.4	aluminum and micarta	305 mm × 305 mm hardrive concrete	44.2	19	2247	24	silty, very fine sand, N = 138
23	open-end diesel	62.4	aluminum and micarta	305 mm × 305 mm hardrive concrete	19.8	34	2230	24	clayey silt till, some sand and gravel, N = 78-44
24	open-end diesel	62.4	aluminum and plywood	305 mm × 305 mm hardrive concrete	32.9	20	2491	27	shale bedrock, N > 100

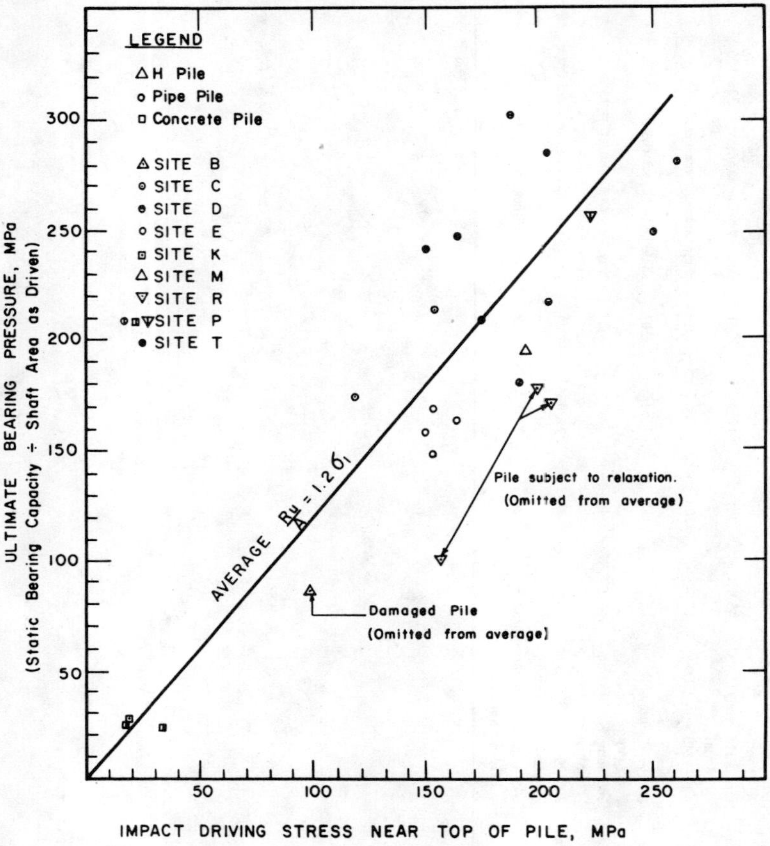

FIG. 5—*Impact driving stress versus ultimate bearing pressure.*

M = mass of pile,
C = wave speed velocity,
L = length of pile,
V_1 = impact velocity at the top of the pile,
V_2 = velocity at the top of the pile at a time interval of $2L/c$ after impact, and
STAT = static force on pile.

$$V = FC/EA \tag{2}$$

$$M = AL\rho \tag{3}$$

$$E = \rho C^2 \tag{4}$$

where

E = modulus of elasticity of pile,

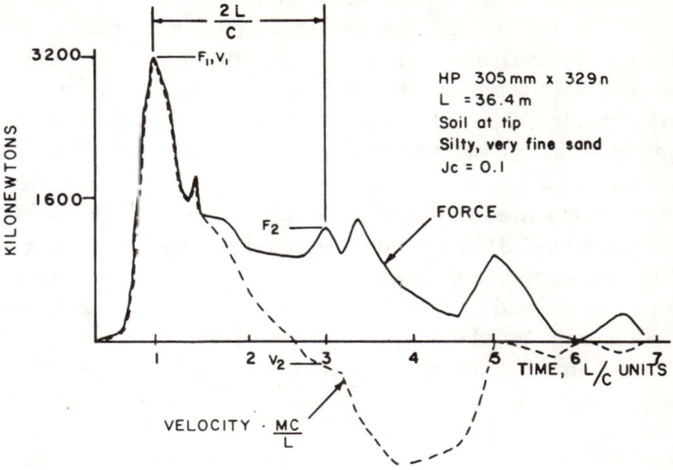

FIG. 6—*Typical force and velocity record.*

A = cross-sectional area of pile shaft, and
ρ = density of pile material.

From Fig. 6 and other observations of the force and velocity parameters for piles bearing in very dense granular soils,

$$F_2 \simeq 0.7F_1 \tag{5}$$

$$V_2 \simeq 0 \tag{6}$$

$$J_c \simeq 0.1 \tag{7}$$

$$J_c\text{STAT} \simeq 0 \tag{8}$$

On this basis, $R_u \simeq 1.3F_1$ or

$$R_u/A \simeq 1.3\sigma_1 \tag{9}$$

where σ_1 = the impact stress near the top of the pile.

Case history A illustrates the benefits of employing impact stress information to increase bearing loads on driven piles. Over a period of 10 years, eight piles were load tested to ultimate failure for plant expansion in relatively uniform soil conditions. Load Tests 1 to 4 were undertaken between 1966 and 1970. There was little apparent difference between the installation of the piles. Consequently, test 4, Table 1, which had the lowest results, controlled the design loading on the piles for many years. While the allowable load of 800 kN per pile, which was employed, was considerably more than values of 530 kN previously used, it was still much less than the

1020 kN, which could have been used on the basis of the highest of the tests. In 1975, a major plant expansion allowed a further series of load tests with dynamic measurements during pile driving. These measurements were correlated with the load test results and clearly indicated for tests 5 to 7 that: (*a*) test pile 6 would have the highest capacity, although test pile 5 had a greater final driving resistance; (*b*) the cap block thickness had a significant effect on the drivability of the piles to high capacities; and (*c*) that impact driving stresses had to be greater than 170 MPa to ensure that ultimate capacities of 3110 kN per pile could be developed. The specifications for the production piling were therefore drawn up to cover these aspects, and a design load of 1560 kN per pile was employed successfully.

In the portion of the plant where test pile 8 was driven and load tested, piles with the same general driving specifications as test piles 1 to 4 were successfully design loaded to 1070 kN each, an increase of one-third over previous loads.

Between 1975 and 1977, 24 pile load tests have been compared with dynamic measurements during pile driving, and these are documented in Table 2 and Fig. 5. (Other tests have been undertaken, but these are not yet available for publication.) The types of pile include: three precast concrete; six steel H and 15 closed-end steel pipe piles with lengths ranging from 7.5 m to 30 m. The soil conditions at the pile tips range from interlayered shale and limestone bedrock through very dense silt, sand, and boulder tills. Some piles derived a large measure of support in friction, while others were end bearing. The hammers included open-end diesel, double and single acting air, and drop hammers. Cap blocks were of durable plastics, hardwood, plywood, steel cable, and rubber belting. The piles were driven to driving resistances of greater than 10 blows per 25 mm. The results of the load tests are plotted in Fig. 5 in terms of the impact driving stress versus the ultimate bearing pressure (ultimate static load divided by shaft area of the pile as driven). The average of the results, excluding tests where relaxation or damaged piles were encountered, is plotted as a straight line through the origin. This average indicates that the ultimate bearing pressure is 1.2 times the impact stress, with a standard deviation of ± 20 percent. On this basis, it is concluded that there is a reasonable correlation with the hypothesis from the Case method wave equation analysis (Eq. 9).

The limitations of determining bearing capacities from Fig. 5 have been examined. The bearing stratum at the pile tip was consistently very dense granular material or bedrock with low damping factors. There was little evidence of soil freeze and only one site on which relaxation occurred. Soil freeze would be expected to occur more widely than what was observed during this testing, usually resulting in greater ultimate bearing pressures at a given impact stress than shown in Fig. 5. Additionally, the wave equation analysis indicates that some deviation from the average should be expected, depending on the nature of the support and soil conditions at the tip. The test results confirmed this, while still indicating that the bearing pressure is a function of the impact stress. Finally, errors can occur in the measurements of both the impact stress and the ultimate static capacity, even when load cells are used. However, it is concluded that the results, which are presented in Fig. 5, clearly indicate that the developed ultimate bearing capacity is a function of the impact stress and that, with due regard to the testing limitations, the relationship may be used to estimate bearing capacities.

Attempts have been made to extend the correlation of impact driving stress with bearing capacity to the more recognized parameters of rated and/or developed energy and driving resistance (final set). Figure 3, case history B, presents the driving resistance, the ultimate capacity, and the impact force as an H-pile penetrates a very dense bouldery silt till. The same hammer-cap block-pile combination applied throughout. The result of an overnight stop was to increase the impact stress by 12 percent, increase the bearing capacity by 7 percent, and decrease the driving resistance by 38 percent. According to conventional approaches, the decrease in driving resistance should have indicated relaxation, but this was not the case. Other similar experience shows that, frequently, the driving resistance is more a function of hammer and cap block performance than of static bearing capacity.

In Fig. 4, case history C, an H-pile was being driven with an open-end diesel hammer through compact to very dense silty sand till. Between levels A and B, elevation 150.4 m approximately, the throttle was opened on the hammer with a resulting large decrease in driving resistance, which correlates with increases in developed energy, force, and capacity. While this correlation was predictable, the throttle operator was not. It was also very difficult to control adjustments to the throttle setting, especially during production piling. This significantly affected driving resistance observations. Similar illustrations could be presented for other hammers such as air and drop type, and for other driving conditions. They all, however, clearly indicate that the impact force or stress is the only parameter that has been observed to consistently correlate with the ultimate bearing capacity.

As higher bearing capacities are employed on driven piles, true factors of safety are decreased. While applied loads were comparatively low, actual factors of safety were usually well in excess of 2. Recent trends to increase loads have resulted in more failures during testing and have called into question the conventional procedures for establishing criteria. Of the dynamic measurements, only impact forces and stresses have consistently correlated with ultimate static capacities in the testing that has been performed during the past 3 years. Consequently, the most suitable procedure to optimize bearing capacities is to establish driving criteria on the basis of impact stress measurements. Figure 5 may be used at the design stage to establish the stress levels required to develop a bearing capacity. Steel piles with yield strengths of 240 to 350 MPa have been driven to impact stresses of more than 80 percent of yield without damage, as long as the seating of the helmet was satisfactory. As the ratio of ultimate bearing pressure to impact stress is 1.2, the ultimate bearing load is of the same order as the yield strength of the steel. For construction control, the hammer should be selected on the basis of a wave equation analysis or experience. The driving criteria should then be established by stress measurements at practical refusal.

Conclusions

The following conclusions are considered applicable for a wide range of hammers, cap blocks, and short to moderate length piles driven to practical refusal to develop high capacities in very dense granular soils and shale bedrock:

1. The ultimate static bearing capacity for 20 load tests has been found to average 1.2 times the impact stress at the top of the pile (Fig. 5). The standard deviation from this value has been observed at ± 20 percent. These values are in close agreement with the Case method wave equation analysis for these piles.
2. Methods of establishing bearing capacities on the basis of rated and/or developed energies have been unsuccessful.
3. The design bearing capacity of steel piles may be established from Fig. 5 for bearing strata capable of supporting the design loads with adequate factors of safety. The needed impact stresses may be assumed to be 80 percent of yield for 240 to 350 MPa yield strength steels, as long as the hammer has a sufficient capability to develop these stresses.
4. The driving criteria for the production piling designed according to Fig. 5 may be determined from impact force measurements near the top of the pile.

References

[*1*] Housel, W. S., *Proceedings*, American Society of Civil Engineers, Vol. 92, No. SM4, 1966, pp. 1-30.

[*2*] Forehand, P. W. and Reese, J. L., *Journal of Soil Mechanics and Foundations Division,* American Society of Civil Engineers, No. SM2, March 1964, pp. 1-25.

[*3*] Smith, E. A. L., *Proceedings*, American Society of Civil Engineers, Vol. 86, No. SM4, 1960, pp. 35-61.

[*4*] Rausche, F., Moses, F., and Goble, G. G., *Proceedings*, American Society of Civil Engineers, Vol. 98, No. SM9, Sept. 1972, pp. 917-937.

[*5*] Lowery, L. L., Edwards, T. C., Hirsch, T. J., "Use of the Wave Equation to Predict Soil Resistance on a Pile During Driving," The Texas Transportation Institute, A, Research Report 33-10.

[6] Thompson, C. D., *Proceedings*, Seminar and Workshop on Design and Installation of Higher Capacity Piles, University of Colorado, Boulder, August 1977.
[7] Davisson, M. T., *Proceedings*, American Society of Civil Engineers' Lecture Series, Innovations in Foundation Construction, Illinois Section, 1972.
[8] *The Ontario Building Code, Section 4.2.5.3(2) Piles*, Ministry of Consumer and Commercial Relations, Ontario, Canada, 1975, p. 184.
[9] Thompson, C. D., *Proceedings*, Regional Technical Conference, Canadian Society for Civil Engineering, Sudbury, Ontario, Canada, February 1977.

M. C. Wang,[1] A. H. Wu,[2] and D. J. Scheessele[3]

Stress and Deformation in Single Piles Due to Lateral Movement of Surrounding Soils

REFERENCE: Wang, M. C., Wu, A. H., and Scheessele, D. J., "**Stress and Deformation in Single Piles Due to Lateral Movement of Surrounding Soils,**" *Behavior of Deep Foundations, ASTM STP 670,* Raymond Lundgren, Ed., American Society for Testing and Materials, 1979, pp. 578–591.

ABSTRACT: This paper presents a case history of structural damage due to lateral soil movement and an analysis of stress and deformation in the supporting piles. The project involved cast-in-place concrete piles cased in steel pipes having 40.6-cm (16-in.) outside diameter and 38.7-cm (15¼-in.) inside diameter. The piles extended through a 7.6-m (25-ft) oyster shell fill and a 17-m (55-ft) soft clay layer and penetrated 7.6 m (25 ft) into a compact sand stratum. Due to overloading at the top of the fill, the soil surrounding the piles underwent lateral movement and caused cracking in the pile caps.

The amount of lateral soil movement was measured using inclinometers. Based upon the measured soil deformation, lateral soil pressures were determined with appropriate moduli of subgrade reaction. The soil pressures were used as lateral loading on a single pile. Using a finite element computer program that was developed based upon the theory of beam on elastic foundation, the deformation of the single pile and the horizontal reaction and resisting moment mobilized in the pile cap were analyzed. By comparing the horizontal reaction and resisting moment with the structurally determined ultimate values, the results of the pile behavior analysis were checked favorably.

KEY WORDS: bearing piles, lateral loads, lateral soil movement, soil modulus, finite element analysis, pile behavior, slope stability evaluation, oyster shell fill, soft clay

When the soil near the ground surface is inadequate to support foundation loads because of either low bearing capacity or high compressibility, piles are used to transfer foundation loads to deeper soil strata that are

[1]Associate professor, Department of Civil Engineering, The Pennsylvania State University, University Park, Pa.

[2]Soil mechanics and foundation specialist, Naval Facilities Engineering Command, Alexandria, Va. 22332.

[3]Structural engineer and computer-aided design specialist, Chesapeake Division, Naval Facilities Engineering Command, Washington, D.C.

stronger and less compressible. Piles are often subjected to lateral loading in addition to the primary loading in the axial direction. The lateral loading may be derived from the superstructure, which is supported by the piles, or from lateral movement of the surrounding soils. Under the combined effect of the axial and lateral loadings, the piles may undergo excessive deformation and cause failure in the superstructure. Thus, pile foundations must be designed so that the stresses and deformation are within tolerable limits. In order that this requirement can be achieved, foundation engineers require knowledge of predicting the behavior of pile foundations subjected to various types of loading.

During the past decades, considerable effort has been directed toward this goal. As a result, various techniques of predicting pile foundation behavior have evolved. While literature is available for piles subjected to lateral loading—for instance, Bowles [*1*],[4] Broms [*2,3*], Davisson [*4*], Davisson and Gill [*5*], Reese and Cox [*6*], Poulos [*7*] and many others—most of them deal with lateral loading derived from the superstructure. Very little information is currently available regarding pile behavior under lateral loading induced by the movement of the surrounding soil.

The lateral load induced by the movement of the surrounding soil is significantly different from that derived from the superstructure. The major difference includes the location and distribution of the loading. In addition, a considerable difficulty is involved in the evaluation of the loading. Thus, the problem of pile foundation subjected to lateral movement of the surrounding soil is rather unique and requires particular attention and treatment. This paper presents a case history of structural damage due to lateral soil movement and an analysis of stress and deformation in the supporting pile.

Description of the Project

The project involved the construction of a multimillion dollar Nimitz Library and Rickover Hall at the U.S. Naval Academy, Annapolis, Md. The original ground elevation at the job site ranged from −1 m (−3 ft) to −7 m (−20 ft). Thus, a landfill of about 1.4 hectares (3.5 acres) was constructed along the Dorsey Creek to support the proposed structures. The landfill was constructed with lightweight oyster shell. At the fill site, the soil deposit contains a soft organic clay layer averaging about 17 m (55 ft) thick followed by a compact sand stratum. To accelerate consolidation of the soft clay deposit, vertical sand drains were installed. The construction of the landfill was started in 1970 and completed in September 1971.

The Nimitz Library was founded on cast-in-place concrete piles. The

[4] Italic numbers in brackets refer to the list of references appended to this paper.

concrete piles were cased in steel pipes having 40.6-cm (16-in.) outside diameter and 38.7-cm (15¼-in.) inside diameter. The piles had a length of about 32 m (105 ft), which extended from the top of the oyster shell fill to a depth of about 7.6 m (25 ft) into the compact sand stratum. Figures 1 and 2 show the aerial view of the landfill area and the elevation view of the project site, respectively.

After completion of the entire 1.4-hectare (3.5-acre) landfill project, the

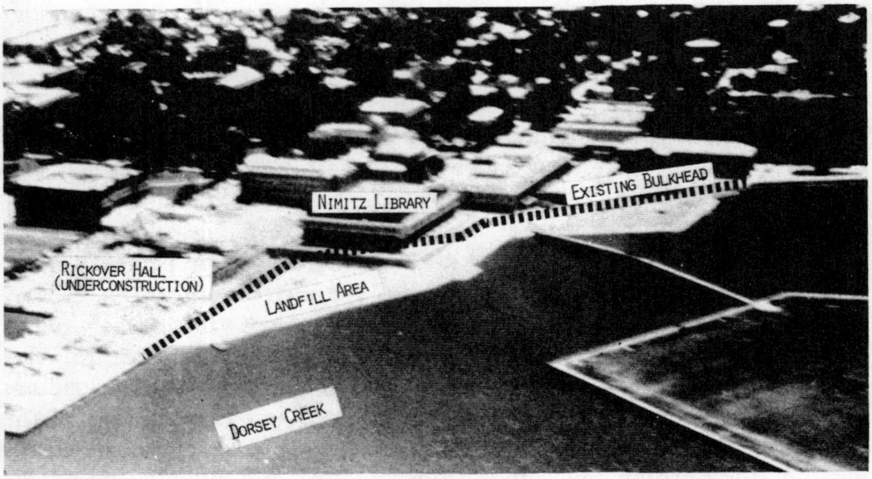

FIG. 1—*Aerial view of the landfill area, 1973.*

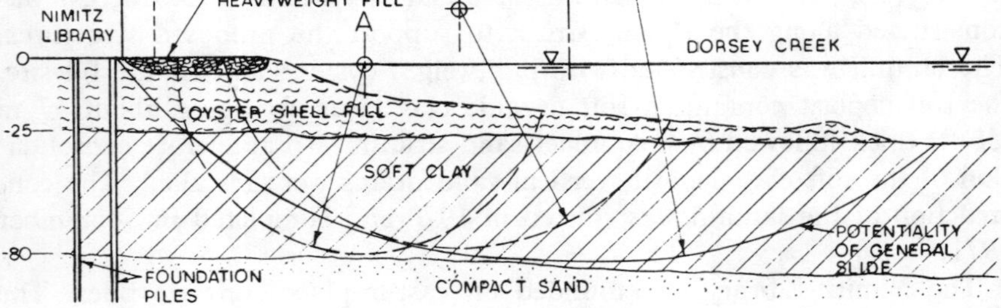

FIG. 2—*Typical landfill section potentiality of instability.*

fill surface showed large and rapid subsidence. The rate of subsidence was far greater than that anticipated from consolidation of the soft clay soil. Meanwhile cracks developed in some of the pile caps and girders of Nimitz Library as shown in Fig. 3. Subsequent investigations further revealed that the excessive fill movements were aggravated by overloading of the upland area during construction of Nimitz Library and Rickover Hall. The lateral soil movement appeared to have caused additional earth pressure to the piles, thus presenting a potential of structural damage to the library building.

Remedial measures were immediately undertaken to prevent the landfill from further subsiding. Detailed information on the various techniques used to stabilize the landfill has been reported by Wu [*8*]. The behavior of two single piles that support the stairway entrance of Nimitz Library is analyzed. These piles were chosen for analysis because cracks developed in the pile cap. Figure 4 shows the locations of the single piles analyzed.

Soil Properties

The soft clay has an average natural water content of 111 percent, an average liquid limit of 75 percent, and an average plastic limit of 42 percent. These data indicate that the clay is highly plastic and sensitive to re-

FIG. 3—*Fracture in pile cap.*

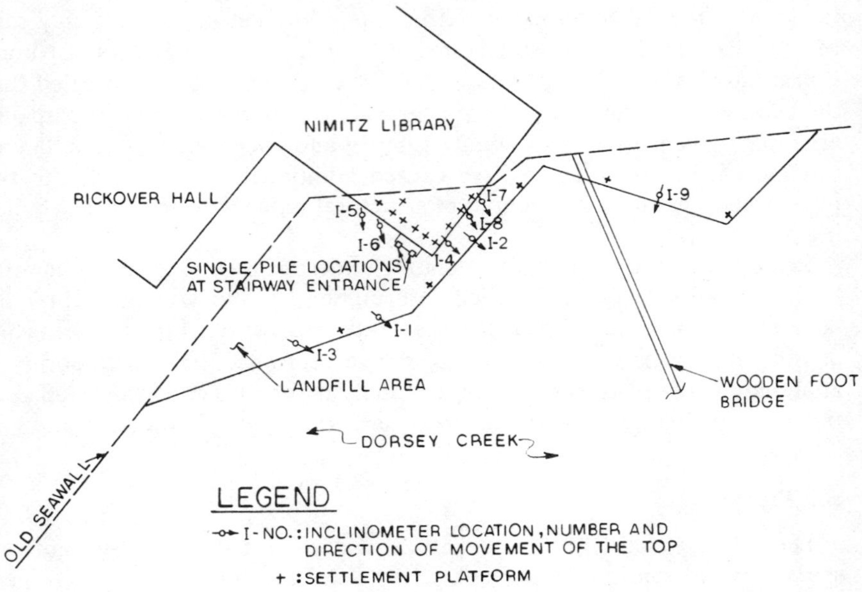

FIG. 4—*Single piles—inclinometers and settlement platform locations.*

molding. The moduli of volume change determined from laboratory consolidation tests equal 0.61 (0.0064), 0.35 (0.0037), and 0.0018 m^2/kn (0.17 ft^2/ton) for pressures of 0.05 to 0.25 (4.78 to 23.91), 0.25 to 0.50 (23.91 to 47.83), and 47.83 to 95.66 kn/m^2 (0.50 to 1.00 tons/ft^2), respectively. Thus, the clay soil is highly compressible, and secondary compression effects are important.

The strength properties of the soft clay were determined both in the laboratory using the triaxial compression test and in the field using the vane shear test. The undrained shear strength ranges from 7.2 kn/m^2 (150 psf) to 16.7 kn/m^3 (350 psf). From the *in situ* vane shear tests, the undisturbed shear strength ranges from 19.2 kn/m^2 (402 psf) to 64.4 kn/m^2 (1346 psf), and the remolded strength ranges from 6.0 kn/m^2 (124 psf) to 28.5 kn/m^2 (595 psf). Some inconsistencies between the laboratory and the field strength data are seen. No attempt was made to determine possible causes for the discrepancy, however.

The density of the oyster shell was determined by filling containers of known volume with shells. The approximate natural density equals about 1.41 g/cm^3 (88 pcf) saturated and 1.06 g/cm^3 (66 pcf) dry. The effective angle of internal friction of the oyster shell determined using the direct shear test averages about 38 deg. The density of the sand stratum was estimated during boring; it equals approximately 2.0 g/cm^3 (125 pcf). The effective angle of internal friction of the sand was evaluated at approximately 36 deg.

Field Instrumentation

In order to monitor the rate of landfill settlement, 15 settlement platforms and nine inclinometers were installed in the fill area as shown in Fig. 4. Included in Fig. 4 are also the direction of movement of top of each inclinometer as designated by arrows. The lateral deformation profiles determined from inclinometers nos. 6 through 8 are given for various dates in Fig. 5. A comparison among the measurements indicated that the data obtained from inclinometer no. 4 are erratic and inconsistent with others. Possible causes for this erratic result were not investigated. Details on the instrumentation program and the results of measurements are available elsewhere [*9*].

Stability Analysis

The preceding inclinometer data indicate that the zone of movement extended throughout the entire clay stratum. The general direction of movement was toward Dorsey Creek. Therefore, a cross section in front of Nimitz Library was used for analysis of the stability of the entire soil mass, including the landfill and the soft clay.

The stability analysis was performed using the procedure of total stress analysis together with the low value of undrained shear strength of 7.2 kn/m^2 (150 psf) for the soft clay. Using a computer program that was developed based upon the Bishop method of analysis, the factors of safety for various possible sliding circles shown in Fig. 2 were computed. Results of the analysis give factors of safety of 1.10, 1.04, and 0.99 for sliding circles A, B, and C, respectively. The results indicate that the most dangerous mode of possible failure is the one involving sliding of the soft clay toward the deepest part of Dorsey Creek some 30 m (100 ft) to 61 m (200 ft) away from the water edge of the fill, that is, sliding circle C.

As indicated by the results of the stability analysis, the weight of the oyster shell fill triggered the sliding of the underlying soft clay soil. Therefore, the subsidence of the landfill was largely caused by a general shear failure rather than merely by consolidation settlement.

Analysis of Pile Behavior

The deformation and stress distribution due to lateral movement of the surrounding soil was analyzed for the single pile shown in Fig. 4. As has been mentioned earlier, a major difficulty in the analysis of pile behavior under lateral soil movement is the evaluation of lateral loading acting on

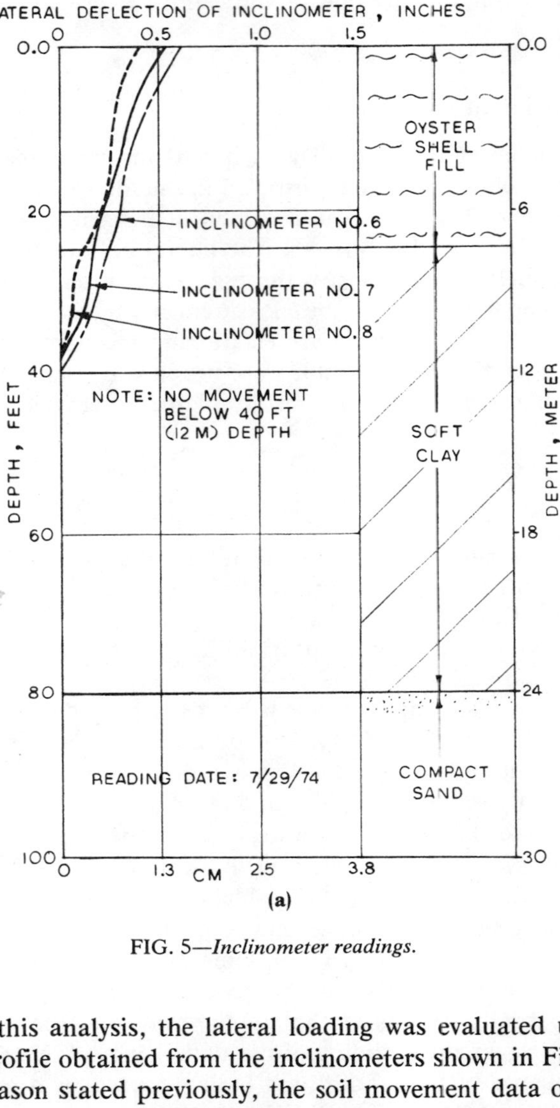

(a)

FIG. 5—*Inclinometer readings.*

the pile. In this analysis, the lateral loading was evaluated using the soil movement profile obtained from the inclinometers shown in Fig. 5*b*.

For the reason stated previously, the soil movement data obtained from inclinometer no. 4 were not used for the analysis. Furthermore, the oyster shell around inclinometer no. 5 was removed in May 1975 for evaluating confinement effect of oyster shell on lateral movement. Since cracking of the pile cap was first noticed in January 1975, the average value of lateral soil movement at this date was computed from inclinometers nos. 6, 7, and 8 and shown in Fig. 5*b*. Then soil pressures required to cause this lateral soil movement were estimated. In the evaluation of soil pressure, the moduli of subgrade reaction of the soils were estimated as follows.

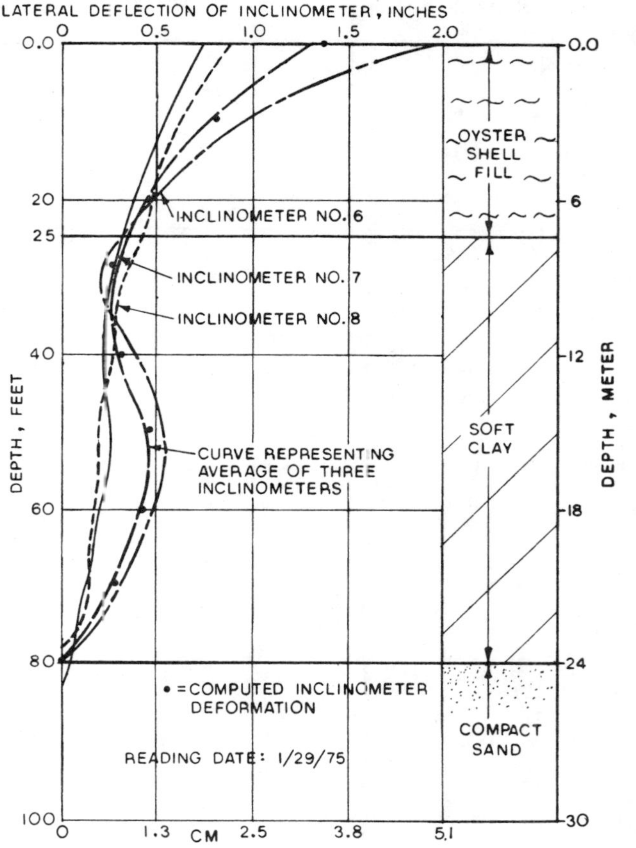

(b)

FIG. 5—(*Continued*).

The value of the soil modulus is not a property of the soil but is a fitting function to correlate the pile deflection with the soil properties. It can only be estimated from pile load tests or from direct observations of pile behavior. Therefore, determination of the exact value of soil modulus is very difficult, if not impossible. Available information [*1,2,4,7*] indicates that for clay soils, the soil modulus can be roughly related with the undrained shear strength of the soil. For instance, the ratio of the soil modulus to the undrained shear strength ranging between 15 and 95 was given by Poulos [*7*], and the modulus of subgrade reaction of clay soils ranging from 238.3 kn/m^3 (1.5 kip/ft^3) to 6354.3 kn/m^3 (40 kip/ft^3) was suggested by Bowles [*1*] and others. For cohesionless materials, Poulos [*7*] used 90 N/cm^2 (130 psi) to 965 N/cm^2 (1400 psi) as the range of soil modulus values, whereas,

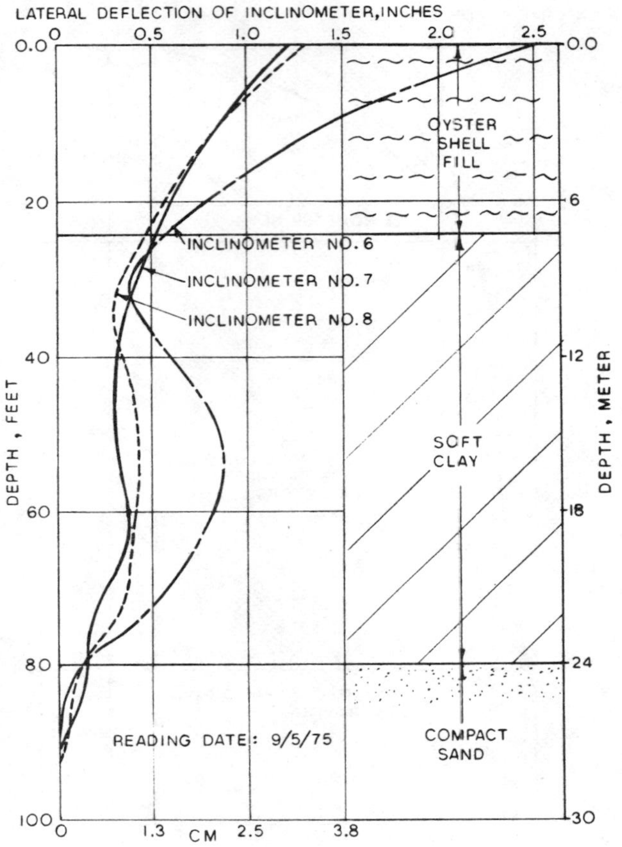

(c)

FIG. 5—(*Continued*).

Bowles [*1*], Davisson [*4*], and others suggested a range of about 1.6 N/cm^3 (6 pci) to 27.1 N/cm^3 (100 pci) for the modulus of subgrade reaction.

It is generally accepted that, for soils having increasing shear strength with depth, the soil modulus values increase linearly with increasing depth. The previous information together with the strength properties of the soils were used as a guide for selection of the modulus of subgrade reaction. Using the selected values of moduli of subgrade reaction and the average lateral soil movements measured from inclinometers nos. 6, 7, and 8 (shown in Fig. 5*b*), the lateral soil pressures were computed. The selected moduli of subgrade reaction are given below, and the computed lateral soil pressure profile is shown in Fig. 6.

Oyster shell	$k = 4.4Z$
Soft clay	$k = 0.6$
Compact Sand	$k = 1.5Z$

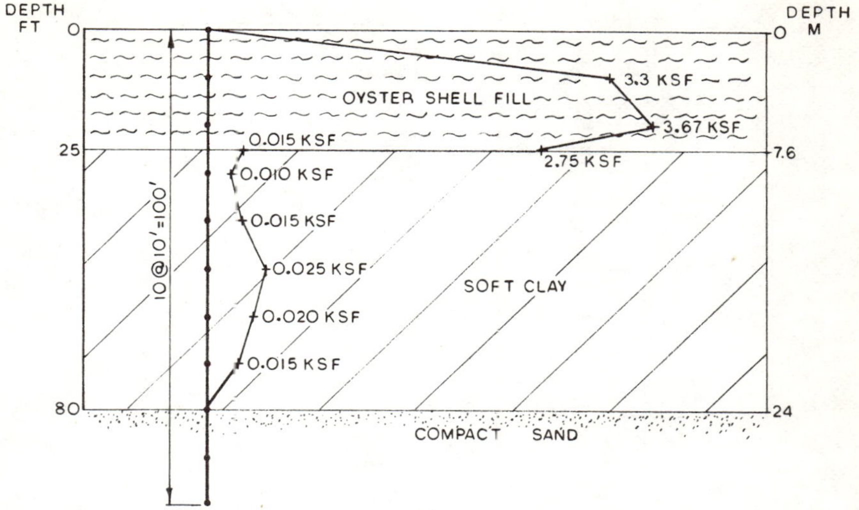

FIG. 6—*Lateral soil pressure diagram.*

Where k is the modulus of subgrade reaction in kcf, and Z in feet is the depth measured from the top of the landfill.

These pressures were then used to calculate nodal loads on the inclinometer tube. In the calculation, the entire 32-m (90-ft) inclinometer tube was divided into nine equal segments. The computed nodal loads shown in Fig. 7*a* along with the selected moduli of subgrade reaction and physical characteristics of the inclinometer were input into a finite element computer program to compute the horizontal deformation of the inclinometer. The finite element program was developed by the Computer Aided Design Group of Chesapeake Division, Naval Facilities Engineering Command, based upon that given by Bowles [*10*]. A salient feature of this newly enhanced program is its capability of solving problems involving piles in layered soil systems. The lateral deformations of the inclinometer computed for the selected moduli of subgrade reaction and soil pressures were compared with the measured data given in Fig. 5*b*. A very close agreement between them was obtained, verifying the validity of the finite element program.

Under the action of the preceding lateral soil pressure, the behavior of the single pile was analyzed using the previously mentioned finite element computer program. In the analysis, the 100-ft (30-m) pile was subdivided into ten 3-m (10-ft) segments. The modulus of elasticity of the pile was equal to 3860 kn/cm^2 (806 400 ksf), and the moment of inertia was 1.34×10^5 cm^4 (0.155 ft^4).

Iterative analyses were performed to determine the moment and horizontal reaction required at the top of the piles to satisfy the boundary conditions of no rotation and no horizontal displacement at the top of the pile as imposed

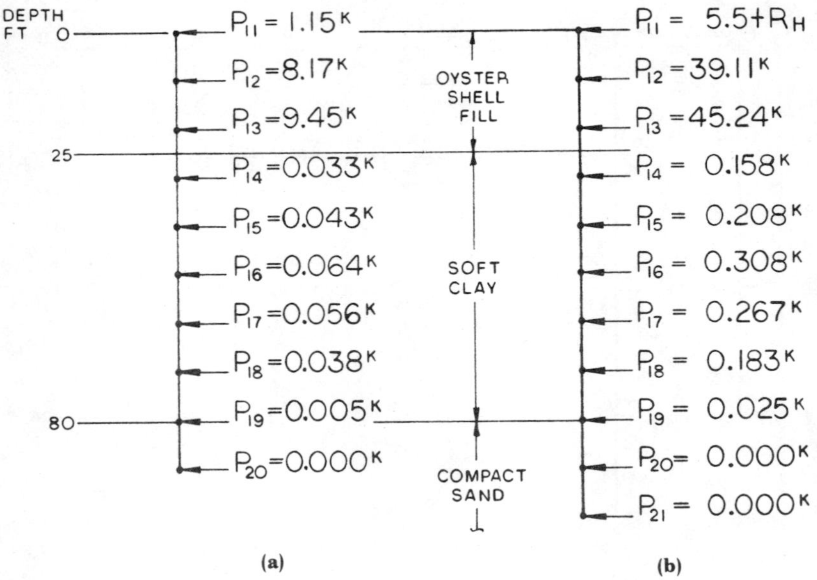

FIG. 7—*Lateral soil loads* (a) *inclinometer and* (b) *pile.*

by the constraint of the grade beam and pile cap structure. The nodal loads on the pile were calculated from the soil pressures given in Fig. 6. Figure 7*b* shows the computed nodal loads on the pile. Results of the finite element analysis give pile deformation, shear, and moment diagrams as shown in Fig. 8. The horizontal reaction 91.2 kN (20.5 kips) and resisting moment 12.9 kN-m (95 ft-kips) at the top of the pile were then used to check the structural integrity of the pile cap.

Structural Analysis of Pile Cap

Figure 9 shows the structural components of the reinforced concrete pile cap. Note that the steel has a yield strength of 413 400 kN/m^2 (60 ksi) and the concrete has a 28-day strength of 20 670 kN/m^2 (3 ksi). The horizontal reaction and resisting moment determined above are applied to the pile cap as shown in the figure. Under the combined action of horizontal force and moment, the structural integrity of the pile cap was analyzed using the ultimate strength method of analysis. It was found that the strength of the pile cap was exceeded by the combined effect of the external force and moment, confirming the development of cracks in the pile cap. Based upon the results of this structural analysis, the distress in the pile cap was checked favorably, and therefore, the behavior of the single pile analyzed using the approach adopted would be satisfactory.

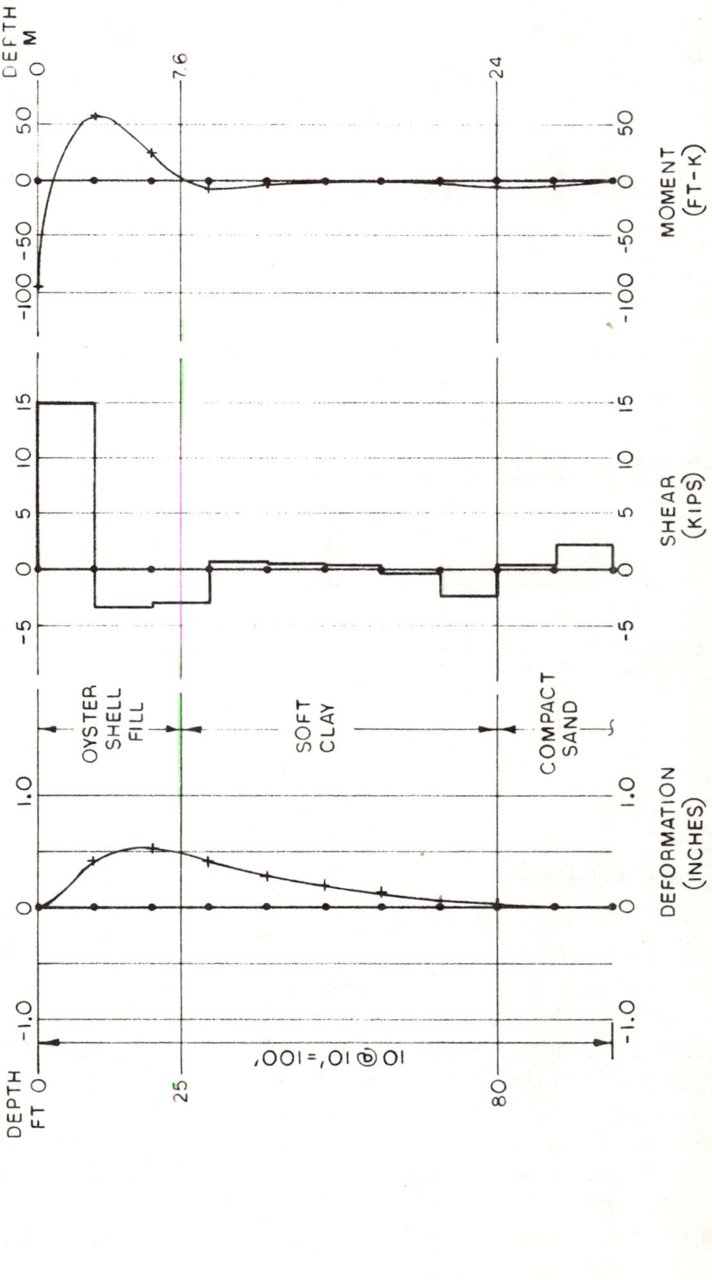

FIG. 8—*Pile deformation, shear, and moment diagrams.*

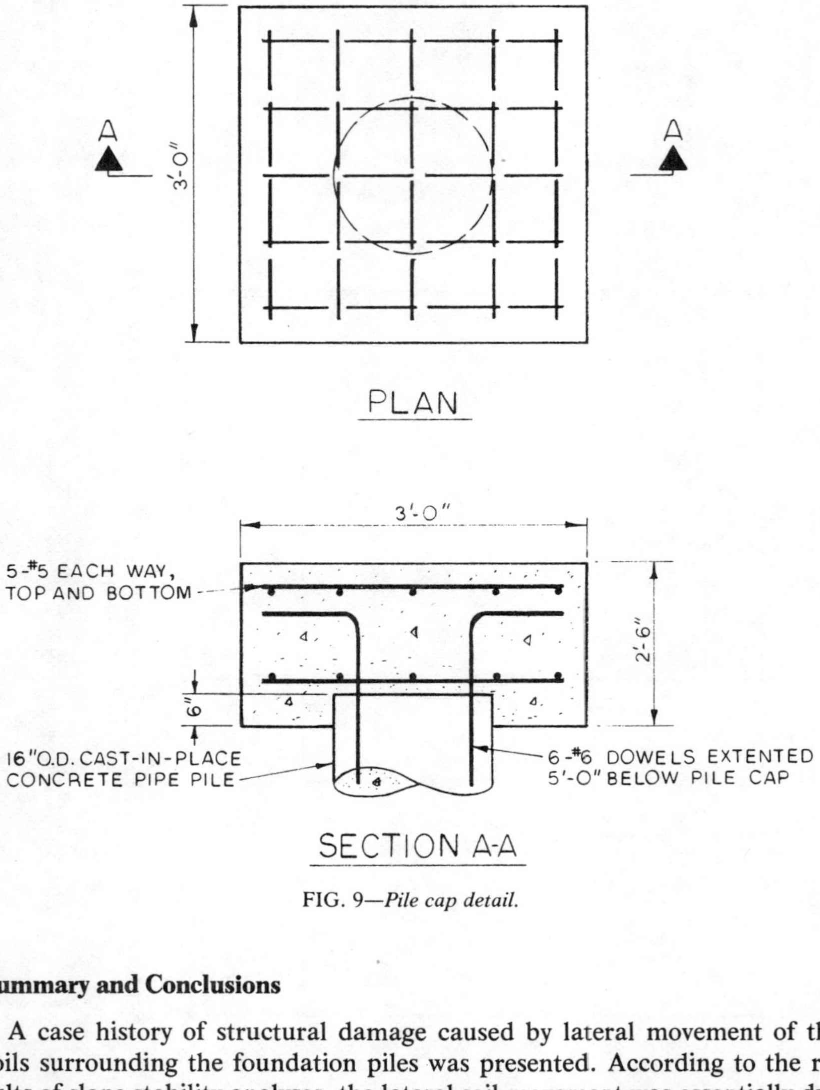

FIG. 9—*Pile cap detail.*

Summary and Conclusions

A case history of structural damage caused by lateral movement of the soils surrounding the foundation piles was presented. According to the results of slope stability analyses, the lateral soil movement was essentially due to a shear failure in the soil mass. The amount of lateral soil movement determined from inclinometers was used to calculate lateral loading on the piles. The behavior of a single pile subjected to this lateral loading was analyzed using a finite element computer program. By checking the pile cap under the combined action of the computed horizontal reaction and resisting moment at the top of the pile, the results of the pile analysis were checked satisfactorily.

In the light of the preceding results of study, it can be concluded that a successful pile foundation design requires due consideration of the effect of lateral soil pressure due to possible movement of the surrounding soil. In this regard, a stability analysis of the supporting soil subjected to overall loading would be needed during the pile foundation design stage. If a possibility of lateral movement of the surrounding soil is indicated, the stress and deformation of the pile induced by the lateral earth pressures should be analyzed. The behavior of single piles subjected to lateral soil movement can be predicted successfully using the finite element computer program together with appropriate properties of the surrounding soil and the pile material, and the magnitude of lateral soil movement.

Acknowledgments

The computer analysis was carried out at Chesapeake Division, Naval Facilities Engineering Command, Washington, D.C. The encouragement and support of R. J. Morony, Manager of Civil/Structural Engineering Branch, Chesapeake Division, during the preparation of this paper is appreciated.

References

[*1*] Bowles, J. E., *Foundation Analysis and Design,* McGraw-Hill, New York, 1968.

[*2*] Broms, B. B., *Journal of the Soil Mechanics and Foundations Division,* Vol. 90, No. SM2, Proc. Paper 3825, American Society of Civil Engineers, Mar. 1974, pp. 27–63.

[*3*] Broms, B. B., *Journal of the Soil Mechanics and Foundations Division,* Vol. 90, No. SM3, Proc. Paper 3909, American Society of Civil Engineers, May 1964, pp. 123–156.

[*4*] Davisson, M. T., "Lateral Load Capacity of Piles," Highway Research Record, No. 333, Highway Research Board, 1970, pp. 104–112.

[*5*] Davisson, M. T. and Gill, H. L., *Journal of the Soil Mechanics and Foundations Division,* Vol. 89, No. SM3, Proc. Paper 3509, American Society of Civil Engineers, May 1963, pp. 63–94.

[*6*] Reese, L. C. and Cox, W. R., in *Performance of Deep Foundation, ASTM STP 444,* American Society for Testing and Materials, 1969, pp. 160–176.

[*7*] Poulos, H. G., "Behavior of Laterally Loaded Piles: I-Single Piles," Proc. Paper 8092, May 1971, pp. 711–731.

[*8*] Wu, A. H., *Navy Civil Engineering,* Vol. 18, No. 3, Fall 1977, pp. 41–42.

[*9*] Law Engineering Testing Company, *Monitoring Services and Engineering Analysis for the Month of September 1975,* Report to the Commanding Officer, Naval Facilities Engineering Command, Washington, D.C., 1975.

[*10*] Bowles, J. E., *Analytical and Computer Methods in Foundation Engineering,* McGraw-Hill, New York, 1974.

A. H. Wu[1] *and R. R. Fox*[2]

Capacity of Axially Loaded Bent Piles in a Bearing Stratum Overlain by a Thick Layer of Soft Clay

REFERENCE: Wu, A. H. and Fox, R. R., **"Capacity of Axially Loaded Bent Piles in a Bearing Stratum Overlain by a Thick Layer of Soft Clay,"** *Behavior of Deep Foundations, ASTM STP 670,* Raymond Lundgren, Ed., American Society for Testing and Materials, 1979, pp. 592–605.

ABSTRACT: This paper presents a case study of four pipe piles, 35.6 cm (14 in.) outside diameter and 34.3 cm (13½ in.) inside diameter driven through a very soft sensitive clay with the pile tip embedded a short distance into a competent gravelly sand stratum. These pipe piles, 9.1 m (30 ft) to 15.2 m (50 ft), long were driven without leads due to the headroom limitation and were filled with high early strength concrete. Load tests were then performed on these piles with a known bend to failure or to the point of yielding. Results of the pile load tests, pile misalignment, soil properties, and pile-driving records were evaluated. A method of predicting the ultimate soil bearing capacity supporting the bent piles is then developed. The computed and measured ultimate pile capacities are compared and commented upon. Graphs are presented to indicate the interrelations among thickness of soft clay, geometric properties of the pile, and reduction of pile end-bearing capacity that likely occur.

KEY WORDS: bent piles, axially loaded piles, load test, ultimate soil bearing capacity, reduction of pile end-bearing capacity, soil pressure distribution against pile, wave equation, *Engineering News* pile-driving formula, soft clay, dense gravelly sand

Pile foundations used to support building additions or renovated buildings are sometimes installed without leads due to restricted headroom. Such piles, driven in short sections through a thick layer of very soft clay and then into a competent granular material, tend to wander and bend so that eccentricity and curvature result. These flaws then contribute to reduced capacity under axial load from that predicted for a concentric load on a straight pile. Therefore, the pile bearing capacity can not be accurately

[1] Soil mechanics and foundation specialist, Chesapeake Division, Naval Facilities Engineering Command, Alexandria, Va. 22332.
[2] Professor of Civil Engineering, George Washington University, Washington, D.C.

predicted by conventional theory or driving formulas. It is the writers' belief that few methods are available for accurately predicting the ultimate bearing capacity of a bent pile. The most reliable method is to conduct a static load test. However, the measurements and analysis of the load test are complicated by the magnitude of pile curvature, pile stiffness, stress distribution along the pile, and lateral earth support.

A project to renovate a hangar building provided the opportunity to evaluate this problem. The hangar building was constructed in about 1940. The interior clearance was about 8.5 m (28 ft). Because of the limited headroom, the contractor selected a hammer that was suspended by a crane, and it was necessary to use short pile sections, as shown by Fig. 1. The steel pipe sections were driven through the sensitive clay and penetrated into a dense gravelly sand in order to form the end-bearing piles. However, the presence of obstructions in the gravel and eccentric blows of the hammer caused the piles to bend to an extent that the misalignment of the piles exceed the specified 2 percent out-of-plumb value.

With the end-bearing piles known to have a gentle bend in the soft clay and a short section of possibly abrupt bend, it became necessary to evaluate the most probable bearing capacity of these piles. The first step taken was to measure the total curvature of the piles. The second step was to load test some piles to failure or to two times the design capacity, whichever was the smaller. The third step was to use the load test results to develop an

FIG. 1—*Pile hammer, pipe piles, crane, and limited headroom conditions.*

analytical method that could be utilized to evaluate the influence of such bends and the soil pressure distribution on the pile bearing capacity.

The method developed herein is then compared with the ultimate bearing capacity calculated from some established methods, including bearing capacity equations, the wave equation, and pile driving formulas. The method developed herein is applicable for an end-bearing pile with a known curvature, fully embedded in the ground, and assumes no lateral support is provided by the sensitive soft clay that overlays the bearing stratum.

Subsoil Conditions and Foundation Requirements

The site was reclaimed from the river about 1900 by being filled with material dredged from the river. Soil borings indicated the site subsoils consist of 9.1 m (30 ft) to 15.2 m (50 ft) of very soft clay, which is generally underlain by a 2.4-m (8-ft) to 4.3-m (14-ft) thick layer of dense gravelly sand. Stiff clay is then encountered below the gravelly layer to the depth of the borings terminated at 27.3 m (90 ft) deep. The soft clay is an alluvial deposit of very soft "river mud" of the Anacostia River of Washington, D.C. The cohesion of the clay is about 6.8 kN/m^2 (140 psf) to 16.0 kN/m^2 (330 psf). Sensitivity of the clay is 4.3 to 6.5. The virgin compression index is 0.44, and the coefficient of consolidation is 1.5×10^{-2} cm^2/min (2.4×10^{-2} ft^2/day). The dense gravelly sand, which has an angle of internal friction of 38 deg, is a deposit of the Pleistocene geologic era, and the lower clay layer is part of the Potomac Group formation of Cretaceous age. Groundwater is about 0.6 m (2 ft) below existing floor grade. Figure 2 shows a typical subsoil profile of the site and the soil test data.

The project was an interior addition forming a mezzanine floor level to the hangar building. The use of deep foundations was considered necessary due to the presence of the thick soft clay layer. Closed-end concrete-filled pipe piles were selected as the most feasible foundations. Pressure-injected footings were not selected, as the driving with high energy for forming the footing bulb might have dislocated the in-place piles. Pipe piles of 35.6-cm (14-in.) diameter driven at least 0.9 m (3 ft) into the gravelly sand layer were initially considered capable of developing the 534-kN (60-ton) design capacity entirely by end bearing.

Pile Capacity

The bearing capacity of an axially loaded pile is generally divided into two components, end-bearing and skin-friction. The general formula for a pile to transfer its load to the soil is

$$\begin{aligned} P_u &= P_p + P_f \\ &= q_p A_p + q_f A_f \end{aligned} \tag{1}$$

where

P_u = ultimate pile bearing capacity
P_p = end bearing capacity
P_f = skin friction
q_p = bearing pressure of soil at pile tip
A_p = pile tip cross-section
q_f = average unit skin friction or adhesion between soil and pile surface
A_f = total surface area of pile in contact with soil

When the pile is driven through a very soft sensitive clay with the pile tip embedded even only a short distance into a competent sand, the frictional capacity is generally negligible compared to the end-bearing capacity. Terzahi's basic bearing capacity equation for a straight end-bearing pile driven a short distance into a dense sand can be expressed by

$$P_u = \gamma D N_q A_p \tag{2}$$

where

P_u = ultimate bearing capacity in the sand stratum,
D = confining depth of the overburden stratum,
γ = effective unit weight of the soil,
N_q = bearing capacity factor of surcharge, and
A_p = pile tip cross section.

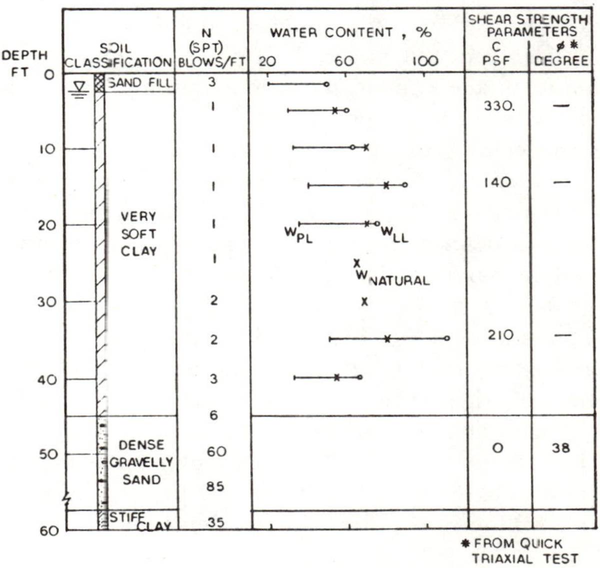

FIG. 2—*Soil profile and soil test data.*

The bearing capacity factor N_q versus angle of internal friction ϕ according to various authors has been summarized by Vesić [*1*].[3] This paper considers a dense sand layer as the bearing stratum, with angle of internal friction values ranging between 30 and 40 deg. The bearing capacity factors for circular deep foundation for the angle of internal friction between 30 and 40 deg is shown in Fig. 3.

When Meyerhof's [*2*] simple formula for computing the bearing capacity of driving piles is adapted to this case, it may be expressed as

$$P_u = 2.5NA_p \qquad (3)$$

where

P_u = ultimate bearing capacity in the sand, tons,
N = standard penetration resistance at the pile tip, blows/ft, and
A_p = pile tip cross section, in 0.09 square metres (square feet).

Since this analysis considers that the pile is embedded only a short distance in the competent sand to develop end-bearing capacity, the writers use the average of the N values recorded in the gravelly sand layer in the application of the above equation.

Parsons and Wilson [*3*] initially attempted to make such an analysis on bent piles that were mainly end-bearing piles. Their Case B was pipe piles driven through fill and organic silt containing a sufficient quantity of boulders to produce some curvature in the pile. The bent pile as well as the assumed force acting on the pile is shown in Fig. 4. This figure refers to the original paper and has only similarity to the soil strata at the site of the hangar considered in this paper. It is presented only to point out the stress distribution in the supporting sand stratum. Parsons and Wilson recognized that some lateral support to the pile is provided by the overlying fill, but it was neglected because of the large restraining forces available in the dense sand and gravel formation immediately underlying the fill. Lateral support from the fill was also discounted in this study, because considerably smaller deformations in sand and gravel than in the overlying fill are required to mobilize the lateral pressure. The authors indicated that, based on these assumptions, the triangular distribution of soil pressure along the pile shown in Fig. 4 is considered reasonable but is recognized to be conservative. No load tests were made on these piles to verify the accuracy of the analysis; however, performance of the piles has demonstrated the method of analysis appeared to be satisfactory.

Johnson [*4*] presented a method of determining the capacity of bent piles that is applicable to both friction and end-bearing piles. It takes into account the pile stiffness, pile load distribution, and soil modulus k. By

[3] The italic numbers in brackets refer to the list of references appended to this paper.

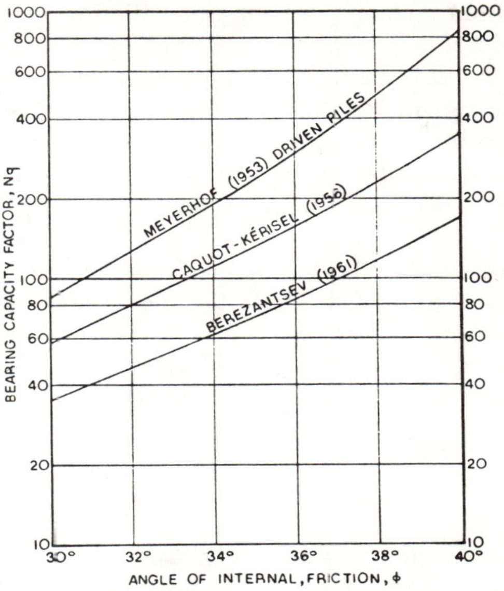

FIG. 3—*Bearing capacity factors for circular deep foundation* (*from Vesić* [1]).

applying Hetenyi's [5] solutions for beams on elastic foundations, one obtains the moment at any point of the piles whose end is free to rotate but has no lateral translation. The solution requires substantial iterations expedited by use of a computer, but the answer is only an estimate of the probable structural capacity, since the method solves for the pile capacity in terms of the pile stresses, which are the sum of the axial and bending stress. The solution does not apply to the determination of the bearing capacity of the soil that supports the bent pile.

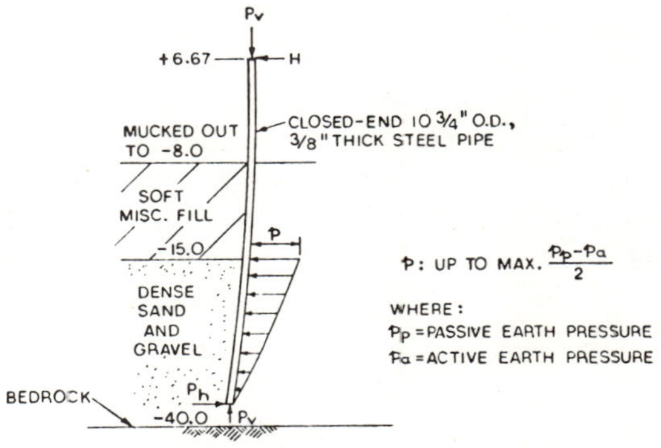

FIG. 4—*Force acting on bent pile* (*from Parsons and Wilson* [3]).

Broms [*6*] presented a method for the calculation of the axial load that can be applied without overstressing the initially bent pile or the soil surrounding the pile. This method is developed by limiting the maximum soil reaction to one-half to one-third the ultimate lateral resistance of the soil and limiting the maximum stress in the pile to one-half to one-fourth the yield strength of the pile material. It was found that this method is not applicable to the bent pile, in which the soil bearing capacity is the main concern.

Hunter and Davisson [*7*] indicated that conventional pile driving in sand may compact the soil beneath the pile tip and improve the tip bearing capacity. In this regard, when the pile is embedded a relatively short distance in an initially dense sand and also has a bend at the tip, the improvement of point bearing resulting from the pile driving may be disregarded.

Analytical Procedure

Consider a pipe pile with known flexural stiffness capable of being driven through a thick soft clay to a dense granular bearing stratum but meanwhile developing curvature in the pile. When the pipe flexural stiffness is much greater than the stiffness of the soft clay, the pile may be considered to have infinite stiffness. When a vertical load is applied to the pile, the pile will deflect laterally because of eccentricity with respect to the support at the lower end of the pile. As the lateral pile displacement occurs, the adjacent clay soil reaction will provide resistance to the lateral movement. When the surrounding soil is a sensitive soft clay, the soil reaction provided by this soft clay may be considered negligible. In this case, it can be assumed that the load is transferred from the relatively straight portion of the pile to the competent bearing soil. With this assumption in mind, we can evaluate the effect of pile curvature on the pile bearing capacity. It should be noted that when the piles are in a cluster, there is a tendency for the pile cap to provide lateral constraint, thus limiting the lateral deflection, which in turn reduces the soil reaction acting on the pile.

To simplify the problem and to give a practical solution, the variation of axial load and earth pressures along the pile length is assumed as in Fig. 5. It assumes that any increase in vertical load due to negative skin friction (down drag) from consolidation of the clay layer is offset by friction (and part of end bearing in special cases) in the supporting sand layer.

From the condition of equilibrium and the end-bearing capacity equation

$$P_{v}e = P_{h}(2d/3) \qquad (4)$$

$$P_{u} = (P_{h}^{2} + P_{v}^{2})^{1/2} \qquad (5)$$

$$P_{u} = \gamma D N_{q} A_{p} \qquad (6)$$

the axial load P_v can therefore be expressed by the equation

$$P_v = \frac{P_u}{[1 + (3e/2d)^2]^{1/2}} \tag{7}$$

where

P_v = ultimate axial load at the top of the pile,
P_h = resultant lateral earth pressure along the pile in sand or granular soil layer, which is equal to but opposed to the lateral force acting at the pile tip,
P_u = ultimate end-bearing capacity of the pile due to curvature from butt to tip,
e = pile eccentricity, and
d = the depth of pile embedment in sand or granular soil layer.

The ultimate axial load predicted from Eq 7 is compared with pile load test results and is discussed in the summary.

Pile Load Test

To verify the design adequacy for the project, pile load tests were performed before driving of the production piles. The pile sections were generally 2.4 m (8 ft) long, allowing work to be done in headroom of less than 8.5 m (28 ft). A Mekiernan-Terry double-acting 11B3 steam hammer rated at 2638 m-kg (19 150 ft-lb) energy per blow was used to drive the 35.6-cm (14-in.) diameter piles. The shell was 0.64 cm (0.25 in.) in thick-

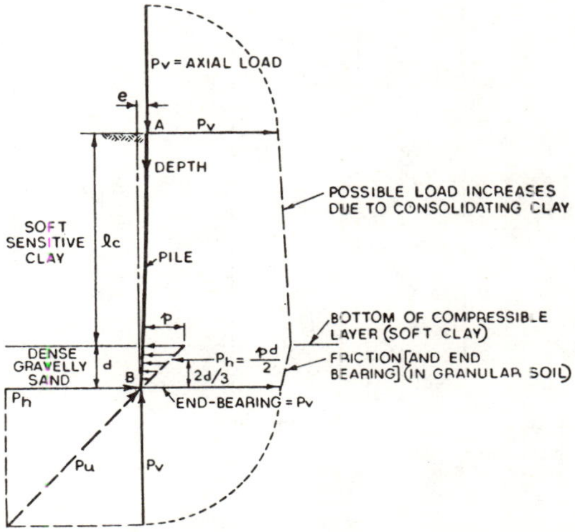

FIG. 5—*Assumed variation of axial load and lateral load along the length of the pile.*

ness. A total of four piles were load tested to failure or to the point of yielding in accordance with ASTM Testing Piles Under Axial Compressive Load (D 1143), which is a slow maintained load test.

Test No. 1

A 15.2-m (50-ft) long pile with the final penetration resistance of 68 blows per foot was tested to failure at a 535-kN (60-ton) load. A soil-boring drilled adjacent to the test pile revealed the pile penetrated 1.3 m (4 ft) below the soft clay, of which 0.6 m (2 ft) was into the medium-dense gravelly sand layer and 0.6 m (2 ft) into a dense granular layer. The pile was out of plumb about 7.9 cm/m (0.66 in./ft), resulting an eccentricity of 0.84 m (2.75 ft) from the butt to the tip. Subsoil conditions, pile driving data, and results are shown in Fig. 6.

Test No. 2

A 15.8-m (52-ft) long pile driven 1.2 m (4 ft) into the medium-dense and an additional 1.2 m (4 ft) into the dense gravelly sand layer was tested and failed at a 579-kN (65-ton) load. Pile data showed increasingly higher penetration resistances for the final 1.2 m (4 ft) of driving. The pile was out of plumb about 6.3 cm/m (0.75 in./ft) or a 1.0-m (3.25-ft) eccentricity. The load test results are shown in Fig. 7.

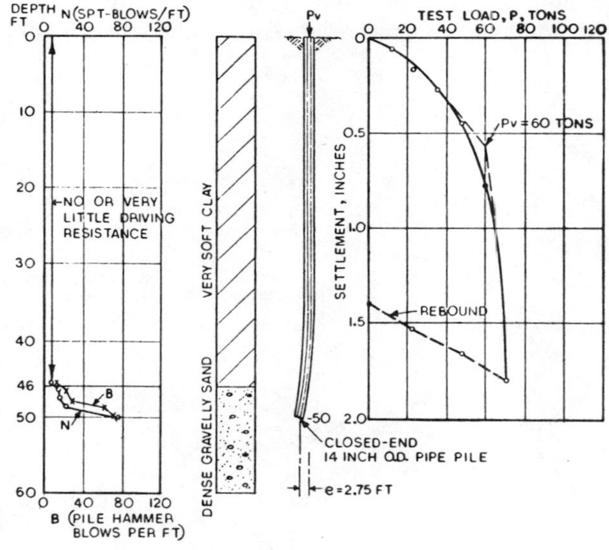

FIG. 6—*Pile Load Test No. 1.*

Test No. 3

A 13.9-m (46-ft) long pile with pile tip driven through 1.2 m (4 ft) of sand and 3.0 m (10 ft) into the stiff clay layer was tested and failed under a 829-kN (93-ton) load. The pile was out of plumb about 5.0 cm/m (0.60 in./ft), a 0.70-m (2.30-ft) eccentricity. The load test results are shown in Fig. 8.

Test No. 4

A 9.1-m (30-ft) long pile with the pile tip driven 1.5 m (5 ft) into the dense gravelly sand was load tested to 1069-kN (120-ton) load. The pile was out of plumb about 4.2 cm/m (0.50 in./ft), a 0.38-m (1.25-ft) eccentricity. This pile appeared to reach the point of yielding at 1069-kN (120-ton) load. The test results are shown in Fig. 9.

The load tests proved that the initial design, which anticipated the dense gravelly sand layer would develop necessary end-bearing design capacity of 535 kN (60 ton), was inaccurate when the soft clay layer was of significant thickness [more than 13 m (44 ft)]. Net settlement for the failed test piles after rebound of the load tests were between 80–90 percent of the total settlement; thus, it appeared that the piles did not fail from structural buckling, although the thick soft clay provided little lateral support.

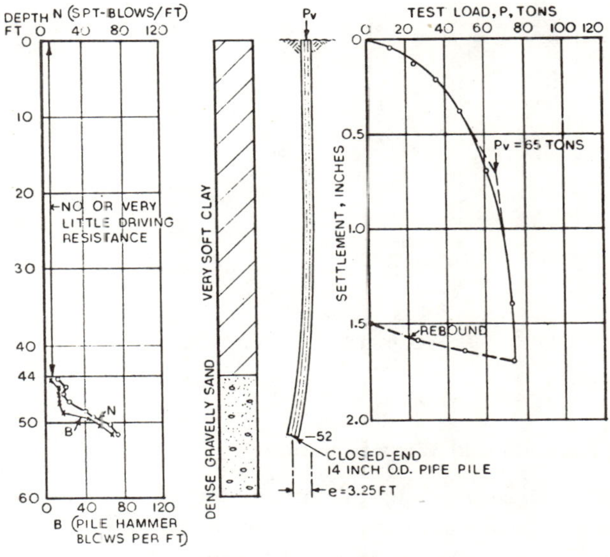

FIG. 7—*Pile Load Test No. 2.*

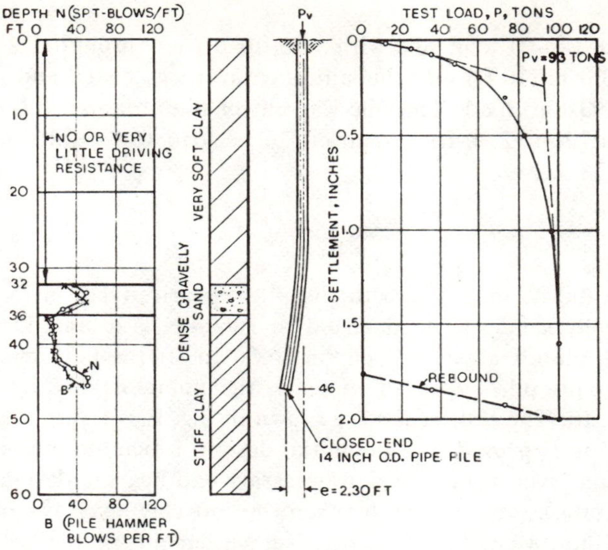

FIG. 8—*Pile Load Test No. 3.*

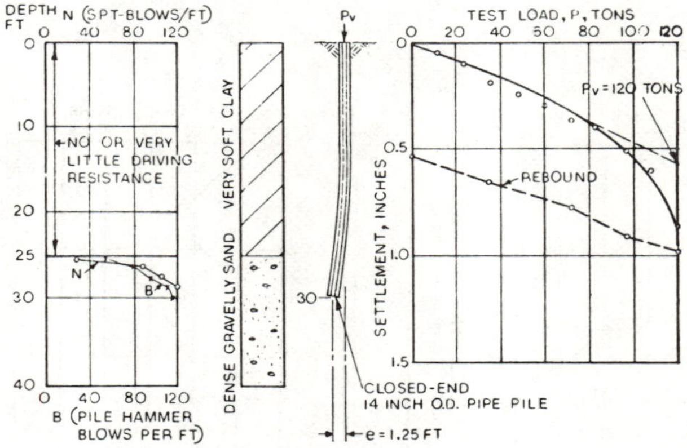

FIG. 9—*Pile Load Test No. 4.*

Pile-Driving Formula and Wave Equation

The simple *Engineering News* pile-driving formula coupled with the load test program was adopted to provide the pile driving criteria for the production piles. The formula was used not in establishing the needed bearing capacity, but was used to ascertain the embedment of the pile in the competent layer. The subsoil conditions and the pile-driving records made it very obvious when the pile penetrated the dense gravelly sand layer. The *Engineering News* formula is especially unreliable when the pile capacity is fully dependent on end bearing and when there is eccentricity that occurred as a result of the limited-headroom driving conditions.

A wave equation analysis by Bowles [*8*] was used to check the driveability of the pipe pile, that is, to check the hammer energy and pile thickness that would permit satisfactory driving in the bearing soils. The wave equation analysis for the end-bearing bent pile indicated the results shown in Table 1.

The wave equation solution also gave ultimate resistance of the piles, which agreed well with the load test results for the two long piles, but underestimated the capacity of one short pile. The wave equation data in the table above also indicated that hard driving or high blow counts (greater than 9 blows per inch) do not result in a proportionately higher bearing capacity.

The wave equation predicted these pipe piles, 35.6 cm (14 in.) outside diameter, 0.64 cm (0.25 in.) wall thickness, would have a maximum compressive stress during driving of 206 700 kN/m^2 (30 000 psi), which is below the yielding stress. However, the damage that occurred on some piles during driving resulted from the fact that the hammer did not deliver a square blow to the pile and/or was not in an axial alignment with the pile.

Summary and Conclusions

Results of the pile load tests, axial pile capacity estimated from the analytical method, the effect of pile misalignment and the thickness of soft clay, analysis of wave equation, and pile-driving formula are summarized in Fig. 10. The graphs are presented to indicate the effect of pile slender-

TABLE 1—*Indicated results of wave equation analysis.*

Blows/Inch	Ultimate Resistance, P_u, tons
2	38
3	48
4	55
5	60
7	66
9	71
15	79

Conversion factors
1 Blows/in. = 2.5 Blows/cm
1 ton = 8.9 kN

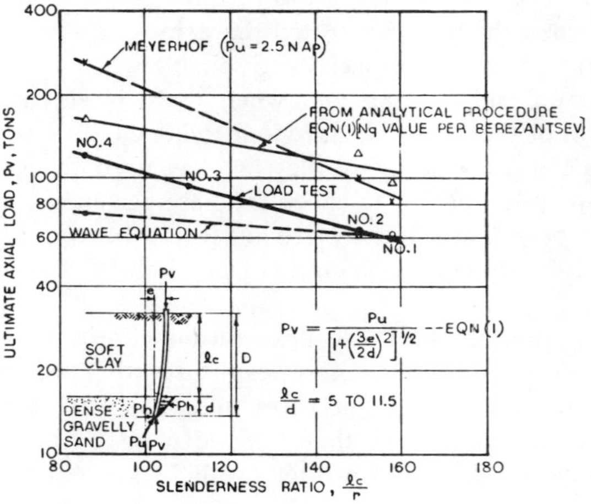

FIG. 10—*Ultimate axial load versus pile slenderness ratio.*

ness ratio on the ultimate axial load. The slenderness ratio used herein is defined as the nondimensional ratio of the soft clay thickness to the radius of gyration of the pile.

The tip of Test Pile No. 3 was not embedded in a dense gravelly sand but in a stiff clay soil. This explains why the results are somewhat inconsistent with the other results except in the pile load tests. The analytical result of Test Pile No. 3 is not plotted in Fig. 10.

Test Pile No. 4 was seated in a very dense gravelly sand, which apparently has a higher angle of internal friction than 38 deg. This can be deduced from the blow counts during standard penetration tests and pile driving. A value of 40 to 42 deg for the friction angle seems appropriate.

It has also been experienced that when pile curvature and damage are likely to occur during driving due to the limited headroom, a thicker pipe wall thickness should be used to lessen the damage resulting from difficult driving and thus avoid delay of work.

The four load test results clearly indicate the effect of the slenderness ratio on the ultimate capacity of axially loaded bent piles embedded a short distance in a dense sand overlain by a thick layer of soft clay. The load test results and the pile foundation construction problems of this project permits the following conclusions:

1. The analytical procedure developed herein show values and trends in Fig. 10 consistent with load test results. A factor of safety of 2 should give reasonable values for designing. The N_q values used in the analytical procedure were obtained from Berezantesv's values shown in Fig. 3.

2. Deficiency of the *Engineering News* driving formula was again

demonstrated in this pile installation. This type of end-bearing piles should not use the *Engineering News* formula in pile capacity quality control.

3. Load test is desirable for pile driving without guide leads through a very soft clay into a dense sand layer because of the potential driving obstructions and the uncertainty of the pile bearing capacity. The load test results may provide justification for using a safety factor of less than 2, probably of the order of 1.5, as demonstrated in this project.

4. The ultimate pile bearing capacities predicted from the wave equation agreed with the load tests no. 1 and no. 2. However, load test no. 4 had much harder pile driving, and the wave equation analysis did not provide as large a capacity as indicated by the load test. Perhaps a modification of the wave equation analysis is needed.

Acknowledgments

We wish to acknowledge the assistance of all those connected with the project. The help and encouragement of R. J. Morony, Manager of Civil/Structural Engineering Branch, Chesapeake Division, Naval Facilities Engineering Command, during preparation of this paper is gratefully acknowledged. R. W. Bell and D. J. Scheessele, both structural engineers of Chesapeake Division, assisted in structural and computer analyses.

References

[*1*] Vesić, A. S., in *Bearing Capacity and Settlement of Foundation*, A.S. Vesić, Ed., Duke University, Durham, N. C., 1965, pp. 53-68.

[*2*] Meyerhof, G. G., "Penetration Tests and Bearing Capacity of Cohesionless Soils," American Society of Civil Engineers, No. SM1, Vol. 82, 1956, p. 866-1.

[*3*] Parson, J. D. and Wilson, S. D., *Transactions*, American Society of Civil Engineers, Vol. 121, 1954, pp. 695-716.

[*4*] Johnson, S. M., *Journal of the Soil Mechanics and Foundations Division, Proceedings*, American Society of Civil Engineers, Vol. 88, December 1962, pp. 65-76.

[*5*] Hetenyi, M., *Beam on Elastic Foundations*, The University of Michigan Press, Ann Arbor, Michigan, 1946.

[*6*] Broms, B. B., *Journal of the Soil Mechanics and Foundations Division, Proceedings*, American Society of Civil Engineers, Vol. 89, September 1973, pp. 73-89.

[*7*] Hunter, A. H. and Davisson, M. T. in *Performance of Deep Foundation, ASTM STP 444*, Raymond Lundgren, Ed., American Society for Testing and Materials, 1969, pp. 106-117.

[*8*] Bowles, J. E., *Analytical and Computer Methods in Foundation Engineering*, McGraw-Hill, New York, 1974.

Index

£8[illegible]